Studies in Classification, Data Analysis, and Knowledge Organization

Titles in the Series

H.-H. Bock and P. Ihm (Eds.)
Classification, Data Analysis and
Knowledge Organization

M. Schader (Ed.)
Analyzing and Modeling Data
and Knowledge

Otto Opitz · Berthold Lausen
Rüdiger Klar (Eds.)

Information and Classification

Concepts, Methods and
Applications

Proceedings of the 16th Annual Conference
of the "Gesellschaft für Klassifikation e.V."
University of Dortmund, April 1-3, 1992

With 165 Figures

Springer-Verlag

Berlin Heidelberg New York
London Paris Tokyo
Hong Kong Barcelona
Budapest

Prof. Dr. Otto Opitz
Lehrstuhl für Mathematische Methoden
der Wirtschaftswissenschaften
Universität Augsburg
Universitätsstr. 2
D-86135 Augsburg, FRG

Dr. Berthold Lausen
Forschungsinstitut
für Kinderernährung
Heinstück 11
D-44225 Dortmund, FRG

Prof. Dr. Rüdiger Klar
Abteilung für Medizinische Informatik
Universitäts-Klinikum Freiburg
Stefan-Meier-Str. 26
D-79104 Freiburg, FRG

ISBN 3-540-56736-4 Springer-Verlag Berlin Heidelberg New York Tokyo
ISBN 0-387-56736-4 Springer-Verlag New York Heidelberg Berlin Tokyo

Preface

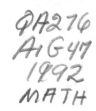

This volume contains refereed and revised versions of 52 papers selected from the contributions presented at the 16th Annual Conference of the "Gesellschaft für Klassifikation e.V." (GfKl), the German Classification Society. The conference stood under the heading "INFORMATION AND CLASSIFICATION: Concepts - Methods - Applications" and took place at the University of Dortmund from April 1-3, 1992. The local organizers were Friedhelm Eicker and Berthold Lausen of the Department of Statistics at the University of Dortmund.

The scientific program of the conference included about 160 presentations which were arranged in plenary, survey or subject sessions, in workshops, software presentations or special tutorial courses. Parts of the scientific program were organized in cooperation with following other scientific institutions and organizations: Deutsche Gesellschaft für Medizinische Informatik, Biometrie und Epidemiologie (GMDS) e.V., Deutsche Verbände für Altertumsforschung, International Biometric Society (German Region), University of Sheffield and with the Dutch-Belgian Vereniging voor Ordinatie en Classificatie (VOC).

The conference owed much to the support by the University of Dortmund and especially its Department of Statistics. On March 31, 1992 the "Festkolloquium" of the Department took place in honour of Professor Friedhelm Eickers 65th birthday. This connection was also fruitful for the overall program of the meeting. Moreover, the international success of the conference was made possible by the generous support of the Deutsche Forschungsgemeinschaft (DFG) and the Bundesland Nordrhein-Westfalen. The kind support by the sponsors Daimler-Benz AG (Stuttgart), Schering AG (Berlin) and the support through the commercial software presentations by DATA VISION (Klosters, Schweiz), GraS (Berlin) and WEKA (Frankfurt) is also gratefully acknowledged.

The editors divided the selected papers into three major sections:

I: **Data Analysis and Classification**
II: **Information Retrieval, Knowledge Processing and Software**
III: **Applications and Special Topics**

Moreover, the papers were grouped and ordered within the major sections according to a common topic covered which is stated in the contents. The editors are aware of the fact, that alternative groupings and orderings are possible, but the structure used reflects the structure of the conference to some extent. Last not least, this hierarchical classification demonstrates the advantages and disadvantages of ordered hierarchical structures as also emphazised in former proceedings of the GfKl and may be regarded as an example for an application of a hierarchical classification.

The editors wish to thank the authors for their contributions and cooperation. Moreover, they have to thank all referees for providing their reports on the submitted papers. Especially the editors are greatly indebted to M. Mißler-Behr, University of Augsburg, for the managing assistance and support during the refereeing process and the technical preparation of this proceedings volume. Finally, the editors have to thank the Springer-Verlag, Heidelberg, for good cooperation in publishing this volume.

Dortmund, December 1992

Otto Opitz
Berthold Lausen
Rüdiger Klar

Contents

Part I: Data Analysis and Classification

Classification Methods

Fuzzy Classification

Conceptual Analysis

Part II: Information Retrieval, Knowledge Processing and Software

Information Retrieval

Neural Networks

Expert Systems and Knowledge Processing

Part III: Applications and Special Topics

Sequence Data and Tree Reconstruction

Data Analysis and Informatics in Medicine

Special Topics - Thesauri, Archaeology, Musical Science and Psychometrics

Part I

Data Analysis and Classification

Hierarchical Clustering of Sampled Functions

G. De Soete[1]

Department of Psychology, University of Ghent,
Henri Dunantlaan 2, B-9000 Ghent, Belgium

Abstract: This paper addresses the problem of performing a hierarchical cluster analysis on objects that are measured on the same variable on a number of equally spaced points. Such data are typically collected in longitudinal studies or in experiments where electro-physiological measurements are registered (such as EEG or EMG). A generalized inter-object distance measure is defined that takes into account various aspects of the similarity between the functions from which the data are sampled. A mathematical programming procedure is developed for weighting these aspects in such a way that the resulting inter-object distances optimally satisfy the ultrametric inequality. These optimally weighted distances can then be subjected to any existing hierarchical clustering procedure. The new approach is illustrated on an artificial data set and some possible limitations and extensions of the new method are discussed.

1 Introduction

In several disciplines, data are collected that can be regarded as samples of continuous functions at discrete points. Following Besse and Ramsay (1986) and Winsberg and Kruskal (1986), such data are called *sampled functions.* Suppose that N objects (or other observational units such as subjects, OTU's, etc.) i $(i = 1, \ldots, N)$ are measured on the same variable at the same M discrete points p $(p = 1, \ldots, M)$, and that the data for each object can be considered as samples of a single underlying function. For the sake of simplicity, it will be assumed that the M discrete points at which the data are observed, are equally spaced on the underlying continuum. The resulting data can be arranged in an $N \times M$ matrix $\mathbf{X} = ((x_{ip}))$. The data for object i will be denoted by the M–component column vector $\mathbf{x}_i = (x_{i1}, \ldots, x_{iM})'$.

When the underlying functions are a function of time, the data are often called *longitudinal data.* Such data are common, for instance, in developmental psychology where subjects are administered a certain test at different stages in their development. Experimental psychologists often collect *learning curves,* which characterize the learning process of individual subjects as a function of time. In other areas, electro-physiological measurements, such as EEG or EMG records, yield sampled functions that can be arranged into a data matrix $\mathbf{X}$.

The underlying continuum on which the functions that are sampled are defined, need not always be time. In psychophysics, for instance, *psychophysical transformations* are studied that relate the psychological sensation of a stimulus to its physical intensity. Other psychophysical experiments yield so-called *psychometric functions* that indicate the probability of judging a variable stimulus as more intense than a fixed reference stimulus as a function of the physical intensity of the variable stimulus.

Often, a researcher wants to study the proximity relations among the objects by performing a hierarchical cluster analysis on inter-object dissimilarities derived from $\mathbf{X}$. An

[1]Supported as "Bevoegdverklaard Navorser" of the Belgian "Nationaal Fonds voor Wetenschappelijk Onderzoek".

experimental psychologist, for instance, can be interested in distinguishing between gradual learners and subjects whose learning curves are step functions. In psychophysics, one might want to group subjects on the basis of their individual psychophysical transformation functions.

While classical linear multivariate statistical methods, such as principal component analysis, have been adapted to deal with functional data (e.g., Besse and Ramsay, 1986; Ramsay, 1982; Winsberg and Kruskal, 1986), there exists no hierarchical clustering technique that explicitly deals with data that are sampled functions. In the absence of such a method, researchers often treat $\mathbf{X}$ as an objects × variables matrix, and compute Euclidean distances, d_{ij}, between the rows of $\mathbf{X}$:

$$d_{ij} = \sqrt{(\mathbf{x}_i - \mathbf{x}_j)'(\mathbf{x}_i - \mathbf{x}_j)}. \tag{1}$$

These Euclidean distances are then subjected to a standard hierarchical clustering procedure. A major drawback of this approach stems from the fact that Euclidean distances only take into account differences between the functions at the sampling points and ignore differences in other characteristics of the underlying functions, such as differences in the first or second order derivatives. To fully characterize the dissimilarity between two objects, the shape of the underlying functions should be taken into account as well. This is illustrated in Fig. 1. The figure shows the data for three objects i, j, and k measured at ten equally spaced points. Although all the underlying functions are linear, the function for object i has a positive slope, while the slopes of the lines for objects j and k are negative. Although most researchers would consider objects j and k to be more similar than objects i and k, it turns out that the Euclidean distance between objects i and k is exactly identical to the Euclidean distance between objects j and k! This example illustrates that Euclidean distances cannot fully quantify the dissimilarities between sampled functions. Consequently, a hierarchical cluster analysis of the Euclidean distances between the rows of $\mathbf{X}$ might result in a quite misleading representation of the data.

In this paper we present an approach to clustering sampled functions that takes the functional aspect of the data (i.e., the fact that the data are sampled functions) explicitly into account. In the next section, a family of metrics is presented that are especially suitable for quantifying the similarity between sampled functions. These metrics are of the form

$$d_{ij} = \sqrt{(\mathbf{x}_i - \mathbf{x}_j)'\mathbf{C}(\mathbf{x}_i - \mathbf{x}_j)}, \tag{2}$$

where $\mathbf{C}$ is a positive (semi-)definite matrix that depends on a small number of parameters. In Section 3, an algorithm is developed for estimating these parameters such that the resulting dissimilarities $\mathbf{D} = ((d_{ij}))$ are optimally suited for a hierarchical cluster analysis. In Section 4, an illustrative application of this method is presented. The final section is devoted to a discussion of some limitations and possible extensions of the present approach.

2   A Distance Metric for Sampled Functions

As argued in the previous section, it is necessary to take the shape of the underlying functions into account to fully quantify the dissimilarity between the functional data of two objects. The shape of such an underlying function can be characterized in term of its first order and higher order derivatives. But, the functions themselves are not observed; only samples at discrete points are available. Therefore, it is not possible to directly base an inter-object

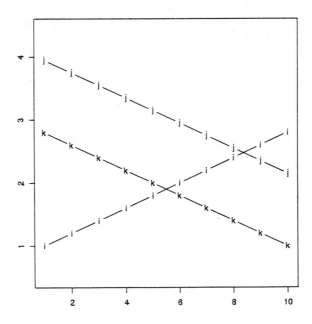

Fig. 1: Example of Functional Data.

dissimilarity metric on differences in function values and in first and higher order derivatives. Instead, we propose to base the dissimilarity metric on squared Euclidean distances between the zero and higher order finite differences in the observed data. The Euclidean distance between the zero order differences for objects i and j, written as $d_{ij}^{(0)}$, is simply equal to the Euclidean distance between $\mathbf{x}_i$ and $\mathbf{x}_j$:

$$
\begin{aligned}
d_{ij}^{(0)^2} &= \sum_{p=1}^{M} (x_{ip} - x_{jp})^2 \\
&= (\mathbf{x}_i - \mathbf{x}_j)'(\mathbf{x}_i - \mathbf{x}_j).
\end{aligned}
\tag{3}
$$

The Euclidean distance between the order t $(0 < t < M)$ finite differences for objects i and j, written as $d_{ij}^{(t)}$, is defined as

$$
d_{ij}^{(t)^2} = \sum_{p=t+1}^{M} \left[(x_{ip} - x_{i(p-t)}) - (x_{jp} - x_{j(p-t)}) \right]^2.
\tag{4}
$$

If we define an $(M - t) \times M$ matrix $\mathbf{A}^{(t)} = ((a_{pq}^{(t)}))$ as follows

$$
a_{pq}^{(t)} = \begin{cases} 1 & \text{if } q = p + t \\ -1 & \text{if } q = p \\ 0 & \text{otherwise,} \end{cases}
\tag{5}
$$

$d_{ij}^{(t)^2}$ can be written as

$$
\begin{aligned}
d_{ij}^{(t)^2} &= (\mathbf{A}^{(t)}\mathbf{x}_i - \mathbf{A}^{(t)}\mathbf{x}_j)'(\mathbf{A}^{(t)}\mathbf{x}_i - \mathbf{A}^{(t)}\mathbf{x}_j) \\
&= (\mathbf{x}_i - \mathbf{x}_j)'\mathbf{A}^{(t)'}\mathbf{A}^{(t)}(\mathbf{x}_i - \mathbf{x}_j).
\end{aligned}
\tag{6}
$$

For $M = 6$, $\mathbf{A}^{(2)}$ has the form

$$\mathbf{A}^{(2)} = \begin{pmatrix} -1 & 0 & 1 & 0 & 0 & 0 \\ 0 & -1 & 0 & 1 & 0 & 0 \\ 0 & 0 & -1 & 0 & 1 & 0 \\ 0 & 0 & 0 & -1 & 0 & 1 \end{pmatrix},$$

and $\mathbf{A}^{(2)}\mathbf{x}_i$ equals

$$\mathbf{A}^{(2)}\mathbf{x}_i = \begin{pmatrix} x_{i3} - x_{i1} \\ x_{i4} - x_{i2} \\ x_{i5} - x_{i3} \\ x_{i6} - x_{i4} \end{pmatrix}.$$

We propose to define the dissimilarity between the sampled functions for objects i and j as the square root of a positive linear combination of the squared distances between the order t differences for $t = 0, 1, \ldots, T$:

$$d_{ij} = \sqrt{\alpha_0 d_{ij}^{(0)2} + \alpha_1 d_{ij}^{(1)2} + \cdots \alpha_T d_{ij}^{(T)2}} \tag{7}$$

with $0 \leq T < M$. The weights $\alpha_0, \alpha_1, \ldots, \alpha_T$ are non-negative

$$\alpha_t \geq 0, \tag{8}$$

and sum to one

$$\sum_{t=0}^{T} \alpha_t = 1. \tag{9}$$

Defining the $M \times M$ matrix $\mathbf{B}^{(t)}$ as

$$\mathbf{B}^{(t)} = \begin{cases} \mathbf{I} & \text{for } t = 0, \\ \mathbf{A}^{(t)\prime}\mathbf{A}^{(t)} & \text{for } 0 < t \leq T, \end{cases} \tag{10}$$

(where $\mathbf{I}$ denotes an identity matrix) allows us to write d_{ij}^2 as

$$d_{ij}^2 = (\mathbf{x}_i - \mathbf{x}_j)'\mathbf{C}(\mathbf{x}_i - \mathbf{x}_j), \tag{11}$$

with

$$\mathbf{C} \equiv \sum_{t=0}^{T} \alpha_t \mathbf{B}^{(t)}. \tag{12}$$

In the case of $M = 6$, $\mathbf{B}^{(2)}$ becomes

$$\mathbf{B}^{(2)} = \begin{pmatrix} 1 & 0 & -1 & 0 & 0 & 0 \\ 0 & 1 & 0 & -1 & 0 & 0 \\ -1 & 0 & 2 & 0 & -1 & 0 \\ 0 & -1 & 0 & 2 & 0 & -1 \\ 0 & 0 & -1 & 0 & 1 & 0 \\ 0 & 0 & 0 & -1 & 0 & 1 \end{pmatrix}.$$

Since the rank of $\mathbf{B}^{(t)}$ equals $M - t$, $\mathbf{B}^{(t)}$ is positive definite for $t = 0$ and positive semi-definite for $t > 0$. Consequently, $\mathbf{C}$ is always positive semi-definite. When

$$\alpha_0 > 0, \tag{13}$$

C is of full rank and C is positive definite. Thus, in the common case where (13) holds, d_{ij}, as defined by (11) and (12), is guaranteed to constitute a metric.

The matrix C on which the dissimilarities $D = ((d_{ij}))$ are based, depends on T and on the weighting coefficients $\alpha_0, \alpha_1, \ldots, \alpha_T$. The data analyst must fix T in advance. From analyses with different values of T the optimal value of T can easily be inferred. While in some situations it might be possible to specify a priori values for the weights $\alpha_0, \alpha_1, \ldots, \alpha_T$, the data analyst usually does not dispose of enough information to fix these weights in advance. In such a case, one would like to determine values of $\alpha_0, \alpha_1, \ldots, \alpha_T$ that result in dissimilarities D that are optimally suited for a hierarchical cluster analysis. In the next section, an algorithm is described for estimating optimal weighting coefficients.

3   Computing Optimal Weights

It is well known that a hierarchical clustering results in an ultrametric tree representation. An ultrametric tree is a terminally labeled rooted tree in which a non-negative weight is attached to each node such that (i) the terminal nodes have zero weight, (ii) the largest weight is assigned to the root, and (iii) the weights attached to the nodes on the path from any terminal node to the root constitute a strictly increasing sequence. In an ultrametric tree, the distance between two terminal nodes i and j, written as δ_{ij}, is defined as the largest weight attached to the nodes on the path that connects i and j. Ultrametric tree distances satisfy the so-called ultrametric inequality:

$$\delta_{ij} \leq \max(\delta_{ik}, \delta_{jk}) \tag{14}$$

for all terminal nodes i, j, and k (see, for instance, Johnson, 1967). An equivalent way of stating this condition is to say that for each triple of terminal nodes i, j, and k, the largest two of δ_{ij}, δ_{ik} and δ_{jk} are equal to each other. The ultrametric inequality is a necessary and sufficient condition that a matrix of distances must satisfy in order to be representable by an ultrametric tree. Moreover, if the condition is met, the ultrametric tree representation is unique. Hence, a matrix of ultrametric distances uniquely determines an ultrametric tree representation and vice versa.

Determining for a given data set X and a fixed value of T, weights $\alpha_0, \alpha_1, \ldots, \alpha_T$ that result in dissimilarities D that are optimally suited for a hierarchical cluster analysis is equivalent to estimating $\alpha_0, \alpha_1, \ldots, \alpha_T$ such that D satisfies the ultrametric inequality as well as possible. Following De Soete (1986), we will adopt a least-squares criterion and solve for the weights that yield distances for which the normalized sum of squared deviations from the ultrametric inequality is minimal. To express the sum of squared deviations from the ultrametric inequality, we define for a matrix of distances D the set Ω_D of ordered triples $\langle i, j, k \rangle$ for which D violates the ultrametric inequality, i.e.,

$$\Omega_D = \{\langle i, j, k \rangle \mid 1 \leq i \leq N, \ 1 \leq j \leq N, \ 1 \leq k \leq N, \ d_{ij} \leq \min(d_{ik}, d_{jk}) \text{ and } d_{ik} \neq d_{jk}\} . \tag{15}$$

According to the ultrametric inequality, for each triple $\langle i, j, k \rangle$ in Ω_D, the equality $d_{ik} = d_{jk}$ should hold. The least-squares function is defined in terms of squared deviations between pairs of distances that should be equal to each other according to the ultrametric inequality. This sum of squared deviations from the ultrametric inequality is normalized by the sum of squared lower-diagonal entries in D in order to prevent certain degenerate solutions. Thus, the optimal values of $\alpha_0, \alpha_1, \ldots, \alpha_T$ are determined by solving the following constrained optimization problem:

$$\underset{\alpha_0, \alpha_1, \ldots, \alpha_T}{\text{minimize}} \quad \left[L(\alpha_0, \alpha_1, \ldots, \alpha_T) = \frac{\underset{\Omega_\mathbf{D}}{\sum}(d_{ik} - d_{jk})^2}{\sum_{i=2}^{N}\sum_{j=1}^{i-1} d_{ij}^2} \right] \qquad (16)$$

subject to constraints (8) and (9), where d_{ij} is defined in (11) and (12). To ensure that the resulting inter-object dissimilarities $\mathbf{D}$ constitute a metric, it is necessary to impose the additional constraint (13).

In order to transform this constrained optimization problem into an unconstrained optimization problem, a transformation of the $T + 1$ constrained variables $\alpha_0, \alpha_1, \ldots, \alpha_T$ into T unconstrained variables $w_1, \ldots, w_T$ is introduced by reparametrizing d_{ij} as follows

$$d_{ij}^2 = \frac{1}{1 + \sum_{s=1}^{T} w_s^2} d_{ij}^{(0)^2} + \sum_{t=1}^{T} \frac{w_t^2}{1 + \sum_{s=1}^{T} w_s^2} d_{ij}^{(t)^2} \qquad (17)$$

(for a description of this transformation of variables technique, see Gill, Murray and Wright, 1981, pp. 270–271). The conversion of $\alpha_0, \alpha_1, \ldots, \alpha_T$ into $w_1, \ldots, w_T$ utilizes the facts that α_0 must be strictly positive (constraint (13)), that $\alpha_1, \ldots, \alpha_T$ must be nonnegative (constraint (8)), and that the $(T + 1)$ α_t coefficients must sum up to one (constraint (9)). The original weights $\alpha_0, \alpha_1, \ldots, \alpha_T$ can be obtained from $w_1, \ldots, w_T$ as follows

$$\alpha_t = \begin{cases} \dfrac{1}{1 + \sum_{s=1}^{T} w_s^2} & \text{for } t = 0, \\[4mm] \dfrac{w_t^2}{1 + \sum_{s=1}^{T} w_s^2} & \text{for } t > 0. \end{cases} \qquad (18)$$

It is apparent from (18) that this transformation of variables technique prevents α_0 from becoming zero and hence ensures that the resulting dissimilarities $\mathbf{D}$ are a metric. The unconstrained minimization problem that must be solved now becomes

$$\underset{w_1, \ldots, w_T}{\text{minimize}} \quad \left[L(w_1, \ldots, w_T) = \frac{\underset{\Omega_\mathbf{D}}{\sum}(d_{ik} - d_{jk})^2}{\sum_{i=2}^{N}\sum_{j=1}^{i-1} d_{ij}^2} \right] \qquad (19)$$

with d_{ij} parametrized in terms of $w_1, \ldots, w_T$ as in (17).

The unconstrained minimization problem (19) can be solved by a variety of iterative numerical methods. We decided to use Powell's (1977) conjugate gradient method with automatic restarts, which was successfully applied for solving similar problems (e.g., De Soete, 1986; De Soete and Carroll, 1988; De Soete, Carroll, and DeSarbo, 1986; De Soete, DeSarbo, and Carroll, 1985). This conjugate gradient procedure requires only the first order

Tab. 1: Artificial Data Set.

Object	Measurement at point									
	1	2	3	4	5	6	7	8	9	10
1	1.02	1.17	1.42	1.63	1.79	2.01	2.18	2.42	2.57	2.78
2	2.10	2.31	2.49	2.71	2.97	3.13	3.30	3.49	3.76	3.97
3	0.93	1.19	1.41	1.67	1.79	1.93	2.18	2.34	2.57	2.81
4	3.92	3.76	3.51	3.33	3.10	2.95	2.74	2.57	2.34	2.11
5	2.83	2.61	2.38	2.21	1.87	1.81	1.63	1.38	1.19	1.03
6	3.50	3.32	3.15	2.90	2.75	2.54	2.36	2.41	2.23	2.16
7	2.27	1.99	1.81	1.59	1.45	1.21	1.01	0.77	0.50	0.41
8	1.70	1.89	2.12	2.33	2.54	2.47	2.31	2.13	1.94	1.68
9	1.25	1.95	2.42	2.63	2.99	2.80	2.55	2.32	1.99	1.47

partial derivatives of the loss function L with respect to the parameters $w_1, \ldots, w_T$. These derivatives are

$$
\frac{\partial L}{\partial w_t} = \frac{2\left[\sum_{\Omega_{\mathbf{D}}}(d_{ik}-d_{jl})\left(\frac{\partial d_{ik}}{\partial w_t}-\frac{\partial d_{jk}}{\partial w_t}\right)\sum_{i=2}^{N}\sum_{j=1}^{i-1}d_{ij}^2-\sum_{i=2}^{N}\sum_{j=1}^{i-1}d_{ij}\frac{\partial d_{ij}}{\partial w_t}\sum_{\Omega_{\mathbf{D}}}(d_{ik}-d_{jk})^2\right]}{\left[\sum_{i=2}^{N}\sum_{j=1}^{i-1}d_{ij}^2\right]^2},
$$

with

$$
\frac{\partial d_{ij}}{\partial w_t} = \frac{w_t}{d_{ij}}\left(\frac{-d_{ij}^{(0)2}}{c^2}+\sum_{s=1}^{T}\frac{c\delta^{st}-w_s^2}{c^2}d_{ij}^{(s)2}\right),
$$

where δ^{st} denotes Kronecker delta and where c is defined as

$$
c = 1 + \sum_{t=1}^{T}w_t^2.
$$

The iterative conjugate gradient method starts from some initial estimates of the parameters $w_1, \ldots, w_T$. Currently, two ways of determining initial parameter estimates have been explored. Either all weights w_t are initially set equal to the same constant T^{-1}, or the initial values of the weights are determined at random. FUNCLUS, a portable Fortran program implementing the algorithm described in this section, allows for a series of runs with different random starts. When the same optimal solution is found in several runs with different random initial parameter estimates, it is not unreasonable to assume that a globally optimal solution was found. In the next section, an illustrative application of FUNCLUS to some artificial data is reported.

4 Illustrative Application

Tab. 1 contains an artificial data set that was constructed by simulating observations of nine objects on a single variable at ten equally spaced points in time. These data are plotted in Fig. 2. As can be seen from the figure, the objects cluster into three groups: $\{1, 2, 3\}$, $\{4, 5, 6, 7\}$, and $\{8, 9\}$. Whereas the functions for objects 1, 2, and 3 are increasing, the functions for objects 4, 5, 6, and 7 are decreasing. Finally, the functions for objects 8 and 9 are single-peaked. Fig. 3 presents a least-squares ultrametric tree representation of the Euclidean distances between the rows of Tab. 1. This least-squares hierarchical clustering was arrived at by means of the LSULT program (De Soete, 1984b) that is based on the mathematical programming procedure by De Soete (1984a). The ultrametric tree in Fig. 3 accounts for 61.3 percent of variance of the Euclidean distances between the rows of Tab. 1. As is apparent from Fig. 3, the ultrametric tree representation of the Euclidean distances does not fully recover the underlying clustering structure. Only the $\{8, 9\}$ cluster is clearly present. This analysis on the Euclidean distances illustrates the point made in the introductory section. Since the Euclidean distances only take into account differences in measurements at the points on which the underlying functions were sampled, they cannot fully quantify the dissimilarity between the objects.

In order to compute inter-object dissimilarities that explicitly deal with the functional nature of the data, FUNCLUS analyses were carried out with $T = 1$, $T = 2$, and $T = 3$. The results of these analyses are summarized in Tab. 2. Column two of Tab. 2 lists the minimum obtained for the loss function L with ten random starts of the FUNCLUS algorithm. The third column gives the goodness of fit—in terms of variance accounted for—of a least squares ultrametric tree representation of the optimal distances $\mathbf{D}$ produced by FUNCLUS. The case $T = 0$ refers to the analysis of the Euclidean distances discussed above. Since the value of FUNCLUS loss function is (after rounding) identical for $T = 2$ and $T = 3$, it can be concluded that $T = 2$ suffices to determine distances that are optimally suited for a hierarchical cluster analysis. The least-squares ultrametric tree representation of the optimal distances for $T = 2$ is displayed in Fig. 4. The ultrametric tree in Fig. 4 clearly recovers the clustering structure in the data. The $\{1, 2, 3\}$, $\{4, 5, 6, 7\}$ and $\{8, 9\}$ clusters are clearly present in this tree representation. This ultrametric tree accounts for 87.8 percent of the variance of the distances computed with FUNCLUS. Tab. 2 shows that the distances obtained by FUNCLUS for $T = 1$ are less optimal than the distances arrived at with $T = 2$. A least-squares ultrametric tree representation of the distances obtained with $T = 1$ is presented in Fig. 5. While this tree reveals the $\{1, 2, 3\}$ and $\{4, 5, 6, 7\}$ clusters, it completely hides the $\{8, 9\}$ grouping. Thus, the representation in Fig. 5 is inferior to the one in Fig. 4.

The analyses on the artificial data set reported in this section demonstrate the superiority of the inter-object dissimilarities computed by means of FUNCLUS over regular Euclidean distances as inter-object dissimilarities. The FUNCLUS distances depend on T. In the example it was illustrated how the most appropriate value of T can be inferred from comparing the values of the FUNCLUS loss function L obtained for various values of T.

5   Discussion

In this paper, a family of metrics is defined that are especially suited for expressing the dissimilarity between sampled functions. This family of metrics depends on a small number of weighting coefficients. A mathematical programming procedure was developed for estimating these coefficients such that the resulting distances satisfy the ultrametric inequality optimally.

The procedure elaborated in this paper can be considered as a special, constrained case of

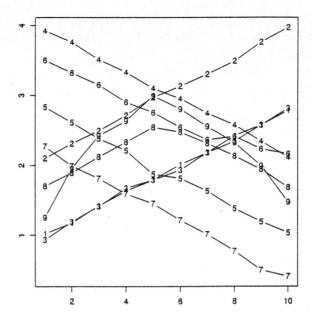

Fig. 2: Data in Tab. 1.

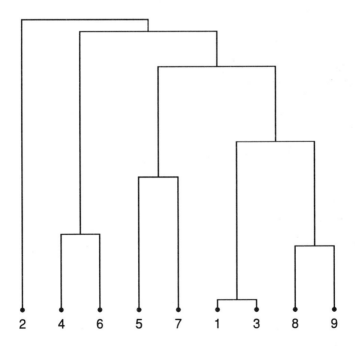

Fig. 3: Least-Squares Ultrametric Tree Representation of the Euclidean Distances

Tab. 2: Goodness-of-fit Results of the FUNCLUS Analyses.

T	FUNCLUS Loss Function L	Percentage VAF by a Least-Squares Ultrametric Tree Representation
0	0.2336	61.33
1	0.0918	81.70
2	0.0849	87.80
3	0.0849	87.99

the general variable weighting method proposed by De Soete (1986) and extensively evaluated by Milligan (1989). Let $\mathbf{Y} = ((y_{ip}))$ denote an $N \times P$ matrix containing measurements of N objects on P variables. The variable weighting method of De Soete (1986) determines P non-negative weights $v_1, \ldots, v_P$ with $v_1 + \cdots + v_P = 1$, such that the weighted Euclidean distances

$$d_{ij} = \sqrt{\sum_{p=1}^{P} v_p (y_{ip} - y_{jp})^2}$$

optimally satisfy the ultrametric inequality (in a least-squares sense). If we define the matrix $\mathbf{Y}$ as

$$\mathbf{Y} = (\mathbf{X} \mid \mathbf{XA}^{(1)'} \mid \mathbf{XA}^{(2)'} \mid \ldots \mid \mathbf{XA}^{(T)'}) \tag{20}$$

with

$$P = \sum_{t=0}^{T} (M - t), \tag{21}$$

then the FUNCLUS procedure is—except for the normalization—equivalent to the general variable weighting method provided the following equality constraints are imposed on the variable weights $v_1, \ldots, v_P$:

$$v_p = v_{p+1} = \ldots = v_q \tag{22}$$

with

$$q = \sum_{s=0}^{t} (M - s); \quad p = q - M - t + 1, \tag{23}$$

for $t = 0, 1, \ldots, T$. In the case of $N = 3$, $M = 4$, and $T = 2$, $\mathbf{Y}$ becomes

$$\mathbf{Y} = \begin{pmatrix} x_{11} & x_{12} & x_{13} & x_{14} & x_{12} - x_{11} & x_{13} - x_{12} & x_{14} - x_{13} & x_{13} - x_{11} & x_{14} - x_{12} \\ x_{21} & x_{22} & x_{23} & x_{24} & x_{22} - x_{21} & x_{23} - x_{22} & x_{24} - x_{23} & x_{23} - x_{21} & x_{24} - x_{22} \\ x_{31} & x_{32} & x_{33} & x_{34} & x_{32} - x_{31} & x_{33} - x_{32} & x_{34} - x_{33} & x_{33} - x_{31} & x_{34} - x_{32} \end{pmatrix},$$

and the sets of equality constraints defined by (22)–(23) are:

$$v_1 = v_2 = v_3 = v_4,$$

$$v_5 = v_6 = v_7$$

$$v_8 = v_9.$$

Of course, the algorithm developed in Section 3 entails a much more direct (and efficient) way of estimating the weighting coefficients in (7).

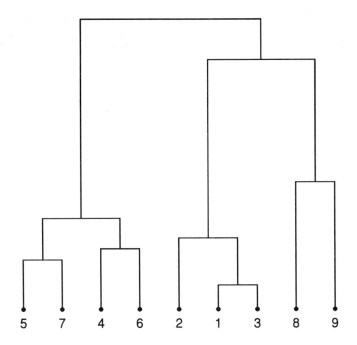

Fig. 4: Least-Squares Ultrametric Tree Representation of the Optimal Distances for $T = 2$.

Like the variable weighting method discussed above, the approach adopted in this paper involves a two-stage process. First, distances $\mathbf{D}$ are computed from the data matrix $\mathbf{X}$ such that $\mathbf{D}$ satisfies the ultrametric inequality as well as possible. Next, in the second step, these optimal distances are subjected to a hierarchical clustering method such as the least-squares ultrametric tree fitting procedure utilized in Section 4. As demonstrated by De Soete et al. (1985), it is in principle possible to combine these two steps into a single minimization problem. Such a combined approach simultaneously estimates from the data matrix $\mathbf{X}$, the weights $\alpha_0, \alpha_1, \ldots, \alpha_T$ and an $N \times N$ matrix of ultrametric distances $\mathbf{\Delta} = ((\delta_{ij}))$, such that the sum of squared deviations between the optimal distances d_{ij} and the corresponding ultrametric distances δ_{ij} is minimal. This amounts to solving the following constrained optimization problem:

$$
\begin{array}{c}
\text{minimize} \\
\alpha_0, \alpha_1, \ldots, \alpha_T \\
\mathbf{\Delta}
\end{array}
\left[L^*(\alpha_0, \alpha_1, \ldots, \alpha_T, \mathbf{\Delta}) = \frac{\displaystyle\sum_{i=2}^{N}\sum_{j=1}^{i-1}(d_{ij} - \delta_{ij})^2}{\displaystyle\sum_{i=2}^{N}\sum_{j=1}^{i-1} d_{ij}^2} \right] \tag{24}
$$

subject to

$$\alpha_0 > 0, \ \alpha_t \geq 0 \quad (\text{for } t = 1, \ldots, T),$$

$$\alpha_0 + \alpha_1 + \cdots + \alpha_T = 1,$$

and

$$\delta_{ij} \leq \max(\delta_{ik}, \delta_{jk}) \quad (\text{for } i, j, k = 1, \ldots, N).$$

14

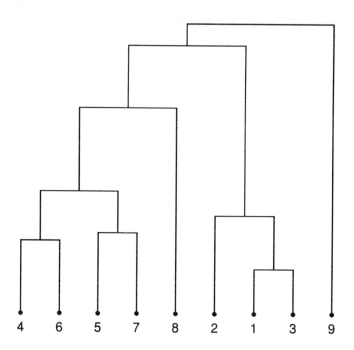

Fig. 5: Least-Squares Ultrametric Tree Representation of the Optimal Distances for $T = 1$.

While this minimization problem can be solved via an alternating least squares algorithm, it is expected that such a procedure will be very computationally intensive. Moreover, as was the case with the general variable weighting problem treated in De Soete (1986) and De Soete et al. (1985), it is very likely that the goodness of fit of the ultrametric tree obtained by solving (24) will not be substantially better in terms of L^* than the ultrametric tree representation arrived at by means of the much simpler two-stage least-squares approach illustrated in Section 4.

In this paper, we restricted ourselves to estimating distances between sampled functions that are optimally suited for an ultrametric tree representation. It should be obvious that the current approach can be easily extended to deal with other types of tree representations, such as additive tree representations (e.g., Sattath and Tversky, 1977). In the case of additive trees, distances between sampled functions are sought that optimally satisfy the additive inequality:

$$d_{ij} + d_{kl} \leq \max(d_{ik} + d_{jl}, d_{il} + d_{jk})$$

for all $i, j, k, l = 1, \ldots, N$ (see, for instance, Dobson, 1974). Such distances can be obtained by re-defining L in (16) as

$$L(\alpha_0, \alpha_1, \ldots, \alpha_T) = \frac{\displaystyle\sum_{\Omega_{\mathbf{D}}} (d_{ik} + d_{jl} - d_{il} - d_{jk})^2}{\displaystyle\sum_{i=2}^{N} \sum_{j=1}^{i-1} d_{ij}^2}$$

where Ω_D is now defined as

$$\Omega_D = \{\langle i, j, k, l \rangle \mid 1 \le i \le N, 1 \le j \le N, 1 \le k \le N, 1 \le l \le N,$$
$$d_{ij} + d_{kl} \le \min(d_{ik} + d_{jl}, d_{il} + d_{jk}) \text{ and } d_{ik} + d_{jl} \ne d_{il} + d_{jk}\}.$$

This modified minimization problem can be solved by an algorithm that is analogous to the one described in Section 3.

The FUNCLUS method assumes that the N objects are measured on a single variable at M equally spaced points. The requirement that the M points are equally spaced on the continuum on which the functions are defined, can be easily relaxed. Let $u_1 < u_2 < \ldots < u_M$ denote the values on the underlying continuum at which the N functions were sampled. The FUNCLUS algorithm as described in the previous sections assumes

$$u_2 - u_1 = u_3 - u_2 = \ldots = u_M - u_{M-1}. \tag{25}$$

If the sampling points are not equally spaced, (25) does not hold. However, in such a case, it suffices to re-define $\mathbf{A}^{(t)}$ as

$$a_{pq}^{(t)} = \begin{cases} \dfrac{1}{u_{p+t} - u_p} & \text{if } q = p + t \\ \dfrac{-1}{u_{p+t} - u_p} & \text{if } q = p \\ 0 & \text{otherwise,} \end{cases}$$

and the FUNCLUS algorithm for estimating the optimal weighting coefficients $\alpha_0, \alpha_1, \ldots, \alpha_T$ can be applied without any further modification.

Finally, in some applications, the N objects are measured on M successive occasions, not on a single variable, but on a number of variables, resulting in a three-way data array. Following the approach of De Soete and Carroll (1988, 1989) to ultrametric tree representations of three-way two-mode data, the FUNCLUS method can be extended to estimate distances that are optimally suited for a two-class ultrametric tree representation (cf. De Soete, DeSarbo, Furnas, and Carroll, 1984).

References

BESSE, P., and RAMSAY, J.O. (1986), Principal Component Analysis of Sampled Functions, *Psychometrika, 51, 285–311.*

DE SOETE, G. (1984a), A Least Squares Algorithm for Fitting an Ultrametric Tree to a Dissimilarity Matrix, *Pattern Recognition Letters, 2, 133–137.*

DE SOETE, G. (1984b), Computer Programs for Fitting Ultrametric and Additive Trees to Proximity Data by Least Squares Methods, *Behavior Research Methods, Instruments, & Computers, 16, 551–552.*

DE SOETE, G. (1986), Optimal Variable Weighting for Ultrametric and Additive Tree Clustering, *Quality & Quantity, 20, 169–180.*

DE SOETE, G., and CARROLL, J.D. (1988), Optimal Weighting for One-Mode and Two-Mode Ultrametric Tree Representations of Three-Way Three-Mode Data, in M. G.H. Jansen and W.H. van Schuur (eds.), *The Many Faces of Multivariate Data Analysis*, RION, Groningen, 16–29.

DE SOETE, G., and CARROLL, J.D. (1989), Ultrametric Tree Representations of Three-Way Three-Mode Data, in: R. Coppi and S. Bolasco (eds.), *Multiway Data Analysis*, North-Holland, Amsterdam, 415–426.

DE SOETE. G., CARROLL, J.D., and DESARBO, W.S. (1986), Alternating Least Squares Optimal Variable Weighting Algorithms for Ultrametric and Additive Tree Representations, in: W. Gaul and M. Schader (eds.), *Classification as a Tool of Research*, North-Holland, Amsterdam, 97–103.

DE SOETE, G., DESARBO, W.S., and CARROLL, J.D. (1985), Optimal Variable Weighting for Hierarchical Clustering: An Alternating Least-Squares Algorithm, *Journal of Classification, 2, 173–192*.

DE SOETE, G., DESARBO, W.S., FURNAS, G.W., and CARROLL, J.D. (1984), The Estimation of Ultrametric and Path Length Trees from Rectangular Proximity Data, *Psychometrika, 49, 289–310*.

DOBSON, A.J. (1974), Unrooted Trees for Numerical Taxonomy, *Journal of Applied Probability, 11, 32–42*.

GILL, P.E., MURRAY, W., and WRIGHT, M.H. (1981), *Practical Optimization*, Academic Press, London.

JOHNSON, S.C. (1967), Hierarchical Clustering Schemes, *Psychometrika, 32, 241–254*.

MILLIGAN, G.W. (1989), A Validation Study of a Variable Weighting Algorithm for Cluster Analysis, *Journal of Classification, 6, 53–71*.

POWELL, M.J.D. (1977), Restart Procedures for the Conjugate Gradient Method, *Mathematical Programming, 12, 241–254*.

RAMSAY, J.O. (1982), When the Data Are Functions, *Psychometrika, 47, 379–396*.

SATTATH, S., and TVERSKY, A. (1977), Additive Similarity Trees, *Psychometrika, 42, 319–345*.

WINSBERG, S., and KRUSKAL, J. (1986), Easy to Generate Metrics for Use with Sampled Functions, in: F. De Antoni, N. Lauro and A. Rizzi (eds.), *COMPSTAT 1986*, Physica-Verlag, Heidelberg, 55–60.

Spatial Clustering of Species Based on Quadrat Sampling

J. Krauth

Department of Psychology, University of Düsseldorf
Universitätsstrasse 1, W-4000 Düsseldorf, F.R.G.

Abstract: The quadrat sampling technique is frequently used in ecology for estimating diversity or analyzing the spatial point pattern of a population. For testing the null hypothesis of spatial randomness against cluster alternatives, Solow and Smith (1991) applied the species-area curve, i.e. the mean number of species in q quadrats. The authors proposed to use a simulation procedure because the calculation of the exact quantiles of the test is computationally costly for large numbers of quadrats, species, and individuals. As an alternative, we propose the use of maximum statistics, and we derive upper and lower bounds for the upper P-values. By combining the upper bounds for the different species we derive tests for spatial clusters. An extension similar to the species-area curve is suggested. The procedures are illustrated by analyzing two data sets from ecological studies.

1 Introduction

Species diversity is an expression community ecologists use when describing community structure. A diversity index seems to have first been proposed by Fisher, Corbet and Williams (1943), and many other indices have been suggested since then. Because the concept became increasingly vague, Hurlbert (1971) claimed that species diversity should be considered as a function of the number of species present (*species richness* or *species abundance*) and of the evenness with which the individuals are distributed among these species (*species evenness* or *species equitability*). A common sampling technique for estimating diversity (Heltshe and Forrester (1983b)), or species richness (Heltshe and Forrester (1983a)), or for analyzing the spatial point pattern of a population (Heltshe and Ritchey (1984)), is *quadrat sampling*. In this method a community containing S species is sampled by choosing at random Q quadrats of unit area. Then, the number of individuals of each species in each quadrat is counted. It is important that the quadrats are disjoint, but it is of course not necessary that they are quadrats in a mathematical sense. All of these disjoint "quadrats" merely have to be of the same size and shape.

A problem with quadrat sampling is that species richness depends on the number (Q) of quadrats sampled. To be able to compare the species richness of different populations estimated from different numbers of quadrats, ecologists therefore use the species-area curve. Let π_i be the probability (independent of the quadrats) that species i occurs in a quadrat and let $S(q)$ be the number of species in q quadrats. The *species-area curve* is defined as the function

$$s(q) = E[S(q)] = \sum_{i=1}^{S}(1 - (1 - \pi_i)^q), \quad q = 1, \ldots, Q - 1. \tag{1}$$

An unbiased estimate of $s(q)$ is given by

$$\hat{s}(q) = S - \sum_{i=1}^{S} \binom{Q - L_i}{q} \bigg/ \binom{Q}{q}, \quad q = 1, \ldots, Q - 1, \tag{2}$$

where L_i is the number of quadrats in which species i is present (Solow and Smith (1991)).

The estimate ($\hat{s}(q)$) of the species-area curve is formally equivalent to the *rarefaction curve* if we assume that S is the number of species in a parent collection, L_i is the number of individuals in the i th species, Q is the total number of individuals in the parent collection, and q is the number of individuals in the rarefied sample (Tipper (1979)). Here, $\hat{s}(q)$ would estimate the species richness a given parent collection of Q individuals would be expected to have were its size to be *rarefied* (restricted) to q individuals.

Solow and Smith (1991) used the species-area curve ($s(q)$) for detecting cluster in the spatial distribution of heterogeneous individuals. They considered the null hypothesis of *spatial randomness* for the spatial distribution of individuals. Let N_i be the total number of individuals of species i present in the Q quadrats for $i = 1, \ldots, S$. Then, conditional on $N_1, \ldots, N_S$, spatial randomness is defined by the following two assumptions (Solow and Smith (1991)):

(i) The individuals of each species are distributed among the Q quadrats independently of the individuals of the other species.

(ii) For each i, each of the Q^{N_i} ways of distributing the N_i individuals among the Q quadrats is equally likely.

The conditional sampling distribution of $\hat{s}(q)$ given $(N_1, \ldots, N_S)$ can be found numerically (Solow and Smith (1991)). For a given number of quadrats (q) the null hypothesis of no cluster can be rejected if $\hat{s}(q)$ is smaller than the lower α-quantile of the null distribution of $\hat{s}(q)$. For, if the individuals of species i are clustered in space, then L_i and therefore $\hat{s}(q)$ tend to be small. Because the calculation of the exact lower α-quantile of $\hat{s}(q)$ is computationally costly for large values of S, Q, and $N_1, \ldots, N_S$, Solow and Smith (1991) proposed to simulate the α-quantiles for $q = 1, \ldots, Q - 1$. For small values of q, $\hat{s}(q)$ is sensitive to spatially common species with π_i near to one, for large values of q, $\hat{s}(q)$ is sensitive to spatially rare species with π_i near to zero.

We propose an alternative procedure based on maximum statistics and offer upper and lower bounds for the upper quantiles that are easy to derive.

2 Bounds for the Upper Tails of a Maximum Statistic for Detecting Clusters

As above, we assume that we have sampled Q quadrats in which S species are observed. In quadrat j, the number of individuals of species i is denoted by N_{ij}, for $i = 1 \ldots, S$, $j = 1, \ldots, Q$. The total number of individuals of species i is denoted by N_i, where $N_i = N_{i1} + \cdots + N_{iQ}$, for $i = 1, \ldots, S$. All calculations are conditional on $N_1, \ldots, N_S$ and are performed under the assumption of spatial randomness. To start with, we shall consider only one single species. This is possible under the assumption of independence (i). From assumption (ii) we conclude that $(N_{i1}, \ldots, N_{iQ})$ follow an equiprobable multinomial distribution $M(N_i, 1/Q, \ldots, 1/Q)$. This implies that N_{ij} follows a binomial distribution $B(N_i, 1/Q)$ and that (N_{ij}, N_{ik}), for $j \neq k$, have a trinomial distribution $T(N_i, 1/Q, 1/Q)$. We now consider the statistic

$$M_i = \max_{1 \leq j \leq Q} N_{ij}. \tag{3}$$

Large values of M_i indicate the possibility of spatial clustering of species i. By defining the events $A_{ij} = \{N_{ij} \geq m_i\}$, we obtain the upper tail probability of M_i,

$$P(M_i \geq m_i) = P\left(\bigcup_{j=1}^{Q} A_{ij}\right). \tag{4}$$

The computation of the exact probability becomes infeasible for large sample sizes. We derive upper bounds for this probability on which we base a conservative test.

Obviously, we have

$$p_{ij} = P(A_{ij}) = \sum_{s=m_i}^{N_i} \binom{N_i}{s} \left(\frac{1}{Q}\right)^s \left(1 - \frac{1}{Q}\right)^{N_i - s} = p_{i1} \quad \text{for } 1 \leq j \leq Q, \tag{5}$$

$$p_{ijk} = P(A_{ij} \cap A_{ik}) = \sum_{s=m_i}^{N_i - m_i} \sum_{t=m_i}^{N_i - s} \frac{N_i!}{s! t! (N_i - s - t)!} \left(\frac{1}{Q}\right)^s \left(\frac{1}{Q}\right)^t \left(1 - \frac{2}{Q}\right)^{N_i - s - t}$$

$$= p_{i12} \quad \text{for } 1 \leq j < k \leq Q. \tag{6}$$

Due to a property of the multinomial distribution (Mallows (1968); Jogdeo and Patil (1975)) we can also conclude that

$$p_{i12} \leq p_{i1} p_{i2}. \tag{7}$$

To derive upper bounds for $P(M_i \geq m_i)$, we consider the Bonferroni bound

$$U_{1i} = \min\{1, Q p_{i1}\} \tag{8}$$

of degree one and the bound

$$U_{2i} = \min\{1, Q p_{i1} - (Q - 1) p_{i12}\} \tag{9}$$

of degree two. The bound U_{1i} is the best linear upper bound of degree one, and the bound U_{2i} is the best linear upper bound of degree two (Kounias and Marin (1974)).

To derive lower bounds for $P(M_i \geq m_i)$, we consider

$$L_{2i} = \frac{Q}{k_i(k_i + 1)}(2k_i p_{i1} - (Q - 1) p_{i12}), \quad k_i = 1 + [(Q - 1) p_{i12}/p_{i1}], \tag{10}$$

where $[x]$ denotes the integer part of x. The bound L_{2i} is the best linear lower bound of degree two (Kounias and Marin (1974)).

From Galambos (1977) it can be concluded that the expression for L_{2i} with arbitrary positive integer values of k_i gives lower bounds for $P(M_i \geq m_i)$. By considering (7) and (10) we therefore derive the lower bound

$$L_{1i} = \frac{Q}{k_i(k_i + 1)}(2k_i p_{i1} - (Q - 1) p_{i1}^2), \quad k_i = 1 + [(Q - 1) p_{i1}] \tag{11}$$

of degree one.

While the bounds of degree two are generally much sharper than the bounds of degree one, the latter ones only require the calculation of a binomial probability while for the former ones a trinomial probability has to be calculated in addition.

If we only consider species i and if the upper bound U_{2i} or even the upper bound U_{1i} is smaller than a given significance level α, we can conclude that this particular species i clusters in space. However, if we are interested in a general test of whether the S species cluster spatially, we should combine the S independent P-values or their upper bounds, respectively, to obtain a total P-value P_T, which is to be compared with α. This global test is significant for $P_T \leq \alpha$. The independence of the P-values follows from the assumption (i) of spatial randomness. If we wish to test whether there is a general tendency of the species to cluster in space, we can consider the sum of the P-values (or of their corresponding upper bounds). If we denote this sum by $D = P_1 + \cdots + P_S$, we can conclude from Edgington (1972) that in this case we have

$$P_T = \frac{D^S}{S!} - \frac{(D-1)^S}{1!(S-1)!} + \frac{(D-2)^S}{2!(S-2)!} \mp \cdots, \tag{12}$$

where for $D \leq 1$ only the first term is considered. For $D > 1$, correction terms are used as far as the difference in the numerator is positive. Edgington (1972) proved that this procedure is conservative if the P-values are calculated for discrete distributions, as in our case.

If we wish to test whether at least one species tends to cluster, or if we want to know which species exhibit spatial clustering, we should use a multiple comparison procedure. For all species with

$$P_i \leq 1 - (1 - \alpha)^{1/S} \tag{13}$$

we can conclude that the null hypothesis of spatial randomness can be rejected. This procedure ensures that the maximum probability of obtaining at least one significant result, though all null hypotheses are valid, will always equal α (Krauth (1988) pp. 35-36).

A problem with the two methods discussed earlier and the test proposed by Solow and Smith (1991) is that they reject the null hypothesis of spatial randomness. This means that a significant result might solely be due to a violation of the assumption of independence (i) while the uniformity assumption (ii) might hold. If we wish to avoid this problem, we can use the Bonferroni procedure

$$P_i \leq \alpha/S \tag{14}$$

or the more powerful Holm (1979) procedure.

For performing the significance tests described above, only upper bounds of the P-values are needed. The lower bounds can be used as a means of evaluating the conservativeness of the tests. The lower bounds might also be useful if we are interested in detecting not spatial clusters of species, but regular spatial patterns (Heltshe and Ritchey (1984)). Small values of the statistics M_i may be caused by such regular patterns. Observing that

$$P(M_i \leq m_i) = 1 - P(M_i \geq m_i + 1) \tag{15}$$

holds, we derive an upper bound for $P(M_i \leq m_i)$ by replacing $P(M_i \geq m_i + 1)$ by a lower bound. However, this test for regular patterns will not be of much use because the distribution of M_i is positively skewed, i.e. small values of M_i occur with a high probability if the null hypothesis holds.

3 Extensions

Similarly as for the species-area curve, we might consider instead of the largest value (M_i) of the N_{ij} the sum $M_i^{(q)}$ of the q largest values of the N_{ij}:

$$M_i^{(q)} = \max_{j_1 < j_2 < \cdots < j_q} \sum_{t=1}^{q} N_{ij_t}, \quad q = 1, \ldots, Q-1. \tag{16}$$

With

$$A_{ij_1 \cdots j_q} = \left\{ \sum_{t=1}^{q} N_{ij_t} \geq m_i \right\} \tag{17}$$

we derive

$$P(M_i^{(q)} \geq m_i) = P\left(\bigcup_{j_1 < j_2 < \cdots < j_q} A_{ij_1 \cdots j_q} \right). \tag{18}$$

Because $N_{ij_1} + \cdots + N_{ij_q}$ follows a binomial distribution $B(N_i, q/Q)$ we obtain

$$p_{ij_1 \cdots j_q}^{(q)} = P(A_{ij_1 \cdots j_q}) = \sum_{t=m_i}^{N_i} \binom{N_i}{t} \left(\frac{q}{Q}\right)^t \left(1 - \frac{q}{Q}\right)^{N_i - t} = p_{i1 \cdots q}^{(q)}. \tag{19}$$

The upper Bonferroni bound of degree one is then given by

$$U_{1i}^{(q)} = \min\left\{ 1, \binom{Q}{q} p_{i1 \cdots q}^{(q)} \right\}. \tag{20}$$

The formulas for the upper bound of degree two and the lower bounds are more complicated and are neglected here.

4 Applications

Solow and Smith (1991) reported counts of $S = 4$ species of seabirds (1 = murre, 2 = crested auklet, 3 = least auklet, 4 = puffin) in $Q = 10$ quadrats (size 250 m by 250 m) in the Anadyr Strait off the coast of Alaska during the summer of 1988. The counts are reproduced in Tab. 1.

Tab. 1: Seabird data

Species	1	2	3	4	5	6	7	8	9	10	N_i	M_i
1	0	0	0	1	1	0	0	1	1	3	7	3
2	0	0	0	2	3	1	5	0	1	5	17	5
3	1	2	0	0	0	0	1	3	2	3	12	3
4	1	0	1	1	0	0	3	1	1	0	8	3

All computations were carried out on a DELL System 325 using Turbo C2.0. The binomial and trinomial probabilities were calculated on a logarithmic scale. By choosing $q = 1$ we derived the upper bounds (9) $U_{21} = .2558$, $U_{22} = .2202$, $U_{23} = 1$, and $U_{24} = .3773$ for the four species. The sum of these values yields $D = 1.8533$, from which $P_T = .4032$ results. From $P_T > \alpha = .05$ we conclude that there is no general tendency of the species to cluster in space. Because all four bounds are larger than $\alpha = .05$, no multiple test procedure will detect deviations from the null hypothesis of spatial randomness. If in this example we consider the extension (16) with $q = 4$, we find $M_1^{(4)} = 6$, $U_{11}^{(4)} = 1$, $M_2^{(4)} = 15$, $U_{12}^{(4)} = .0120$, $M_3^{(4)} = 10$, $U_{13}^{(4)} = .5901$, $M_4^{(4)} = 6$, $U_{14}^{(4)} = 1$. This leads to $D = 2.6021$ and $P_T = .8451$, i.e. there is again no general tendency of the species to cluster. However, for the second species we find $P_2 < 1 - (1 - .05)^{1/4} = .0127$ and $P_2 < .05/4 = .0125$ which indicates spatial clustering of the second species. Of course, the choice of $q = 4$ should be made before analyzing the data. In this example, we would have obtained the same result with $q = 3$.

A larger set of data was examined in Heltshe and Forrester (1983a), where $S = 14$ species (1 = Streblospio benedicti, 2 = Nereis succines, 3 = Polydora ligni, 4 = Scoloplos robustus, 5 = Eteone heteropoda, 6 = Heteromastus filiformis, 7 = Capitella capitata, 8 = Scolecolepides viridis, 9 = Hypaniola grayi, 10 = Branis clavata, 11 = Macoma balthica, 12 = Ampelisca abdita, 13 = Neopanope texana, 14 = Tubifocodies sp.) were observed in $Q = 10$ quadrats. This benthic infaunal sample was collected from a subtidal marsh creek in the Pettaquamscutt River in southern Rhode Island in April 1978. These data are reproduced in Tab. 2.

Tab. 2: Subtidal marsh creek data

Species	Quadrats 1	2	3	4	5	6	7	8	9	10	N_i	M_i	U_{2i}
1	0	13	21	14	5	22	13	4	4	27	123	27	.0000
2	2	2	4	4	1	1	1	0	1	6	22	6	.0510
3	0	1	0	0	0	0	0	1	0	0	2	1	1
4	1	0	1	2	0	6	0	0	1	2	13	6	.0021
5	0	0	1	2	0	0	1	0	0	1	5	2	.6084
6	1	1	2	1	0	1	0	0	1	5	12	5	.0134
7	1	0	0	0	0	0	0	0	0	0	1	1	1
8	2	0	0	0	0	0	0	0	0	0	2	2	.0714
9	0	1	0	0	0	0	0	0	0	0	1	1	1
10	0	0	1	0	0	0	0	0	0	0	1	1	1
11	0	0	3	0	0	0	0	0	0	2	5	3	.0457
12	0	0	5	1	0	2	0	0	0	3	11	5	.0083
13	0	0	0	0	0	0	0	1	0	0	1	1	1
14	8	36	14	19	3	22	6	8	5	41	162	41	.0000

Here we find $D = 5.8004$ and $P_T = .1349$, indicating that there is no general tendency of the species to cluster spatially. Using multiple comparison procedures, we obtain from (13) $1 - (1 - .05)^{1/14} = .0037$, indicating that species 1, 4, and 14 exhibit spatial clustering. If independence between the species is not assumed, we can consider the Bonferroni bound (14) $.05/14 = .0036$, from which it can again be concluded that species 1, 4, and 14 cluster.

To apply the Holm procedure, we compare the smallest bound (species 14) with $.05/14$, the second smallest bound (species 1) with $.05/13$, the third smallest bound (species 4) with $.05/12$, and the fourth smallest bound (species 12) with $.05/11$. We thus obtain $.0000 < .0036$, $.0000 < .0038$, $.0021 < .0042$, $.0083 > .0045$. Again, only species 1, 4, and 14 are found to cluster, while for species 12, and thus for all species with even higher bounds, the null hypothesis of spatial randomness cannot be rejected.

References

EDGINGTON, E.S. (1972), An Additive Method for Combining Probability Values for Independent Experiments, *Journal of Psychology, 80*, 351-363.

FISHER, R.A., CORBET, A.S. and WILLIAMS, C.B. (1943), The Relation Between the Number of Species and the Number of Individuals in a Random Sample of an Animal Population, *Journal of Animal Ecology, 12*, 42-58.

GALAMBOS, J. (1977), Bonferroni Inequalities, *Annals of Probability, 5*, 577-581.

HELTSHE, J.F. and FORRESTER, N.E. (1983a), Estimating Species Richness Using the Jackknife Procedure, *Biometrics, 39*, 1-11.

HELTSHE, J.F. and FORRESTER, N.E. (1983b), Estimating Diversity Using Quadrat Sampling, *Biometrics, 39*, 1073-1076.

HELTSHE, J.F. and RITCHEY, T.A. (1984), Spatial Pattern Detection Using Quadrat Samples, *Biometrics, 40*, 877-885.

HOLM, S. (1979), A Simple Sequentially Rejective Multiple Test Procedure, *Scandinavian Journal of Statistics, 6*, 65-70.

HURLBERT, S.H. (1971), The Nonconcept of Species Diversity: a Critique and Alternative Parameters, *Ecology, 58*, 577-586.

JOGDEO, K. and PATIL, G.P. (1975), Probability Inequalities for Certain Multivariate Discrete Distributions, *Sankhyā, Series B, 37*, 158-164.

KOUNIAS, E. and MARIN, D. (1974), Best Linear Bonferroni Bounds, in: *Proceedings of the Prague Symposium on Asymptotic Statistics*, Volume II, Charles University, Prague, 179-213.

KRAUTH, J. (1988), *Distribution-Free Statistics: An Application-Oriented Approach*, Elsevier, Amsterdam.

MALLOWS, C.L. (1968), An Inequality Involving Multinomial Probabilities, *Biometrika, 55*, 422-424.

SOLOW, A.R. and SMITH, W. (1991), Detecting Cluster in a Heterogeneous Community Sampled by Quadrats, *Biometrics, 47*, 311-317.

TIPPER, J.C. (1979), Rarefaction and Rarefiction - the Use and Abuse of a Method in Paleoecology, *Paleobiology, 5*, 423-434.

A k_n-Nearest Neighbour Algorithm for Unimodal Clusters

A. Kovalenko

Russian Transputer Association, 4-th Likhachevsky Lane 15,
125438, Moscow, RUSSIA

Abstract: This paper is devoted to the problem of construction of unimodal clusters relative to local modes of density. Let us consider a given set of multidimensional observations $x_1, \ldots, x_n \in \mathbf{R}^d$ as realizations of n independent d-dimensional random vectors, all having the same distribution density f. High-density clusters are the connected components of the set $\{x \mid x \in \mathbf{R}^d,\ f(x) \geq c\}$, where the (continuous) density f exceeds some given threshold $c > 0$. The high-density cluster $\mathcal{B}(c)$ is said to be an *unimodal cluster*, if for any $c' \geq c$ there is no more than one high-density cluster $\mathcal{B}(c') \subseteq \mathcal{B}(c)$. The proposed k_n-nearest neighbour algorithm for an unimodal cluster is a modification of Wishart's Mode-Analysis-Algorithm. It is based on the uniformly consistent k_n-nearest neighbour density estimate. It is proved that the proposed algorithm is asymptotically consistent for unimodal clusters.

1 Introduction

One of the cluster analysis methods provides a selection of multivariate density function mode regions based on a given random sample. A review of works described mode analysis algorithms and proof of their consistency can be found in Bock (1985, 1989). An hierarchical clustering algorithm using a k_n-nearest neighbours density estimate is given in Wong and Lane (1981). This algorithm is based on a high-density clusters model (Bock (1974), Hartigan (1985)). A proof of consistency for this algorithm is based on the strong uniform convergency of the k_n-nearest neighbour density estimate (Devroye/Wagner (1977)). An unimodal clustering model and an algorithm for their creation that is a modification of Wishart's Algorithm (Wishart (1969)) are provided in this paper.

2 The Unimodal Clustering Model

Suppose a set of data vectors is randomly sampled and characterized by a probability density. Let an algorithm for partitioning a set of data vectors into a number of classes be given. The problem of consistency of the algorithm consists in setting up a conformity between classes of data and theoretical clusters. Therefore, a research of consistency of an algorithm must be started from which the definition of theoretical clusters, which the algorithm will partition.

We call maximal connected subsets of a set $\{x \mid x \in \mathbf{R}^d,\ f(x) \geq c\}$ the high-density clusters or c-clusters (Bock (1974), Hartigan (1975)), where $\mathbf{R}^d$ is a d-dimensional space, and $f(x)$ is the positive, uniformly continuous density on $\mathbf{R}^d$ with a finite number m of several local modes (local maxima of f). The family of c-clusters built on $f(x)$ for any $c > 0$ forms a hierarchical clustering. The asymptotical consistent hierarchical clustering procedure is given by Wong/Lane (1981).

We will study unimodal clusters. Let $\mathcal{B}(c)$ be a high-density cluster at given threshold $c > 0$. The set $\mathcal{B}(c)$ is called the *unimodal cluster* if for any $c' \geq c$ there exists no more than one high-density cluster $\mathcal{B}(c') \subseteq \mathcal{B}(c)$. Let $\mathbf{B} = \{\mathcal{B}_1(c_1), \ldots, \mathcal{B}_m(c_m)\}$ be the set of unimodal clusters built by the density $f(x)$, such that $\mathcal{B}_i \cap \mathcal{B}_j = \emptyset$ for all $1 \leq i < j \leq m$ and $\mu(\mathcal{B}_i) > 0$ is the Lebesgue measure. Let

$$H_i := \max_{x \in \mathcal{B}_i}\{f(x) - c_i\}$$

be the height of cluster $\mathcal{B}_i$. Let $H := \min_i\{H_i\}$.

Let $x_1, \ldots, x_n \in \mathbf{R}^d$ be n d-dimensional observations considered as realizations of n independent d-dimensional random vectors $X_1, \ldots, X_n$, all with the same distribution density $f(x)$. Suppose that we have partitioned with the help of the Mode-Analysis-Algorithm the data vectors $\{x_1, \ldots, x_n\}$ for $\nu_n \geq 1$ classes $\mathbf{B}_n = \{\mathcal{B}_{n1}, \ldots, \mathcal{B}_{n\nu_n}\}$ and a background-class $\mathcal{B}_{n0}$. (By a background-class we mean the elements of $\{x_1, \ldots, x_n\}$ that are not included in $\mathbf{B}_n$.) The partition $\mathbf{B}_n$ is said to be *strongly consistent* for unimodal clustering $\mathbf{B}$ if

$$\mathbf{P}\{\nu_n \to m \text{ as } n \to \infty\} = 1$$

and if with probability 1 for n large for any pair $\mathcal{B}_i, \mathcal{B}_j \in \mathbf{B}$, $i \neq j$, there exist $\mathcal{B}_{ns}, \mathcal{B}_{nt} \in \mathbf{B}_n$ with $\mathcal{B}_{ns} \supset \mathcal{B}_i \cap \{x_1, \ldots, x_n\}$, $\mathcal{B}_{nt} \supset \mathcal{B}_j \cap \{x_1, \ldots, x_n\}$ and $\mathcal{B}_{ns} \cap \mathcal{B}_{nt} = \emptyset$.

3 A k_n-Nearest Neighbour Algorithm for Unimodal Clusters

The proposed algorithm for unimodal clusters is based on the uniformly consistent k_n nearest neighbour density estimate: the estimated density at point x is

$$f_n(x) := k_n/(nV_{k_n}(x)),$$

where $V_{k_n}(x)$ is the volume of the closed sphere centered at x containing k_n sample points. We denote the radius of $V_{k_n}(x)$ by $\rho_{k_n}(x)$ and the Euclidean distance by $\rho(\cdot, \cdot)$.

We define by a graph the relation of "neighbourhood" $G(S_n, U_n)$ with the set $S_n = \{x_1, \ldots, x_n\}$ as vertices and the set

$$U_n = \{(x_i, x_j) \mid \rho(x_i, x_j) \leq \rho_{k_n}(x_i), \; i, j = 1, \ldots, n, \; i \neq j\}$$

as edges. We also define the family of subgraphs $G(S_i, U_i)$, $1 \leq i \leq n$, where $S_i = \{x_1, \ldots, x_i\}$, $U_i = U_n \cap (S_i \times S_i)$.

Suppose class $\mathcal{B}_{ns}$ has a significant height relative to the level $h > 0$ if

$$H_{ns} := \max_{x_i, x_j \in \mathcal{B}_{ns}} \{|f_n(x_i) - f_n(x_j)|\} \geq h.$$

The proposed algorithm consists of four stages:

The Choice of Parameters Stage. Take the k_n number of nearest neighbours and level h $(0 < h < H)$.

The Calculation of Distances to the k_n-th Nearest Neighbour Stage. For any $i = 1, \ldots, n$ find distance $\rho_{k_n}(x_i)$ to the k_n-the nearest neighbour from set $\{x_1, \ldots, x_n\}$.

The Sorting and Renumbering Stage. Number the observations $x_1, \ldots, x_n$ in increasing order of $\rho_{k_n}(x_i)$.

The Unimodal Clustering Stage. At this stage we number the classes using their first elements as an estimate of the corresponding local mode. Let us define the background-class as having unsignificant height. We will distinguish classes that have been formed up to this step (formed classes) and classes that are being formed now. Let $\omega(x_i)$ be the number of the class in which the observation x_i exists and let $i = 1$.

Step 1. Consider subgraph $G(S_i, U_i)$. The following cases are possible:

1.1. The vertice x_i is isolated in subgraph $G(S_i, U_i)$. Begin to form a new class $\mathcal{B}_{ni}$, where $\omega(x_i) = i$. Go to step 2.

1.2. The vertice x_i is connected only with vertices from $\mathcal{B}_{ns}$. If the s-th class has already been formed, let $\omega(x_i) = 0$. Otherwise, $\omega(x_i) = s$. Go to step 2.

1.3. The vertice x_i is connected with vertices from classes $\mathcal{B}_{ns_1}, \ldots, \mathcal{B}_{ns_t}$, where $s_1 < s_2 < \ldots < s_t$, $t \geq 2$.

1.3.1. If all t classes are formed, then $\omega(x_i) = 0$. Go to step 2.

1.3.2. If unformed classes exist, verify the significance of heights of classes $\mathcal{B}_{ns_1}, \ldots, \mathcal{B}_{ns_t}$ relative to level h. Let $z(h)$ be the number of significant classes.

(a) If $z(h) > 1$ or $s_1 = 0$, then set $\omega(x_i) = 0$ and classes with the significant height must be marked as formed and the classes with unsignificant height must be removed, with $\omega(x_j) = 0$ for all included x_j.

(b) If $z(h) \leq 1$ and $s_1 > 0$, then classes $\mathcal{B}_{ns_2}, \ldots, \mathcal{B}_{ns_t}$ must be included into class $\mathcal{B}_{ns_1}$, where $\omega(x_j) = s_1$ for all included x_j. We set $\omega(x_i) = s_1$.

Step 2. Let $i = i + 1$. If $i \leq n$ then repeat step 1. Otherwise go to next stage.

The Renumbering of Classes. Let ν_n be equal to the number of classes. Renumber classes (except the background-class) in the increasing row of numbers $1 \leq s \leq \nu_n$.

The main difference between the given algorithm and Wishart's Mode-Analysis-Algorithm (Wishart (1969)) consists of checking the significance of the height of the formed clusters. This algorithm is easily transformed into Wishart's algorithm if the conditions of checking the significance of the height and the formed classes and the corresponding steps be excluded.

4   The Strong Consistency of Unimodal Clustering

Suppose, that

(i) $f(x)$ is a positive uniformly continuous function on $\mathbf{R}^d$ with a finite number m of several local modes;

(ii) $x_1, \ldots, x_n \in \mathbf{R}^d$ are realizations of n independent d-dimensional random vectors $X_1, \ldots, X_n$, all with the same distribution density $f(x)$;

(iii) k_n is a sequence of positive integers such that $k_n/n \to 0$ and $k_n/\log n \to \infty$ as $n \to \infty$.

Lemma 1: If (i)-(iii) are satisfied, then $\sup_x |f_n(x) - f(x)| \to 0$ with probability 1.

The proof of Lemma 1 is given by Devroye/Wagner (1977).

Let A be an unimodal c-cluster with density $f(x)$ and $\mu(A) > 0$. Consider the subgraph $G(S_A, U_A)$ with the set of vertices $S_A = A \cap \{x_1, \ldots, x_n\}$ and the set of edges $U_A = U_n \cap (S_A \times S_A)$, with the subgraph corresponding to the relation of the "neighbourhood" between points of S_A.

Lemma 2: *If (i)–(iii) are satisfied, then*

$$\mathbf{P}\{G(S_A, U_A) \ \text{is connected as } n \to \infty\} = 1.$$

Proof: Function $f(x)$ is integrable and uniformly continuous. Therefore, it is limited on A and hence $0 < M < \infty$ exists such that $f(x) < M$ everywhere on A. Fix $0 < \epsilon < c$. By Lemma 1, with probability 1 there exists an integer n_0 such that for $n > n_0$ and any $x \in A$

$$\rho_{k_n}(x) > \left(\frac{k_n}{n(M + \epsilon)V_d}\right)^{1/d} := 4\delta_n,$$

where V_d is the volume of the unit sphere. Therefore, if a distance between any $x', x'' \in S$ is equal to, or less than, $4\delta_n$, they must be related as neighbours, viz. $\rho(x', x'') \le \rho_{k_n}(x')$ or $\rho_{k_n}(x'')$.

Sequence δ_n satisfies conditions $\delta_n^d \to 0$, and $n\delta_n^d/\log n \to \infty$, as $n \to \infty$. The function $g_n(x) = K(x)/nV_d\delta_n^d$, where $K(x)$ is the number of points from $\{x_1, \ldots, x_n\}$ situated in the sphere centered at x with the radius δ_n, is the kernel estimate of density $f(x)$. From Devroye/Wagner (1980) it follows that with probability 1 $\sup_x |g_n(x) - f(x)| \to 0$ as $n \to \infty$. Therefore, for all $x \in A$ there exists, with probability 1, an integer $n_1 > n_0$ such that for with $n > n_1$

$$K(x) > (c - \epsilon)nV_d\delta_n^d \ge 1.$$

It means that in δ_n-neighbourhood $\Omega_{\delta_n}(x)$ of any point $x \in A$ there is at least one point from set $\{x_1, \ldots, x_n\}$.

From the connectivity and unimodality of cluster A it follows that there exists, with probability 1, an integer $n_2 > n_1$ such that for $n > n_2$ there are any two points $x', x'' \in S$, $\Omega_{\delta_n}(x') \subset A$ and $\Omega_{\delta_n}(x'') \subset A$ that can be connected with the finite path $L_{\delta_n}(x', x'')$, lying in A with its own δ_n-neighbourhood.

Let us divide the path $L_{\delta_n}(x', x'')$ on intervals with length $2\delta_n$. At least one point of the set $\{x_1, \ldots, x_n\}$ exists in δ_n-neighbourhood of the beginning and ending points of every interval. The distance between such points is not longer than $4\delta_n$. Therefore, these points are connected as neighbours, that proves connectivity of graph $G(S_A, U_A)$. Proof of Lemma 2 is completed.

Theorem: *Let* $\mathbf{B} = \{\mathcal{B}_1(c_1), \ldots, \mathcal{B}_m(c_m)\}$ *be the set of unintersecting unimodal clusters with a non-zero measure built by density* $f(x)$ *and let* $\mathbf{B}_n = \{\mathcal{B}_{n1}, \ldots, \mathcal{B}_{n\nu_n}\}$ *be the partition of set* $\{x_1, \ldots, x_n\}$ *specified by the unimodal clustering algorithm. Then, provided that (i)–(iii) satisfied, the partition* $\mathbf{B}_n$ *is strongly consistent for unimodal clustering* $\mathbf{B}$.

Proof: By the above definition, for any pair of unimodal clusters $\mathcal{B}_i, \mathcal{B}_j \in \mathbf{B}$, $1 \le i < j \le m$, there exist $0 < \epsilon < \min\{(H - h)/2, h/2\}$, $\lambda_{ij} > 0$ and n_0, such that for $n > n_0$,

(a) $f(x) \ge \lambda_{ij}$ for all $x \in \mathcal{B}_i \cup \mathcal{B}_j$, and

(b) each rectilinear path between $\mathcal{B}_i$ and $\mathcal{B}_j$ contains a segment with a length greater than $\delta_n(\epsilon, \lambda_{ij})$,

$$\delta_n(\epsilon, \lambda_{ij}) > 2\left(\frac{k_n}{n(\lambda_{ij} - \epsilon)V_d}\right)^{1/d},$$

along which the density $f(x) < \lambda_{ij} - 3\epsilon$.

By Lemma 1, there exists with probability 1 an integer $n_1 > n_0$ such that $\sup_x |f_n(x) - f(x)| < \epsilon$ for $n > n_1$. Thus from (a) and (b) it follows that with probability 1 for $n > n_1$,

(c) $f_n(x) > \lambda_{ij} - \epsilon$ for all $x \in \mathcal{B}_i \cup \mathcal{B}_j$, and

(d) each rectilinear path betwen $\mathcal{B}_i$ and $\mathcal{B}_j$ contains a segment of a length greater than $\delta_n(\epsilon, \lambda_{ij})$ along which the density estimate $f_n(x) < \lambda_{ij} - 2\epsilon$.

Since $\mathcal{B}_i$ and $\mathcal{B}_j$ are disjoint, from (c) and (d) it follows that points from $\mathcal{B}_i$ and $\mathcal{B}_j$ do not have mutual neighbours among points from $\{x_1, \ldots, x_n\}$ in which the density estimate is greater than $\lambda_{ij} - \epsilon$. Therefore, the points from sets $\mathcal{B}_i \cap \{x_1, \ldots, x_n\}$ and $\mathcal{B}_j \cap \{x_1, \ldots, x_n\}$ will be separated into different classes from $\mathcal{B}_n$.

By Lemma 2, we have for n large the points of sets $\mathcal{B}_i \cap \{x_1, \ldots, x_n\}$, $1 \leq i \leq m$, form, with probability 1, connected graphs of the relation of "neighbourhood". From condition $h < H$ by Lemma 1 for $n > n_1$ the clustering of points from $\mathcal{B}_i \cap \{x_1, \ldots, x_n\}$ can lead to the building of no more than one $\mathcal{B}_{ns} \in \mathbf{B}_n$ of significant height such that $\mathcal{B}_{ns} \supset \mathcal{B}_i \cap \{x_1, \ldots, x_n\}$. By Lemma 1, with probability 1 the number of classes with the significant height, ν_n, tends to m. The proof of Theorem is complete.

5 Clustering of Points from Background-Class

In addition to unimodal clusters, the algorithm also builds background-class $\mathcal{B}_{n0}$ that contains "rejected" points. It is often necessary to build the clustering of all the observations. In that case the algorithm can be added by the stage of point classification from the background-class using the rule of the nearest neighbour. The classification should be done in order of increasing point numbers from the background-class.

6 Numerical Example

Let the random sample $n = 50$ from a mixture of two bivariate normal populations

$$(1/2)\mathcal{N}(\mu_1, \Sigma_1) + (1/2)\mathcal{N}(\mu_2, \Sigma_2)$$

be given, where the mean vectors $\mu_1 = (0, 0)$, $\mu_2 = (0, 6)$ and the covariance matrices

$$\Sigma_1 = \begin{pmatrix} 2 & 0 \\ 0 & 1 \end{pmatrix}, \quad \Sigma_2 = \begin{pmatrix} 1 & 0 \\ 0 & 2 \end{pmatrix}.$$

The scatter-plot of this sample set is shown in Figure 1, in which the observation numbers are plotted next to the observations. Let $k_n = 7$ and $h = 0.01$. The above algorithm has builded two classes as shown in Figure 1 by continuous lines. For the first class $H_{n1} \approx 0.038$ and for second class $H_{n2} \approx 0.036$. Hence, the result of clustering with chosen parameters is near the expected one.

The point clustering from the background-class using the rule of the nearest neighbour leads to the results shown in Figure 1 with the dotted arrows.

7 Choice of Parameters

To use this algorithm we must set parameters k_n and h. Using the condition *(iii)* we can set, for example, $k_n \sim O(n^b)$, where $0 < b < 1$. The value of h depends on n and is defined by the speed of convergence of the k_n nearest neighbours density estimate. There is a problem of choosing the value of h for finite n. If we choose h small enough, we will probably get "accidental" clusters. With big enough h we can lose real clusters.

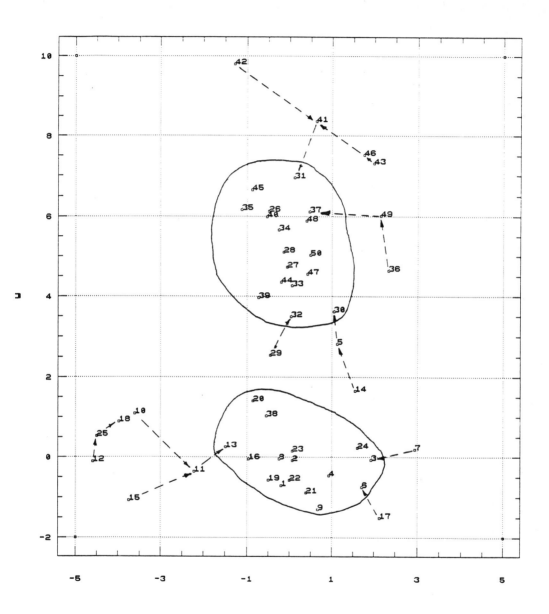

Figure 1: Plot of y vs. x

Acknowledgements

The author is grateful to Prof. Dr. H.H. Bock of the Institut für Statistik und Wirtschafts-mathematik, Reinisch-Westfälische Technische Hochschule Aachen, F.R.G., for useful discussions of a preliminary version of this paper.

References

BOCK, H.H. (1974), *Automatische Klassifikation. Theoretische und praktische Methoden zur Gruppierung und Strukturierung von Daten (Clusteranalyse)*, Vandenhoeck&Ruprecht, Göttingen.

BOCK, H.H. (1985), On some significance tests in cluster analysis, *Journal of Classification*, 2, 77–108.

BOCK, H.H. (1989), Probabilistic Aspects in Cluster Analysis, in: O. Opitz (ed.), *Conceptual and Numerical Analysis of Data*, Springer, Berlin, 12–44.

DEVROYE, L., WAGNER, T.J. (1977), The Strong Uniform Consistency of Nearest Neighbor Density Estimates, *Annals of Statistics, 5, No. 3, 536–540.*

DEVROYE, L., WAGNER, T.J. (1980), The Strong Uniform Consistency of Kernel Density Estimates, in: P.R. Krishnaiah (ed.): *5th Int. Symp. on Multivariate Analysis*, New York, 59–77.

HARTIGAN, J.A. (1975), *Clustering Algorithms*, John Wiley, New York.

WISHART, D. (1969), Mode Analysis: A generalization of nearest neighbor which reduces chaining effects, in: A. Cole (ed.): *Numerical Taxonomy*, Academic Press, New York, 282–319.

WONG, M.A., LANE, T. (1981), A kth Nearest Neighbour Clustering Procedure, in: *Comput. Sci. and Statist. Proc. 13th Symp. Interface*, Pittsburg, New York, 308–311.

Asymptotic Robustness in Cluster-Analysis for the Case of Tukey-Huber Distortions

Yu. Kharin, E. Zhuk

Department of Mathematical Modeling and Data Analysis, Byelorussian State University,
pr. Franciska Skariny 4, 220050, Minsk-50, Republic of Byelarus

Abstract: We consider the problem of cluster-analysis of multivariate observations in the situation when their probabilistic models are subjected to Tukey-Huber distortions and parametric plug-in decision rules with minimum contrast estimates for unknown parameters are used. We construct asymptotic expansions for the risk (the expected losses) and the robustness coefficient for classical rules and for rules used "truncated" contrast functions. The case of the "contaminated" Fisher model is investigated.

1 Introduction

Real data, subject to cluster-analysis, are often not fully adequate for using traditional probabilistic models in practice (Aivazyan at al. (1989), Huber (1981)). Therefore probabilistic distributions, which define the classes, should allow for these distortions. The risk of the classical decision rules (DR) is minimal only for the "ideal" ("not distorted") model (see, for example, Aivazyan at al. (1989), Bock (1989)). Therefore the problem of the evaluation of the classical DR stability (robustness) and the construction of a stable (robust) DR in the sense of the risk functional are very topical. In this paper the method of asymptotic expansions of the risk is developed, which was used earlier for the construction of the robust DR in discriminant analysis by Kharin (1982, 1983).

2 Mathematical Model of Data and Decision Rules

Let us consider n jointly independent random observations $x_1, \ldots, x_n$ in R^N from $L \geq 2$ classes $\{\Omega_1, \ldots, \Omega_L\}$. Unlike the classical model (Aivazyan at al. (1989)), an observation from Ω_i is a random N-vector with probability density $p_i(x; \theta_i^o), x \in R^N$, which may be a distorted one:

$$p_i(\cdot; \theta_i^o) \in \mathcal{P}_i(\varepsilon_{+i}), \quad i \in S = \{1, \ldots, L\}; \tag{1}$$

$\mathcal{P}_i(\varepsilon_{+i})$ is the set of admissible densities for Ω_i and ε_{+i} is the level of distortions for Ω_i: If $\varepsilon_{+i} = 0$, then there is no distortion in Ω_i, $p_i(\cdot; \theta_i^o) \equiv q(\cdot; \theta_i^o)$ is the "not distorted" parametric probability distribution density; for this case $\mathcal{P}_i(0)$ contains the only element $q(\cdot; \theta_i^o)$); $\theta_i^o \in \Theta \subseteq R^m$ is the true unknown value of the parameter for the i-th class.

Let us introduce some notation: $d_t^o \in S$ is the unknown true index of the class to which the observation x_t belongs; $D^o = (d_1^o, \ldots, d_n^o)^T$ is the true classification vector of the sample $X = (x_1, \ldots, x_n)^T$, where "T" is the transposition symbol. A priori $\{d_t^o\}$ are jointly independent, identically distributed discrete random values with the distribution $\pi_i = \mathbf{P}\{d_t^o = i\}$, $i \in S$, ($\sum_{i \in S} \pi_i = 1$). We consider a given loss matrix $W = (w_{ik})$, where $w_{ik} \geq 0$ is the loss when assigning an observation from Ω_i to Ω_k ($i, k \in S$). The problem of cluster-analysis consists in the construction of DR for the classification of the sample X, i.e. in construction of the estimate $D = (d_1, \ldots, d_n)^T$ for D^o on X.

Let us consider the following model of type (1) which is most often used in practice: It is the Tukey-Huber distortion model (Huber (1981)):

$$\mathcal{P}_i(\varepsilon_{+i}) = \{p(x), x \in R^N : p(x) = (1 - \varepsilon_i)q(x; \theta_i^o) + \varepsilon_i h_i(x), 0 \le \varepsilon_i \le \varepsilon_{+i} < 1\}, i \in S, \quad (2)$$

where $h_i(\cdot)$ is the arbitrary density of the "contaminating" distribution; ε_i is the probability of the "contamination" in the class Ω_i. According to (2) the class Ω_i is generated by two subclasses $\Omega_i^o, \Omega_i^h \subseteq \Omega_i$, $\Omega_i^o \cap \Omega_i^h = \emptyset$. An observation from Ω_i^o is described by the "ideal" density $q(\cdot; \theta_i^o)$, and an observation from Ω_i^h is determined by the unknown density of the "contaminating" distribution $h_i(\cdot)$, which may correspond, for example, to the "outlying" observations. When sampling an observation from Ω_i, this observation corresponds to Ω_i^o with probability $1 - \varepsilon_i$ and it corresponds to Ω_i^h with probability ε_i.

It is known (see, for example, Aivazyan at al. (1989)), that if distortions are absent ($\varepsilon_+ = \max_{i \in S} \varepsilon_{+i} = 0$) and the value of the parameter vector $\theta^o = (\theta_1^{o^T}, \ldots, \theta_L^{o^T})^T \in \Theta^L \subseteq R^{Lm}$ is a priori known, then the classical DR (Bayes DR)

$$d = d(x; \theta^o) = \arg \min_{i \in S} \sum_{j \in S} \pi_j w_{ji} q(x; \theta_j^o), x \in R^N \quad (3)$$

has the minimal risk (expected loss):

$$
\begin{aligned}
r_o &= R(\theta^o; \theta^o), \quad &(4)\\
\text{where} \quad R(\theta; \theta^o) &= \sum_{i,j \in S} \pi_i w_{ij} \int_{d(x;\theta)=j} q(x; \theta_i^o) dx \ge 0, \ \theta \in \Theta^L.
\end{aligned}
$$

As θ^o is unknown, its statistical estimate $\hat{\theta}$ under the "contaminated" sample X is used. To solve the classification problem plug-in DR $d(\cdot; \hat{\theta})$ will be used, which is obtained by substituting $\hat{\theta}$ for the true value θ^o. This DR is characterized by the expected risk

$$r_\varepsilon = \mathbf{E}_{\theta^o}\{R(\hat{\theta}; \theta^o)\}, \quad (5)$$

where $\mathbf{E}_{\theta^o}\{\cdot\}$ is the expectation symbol under the density $q_\pi^\varepsilon(x; \theta^o) = \sum_{i \in S} \pi_i p_i(x; \theta_i^o)$. As a stability measure for DR $d(\cdot; \hat{\theta})$ let us define the robustness coefficient (the relative risk bias if $r_o > 0$):

$$\Re_\varepsilon = \frac{r_\varepsilon - r_o}{r_o}. \quad (6)$$

The smaller the robustness coefficient $\Re_\varepsilon$, the more stable is the DR $d(\cdot; \hat{\theta})$.

3   Asymptotic Risk Expansions

Let $\hat{\theta}$ be the minimum contrast estimate (MCE) (Pfanzagl (1969)), which is determined by the contrast function (CF). In the model (2) the random observation $x \in R^N$ has the density

$$q_\pi^\varepsilon(x; \theta^o) = \sum_{i \in S} \pi_i p_i(x; \theta_i^o) = q_\pi^o(x; \theta^o) + \sum_{i \in S} \varepsilon_i \pi_i (h_i(x) - q(x; \theta_i^o)), \quad (7)$$

where

$$q_\pi^o(x; \theta^o) = \sum_{i \in S} \pi_i q(x; \theta_i^o) \tag{8}$$

is the "ideal" density (when $\varepsilon_+ = 0$), for which, according to Chibisov (1973), the CF $f(x; \theta)$ satisfies the following inequality (optimality property):

$$b(\theta^o; \theta^o) < b(\theta; \theta^o), \text{ with } b(\theta; \theta^o) := \int_{R^N} f(x; \theta) q_\pi^o(x; \theta^o) dx, \text{ for } \theta^o \in \Theta^L, \theta \in \overline{\Theta}^L, \theta^o \neq \theta, \tag{9}$$

(here $\overline{\Theta}$ is the closure of Θ). If $\varepsilon_+ = 0$ then the classical MCE applies:

$$\tilde{\theta} := \arg \min_{\theta \in \overline{\Theta}^L} \tilde{L}_n(\theta), \quad \text{where } \tilde{L}_n(\theta) = n^{-1} \sum_{t=1}^n f(x_t; \theta); \tag{10}$$

in particular, if $f(x; \theta) = -\ln q_\pi^o(x; \theta)$ then $\tilde{\theta}$ is the maximum likelihood estimate. If $\varepsilon_+ > 0$ then the use of the mixture (7) for the construction of CF is impossible, because $\{h_i(\cdot)\}$ are arbitrary and not known. Therefore let us use the "truncation" principle from Huber (1981) and define the estimate $\hat{\theta}$:

$$\hat{\theta} := \arg \min_{\theta \in \overline{\Theta}^L} L_n(\theta), \quad \text{where } L_n(\theta) = n^{-1} \sum_{t=1}^n \psi(x_t; \theta), \tag{11}$$

with

$$\psi(x; \theta) = f(x; \theta) - (f(x; \theta) - c) \cdot \mathbf{H}(f(x; \theta) - c), \tag{12}$$
$$\mathbf{H}(z) = \{1, \text{ if } z \geq 0; 0, \text{ if } z < 0\}.$$

Obviously, if $c = +\infty$ then $\psi(\cdot) \equiv f(\cdot)$. Let us construct $\psi(x; \theta)$ as CF for the distorted densities (7) by a special choice of the "truncation" constant $c \in R^1$.

Theorem 1. The function $\psi(x; \theta)$ from (12) is CF for the family (7), if the following regularity conditions are satisfied:

C1) $f(x; \theta)$ is bounded from below and differentiable with respect to $\theta \in \overline{\Theta}^L$;

C2) $f(x; \theta)$ is integrable with respect to the densities $\{h_i(x)\}$;

C3) the distortion level $\varepsilon_+ = \varepsilon_+(n)$ and the "truncation" constant $c = c(\varepsilon_+)$ satisfy the asymptotic conditions ($n \to +\infty$):

$$\varepsilon_+(n) \to 0, \tag{13}$$

$$\sup_{\theta^* \in \Theta, \theta \in \overline{\Theta}^L} | \int_{R^N} f(x; \theta) \cdot \mathbf{H}(f(x; \theta) - c) \cdot q(x; \theta^*) dx | = O(\varepsilon_+). \tag{14}$$

Proof. The integrability of $\psi(x; \theta)$:

$$\int_{R^N} \psi(x; \theta) q_\pi^e(x; \theta^o) dx < +\infty$$

is proved by the conditions C1, C2 and by the definition (12). The inequality, which is analogous to (9) and defines CF $\psi(\cdot)$ for the family (7), is checked by the regularity conditions C1, C3. $\square$

Let us construct the asymptotic expansion for $\hat{\theta}$, which is defined by (11). Let us introduce the notations: $M = m \cdot L$; $\mathbf{0}_M$ is the M-column-vector the elements of which are all equal to 0; ∇_θ^k is the differential operator for the calculation of the vector ($k = 1$) and the matrix ($k = 2$) of k-th order partial derivatives with respect to $\theta = (\theta_1, \ldots, \theta_M)^T$; $o_1(Z_n) \in R^K$ is the random sequence which depends on the random sequence $Z_n \in R^M$ and fulfils

$$\frac{|\, o_1(Z_n)\,|}{|\,Z_n\,|} \longrightarrow 0 \quad \text{almost surely for } n \longrightarrow \infty,$$

where convergence takes place with probability 1 and $|\cdot|$ is Euclidian norm; $\mathbf{1}_M$ is the M-column-vector the elements of which are all equal to 1; the $(M \times M)$-matrix $\mathbf{1}_{M \times M}$ is determined analogously.

Theorem 2. Let the regularity conditions C1-C3 be satisfied and let the conditions C4-C6 take place:

C4) the function $f(x; \theta)$, $x \in R^N$, is twice differentiable with respect to $\theta \in \overline{\Theta}^L$ so that the generalized functions

$$\frac{\partial \psi(x; \theta)}{\partial \theta_i}, \quad \frac{\partial \psi(x; \theta)}{\partial \theta_i} \cdot \frac{\partial \psi(x; \theta)}{\partial \theta_j}, \quad \frac{\partial^2 \psi(x; \theta)}{\partial \theta_i \partial \theta_j}, \quad \frac{\partial \psi(x; \theta)}{\partial \theta_i} \cdot \frac{\partial \psi(x; \theta)}{\partial \theta_j} \cdot \frac{\partial \psi(x; \theta)}{\partial \theta_k}, \quad x \in R^N,$$

are uniformly integrable in R^N with respect to $q(x; \theta^*)$, $\{h_l(x)\}$, $\theta \in \overline{\Theta}^L$, $\theta^* \in \Theta$, $i, j, k = \overline{1, M}$;

C5) the following integration and differentiation operations may be permuted:

$$\nabla_\theta^k \int_{R^N} \psi(x; \theta) q(x; \theta^*) dx = \int_{R^N} \nabla_\theta^k \psi(x; \theta) q(x; \theta^*) dx,$$

$$\nabla_\theta^k \int_{R^N} \psi(x; \theta) h_l(x) dx = \int_{R^N} \nabla_\theta^k \psi(x; \theta) h_l(x) dx, \quad l \in S,$$

for $\theta \in \overline{\Theta}^L$, $\theta^* \in \Theta$, $k = 1, 2$;

C6) for the functional $b(\theta; \theta^o)$ and its gradient vector $B(\theta; \theta^o) = \nabla_\theta b(\theta; \theta^o)$ we have

$$B(\theta^o; \theta^o) = \mathbf{0}_M,$$

and the matrix of second order partial derivatives $a(\theta; \theta^o) = \nabla_\theta^2 b(\theta; \theta^o)$ is positively definite for $\theta = \theta^o$:

$$A(\theta^o) = a(\theta^o; \theta^o) \succ 0.$$

Then the following asymptotic expansion for the estimated deviation (11) holds for $n \longrightarrow \infty$:

$$\triangle \theta = \hat{\theta} - \theta^o = -A^{-1}(\theta^o) \nabla_{\theta^o} L_n(\theta^o) + o_1(A^{-1}(\theta^o) \nabla_{\theta^o} L_n(\theta^o)). \tag{15}$$

Proof. Under the conditions of Theorem 2 it follows from (11) that $\hat{\theta}$ is the root of the equation

$$\nabla_\theta L_n(\theta) = \mathbf{0}_M. \tag{16}$$

From the relation
$$\nabla_{\theta^o} L_n(\theta^o) \longrightarrow 0_M, \quad \text{a. s. for } n \longrightarrow \infty$$

which is based on the strong law of large numbers and on the following relation

$$\mathbf{E}_{\theta^o}\{\nabla_{\theta^o} L_n(\theta^o)\} = B(\theta^o; \theta^o) + O(\varepsilon_+) = O(\varepsilon_+)$$

we obtain with the help of the known result from Borovkov (1984):

$$\hat{\theta} \longrightarrow \theta^o, \quad \text{a. s. for } n \longrightarrow \infty.$$

This fact permits to use the Taylor formula in the neighbourhood of θ^o:

$$
\begin{aligned}
\nabla_\theta L_n(\theta) &= \nabla_{\theta^o} L_n(\theta^o) + \nabla_{\theta^o}^2 L_n(\theta^o)(\theta - \theta^o) + 1_M O(|\theta - \theta^o|^2) \qquad (17) \\
&= \nabla_{\theta^o} L_n(\theta^o) + A(\theta^o)(\theta - \theta^o) + (\nabla_{\theta^o}^2 L_n(\theta^o) - \mathbf{E}_{\theta^o}\{\nabla_{\theta^o}^2 L_n(\theta^o)\})(\theta - \theta^o) + \\
&\quad + (\theta - \theta^o) O(\varepsilon_+) + 1_M O(|\theta - \theta^o|^2),
\end{aligned}
$$

where the following relation has been used:

$$\mathbf{E}_{\theta^o}\{\nabla_{\theta^o}^2 L_n(\theta^o)\} = A(\theta^o) + O(\varepsilon_+).$$

Then the asymptotic expansion (15) is obtained by solving for the difference $\Delta\theta = \hat{\theta} - \theta^o$ when we use the expansion (17) in (16). $\square$

Let us introduce the matrix

$$I_o(\theta^o) = A^{-1}(\theta^o) \int_{R^N} \nabla_{\theta^o} f(x; \theta^o)(\nabla_{\theta^o} f(x; \theta^o))^T q_\pi^o(x; \theta^o) dx \cdot A^{-1}(\theta^o)$$

in analogy to the inverse Fisher information matrix (when $\varepsilon_+ = 0$, $f(\cdot) \equiv -\ln q_\pi^o(\cdot)$).

<u>Theorem 3.</u> Under the conditions of Theorem 2 the following asymptotic expansions are true:

a) for the estimate $\hat{\theta}$ bias:
$$\mathbf{E}_{\theta^o}\{\Delta\theta\} = \beta(\theta^o) + 1_M o(\varepsilon_+);$$

with

$$
\begin{aligned}
\beta(\theta^o) &:= A^{-1}(\theta^o) \int_{R^N} \nabla_{\theta^o} f(x; \theta^o)(\mathbf{H}(f(x; \theta^o) - c) q_\pi^o(x; \theta^o) - \\
&\quad - \sum_{i \in S} \varepsilon_i \pi_i (h_i(x) - q(x; \theta_i^o)) \mathbf{H}(c - f(x; \theta^o))) dx;
\end{aligned}
$$

b) for the variance matrix of $\hat{\theta}$:

$$\mathbf{E}_{\theta^o}\{\Delta\theta(\Delta\theta)^T\} = I_o(\theta^o) n^{-1} + \beta(\theta^o)\beta^T(\theta^o) + 1_{M \times M} o(\varepsilon_+^2 + n^{-1}).$$

<u>Proof:</u> It is based on the use of (15), (7), (11), (12) and the regularity conditions C1-C6.
$\square$

Now let us construct the asymptotic expansion for the risk r_ε of DR $d(\cdot; \hat{\theta})$ with the help of the method from Kharin (1982, 1983) in the case of $L = 2$ classes. But the results can be analogously obtained in the general case ($L \geq 2$). The DR for $L = 2$ has the form

$$d = d(x; \hat{\theta}) = \mathbf{H}(G(x; \hat{\theta})) + 1 \tag{18}$$
$$\text{with } G(x; \theta^o) = a_2 q(x; \theta_2^o) - a_1 q(x; \theta_1^o), a_i = \pi_i(w_{i,3-i} - w_{ii}),$$

and the following relation for the risk (5) is true:

$$r_\varepsilon = \pi_1 w_{11} + \pi_2 w_{21} - \mathbf{E}_{\theta^o}\{\int_{R^N} \mathbf{H}(G(x; \hat{\theta}))G(x; \theta^o)dx\}. \tag{19}$$

<u>Theorem 4.</u> Under the conditions of theorem 2 let the density $q(x; \theta^*), \theta^* \in \Theta$, be differentiable with respect to $x \in R^N$ and the integrals

$$I_1 = \frac{1}{2}\int_\Gamma (\nabla_{\theta^o} G(x; \theta^o))^T I_o(\theta^o) \nabla_{\theta^o} G(x; \theta^o) \mid \nabla_x G(x; \theta^o) \mid^{-1} d\mathcal{S}_{N-1},$$
$$I_2 = \frac{1}{2}\int_\Gamma ((\nabla_{\theta^o} G(x; \theta^o))^T \beta(\theta^o))^2 \mid \nabla_x G(x; \theta^o) \mid^{-1} d\mathcal{S}_{N-1}$$

be finite along the discriminant surface $\Gamma = \{x : G(x; \theta^o) = 0\}$. Then the risk r_ε allows the asymptotic expansions $(n \to +\infty)$:

A1) in the asymptotics $\varepsilon_+ = o(n^{-1/2})$:

$$r_\varepsilon = r_o + I_1/n + o(n^{-1}); \tag{20}$$

A2) in the asymptotics $\varepsilon_+ = O(n^{-1/2})$:

$$r_\varepsilon = r_o + I_1/n + I_2 + o(n^{-1}); \tag{21}$$

A3) in the asymptotics $\varepsilon_+/n^{-1/2} \to +\infty, \varepsilon_+ = \varepsilon_+(n) \longrightarrow 0$:

$$r_\varepsilon = r_o + I_2 + o(\varepsilon_+^2). \tag{22}$$

<u>Proof</u> is based on the use of the Taylor formula in the neighbourhood of θ^o with respect to $\Delta\theta = \hat{\theta} - \theta^o$ to the integral from (19) with the help of Theorem 3 and the theory of generalized functions as in Kharin (1982, 1983). $\square$

<u>Corollary.</u> The risk $\tilde{r}_\varepsilon$ of the DR $d(\cdot; \tilde{\theta})$ which uses the classical estimate $\tilde{\theta}$ from (10) allows the asymptotic expansions:
in the asymptotics A1:
$$\tilde{r}_\varepsilon = r_o + I_1/n + o(n^{-1}); \tag{23}$$

in the asymptotics A2:
$$\tilde{r}_\varepsilon = r_o + I_1/n + \tilde{I}_2 + o(n^{-1}); \tag{24}$$

in the asymptotics A3:
$$\tilde{r}_\varepsilon = r_o + \tilde{I}_2 + o(\varepsilon_+^2), \tag{25}$$

where

$$\tilde{I}_2 = \frac{1}{2} \int_\Gamma ((\nabla_{\theta^\circ} G(x; \theta^\circ))^T \tilde{\beta}(\theta^\circ))^2 \mid \nabla_x G(x; \theta^\circ) \mid^{-1} d\mathcal{S}_{N-1},$$

$$\tilde{\beta}(\theta^\circ) = A^{-1}(\theta^\circ) \sum_{i \in S} \varepsilon_i \pi_i \int_{R^N} \nabla_{\theta^\circ} f(x; \theta^\circ)(h_i(x) - q(x; \theta_i^\circ)) dx.$$

Proof. If $c = +\infty$ then the condition (14) is satisfied, and the estimate $\hat{\theta}$, introduced in (11), is replaced by $\tilde{\theta}$, introduced by means of (10). The end of proof is obtained by substitution of $c = +\infty$ in (20)-(22). $\square$

From (20) and (23) it follows that in the asymptotics A1 the use of $\hat{\theta}$ instead of its classical version $\tilde{\theta}$ is useless, because the risks $r_\varepsilon, \tilde{r}_\varepsilon$ (and, consequently, the robustness coefficients $\mathfrak{R}_\varepsilon, \tilde{\mathfrak{R}}_\varepsilon$) coincide in the main terms of asymptotic expansions with the remainder $o(n^{-1})$:

$$\tilde{r}_\varepsilon - r_\varepsilon = o(n^{-1});$$

$$\lim_{n \to +\infty} n\mathfrak{R}_\varepsilon = \lim_{n \to +\infty} n\tilde{\mathfrak{R}}_\varepsilon = I_1/r_o.$$

The results of the Theorem 4 and its Corollary allow to indicate the situations when in the asymptotic cases A2 and A3 the use of $d(\cdot; \hat{\theta})$ instead of its classical version $d(\cdot; \tilde{\theta})$ gives an considerable increase in the robustness coefficient (i.e. when $\mathfrak{R}_\varepsilon < \tilde{\mathfrak{R}}_\varepsilon$). Particularly, if the following norm

$$\parallel \nabla_{\theta^\circ} f(x; \theta^\circ) \parallel = \sqrt{(\nabla_{\theta^\circ} f(x; \theta^\circ))^T V(\theta^\circ) \nabla_{\theta^\circ} f(x; \theta^\circ)}$$

is used where

$$V(\theta^\circ) = A^{-1}(\theta^\circ) \int_\Gamma \nabla_{\theta^\circ} G(x; \theta^\circ)(\nabla_{\theta^\circ} G(x; \theta^\circ))^T \mid \nabla_x G(x; \theta^\circ) \mid^{-1} d\mathcal{S}_{N-1} A^{-1}(\theta^\circ) \succ 0$$

is not bounded in the region $U = \{x : f(x; \theta^\circ) \geq c\}$, then, obviously, there exist such densities $\{h_i(\cdot)\}$ which are concentrated in U:

$$\int_U h_i(x) dx \to 1, \ i \in S,$$

and for which $r_\varepsilon < \tilde{r}_\varepsilon$ ($\mathfrak{R}_\varepsilon < \tilde{\mathfrak{R}}_\varepsilon$). In this situation such densities $\{h_i(\cdot)\}$ can describe the "outlying" observations.

4 Robustness for the "Contaminated" Fisher Model

Let us illustrate the obtained results for the well known "contaminated" normal distribution Fisher model:

$$q(x; \theta_i^\circ) = n_N(x \mid \theta_i^\circ, \Sigma),$$
$$h_i(x) = n_N(x \mid \theta_i^+, \Sigma), \qquad i = 1, 2 \quad (L = 2),$$

where

$$n_N(x \mid \theta, \Sigma) = (2\pi)^{-\frac{N}{2}} (\det \Sigma)^{-\frac{1}{2}} \exp(-\frac{1}{2}(x - \theta)^T \Sigma^{-1}(x - \theta)), \ x \in R^N,$$

is the N-variate Gaussian density with expectation vector θ and nonsingular covariance $(N \times N)$-matrix Σ ($\det \Sigma > 0$).

Let the classes Ω_1, Ω_2 be a priori equiprobable ($\pi_1 = \pi_2 = 0.5$) and be "equicontaminated" ($\varepsilon_1 = \varepsilon_2 = \varepsilon \le \varepsilon_+ < 1$); $w_{ij} = \{1, \text{ if } i \ne j; 0, \text{ if } i = j\}$ (in this case the risk r_ε is the classification error probability). Let us introduce the notations: $\varphi(z) = n_1(z \mid 0, 1)$, $\Phi(z)$ are the standard normal distribution density and distribution function, respectively, $\rho(z) = \Phi(-z)/\varphi(z)$ is Mills ratio,

$$\Delta = \sqrt{(\theta_1^o - \theta_2^o)^T \Sigma^{-1}(\theta_1^o - \theta_2^o)}$$

is the Mahalanobis interclass distance. Note that if $\varepsilon_+ = 0$ and $\theta^o = (\theta_1^{o^T} : \theta_2^{o^T})^T$ are known a priori then the Bayes (classical) DR error probability is $r_o = \Phi(-\Delta/2)$.

Let us investigate the stability of DR $d(\cdot; \tilde{\theta})$ where $\tilde{\theta}$ is MCE with CF:

$$f(x; \theta^o) = -\ln \sum_{i=1}^{2} \exp(-\frac{1}{2}(x - \theta_i^o)^T \Sigma^{-1}(x - \theta_i^o)).$$

In the asymptotics A3 ($\varepsilon_+ = n^{-0.5+\nu}, 0 < \nu \le 0.4$) and the situation, when

$$(\theta_i^+ - \theta_i^o)^T \Sigma^{-1}(\theta_i^+ - \theta_i^o) \le \delta^2 \ (i = 1, 2),$$

with the help of the Corollary of Theorem 4, the guaranteed value of robustness coefficient (6) is obtained:

$$\tilde{\Re}_\varepsilon \le \tilde{\Re}_\varepsilon^+ + o(\varepsilon_+^2), \tag{26}$$

$$\tilde{\Re}_\varepsilon^+ = \frac{\varepsilon_+^2 \delta^2(4 + \Delta^2)}{\Delta \rho(\Delta/2)}.$$

The smaller ε_+, the more stable is DR $d(\cdot; \tilde{\theta})$. From the condition $\tilde{\Re}_\varepsilon^+ \le \gamma$ ($\gamma > 0$ is the predetermined value of the robustness coefficient) with the help of the relation (26) the γ-admissible sample size is evaluated:

$$n^* = n^*(\gamma) = [Z^{\frac{1}{2\nu-1}}] + 1,$$

$$Z = \frac{\gamma \Delta \rho(\Delta/2)}{\delta^2(4 + \Delta^2)},$$

where $[x]$ is the entire part of x.

The relation (26) helps us to determine the "breakdown point" (see Huber (1981)):

$$\varepsilon_+^* = \sqrt{\Delta(0.5 - \Phi(-\Delta/2))/(\delta^2 \varphi(\Delta/2)(4 + \Delta^2))};$$

if $\varepsilon_+ > \varepsilon_+^*$ then the error probability of DR $d(\cdot; \tilde{\theta})$ can increase up to the "breakdown value" $\tilde{r}_\varepsilon^* = 0.5$, which corresponds to the equiprobable coin tossing.

If we use the estimate $\hat{\theta}$ with "truncated" CF (12), then the "truncation" constant is given by the following relation which is obtained by the condition C3:

$$c = c(\varepsilon_+) = \frac{N}{2}(\Phi^{-1}(y))^2,$$

$$y = \frac{1 + (1 - \varepsilon_+)^{1/N}}{2},$$

where $\Phi^{-1}(y)$ is the y-level quantile of the standard normal distribution.

References

AIVAZYAN, S.A., BUCHSTABER, V.M., YENYUKOV, I.S., MESHALKIN, L.D. (1989), *Applied Statistics: Classification and Dimensionality Reduction*, Finansy i Statistika, Moscow.

BOCK, H.H. (1989), Probabilistic Aspects in Cluster Analysis, in: O. Opitz (ed.), *Conceptual and Numerical Analysis of Data*, Proceedings of the 13-th Conference of the Gesellschaft fr Klassifikation, Springer, Heidelberg et. al., 12–44.

BOROVKOV, A.A. (1984), *Mathematical statistics: Parameter Estimation and Hypotheses Testing*, Nauka, Moscow.

CHIBISOV, D.M. (1973), An Asymptotic Expansion for a Class of Estimators containing ML-Estimators, *Theory of Probability and its Applications*, 18, 295–303.

HUBER, P.J. (1981), *Robust Statistics*, John Wiley and Sons, New York.

KHARIN, Yu.S. (1982), Asymptotic Expansion for the Risk of Parametric Decision Functions, *Asymptotic Statistics (by Mandl)*, North Holland, Amsterdam.

KHARIN, Yu.S. (1983), Investigation of Risk for Statistical Classifiers Using Minimum Contrast Estimates, *Teor. Ver. Prim.*, 28, No. 3, 592–598.

PFANZAGL, J. (1969), On the measurability and consistency of minimum contrast estimates, *Metrika*, 14, 249–272.

Choosing the Number of Component Clusters in the Mixture-Model Using a New Informational Complexity Criterion of the Inverse-Fisher Information Matrix

Hamparsum Bozdogan
Department of Statistics
The University of Tennessee
Knoxville, TN 37996-0532, USA

Abstract: This paper considers the problem of choosing the number of component clusters of individuals within the context of the standard mixture of multivariate normal distributions. Often the number of mixture clusters K is unknown, but varying and needs to be estimated. A two-stage iterative maximum-likelihood procedure is used as a clustering criterion to estimate the parameters of the mixture-model under several different covariance structures. An approximate component-wise inverse-Fisher information (IFIM) for the mixture-model is obtained. Then the informational complexity (ICOMP) criterion of IFIM of this author (Bozdogan 1988, 1990a, 1990b) is derived and proposed as a new criterion for choosing the number of clusters in the mixture-model. For comparative purposes, Akaike's (1973) information criterion (AIC), and Rissanen's (1978) minimum description length (MDL) criterion are also introduced and derived for the mixture-model. Numerical examples are shown on simulated multivariate normal data sets with a known number of mixture clusters to illustrate the significance of ICOMP in choosing the number of clusters and the best fitting model.

1 Introduction and Statement of the Problem

A general common problem in all clustering techniques is the difficulty of deciding on the number of clusters present in a given data set, cluster validity, and the identification of the approximate number of clusters present. This paper considers the problem of choosing the number of clusters within the context of multivariate mixture-model cluster analysis, known as the standard normal mixture model, unsupervised pattern recognition, or "learning without a teacher" (Bozdogan 1981, 1983, and 1992).

In the mixture-model for cluster analysis, often the number of component clusters K is not known and needs to be estimated and determined. Under this model conventional inferential procedures suffer from sampling distributional problems. If the data actually contain K clusters, the null distribution of the usual likelihood ratio is still not known, and remains largely unresolved.

In the statistical literature, despite the increased number of books appearing on finite mixture of distributions such as Everitt and Hand (1981), Titterington et al. (1985), and McLachlan and Basford (1988), a relatively little work has been done concerning the choice of the number of component mixture clusters.

In this paper, our objective from a clustering viewpoint is to identify and describe the class distribution using a sample drawn from the mixture-model, and estimate K, the number of mixture clusters such that $\hat{K} < K$ for $k = 1, 2, \ldots, K$. To achieve this, we will use a two-stage iterative maximum-likelihood procedure to estimate the parameters in the mixture-model, and develop an informational complexity (ICOMP) criterion of this author (Bozdogan 1988,1990a,1990b) as a new criterion for choosing the number of clusters. For comparative purposes, Akaike's (1973) information criterion (AIC), and Rissanen's (1978) minimum description length (MDL) criterion based on coding theory are also introduced and derived.

We give numerical examples based on simulated multivariate normal data sets with a known true number of mixture clusters to illustrate the significance of ICOMP in choosing the number of clusters and the best fitting model. ICOMP takes into account simultaneously the badness-of-fit (or lack-of-fit) of a cluster, the number of parameters, the sample size, and the complexity of the increased number of clusters to achieve the best fit explicitly in one criterion function.

2 The Standard Mixture-Model Cluster Analysis

2.1 The Model

The problem of clustering of n individuals on the basis of p-dimensional observation vectors x_1, x_2, ..., x_n will be studied using a mixture of normal probability density functions (p.d.f.'s) without being told of their classification. In this method, a priori we do not know K the number of mixture clusters, the mixing proportions, the mean vectors, and the covariance matrices of the class distributions. If we assume that each observation vector x_i has probability π_k of coming from the k-th population $k \in (1, 2, \ldots, K)$, then $x_1, x_2, \ldots, x_n$ is a sample from

$$f(x) \equiv f(x; \pi, \mu, \Sigma) = \sum_{k=1}^{K} \pi_k g_k(x; \mu_k, \Sigma_k), \qquad (1)$$

where $\pi = (\pi_1, \pi_2, \ldots, \pi_{K-1})$ are $K - 1$ independent mixing proportions such that

$$0 \leq \pi_k \leq 1 \quad \text{for } k = 1, 2, \ldots, K \text{ and } \pi_k = 1 - \sum_{k=1}^{K-1} \pi_k, \qquad (2)$$

and where $g_k(x; \mu_k, \Sigma_k)$ is the k-th component multivariate normal density function given by

$$g_k(x; \mu_k, \Sigma_k) = (2\pi)^{-p/2} |\Sigma_k|^{-1/2} exp\{-1/2(x - \mu_k)'\Sigma_k^{-1}(x - \mu_k)\}. \qquad (3)$$

The model given by the p.d.f. in (1) is called the standard multivariate normal mixture model to distinguish it from the modified conditional mixture model considered by Symons (1981), Sclove (1977, 1982), Scott and Symons (1971), and John (1970).

In the statistical literature, several authors, including Wolfe (1967, 1970), Day (1969), Binder (1978), Hartigan (1977), and others, have considered clustering problems in which standard of multivariate normal mixture-model is used as a statistical model in (1).

2.2 The Covariance Structures of the Mixture-Model

We note that the mixture model in (1) is a highly overparameterized model since there are $1/2(p + 1)(p + 2)K$ parameters to estimate, where p is the number of variables, and $K \geq 1$

is the hypothesized number of components. To be able to fit the mixture model, following Hartigan (1975, p.116), we require that the total sample size $n > 1/2(p+1)(p+2)K$ as the rule of thumb. In terms of the covariance structures, based on the suggestions of Hartigan (1975) and Bock (1981), we shall consider the following four covariance structures between the component mixture clusters. We denote these models by M_1, M_2, M_3 and M_4 corresponding to their covariance structures. These are:

- M_1 = General covariances, that is, covariance matrices are different between component mixture clusters. The parameter space for this model is:

$$\Theta = \{\theta : \theta = (\pi_1, \mu_1, \Sigma_1, \pi_2, \mu_2, \Sigma_2, \ldots, \pi_K, \mu_K, \Sigma_K)\} \tag{4}$$

- M_2 = Covariance matrices are equal between component mixture clusters, $\Sigma_k = \Sigma$. The parameter space for this model is:

$$\Theta = \{\theta : \theta = (\pi_1, \mu_1, \Sigma, \pi_2, \mu_2, \Sigma, \ldots, \pi_K, \mu_K, \Sigma)\} \tag{5}$$

- M_3 = Covariance matrices are equal and diagonal between component mixture clusters, $\Sigma_k = diag(\sigma_1^2, \ldots, \sigma_p^2)$. The parameter space for this model is:

$$\Theta = \{\theta : (\pi_1, \mu_1, diag(\sigma_1^2, \ldots, \sigma_p^2), \ldots, \pi_K, \mu_K, diag(\sigma_1^2, \ldots, \sigma_p^2)\} \tag{6}$$

- M_4 = All variables have the same variance and are pairwise independent between component mixture clusters (spherical model), $\Sigma_k = \sigma^2 I$. The parameter space for this model is:

$$\Theta = \{\theta : \theta = (\pi_1, \mu_1, \sigma^2 I, \pi_2, \mu_2, \sigma^2 I, \ldots, \pi_K, \mu_K, \sigma^2 I)\} \tag{7}$$

The choice between these covariance structures is very important since we can achieve simplification of the model, protect the model from overparameterization, specially when the sample size is not large, and help to eliminate singular solutions and other anomalies in the mixture model.

2.3 Estimating the Unknown Parameters of the Mixture-Model

To estimate the parameters of each of the models M_1, M_2, M_3, and M_4 we write the log likelihood function of the data $x_1, x_2, \ldots, x_n$ as

$$l(\theta) \equiv \log L(\theta|X) = \sum_{i=1}^{n} \log \left[\sum_{k=1}^{K} \pi_k (2\pi)^{-p/2} |\Sigma_k|^{-1/2} exp\{-\frac{1}{2}(x_i - \mu_k)'\Sigma_k^{-1}(x_i - \mu_k)\} \right] \tag{8}$$

To obtain the maximum likelihood estimators (MLE's) of the unknown parameters, we use matrix differential calculus and compute the partial derivatives of the log likelihood function $l(\theta)$ with respect to π_k, the mean vector μ_k, and Σ_k^{-1}, respectively, and set these equal to zero. After some algebra, for model M_1, we obtain the ML equations:

$$\hat{P}(k|x_i) = \frac{\hat{\pi}_k g_k(x_i; \hat{\mu}_k, \hat{\Sigma}_k)}{\sum_{i=1}^{K} \hat{\pi}_k g_k(x_i; \hat{\mu}_k, \hat{\Sigma}_k)} \qquad k = 1, 2, \ldots, K \tag{9}$$

$$\hat{\pi}_k = \frac{1}{n} \sum_{i=1}^{n} \hat{P}(k|x_i) \qquad k = 1, 2, \ldots, K \tag{10}$$

$$\hat{\mu}_k = \frac{1}{n\hat{\pi}_k} \sum_{i=1}^{n} x_i \hat{P}(k|x_i) \qquad k = 1, 2, \ldots, K \tag{11}$$

$$\hat{\Sigma}_k = \frac{1}{n\hat{\pi}_k} \sum_{i=1}^{n} \hat{P}(k|x_i)(x_i - \hat{\mu}_k)(x_i - \hat{\mu}_k)' \qquad k = 1, 2, \ldots, K \tag{12}$$

where $\hat{\pi}_k$ is the estimated mixing proportion π_k, $\hat{\mu}_k$ is the estimated mean vector μ_k, $\hat{\Sigma}_k$ is the estimated covariance matrix Σ_k, and $\hat{P}(k|x_i)$ is the estimated posterior probability of group membership of the observation vector x_i in the cluster k.

In the literature, there are several methods for solving the mixture likelihood equations by the Newton-Raphson method or the EM algorithm, with many small modifications. Following Wolfe (1970) and Hartigan (1975), here we use a modified two-stage iterative Newton-Raphson method and obtain the solutions for the above recursive equations (9)-(12). For this purpose we use our program package called NEWMIXTURE. The steps followed in the NEWMIXTURE program is as follows.

First Step: We choose initial values of the parameters of the mixture model denoted by $\pi_k^{(0)} = 1/K, \mu_k^{(0)} = \hat{\mu}_k, \Sigma_k^{(0)} = \hat{\Sigma}_k$, for $k = 1, 2, \ldots, K$. To initialize the mean vectores, we provide three options: $0 =$ default means after splitting data up in the order entered, $1 =$ user inputed initial means, and $2 =$ Bozdogan's (1983) mean initialization scheme. For the covariance matrix a user defined initial covariance matrix is utilized. At this step, a user can also initialize the mixture algorithm by using the results from the "k-means" algorithm. See, e.g., Hartigan (1975, p.124).

Iteration Step: We calculate $\hat{P}(k|x_i), \hat{\pi}_k, \hat{\mu}_k,$ and $\hat{\Sigma}_k$ in turn according to (9), (10), (11) and (12). The iteration cycle continues until the value of $\pi_k, \mu_k,$ and Σ_k all converge.

Stop Criterion: We stop the algorithm when the log likelihood increases by less than 0.01, after the two-stage process iterates $r = 1, 2, \ldots, 40, \ldots$ times.

Clustering Step: Once the MLE's are known, we regard each distribution as indicating a separate cluster. Individuals are then assigned by the Bayes allocation rule. That is, assign x_i to the k-th mixture cluster when

$$\hat{\pi}_l g_l(x_i; \hat{\mu}_l, \hat{\Sigma}_l) \le \hat{\pi}_k g_k(x_i; \hat{\mu}_k, \hat{\Sigma}_k) \quad \text{for all } l \ne k. \tag{13}$$

Another way to put it, individual i is assigned to that mixture cluster k for which the estimated posterior probability of group membership, $\hat{P}(k|x_i)$, is the largest.

Similarly, we obtain the MLE's of the models $M_2, M_3,$ and M_4, respectively.

3 AIC and MDL for the Mixture-Model

For the mixture-model, AIC in Bozdogan (1981, 1983) is given by

$$AIC = -2 \log L(\hat{\theta}) + 3m. \tag{14}$$

In the literature of model selection, as n, the sample size, gets larger and larger, other penalties have been proposed by many other researchers to penalize overparameterization more stringently to pick only the simplest of the best approximating models. For this purpose, for example, Rissanen (1978, 1989) proposed a criterion based on coding theory using statistical information in the data and the parameters. His criterion is called the *minimum description length (MDL)* and is defined by

$$MDL = -2 \log L(\hat{\theta}) + m \log(n), \tag{15}$$

where $\log(n)$ denotes the "natural logarithm" of the sample size n.

We note that AIC and MDL criteria differ from one another in terms of their penalty, that is, in their second components in order to achieve the *principle of parsimony* in model selection.

From Bozdogan (1981,1983) for the mixture-model when the general covariances are used between the component clusters, that is, under model M_1, the derived analytical form of Akaike's (1973) information criterion (AIC) is given by

$$AIC(M_1) = -2 \sum_{i=1}^{n} \log[\sum_{k=1}^{K} \hat{\pi}_k g_k(x_i; \hat{\mu}_k, \hat{\Sigma}_k)] + 3[kp + (k-1) + kp(p+1)/2] \tag{16}$$

We note that the marginal cost per parameter, that is, the so called "magic number 2" in the original definition of AIC is not adequate for the mixture-model. In Wolfe (1970), if we consider the fact that the distribution of the usual likelihood ratio for the mixture-model, that is , if

$$- 2 \log \lambda \overset{a.d.}{\sim} \chi_{\nu*}'^2(\delta) \quad \text{(noncentral chi-square)} \tag{17}$$

with $\nu^* = 2(M - m)$ degrees of freedom, and where *a.d.* denotes asymptotically distributed, then in AIC we obtain 3 as the "magic number in (16). In the literature, this point has been confused by many authors. For other generalizations of AIC, we refer the reader to Bozdogan (1987).

For models M_2, M_3 and M_4, we similarly obtain the AIC's using (16). The number of estimated parameters under these models which are penalized in AIC, are given as follows:

$$M_2 : m \equiv m(k) = kp + (k-1) + p(p+1)/2, \tag{18}$$
$$M_3 : m \equiv m(k) = kp + (k-1) + p, \tag{19}$$
$$M_4 : m \equiv m(k) = kp + (k-1) + 1, \tag{20}$$

Likewise, Rissanan's (1978, 1989) minimum description length (MDL) criterion for the mixture model when the general covariances are used between component clusters is given by

$$MDL(M_1) = -2 \sum_{i=1}^{n} \log[\sum_{k=1}^{K} \hat{\pi}_k g_k(x_i; \hat{\mu}_k, \hat{\Sigma}_k)] + m \log(n) \tag{21}$$

This is modified accordingly for models M_2, M_3 and M_4 with m given in (18)-(19) above for each of these models.

4 ICOMP Model Selection Criterion For the Mixture-Model

As an alternative to AIC and MDL, this author (Bozdogan 1988, 1990a, 1990b) developed a new entropic statistical complexity criterion called ICOMP for model selection for general multivariate linear and nonlinear structural models. Analytic formulation of ICOMP takes the "spirit" of AIC, but it is based on the generalization and utilization of an entropic covariance complexity index of van Emden (1971) for a multivariate normal distribution in parametric estimation. For exampes, for a general multivariate linear or nonlinear structural model defined by

$$\text{Statistical model} = \text{Signal} + \text{Noise} \qquad (22)$$

ICOMP is designed to estimate a loss function:

$$\text{Loss} = \text{Lack of fit} + \text{Lack of Parsimony} + \text{Profusion of Complexity} \qquad (23)$$

in two ways using the additivity property of information theory, and the developments of Rissanen (1976) in his Final Estimation Criterion (FEC) for estimation and model identification problems, as well as Akaike's (1973) AIC, and its analytical extensions in Bozdogan (1987). Here, we shall discuss the most general second approach which uses the *complexity of the estimated inverse-Fisher information matrix (IFIM)* $\hat{F}^{-1}$, of the entire parameter space of the model. In this case ICOMP is defined as

$$ICOMP(\text{Overall Model}) = -2\log L(\hat{\theta}) + C_1(\hat{F}^{-1}) \qquad (24)$$

where C_1 denotes the maximal information complexity of $\hat{F}^{-1} \equiv Est.Cov(\hat{\theta})$, the estimated IFIM of the model given by

$$C_1(\hat{F}^{-1}) = \frac{s}{2}\log[tr(\hat{F}^{-1})/s] - \frac{1}{2}\log|tr(\hat{F}^{-1})|, \qquad (25)$$

amd where $s = dim(\hat{F}^{-1}) = rank(\hat{F}^{-1})$.

For more on the C_1 measure as a "scalar measure" of a non-singular covariance matrix of a multivariate normal distribution, we refer the reader to the original work of van Emden (1971). C_1 measure also appears in Maklad and Nichols (1980, p.82) with an incomplete discussion based on van Emden's (1971) work in estimation. However, we further note that these authors abandoned the C_1 measure, and they never used it in their problem. See, e.g., Bozdogan (1990).

The first component of ICOMP in (24) measures the lack of fit of the model, and the second component measures the complexity of the estimated inverse-Fisher information matrix (IFIM), which gives a scalar measure of the celebrated *Cramér-Rao lower bound* matrix of the model.

ICOMP controls the risks of both insufficient and overparameterized models. It provides a criterion that has the virtue of *judiciously* balancing between *lack of fit* and the *model complexity* data adaptively. A model with minimum ICOMP is chosen to be the best model among all possible competing alternative models. ICOMP removes researcher from any need to consider the parameter dimension of a model explicitly and adjusts automatically for the sample size.

The theoretical justification of this approach is that it combines all three ingredients of statistical modeling in (23) based on the joining axiom of complexity of a system. Also, it refines futher the derivation of AIC, and represents a compromise between AIC and MDL, or CAIC of Bozdogan (1987, p.358). For more on this, see, Bozdogan (1992).

With ICOMP, complexity is viewed not as the number of parameters in the model, but as the degree of interdependence among the components of the model. By defining complexity in this way, ICOMP provides a more judicious penalty term than AIC and MDL (or CAIC). The lack of parsimony is automatically adjusted by $C_1(\hat{F}^{-1})$ across the competing alternative models as the parameter spaces of these models are constrained in the model fitting process.

The derivation of ICOMP in (24) requires the knowledge of the inverse-Fisher information matrix (IFIM) of the mixture-model in (1), and its sample estimate. Let

$$F^{-1} = -\left[E\left(\frac{\partial^2 \log L(\theta)}{\partial\theta\partial\theta'}\right)\right]^{-1} \tag{26}$$

denote the IFIM of the mixture-model in (1). As is well known the expectations in IFIM in (26) involve multiple integrals which are impossible to obtain in closed form, and moreover the second partial derivatives of the log likelihood function will involve complicated nonlinear expressions, whose exact expected values are unknown and difficult to evaluate. Therefore, presently the derivation of IFIM in a closed form expression using directly (1) is an unsolved and a very hard problem (Magnus 1989). This is currently being investigated by this author and Magnus, and will be reported elsewhere.

However, to remedy this existing difficulty, we propose an approximate solution and define the IFIM of (1) after we obtain the K component mixture clusters for $k = 1, 2, \ldots, K$ from the mixture-model when all the observations are classified in their own clusters or when all the group memberships are known. The situation now becomes analogous to that of multivariate analysis of variance (MANOVA) model, or the conditional mixture model with varying mean vectors and covariance matrices. In this case after some work, it becomes relatively easy to produce the IFIM of the mixture-model. We call this approach an approximate component-wise derivation of IFIM. We intuitively explain the process of fitting K component mixture clusters for $k = 1, 2, \ldots, K$ as follows.

Number of Mixture Types k	Estimated Clusters	Approximated Est. IFIM
1	$C_1 \sim N_p(\hat{\mu}_1, \hat{\Sigma}_1)$ with $\hat{\pi}_1$	$\hat{F}_1^{-1}$
2	$C_1 \sim N_p(\hat{\mu}_1, \hat{\Sigma}_1)$ with $\hat{\pi}_1,$ $C_2 \sim N_p(\hat{\mu}_2, \hat{\Sigma}_2)$ with $\hat{\pi}_2$	$\hat{F}_1^{-1}, \hat{F}_2^{-1}$
$\vdots$	$\vdots$	$\vdots$
K	$C_1 \sim N_p(\hat{\mu}_1, \hat{\Sigma}_1)$ with $\hat{\pi}_1, \ldots,$ $C_K \sim N_p(\hat{\mu}_K, \hat{\Sigma}_K)$ with $\hat{\pi}_K$	$\hat{F}_1^{-1}, \ldots, \hat{F}_K^{-1}$

Following the notation in Magnus (1988, p.169), Magnus and Neudecker (1988, p.319), we derive the estimated inverse Fisher information matrix (IFIM) of the mixture-model (1) under M_1 as

$$\hat{F}^{-1} = diag(\hat{F}_1^{-1}, \dots, \hat{F}_K^{-1}) \tag{27}$$

with

$$\hat{F}_k^{-1} = \begin{bmatrix} \frac{1}{\hat{\pi}_k}\hat{\Sigma}_k & 0 \\ 0 & 2D_p^+(\hat{\Sigma}_k \otimes \hat{\Sigma}_k)D_p^{+\prime} \end{bmatrix} \tag{28}$$

This is a block diagonal matrix with the diagonal block given by the estimated asymptotic covariance matrices $\hat{F}_k^{-1}$ for the k-th mixture cluster or type, and $\otimes$ denotes the Kronecker product. In (28), $D_p^+ = (D_p'D_p)^{-1}D_p'$ is the Moore-Penrose inverse of the duplication matric D_p. A duplication matrix is a unique $p^2 \times 1/2p(p+1)$ matrix which transforms $v(\Sigma_k)$ into $vec(\Sigma_k)$. $v(\Sigma_k)$ denotes the $1/2p(p+1)$-vector that is obtained from $vec(\Sigma_k)$ by eliminating all supradiagonal elements of Σ_k and stacking the remaining columns one underneath the other.

Our main result is given by

Proposition 1 *The maximal informational complexity of the estimated inverse-Fisher information (Est. IFIM) for the mixture model with general covariances is*

$$
\begin{aligned}
ICOMP(IFIM) &= -2\log L(\hat{\theta}) + \frac{s}{2}\log[tr(\hat{F}^{-1})/s] - \frac{1}{2}\log|\hat{F}^{-1}| \tag{29} \\
&= -2\sum_{i=1}^{n}\log\left[\sum_{k=1}^{K}\hat{\pi}_k g_k(x_i; \hat{\mu}_k, \hat{\Sigma}_k)\right] + \frac{kp + kp(p+1)/2}{2} \tag{30} \\
&\times \log\left[\frac{\sum_{k=1}^{K}\{1/\hat{\pi}_k tr(\hat{\Sigma}_k) + 1/2tr(\hat{\Sigma}_k^2) + 1/2tr(\hat{\Sigma}_k)^2 + \sum_{j=1}^{p}(\hat{\sigma}_{kjj})^2\}}{kp + kp(p+1)/2}\right] \tag{31} \\
&\quad -\frac{1}{2}\{(p+2)\sum_{k=1}^{K}\log|\hat{\Sigma}_k| - p\sum_{k=1}^{K}\log(\hat{\pi}_k n)\} - kp/2\log(2n) \tag{32}
\end{aligned}
$$

for $k = 1, 2, \dots, K$, where $s = dim(\hat{F}^{-1})$.

Proof - See Bozdogan (1990a, 1990b).

We note that in ICOMP(M_1) above, the effects of the parameter eatimates and their accuracy is taken into account. Thus, counting and penalizing the number of parameters are eliminated in the model fitting process.

In a similar fashion as above, we derive ICOMP(IFIM) for models M_2, M_3, and M_4 by simply substituting their corresponding covariance structures given in (5)-(7) into (29).

We note that ICOMP(IFIM), AIC, and MDL differ from one another in terms of their penalty, that it, in their second components in order to achieve the principle of parsimony in model selection. According to the experience of this author, in many Monte Carlo simulation studies, MDL penalizes models too stringently to the point that it underfits the true model. ICOMP(IFIM), however, achieve a more satisfactory compromise between the two extreme of AIC and MDL by the judicious application of a penalty term based upon the interdependencies among the parameter estimates of the model, rather than penalizing the number parameters estimated within the model. Penalizing the number of parameters is necessary, but not suffficient in general.

Next we show numerical examples on simulated multivariate normal data sets with a known number of mixture clusters to illustrate the working of these new model selection procedures in a difficult problem such as the mixture-model in choosing the number clusters and the best fitting model.

5 Numerical Examples

We simulated two different configurations of data sets to determine and estimate the appropriate choices for the number K of mixture clusters and for fitting the model using ICOMP(IFIM), AIC and MDL model selection criteria.

Example 5.1 Cluster of different shape and seperation: In this example, we simulated a total of $n = 450$ observations from a three-dimensional multivariate normal distribution with $K = 3$ clusters. Table 5.1 shows the specific parameters for this data set.

Tab. 5.1 Specific Parameters for Clusters of Different Shape
and Separation and the Sample Sizes

$$\mu_1 = \begin{bmatrix} 0 \\ 0 \\ 0 \end{bmatrix} \quad \Sigma_1 = \begin{bmatrix} 9 & 0 & 0 \\ 0 & 4 & 0 \\ 0 & 0 & 1 \end{bmatrix} \quad n_1 = 100$$

$$\mu_2 = \begin{bmatrix} -6 \\ 3 \\ 6 \end{bmatrix} \quad \Sigma_1 = \begin{bmatrix} 4 & 0 & 0 \\ 0 & 4 & 0 \\ 0 & 0 & 1 \end{bmatrix} \quad n_1 = 150$$

$$\mu_3 = \begin{bmatrix} 6 \\ 6 \\ 4 \end{bmatrix} \quad \Sigma_1 = \begin{bmatrix} 9 & 0 & 0 \\ 0 & 4 & 0 \\ 0 & 0 & 4 \end{bmatrix} \quad n_1 = 200$$

. We note that the mean vectors and the covariance matrices between these $K = 3$ populations are different from one another. The Fig. 5.1, 5.2, and 5.3 depict the three-dimensional mesh plots of each of the variable pairs and their contour plots to show the separation of $K = 3$ clusters on the direction of each of the variable pairs.

Examining the results in Tab. 5.2, we see that all three criteria achieve their minimum at $\hat{K} = 3$ mixture clusters, indicating the recovery of the true number of clusters. Looking at the confusion matrix, we see that 2 of the observations from the third cluster are classified with the first cluster. So the error of missclassification is 0.4%, which is quite small. This is not unexpected since the original clusters were chosen to have different shape and separation. Although, the three-dimensional mesh plots by taking the variables two at a time indicate the presence of $K = 2$ mixture clusters in the directions of variable one and two (Fig. 5.1); in the directions of variable two and three (Fig. 5.3); but there is a clear indication of the presence of $K = 3$ mixture clusters in the directions of variable one and three (Fig. 5.2) which appears to be dominant.

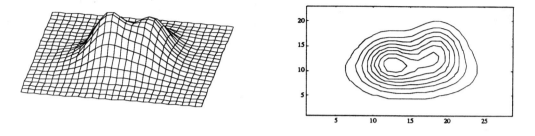

Fig. 5.1 Three-Dimensional Mesh and Contour Plots for Variable 1 & 2 Pair.

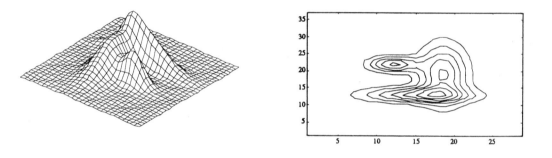

Fig. 5.2 Three-Dimensional Mesh and Contour Plots for Variable 1 & 3 Pair.

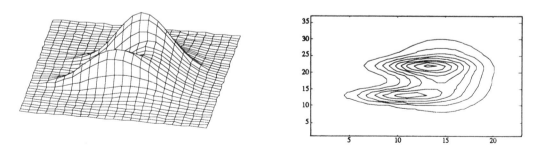

Fig. 5.3 Three-Dimensional Mesh and Contour Plots for Variable 2 & 3.

Tab. 5.2 ICOMP, AIC$_3$, and MDL Values for the Mixture Cluster Analysis
When the Covariances are General Between the Component Clusters

k	$-2\log L(\hat{\theta})$	m	$C_1(\hat{F}^{-1})$	$ICOMP(\hat{F}^{-1})$	AIC	MLD
1	7053.0	9	7.4300	7060.5	7080.0	7108.0
2	6651.4	19	9.8957	6661.3	6708.4	6767.5
3	6331.2	29	9.0442	6340.2*	6418.2*	6508.3*
4	6335.5	39	15.4289	6350.9	6452.5	6573.7
5	6319.6	49	33.7812	6353.4	6466.6	6618.9

Note: n=625 Obcrscrvations, p=3 Variables, *Minimum of criterion value

Confusion Matrix

K	1	2	3
1	100	0	0
2	0	150	0
3	2	0	198

Estimated Mixture Probabilities π_k, $k = 1, 2, \ldots, K$

K	1	2	3	4	5
2	0.56	0.44			
3	0.23	0.33	0.44		
4	0.23	0.33	0.12	0.32	
5	0.22	0.31	0.02	0.39	0.06

Example 5.2 Overlapping clusters of different compactness: In this example, we carried out a Monte Carlo study and replicated the experiment 50 times. In each replication, we generated a total of $n = 625$ observations each from a four-dimensional multivariate normal distribution with $K = 5$ clusters. We configured the structure of the original clusters to be highly overlapping with different compactness, so that it would be particularly challenging to the mixture clustering algorithm to identify even three clusters. It is not so obvious from this data set that there are $K = 5$ clusters. Tab. 5.3 shows the specific parameters and the sample sizes for each of the $K = 5$ clusters.

Tab. 5.3 Specific Parameters for Overlapping Clusters of
Different Compactness and the Sample Sizes

$$\mu_1 = \begin{bmatrix} 10 \\ 12 \\ 10 \\ 12 \end{bmatrix} \quad \Sigma_1 = \begin{bmatrix} 1 & 0 & 0 & 0 \\ 0 & 1 & 0 & 0 \\ 0 & 0 & 1 & 0 \\ 0 & 0 & 0 & 1 \end{bmatrix} \quad n_1 = 75 \quad \mu_2 = \begin{bmatrix} 8.5 \\ 10.5 \\ 8.5 \\ 10.5 \end{bmatrix} \quad \Sigma_2 = \begin{bmatrix} 1 & 0 & 0 & 0 \\ 0 & 1 & 0 & 0 \\ 0 & 0 & 1 & 0 \\ 0 & 0 & 0 & 1 \end{bmatrix} \quad n_2 = 100$$

$$\mu_3 = \begin{bmatrix} 12 \\ 14 \\ 12 \\ 14 \end{bmatrix} \quad \Sigma_3 = \begin{bmatrix} 1 & 0 & 0 & 0 \\ 0 & 1 & 0 & 0 \\ 0 & 0 & 1 & 0 \\ 0 & 0 & 0 & 1 \end{bmatrix} \quad n_3 = 125 \quad \mu_4 = \begin{bmatrix} 13 \\ 15 \\ 7 \\ 9 \end{bmatrix} \quad \Sigma_4 = \begin{bmatrix} 4 & 0 & 0 & 0 \\ 0 & 4 & 0 & 0 \\ 0 & 0 & 4 & 0 \\ 0 & 0 & 0 & 4 \end{bmatrix} \quad n_4 = 150$$

$$\mu_5 = \begin{bmatrix} 7 \\ 9 \\ 13 \\ 15 \end{bmatrix} \quad \Sigma_5 = \begin{bmatrix} 9 & 0 & 0 & 0 \\ 0 & 9 & 0 & 0 \\ 0 & 0 & 9 & 0 \\ 0 & 0 & 0 & 9 \end{bmatrix} \quad n_5 = 175 \qquad \text{Total n=625}$$

We note that in terms of separation, examining the bivariate scatter plots, we see that each of the four cluster 2, 3, 4 and 5 all overlap slightly with cluster 1. Cluster 2 has the most overlap. In terms of size and compactness, we see that cluster 1 is very dense, and cluster 2 and 3 cover the same sized area similar to cluster 1. This can be very readily seen from the equality of the covariance matrices between these clusters. If there are any differences, they are in their mean vectors. Clusters 4 and 5 cover much large areas, and they are not very compact. We note that both the mean vectors and the covariance matrices for this clusters are differnent as shown in Tab. 5.3 above. Tab. 5.4 shows our results for this experiment when we use the general covaiances between the component clusters.

Tab. 5.4 Percent Frequency of Choosing K-Mixture Clusters
in 50 Runs of the Monte Carlo Experiment

No. of Cluster K	m	ICOMP($\hat{F}^{-1}$)	AIC	MDL
1	14	0	0	0
2	29	0	0	0
3	44	92%	94%	100%
4	59	8%	6%	0
5	74	0	0	0
6	89	0	0	0

Tab. 5.5 ICOMP, AIC$_3$, and MDL Values From One Run When the Covariance
Matrices are Different Between the Component Clusters

k	$-2\log L(\hat{\theta})$	m	$C_1(\hat{F}^{-1})$	$ICOMP(\hat{F}^{-1})$	AIC	MLD
1	12034.233	14	10.39	12044.631	12076.23	12124.362
2	11515.669	29	32.10	11547.773	11602.66	11702.363
3*	11325.561	44	22.45	11348.020*	11457.56*	11608.821*
4	11374.919	59	38.84	11413.766	11551.91	11754.746
5	11577.804	74	64.37	11642.180	11799.80	12054.197
6	11639.244	89	68.37	11707.6215	11906.24	12212.204

Note: n=625 Oberservations, p=4 Variables. *Minimum criterion value

Confusion Matrix

K	{ 1,2,3 }	4	5
1	73	1	1
2	99	0	1
3	124	0	1
4	10	140	0
5	7	0	168

Estimated Mixture Probabilities π_k,
$k = 1, 2, \ldots, K$

K	1	2	3	4	5	6
2	0.41	0.59				
3	0.50	0.23	0.27			
4	0.33	0.13	0.24	0.26		
5	0.17	0.19	0.26	0.05	0.20	
6	0.08	0.25	0.04	0.21	0.05	0.19

Examining the results in Tab. 5.4, we see that ICOMP(IFIM), AIC, and MDL all choose $\hat{K} = 3$ mixture clusters with high relative frequencies each. Recall that the true number of original clusters is $K = 5$ for this data set and the observations are highly overlapped. It would be difficult to identify $K = 5$ clusters. However, we note that ICOMP(IFIM) chooses

$\hat{K} = 4$ 8% of the time, compared to AIC which chooses 6% of the time, indicating toward the direction of the true number of clusters $K = 5$. On the other hand, MDL does not have this behavior and fully automatically chooses $\hat{K} = 3$ 100% of the time. Looking at the confusion matrix, we note that we can practically collapse the original clusters 1, 2, and 3 together to form one cluster $\{1, 2, 3\}$. Cluster 4 and 5 remain separate in their own clusters, except that 10 of the observations from cluster 5, are classified with that of $\{1, 2, 3\}$. This indicates that clearly $\hat{K} = 3$ is a very satisfactory solution. Tab. 5.5 gives the values of each of the criteria for one run of the experiment. All criteria are minimized at $\hat{K} = 3$ mixture clusters.

In concluding, we see that our numerical examples clearly demonstrate the utility of ICOMP(IFIM), AIC, and MDL model selection criteria in identifying the existing structure in the simulated data sets, and estimating the number of component clusters present in the mixture-model. Other numerical examples on the performance of these new procedures on real as well as on simulated data set in the mixture-model cluster analysis are illustrated in Bozdogan (1992).

Acknowledgements

The author is grateful to the detailed positive and many helpful comments of two referees. I wish to thank Dr. B. Lausen for inviting me to present his paper at the 16th Annual Meeting of the German Classification Society at the University of Dortmund, Germany during April 1-3, 1992, I wish also especially to thank my graduate student Chen Xi for transferring the existing MathWriter files into LaTeX form for the publication of this Manuscript.

This research was partially supported by the Institute of Statistical Mathematics, Tokyo, Japan, during January-August, 1988, as a Research Fellow, and Faculty Research Fellowship Award at the University of Tennessee during the summers of 1991 and 1992.

References

AKAIKE, H. (1973), Information Theory and an Extension of the Maximum Likelihood Principle, in: B. N. Pretrov and F. Csaki (eds.), *Second International Symposium on Information Theory*, Academiai Kiado Budapest, 267-281.

BINDER, D.A. (1978), Bayesian Cluster Analysis, *Biometrika*, 65, 31-38.

BOCK, H. H. (1981), Statistical Testing and Evaluation Methods in Cluster Analysis, in the Proceedings of the Indian Statistical Institute Golden Jubilee International Conference on: *Statistics: Applications and New Directions*, J. K. Gosh and J. Roy (eds.) December 16-19, Calcutta, 116-146.

BOZDOGAN, H. (1981), *Multi-Sample Cluster Analysis and Approaches to Validity Studies in Clustering Individuals*, Ph.D. thesis, Department of Mathematics, University of Illinois at Chicago, Chicago, Illinois 60680.

BOZDOGAN, H. (1983), Determining the Number of Component Clusters in the Standard Multivariate Normal Mixture Model Using Model-Selection Criteria, Technical Report No. UIC/DQM/A83-1, June 16, 1983, ARO Contract DAAG29-82-k-0155, Quantitative Methods Department, University of Illinois at Chicago, Chicago, Illinois 60680.

BOZDOGAN, H. (1987), Model Slection and Akaike's Information Criterion (AIC): The General Theory and Its Analytical Extensions, *Psychometrika*, Vol. 52, No. 3, 1987, Special Section (invited paper), 345-370.

BOZDOGAN, H. (1988), ICOMP: A New Model Selection Criterion, in: Hans H. Bock (ed.), *Classification and Related Methods of Data Analysis*, North-Holland, Amsterdam, April, 599-608.

BOZDOGAN, H. (1990a), On the Information-Based Measure of Covariance Complexity and its Application to the Evaluation of Multivariate Linear Models, *Communications in Statistics, Theory and Methods*, 19(1), 221-278.

BOZDOGAN, H. (1990b), Multisample Cluster Analysis of the Common Principle Component Model in K Groups Using an Entropic Statistical Complexity Criterion, invited paper presented at the International Symposium on Theory and Practice of Classification, December 16-19, Puschino, Soviet Union.

BOZDOGAN, H. (1992), Mixture-Model Cluster Analysis and Choosing the Number of Clusters Using a New Informational Complexity ICOMP, AIC, and MDL Model-Selection Criteria, invited paper presented at the First US/Japan Conference on the Frontiers of Statistical Modeling: An Informational Approach, May 24-29, The University of Tennessee, Knoxville, TN 37996, USA. To appear in: H. Bozdogan (ed.), Multivariate Statistical Modeling, Vol. II, Kluwer Academic Publishers, Holland, Dordrecht.

DAY, N.E. (1969), Estimating the Components of a Mixture of Normal Distributions, *Biometrika*, 11, 235-254.

EVERITT, B.S., and HAND, D.J. (1981), *Finite Mixture Distributions*, Champman and Hall, New York.

HARTIGAN, J.A. (1975), *Clustering Algorithms*, John Wiley & Sons, New York.

HARTIGAN, J.A. (1977), Distribution Problems in Clustering, in: J. Van Ryzin (ed.), *Classification and Clustering*, Academic Press, New York, 45-71.

JOHN, S. (1970), On Identifying the Population of Origin of Each Observation in a Mixture of Observation from Two Normal Populations, *Technometrics*, 12, 553-563

KULLBACK, S., and LEIBLER, R.A. (1951), On Information and Sufficiency, *Ann. Math. Statist.*,22, 79-86.

MAGNUS, J.R. (1989), *Linear Structures*, Oxford University Press, New York.

MAGNUS, J.R. (1989), Personal correspondence.

MAGNUS, J.R., and NEUDECKER, H. (1988), *Matrix Differential Calculus with Applications in Statistics and Economitrics*, John Wiley & Sons, New York.

MAKLAD, M.S., and NICHOLS, T. (1980), A New Approach to Model Structure Discrimination, *IEEE Trans. on Systems, Man, and Cybernetics*, SMC-10, No. 2, 78-84.

MCLACHLAN, G.L., and BASFORD, K.E. (1988), *Mixture Models: Inference and Applications to Clustering*, Marcel Dekker, Inc., New York.

RISSANEN, J. (1976), Minmax Entropy Estimation of Models for Vector Processes, in: R.K.Mehra and D.G. Lainiotis (eds.), *System Identification*, Academic Press, New York, 97-119.

RISSANEN, J. (1978), Modeling by Shortest Data Description, *Automatica*, Vol.14, 465-471.

RISSANEN, J. (1989), *Stochastic Complexity in Statistical Inquiry*, World Scientific Publishing Company, Teaneck, New Jersey.

SCLOVE, S.L. (1977), Population Mixture Models and Clustering Algorithms, *Communications in Statistics, Theory and Methods*, A6, 417-434.

SCLOVE, S.L. (1982), Application of the Conditional Population Mixture Model to Image Segmentation, Technical Report A82-1, 1982, ARO Contract DAAG29-82-K-0155, University of Illinois at Chicago, Chicago, Illinois 60680.

SCOTT, S.L., and SYMONS, M.J. (1971), Clustering Methods Based on Likelihood Ratio Criteria, *Biometrics*, 27, 389-397.

SYMONS, M. J. (1981), Clustering Criteria and Multivariate Normal Mixtures, *Biometrics*, 37, 35-43.

TITTERINGTON, D.M., SMITH, A.M.F., and MARKOV, U.E. (1985), *Statistical Analysis of Finite Mixture Distributions*, John Wiley & Sons, New York.

VAN EMDEN, M.H. (1971), *An Anlysis of Complexity*, Mathematical Center Tracts, 35, Amsterdam.

WOLFE, J.H. (1967), NORMIX: Computational Methods for Estimating the Parameters of Multivariate Normal MIxtures of Distributions, Research Memorandum, SRM 68-2, U.S. Naval Personnel Research Activity, San Diego, California.

WOLFE, J.H. (1970), Pattern Clustering by Multivariate Mixture Analysis, *Multivariate Behavioral Res.*, 5, 329-350.

Fuzzy Sets and Fuzzy Partitions

Slavka Bodjanova

Department of Mathematics, Texas A&I University
Kingsville, TX 78363, U.S.A.

Abstract: In this paper some connections between a characterization of fuzzy sets and a characterization of fuzzy partitions are explored. We extend the notion of α-level set and sharpness of fuzzy sets to fuzzy partitions. Then we define α-level equivalence and relation of sharpness on the set of all fuzzy partitions of a finite set of objects into k clusters. We define sharpnesshood and complementhood of fuzzy partitions and then we show how to find the complement of a fuzzy partition.

1   Introduction

Let $X = \{x_1, x_2, \ldots, x_n\}$ be a given set of objects. Denote by $F(X)$ the set of all fuzzy sets defined by X. Let X be partitioned into k clusters. The fuzzy k-partition space associated with X was defined by Bezdek (1981) as follows:

$$P_{fko} = \{U \in V_{kn}; u_{ij} \in [0,1]; \sum_i u_{ij} = 1 \text{ for all } j\} \tag{1}$$

Here u_{ij} is the grade of membership $x_j \in X$ in the fuzzy set $u_i : X \to [0,1]$, hence $u_i \in F(X)$; and V_{kn} is the usual vector space of real $k \times n$ matrices.

The hard k-partition space was defined by

$$P_{ko} = \{U \in V_{kn}; u_{ij} \in \{0,1\}; \sum_i u_{ij} = 1 \text{ for all } j\} \tag{2}$$

Here u_{ij} is the value of a characteristic function $u_i : X \to \{0,1\}$, which specifies the membership of $x_j \in X$ in a hard subset $Y \subset X$.

Bezdek has shown that P_{fko} is the convex hull of P_{ko},

$$P_{fko} = \text{conv}(P_{ko}) = \text{conv}(B_k)^n \tag{3}$$

where B_k is the standard basis of $\Re^k$.

2   Alpha-level fuzzy partitions

It is well known that every fuzzy set $u \in F(X)$ can be approximated by an α-level set u_α defined by

$$u_\alpha(x) = \begin{cases} 1 & \text{if and only if } u(x) \geq \alpha \\ 0 & \text{otherwise.} \end{cases} \tag{4}$$

The above definition implies that u_α is a hard (ordinary) subset of X. We can define the α-level fuzzy partition corresponding to $U \in P_{fko}$ as follows:

Definition 1 *Let $U \in P_{fko}$, $\alpha \in (0, \min_j\{\max_i u_{ij}\})$ and let $c_j = card\{u_{ij} : u_{ij} \geq \alpha\}$. Then the α-level fuzzy partition U^α associated with U is given by*

$$u_{ij}^\alpha = \begin{cases} \frac{1}{c_j} & \text{if and only if } u_{ij} \geq \alpha \\ 0 & \text{elsewhere} \end{cases} \tag{5}$$

Note 1. If $c_j = 1$ for all $x_j \in X$, then U^α is a hard approximation of the fuzzy partition U according to the well known method of maximum membership (MM method).

    Note 2. If for any $x_j \in X$, $c_j = m (1 \leq m \leq k)$, then there are m values $u_{i_r j}, r = 1, \ldots, m$ such that $u_{i_r j} \geq \alpha$. Therefore we can approximate the partitioning of object $x_j \in X$ (that is column $U(x_j)$ from $U \in P_{fko}$) by m different hard partitionings $U_r^\alpha(x_j), r = 1, \ldots, m$, defined as follows

$$(u_r^\alpha)_{ij} = \begin{cases} 1 & \text{if } i = i_r \\ 0 & \text{if } i \neq i_r \end{cases} \tag{6}$$

$U^\alpha(x_j)$ is then the convex combination of $U_r^\alpha(x_j), r = 1, \ldots, m$, defined as follows

$$U^\alpha(x_j) = \frac{1}{m} \sum_r U_r^\alpha(x_j) \tag{7}$$

We can also define the complementary α-level fuzzy partition $\sim U^\alpha$ associated with $U \in P_{fko}$ as follows:

Definition 2 *Let $U \in P_{fko}$, $\alpha \in (0, \min_j\{\max_i u_{ij}\})$. Let $c_j = card\{u_{ij}; u_{ij} \geq \alpha\}$. Then the complementary α-level fuzzy partition $\sim U^\alpha$ is given by*

$$\sim u_{ij}^\alpha = \begin{cases} \frac{1}{k - c_j} & \text{if } k > c_j \text{ and } u_{ij} < \alpha \\ \\ \frac{1}{k} & \text{if } k = c_j \\ 0 & \text{otherwise} \end{cases} \tag{8}$$

Example 1. Let $X = \{x_1, x_2, x_3, x_4\}$ and $U \in P_{f3o}$ be given by the matrix

$$U = \begin{bmatrix} 0.5 & 0.7 & 0.3 & 0 \\ 0.4 & 0.2 & 0.4 & 0.1 \\ 0.1 & 0.1 & 0.3 & 0.9 \end{bmatrix}$$

If $\alpha = 0.3$, then

$$U^\alpha = \begin{bmatrix} 0.5 & 1 & \frac{1}{3} & 0 \\ 0.5 & 0 & \frac{1}{3} & 0 \\ 0 & 0 & \frac{1}{3} & 1 \end{bmatrix} \text{ and } \sim U^\alpha = \begin{bmatrix} 0 & 0 & \frac{1}{3} & 0.5 \\ 0 & 0.5 & \frac{1}{3} & 0.5 \\ 1 & 0.5 & \frac{1}{3} & 0 \end{bmatrix}$$

Definition 3 *Let $U, V \in P_{fko}$ and $\alpha \in (0, t]$, $0 < t \leq 1$. We say that U and V are α-equivalent, denoted by $U =_\alpha V$, if and only if $U^\alpha = V^\alpha$.*

Theorem 1 *The relation $=_\alpha$ is an equivalence relation on P_{fko}.*

With respect to the equivalence relation $=_\alpha$ the partition space P_{fko} is divided into the classes

$$U_\alpha = \{W \in P_{fko}; W^\alpha = U^\alpha\}. \tag{9}$$

That is, U_α is the subset of all fuzzy partitions from P_{fko} that can be approximated by the same α-level fuzzy partition U^α. We will call U_α the "α-levelhood of U."

The subset

$$\sim U_\alpha = \{W \in P_{fko}; W^\alpha =\sim U^\alpha\} \tag{10}$$

is called the "α-complementhood of U."

For every $V \in P_{fko}$, we can measure the degree to which V can be approximated by U^α, that is, the degree to which V belongs to the class U_α. This measure is given in Definition 4.

Definition 4 *Let* $U, V \in P_{fko}$. *Let* $M_1 = \{(i,j) : u_{ij} \geq \alpha\}$, *and* $M_2 = \{(i,j) : u_{ij} < \alpha\}$. *Write* $\text{card } M_1 = |M_1|$. *Then the degree to which* V *can be approximated by* U^α *is measured by the operator*

$$\mathcal{E}_\alpha(V,U) = 1 - \frac{\sum_{M_1} \max\{0, \alpha - v_{ij}\} + \sum_{M_2} \max\{0, v_{ij} - \alpha\}}{|M_1|\alpha + (nk - |M_1|)(1-\alpha)} \tag{11}$$

Theorem 2 *Let* $U, V \in P_{fko}$. *Then*

1. $0 \leq \mathcal{E}_\alpha(V,U) \leq 1$,

2. If $V =_\alpha U$, *then* $\mathcal{E}_\alpha(V,U) = 1$.

3 Sharpness of fuzzy partitions

De Luca and Termini (1972) have introduced the relation of sharpness on $F(X)$ as follows. Let $u, v \in F(X)$. Then v is a sharpened version of u, denoted by $v \prec u$, if and only if

$$v(x) \leq u(x), \qquad \text{for } u(x) \leq \frac{1}{2}, \text{ and} \tag{12}$$

$$v(x) \geq u(x), \qquad \text{for } u(x) \geq \frac{1}{2} \tag{13}$$

We can generalize the relation of sharpness to fuzzy partitions as follows:

Definition 5 *Let* $U, V \in P_{fko}$. *Then* V *is a sharpened version of* U, *denoted by* $V \prec U$, *if and only if*

$$v_{ij} \leq u_{ij}, \qquad \text{for } u_{ij} \leq \frac{1}{k}, \text{ and} \tag{14}$$

$$v_{ij} \geq u_{ij}, \qquad \text{for } u_{ij} \geq \frac{1}{k} \tag{15}$$

Example 2. Let $U, V, W \in P_{f3o}$ be given by the matrices

$$U = \begin{bmatrix} 0.1 & 0.2 & 0.6 \\ 0.5 & 0.4 & 0.3 \\ 0.4 & 0.4 & 0.1 \end{bmatrix} V = \begin{bmatrix} 0 & 0.1 & 0.7 \\ 0.6 & 0.4 & 0.2 \\ 0.4 & 0.5 & 0.1 \end{bmatrix} W = \begin{bmatrix} 0.5 & 0 & 0.35 \\ 0 & 0.4 & 0.65 \\ 0.5 & 0.6 & 0 \end{bmatrix}$$

It is obvious that $V \prec U$ and that there is no fuzzy partition $S \in P_{f3o}$, $S \neq W$ such that $S \prec W$.

Theorem 3 *Let $U \in P_{fko}$. Then there exists $V \in P_{fko}$, $V \neq U$, $V \prec U$, if and only if the set $\{u_{ij} : 0 < u_{ij} \leq \frac{1}{k}\}$ is nonempty.*

Theorem 4 *P_{fko} is partially ordered by the relation $\prec$.*

For every $U \in P_{fko}$, we have the subset

$$U_{\prec} = \{W \in P_{fko} : W \prec U\} \tag{16}$$

which we call the "sharpnesshood of U."

For every $V \in P_{fko}$ we can measure the degree to which V is a sharpened version of U, that is the degree to which V belongs to $U_{\prec}$. This measure is given in Definition 6.

Definition 6 *Let $U, V \in P_{fko}$. Let $K_1 = \{(i,j) : u_{ij} > \frac{1}{k}\}$, and $K_2 = \{(i,j) : u_{ij} < \frac{1}{k}\}$. Then the degree to which V is a sharpened version of U is measured by the operator*

$$\mathcal{H}(V,U) = 1 - \frac{\sum_{K_1} \max\{0, u_{ij} - v_{ij}\} + \sum_{K_2} \max\{0, v_{ij} - u_{ij}\}}{2n} \tag{17}$$

Theorem 5 *Let $U, V \in P_{fko}$. Then*

1. $0 \leq \mathcal{H}(V,U) \leq 1$

2. $\mathcal{H}(V,U) = 1$ if and only if $V \prec U$.

Let $\overline{U_j}$ denote the partitioning of object $x_j \in X$ such that $u_{ij} = \frac{1}{k}$ for all i. Let $\alpha = \frac{1}{k}$. From the geometrical point of view, we can find for every $U_j \neq \overline{U_j}$ the approximation $\gg U_j \in U_{\alpha j}$ and the approximation $\ll U_j \in (\sim U_{\alpha j})$ such that $\gg U_j$ is the sharpest fuzzy partition of object x_j in $U_{\alpha j}$ on the line joining U_j to $\overline{U_j}$, and $\ll U_j$ is the sharpest fuzzy partition of the object x_j in $\sim U_{\alpha j}$ on the line joining U_j to $\overline{U_j}$.

We will say that $\gg U_j$ is the "maximal linear sharpened" (MLS) α-level fuzzy approximation associated with U_j, and $\ll U_j$ is the complementary MLS α-fuzzy approximation associated with U_j.

If $U_j = \overline{U_j}$ then we define $(\ll \overline{U_j}) = (\gg \overline{U_j}) = \overline{U_j}$.

Definition 7 *Let $U \in P_{fko}$. The MLS α-level fuzzy partition of U for $\alpha = \frac{1}{k}$ is a fuzzy partition $\gg U \in U_\alpha$ such that the matrix $\gg U$ consists of columns $\gg U_j$, where every column $\gg U_j$ is the MLS $\frac{1}{k}$-level fuzzy approximation associated with U_j (column of U).*

Definition 8 *Let $U \in P_{fko}$. The complementary MLS α-level fuzzy partition of U for $\alpha = \frac{1}{k}$ is the fuzzy partition $\ll U \in (\sim U_{\alpha j})$ such that the matrix $\ll U$ consists of columns $\ll U_j$, where every column $\ll U_j$ is the complementary MLS $\frac{1}{k}$-level approximation associated with U_j (column of U).*

Theorem 6 *Let $U \in P_{fko}$, $U \neq \overline{U}$. Let $u_{rj} = \min_i\{u_{ij}\}$ and $u_{sj} = \max_i\{u_{ij}\}$. Then*

$$\gg u_{ij} = \frac{1}{k} + \frac{1}{1 - ku_{rj}}(u_{ij} - \frac{1}{k}), \tag{18}$$

$$\ll u_{ij} = \frac{1}{k} + \frac{1}{1 - ku_{sj}}(u_{ij} - \frac{1}{k}). \tag{19}$$

4 Complement of fuzzy partition

Definition 9 *Let $U \in P_{fko}$. The complement of U is a fuzzy partition $U^c \in P_{fko}$ given by*

$$u_{ij}^c = \frac{u_{ij} - \frac{\lambda_j}{k}}{1 - \lambda_j}, \tag{20}$$

where $\lambda_j = \frac{k(\max_i u_{ij} - \min_i u_{ij})}{1 - k \min_i u_{ij}}$ if $U_j \neq \overline{U}_j$, and $\lambda_j = 0$ otherwise.

Note. For $k = 2$ (that is, the partition consists of a fuzzy set and its complement) we get Zadeh's usual definition of complement, that is $u_{ij}^c = 1 - u_{ij}$. So we can say that our definition of complementation is a generalization of Zadeh's complementation for fuzzy partitions into $k \geq 2$ clusters.

Theorem 7 *Let $U \in P_{fko}$. Then*

$$(\gg U)^c = (\gg U^c) = (\ll U) \tag{21}$$
$$(\ll U)^c = (\ll U^c) = (\gg U) \tag{22}$$

Corollary 1 *Let $U \in P_{fko}$, $\alpha = \frac{1}{k}$. Then $(U^\alpha)^c = (\sim U^\alpha)$, and $(\sim U^\alpha)^c = U^\alpha$.*

Theorem 8 *Let $U \in P_{fko}$. Then $(U^c)^c = U$.*

Example 3. Let $U \in P_{f3o}$ be given by the matrix

$$U = \begin{bmatrix} 0.8 & 0.2 & 0.1 & 0.5 \\ 0.1 & 0.6 & 0.6 & 0.2 \\ 0.1 & 0.2 & 0.3 & 0.3 \end{bmatrix} \quad \text{Then the complement of } U \text{ is } U^c = \begin{bmatrix} 0.10 & 0.4 & 0.538 & 0.20 \\ 0.45 & 0.2 & 0.099 & 0.40 \\ 0.45 & 0.4 & 0.363 & 0.36 \end{bmatrix}$$

5 Conclusion

One of the important practical problems connected with fuzzy clustering and fuzzy classification is the comparison and evaluation of fuzzy partitions. In this paper we proposed a comparison and evaluation based on the so called α-level approximation (α-level fuzzy partitions) and the sharpness of fuzzy partitions. We have shown how α-level approximation and sharpness can be used in order to find the complement of a fuzzy partition.

For every fuzzy partition $U \in P_{fko}$, we can find the following associated partitions: U^α, $\gg U$, $\sim U^\alpha$, $\ll U$ and U^c. Our definition of complement U^c can be considered as a generalization of Zadeh's complementation. U^α and $\gg U$ represent approximations of U. $\sim U^\alpha$ and $\ll U$ represent approximations of U^c. We have shown that the partitions from P_{fko} can be compared (evaluated) using the relation of α-equivalence or the relation of sharpness. The degree of α-levelhood and the degree of sharpnesshood can be computed for every pair of partitions $U, V \in P_{fko}$.

References

BACKER, E. (1978), *Cluster Analysis by Optimal Decomposition of Induced Fuzzy Sets*, Delftse Universitaire Pers, Delft.

BEZDEK, J. (1981) *Pattern Recogition with Fuzzy Objective Function Algorithms*, Plenum Press, New York.

BEZDEK, J., and HARRIS, J. (1979), Convex Decomposition of Fuzzy Partitions, *Journal of Math. Anal. and Applications, 67, 490–512.*

DE LUCA, A., and TERMINI, S. (1972), A Definition of a Nonprobabilistic Entropy in the Setting of Fuzzy Sets Theory, *Information and Control, 20, 301–312.*

DUBOIS, D., and PRADE, H. (1980), *Fuzzy Sets and Systems: Theory and Applications*, Academic Press, New York.

KOSKO, B. (1992), *Neural Networks and Fuzzy Systems*, Prentice Hall, Englewood Cliffs.

ZADEH, L.A. (1965), Fuzzy Sets, *Information and Control, 8, 338–353.*

Fuzzy Clustering by Minimizing the Total Hypervolume

E. Trauwaert
Belgoprocess, Gravenstraat 73, B-2480 Dessel

P. Rousseeuw
University of Antwerp, WISINF, Universiteitsplein 1, B-2610 Wilrijk

L. Kaufman
VUB, STOO, Pleinlaan 2, B-1050 Brussels

Abstract: The total hypervolume of a clustering, measured by the sum of the square root of the determinants of the fuzzy covariance matrices, is proposed as objective function. It is shown that this criterion, unlike most other well-known methods, is independent of differences in shape, size and cardinality between the different clusters.

1   Introduction

For partitioning a data set in groups of similar objects, a wide variety of methods seems to be available. However, when the required clusters can be different in shape (although of approximate hyperellipsoidal form) the number of possible methods is largely reduced. If furthermore the different clusters are possibly unequal in size, there is hardly any published method able to identify this stucture adequately and without prior knowledge of the sizes of the clusters.

Gustafson and Kessel [1] have developed a method based on a generalization of the **fuzzy** $k-$**means** algorithm [2], which is certainly one of the most advanced methods in this field. However, although it accepts clusters of different shapes, they must be of known relative sizes; clusters with unknown sizes will normally not be satisfactorily detected by this method. The product of fuzzy determinants criterion developed by Trauwaert, Kaufman and Rousseeuw [3] avoids this restriction, but at the price of other limitations and inherent bias.

Overall, the problem appears to be the choice of an adequate objective function which is sufficiently versatile to describe the various clusters with all their potential differences, and without bias.

In this paper we propose an objective function based on the total volume of the clusters. In this approach the clusters are described by their covariance matrices as in [3], whereas the hypervolume of a cluster is measured by the square root of the determinant of its covariance matrix. The total hypervolume of the clusters corresponds with the sum, over all clusters, of the square roots of the determinants of the covariance matrices. Furthermore, the residual fuzziness bias present in this method (as indeed in all other fuzzy methods) can be avoided by adopting a **semi-fuzzy** algorithm. This approach uses the fuzzy methodology during the iterative calculations but tends to converge to a mainly crisp result as final solution.

In the next section the hypervolume method will be described in both its fuzzy and semi-fuzzy mode. This section will be followed by the description of the algorithm adapted from **fuzzy** $k-$**means**. Section 4 will illustrate the qualities of the method on some classical examples, as well as on some generated data sets. The conclusions are gathered in the last section.

2 Description of the hypervolume model

2.1 General elements

Let us suppose a data set is to be inspected for the presence of clusters and consists of n objects x_i described by m variables $(x_{i1}, \ldots, x_{ip}, \ldots, x_{im})$. The data set has some local concentrations of objects, which corresponds to the k clusters[1]. Each of the clusters is defined by its center and its covariance matrix, whereas the membership of each object i to each cluster t is measured by a function u_{it}. This membership function can vary between 1, for full membership, and 0, for absence of membership. In the crisp mode the membership function can take only one of the two extreme values $\{0, 1\}$, whereas in the fuzzy or semi-fuzzy mode all values between zero and one are also allowed. In all cases, the sum over all clusters of the membership functions related to any object must always be equal to one; hence

$$\sum_t u_{it} = 1 \qquad \text{for all } i \qquad (1)$$

$$\text{with} \quad u_{it} \geq 0 \qquad \text{for all } i \text{ and for all } t . \qquad (2)$$

The constraint that each membership should be less than or equal to 1 is already implicitly included in the two previous conditions. The number of objects belonging to a cluster is defined as :

$$n_t = \sum_i u_{it} \qquad \text{for all } t \qquad (3)$$

Through (1) and (3) we have :

$$\sum_t n_t = \sum_t \sum_i u_{it} = \sum_i \sum_t u_{it} = n \qquad (4)$$

The concentrations of objects around the k centers can be described by the covariance matrices, defined as follows :

$$S_t = \frac{\sum_i \Phi(u_{it})(x_i - \mu_t)(x_i - \mu_t)'}{n_t} \qquad \text{for all } t \qquad (5)$$

supposing that the clusters are not empty $(n_t > 0)$. In this equation $\Phi(u_{it})$ is a function of the memberships, taking one of the three following forms :

$$\text{crisp mode} \quad : \quad \Phi(u_{it}) \equiv u_{it} \qquad (6a)$$

$$\text{fuzzy mode} \quad : \quad \Phi(u_{it}) \equiv u_{it}^\beta \quad \text{with } \beta > 1 \qquad (6b)$$

$$\text{semi-fuzzy mode} \quad : \quad \Phi(u_{it}) \equiv \alpha u_{it} + (1 - \alpha)u_{it}^\beta \quad \text{with } 0 < \alpha < 1 \text{ and } \beta > 1 \qquad (6c)$$

It can be seen that each term of the covariance matrix consists of the average cross product of the difference between each object and the cluster center, with $\Phi(u_{it})$ as weighing factors.

---

[1] Unless otherwise mentioned, i will always vary from 1 to n, p from 1 to m and t from 1 to k; hence $\sum_t$ stands for $\sum_{t=1}^k$ and so on.

In the crisp mode the covariance matrix has its classical form. To get a fuzzy approach it is not enough to replace the boolean u_{it} by a continuous function with values in the range [0,1]; one must also adopt an exponent β strictly larger than 1 (see Dunn [2], Bock [4]). From now on we will restrict ourselves to the value $\beta = 2$.

From the definition of S_t one can further infer that this matrix is always symmetric; it is also at least semi-definite. Moreover the matrix S_t can generate a quadratic equation

$$x'S_t^{-1}x \;=\; 1 \tag{7}$$

which represents a hyperellipsoid of rank m (when S_t is positive definite) or lower (when S_t is positive semi-definite) centered at the origin of the axes of coordinates. The axes of the hyperellipsoid are the square roots of the eigenvalues of the original matrix S_t. Note that the hypervolume of (7) is proportional to the square root of the determinant of S_t (where the proportionality factor depends only on m, the number of variables).

2.2 The objective function and its derivatives

Minimizing the total hypervolume of the clusters is equivalent to minimizing the objective function

$$\Theta \;=\; \sum_t |S_t|^{\frac{1}{2}} \tag{8}$$

The minimum of this objective function will be reached for some optimal values of the centers and the memberships. We will not tackle the problem in the crisp case, with the boolean u_{it}, for which a solution may be found using an exchange procedure. The fuzzy and semi-fuzzy problems on the contrary are continuous, and can be solved by searching for the values of the memberships and of the centers for which the derivatives of the objective function (8) equal zero.

Hence for the partial derivative, with respect to the center of any cluster t, we have

$$\frac{\partial \Theta}{\partial \mu_t} \;=\; \frac{1}{2}|S_t|^{\frac{1}{2}}\frac{\partial |S_t|}{\partial \mu_t}$$

with

$$\frac{\partial |S_t|}{\partial \mu_t} \;=\; 2\frac{|S_t|}{n_t}\sum_i \left[\Phi(u_{it})(x_i - \mu_t)'S_t^{-1}\right]$$

hence

$$\frac{\partial \Theta}{\partial \mu_t} \;=\; \frac{|S_t|^{\frac{1}{2}}}{n_t}\sum_i \left[\Phi(u_{it})(x_i - \mu_t)'S_t^{-1}\right] \tag{9}$$

This partial derivative becomes zero when

$$\mu_t \;=\; \frac{\sum_i \Phi(u_{it})x_i}{\sum_i \Phi(u_{it})} \tag{10}$$

in which the denominator is nonzero under the same conditions as n_t. This is the only solution provided S_t is strictly positive definite, as in that case the determinant is also strictly positive.

For the partial derivative with respect to the memberships u_{it}, due account has to be taken of the two constraints (1) and (2), with which a Lagrangean objective function can be constructed

$$\Theta' = \sum_t |S_t|^{\frac{1}{2}} - \sum_i \Gamma_i(\sum_t u_{it} - 1) - \sum_i \sum_t \tau_{it} u_{it} \tag{11}$$

Its partial derivative can be written

$$\frac{\partial \Theta'}{\partial u_{it}} = \frac{1}{2}|S_t|^{-\frac{1}{2}}\frac{\partial S_t}{\partial u_{it}} - \Gamma_i - \tau_{it}$$

with

$$\frac{\partial S_t}{\partial u_{it}} = \frac{|S_t|}{n_t}[\Phi'(u_{it})(x_i - \mu_t)'S_t^{-1}(x_i - \mu_t) - m]$$

hence

$$\frac{\partial \Theta'}{\partial u_{it}} = \frac{1}{2}\frac{|S_t|^{\frac{1}{2}}}{n_t}[\Phi'(u_{it})(x_i - \mu_t)'S_t^{-1}(x_i - \mu_t) - m] - \Gamma_i - \tau_{it} \tag{12}$$

which in the general semi-fuzzy case, with $\Phi'(u_{it}) = \alpha + 2(1-\alpha)u_{it}$, becomes

$$\frac{\partial \Theta'}{\partial u_{it}} = B_{it}u_{it} - A_{it} - \Gamma_i - \tau_{it} = 0 \tag{13}$$

$$\text{with} \quad A_{it} = \frac{1}{2}\frac{|S_t|^{\frac{1}{2}}}{n_t}[m - \alpha(x_i - \mu_t)'S_t^{-1}(x_i - \mu_t)]$$

$$\text{and} \quad B_{it} = \frac{|S_t|^{\frac{1}{2}}}{n_t}(x_i - \mu_t)'S_t^{-1}(x_i - \mu_t)(1 - \alpha)$$

The corresponding expression in the fuzzy case can be deduced from the above expression by putting $\alpha = 0$. It can be seen that in that case A_{it} is no longer function of the objects so that consequently A_{it} can be identified as A_t.

2.3 The optimal solution

To solve equation (13) different cases have to be considered.

2.3.1 Coincidence of a cluster center with an object

Let us first suppose that for an object i some of the B_{it} are zero, resulting from the coincidence of the object i with one or more cluster centers μ_t. In this case the solution of (13) is obtained by assigning the value zero to the u_{it} for which the B_{it} are not zero, and sharing a value one, in whatever way one likes, between the u_{it} for which the B_{it} are zero. Proper values for Γ_i and τ_{it} can ensure that (13) is verified, although these values are without any further importance.

2.3.2 Solution without positiveness constraint

A more general case is obtained when supposing that all the τ_{it} are equal to zero, at least for some object i and all clusters t. According to the Lagrangean conditions this is only possible if the corresponding u_{it} are not negative.

Under this condition, equation (13) can easily be solved for the unknown values of the memberships, after elimination of Γ_i by the condition (1); one finds

$$u_{it} = \frac{\frac{1}{B_{it}}}{\sum_r \frac{1}{B_{ir}}} - \frac{1}{B_{it}} \left[\frac{\sum_r \frac{A_{ir}}{B_{ir}}}{\sum_r \frac{1}{B_{ir}}} - A_{it} \right] \tag{14}$$

Two remarkable forms of this expression deserve special attention. Let us first observe that the first two terms of (14) define the Lagrangean constant Γ_i, necessary to ensure that (1) is verified. Hence, if these two terms happen to be identical, we have $\Gamma_i = 0$, so that (13) and (14) reduce to

$$u_{it} = \frac{A_{it}}{B_{it}} \tag{15}$$

which is the unconstrained result, expressed in function of the dimension of the variable space m, of the semi-fuzziness factor α and of the Mahalanobis distance of object i to cluster center μ_t.

Another special form of expression (14) is obtained when all A_{it} are independent of t. In that case the memberships reduce to

$$u_{it} = \frac{\frac{1}{B_{it}}}{\sum_r \frac{1}{B_{ir}}} \tag{16}$$

which evidently satisfies (1). This expression is function of the Mahalanobis distance of each object to the center of each cluster, of the fuzziness factor and of the specific density of the clusters.

As the sum over all clusters of all terms of (16) and of (14) both equal one, the sum of the second and third terms of (14) must equal zero. Hence, in the general case, the first term of (14) can be considered as the main membership term; the second and third terms, which in the fuzzy mode are generally small, can thus be considered as corrections, accounting for a difference in specific cluster density. In the case of the semi-fuzzy mode this correction term is also function of the Mahalanobis distance of an object with respect to the center of each cluster.

2.3.3 General solution

In the last particular case above (16) all u_{it} are always positive. This is in general not necessary, as both equations (14) and (15) can become negative, which is in contradiction with (2). To avoid these negative values of u_{it} the corresponding values of τ_{it} must become strictly positive with a value that renders the negative u_{it} equal to zero. From (13) it can be seen that the τ_{it} act as corrections rendering the A_{it} more positive or less negative. In doing so all other memberships u_{ir} of the same object i to the other clusters r will be reduced, as the constraint (1) must always be satisfied. This effect of the τ_{it} on the other memberships can be deduced from the role of A_{it} in (14). Being reduced, some new negative memberships could appear needing some new corrections of the negative memberships and new reductions of the strictly positive memberships. To show that this procedure converges to a solution in which no memberships are strictly negative and for each object i at least some memberships are strictly positive, the following reasoning can be made.

Let us first observe that the τ_{it} are allowed to become strictly positive only to avoid that some u_{it} would become negative; according to (14), this condition can be expressed as follows

$$\tau_{it} \ > \ 0 \quad \text{only if}$$

$$0 \ > \ u_{it} = \frac{\frac{1}{B_{it}}}{\sum_r \frac{1}{B_{ir}}} - \frac{1}{B_{it}} \left[\frac{\sum_r \frac{A_{ir}}{B_{ir}}}{\sum_r \frac{1}{B_{ir}}} - A_{it} \right] \tag{17}$$

or, as $B_{it} > 0$ (otherwise we have the solution of section 2.3.1), only if

$$A_{it} < \frac{\sum_r \frac{A_{ir}}{B_{ir}} - 1}{\sum_r \frac{1}{B_{ir}}} \tag{18}$$

The summations in equations (17) and (18) originate from equation (1). However, it can be seen that (1) must still be satisfied even if the zero memberships are deleted from the sums. The problem can hence be reformulated as follows: find, for an object i, a partition Θ_i of the total set of cluster indices T, such that, when the summations of the right hand terms of (18) are performed over all $r \in \Theta(i) \subset T$, the inequality (18) is verified for all $t \notin \Theta(i)$, but is contradicted for all $t \in \Theta(i)$. So, we can classify, for each object i, the left hand term of (18), A_{it}, in decreasing order of magnitude. We will designate a variable term of this series by A_{is}, for $s = 1$ to k, with the condition that

$$A_{is} > A_{i(s+1)} \tag{19}$$

If some A_{is} satisfies (18), any $A_{is'}$ in which $s' \geq s$ will also satisfy (18); on the other hand if some A_{is} contradicts (18), any $A_{is'}$ in which $s' \leq s$ will also contradict (18). Hence, the problem reduces to the search for an s such that

$$A_{is} \sum_{r=1}^{s} \frac{1}{B_{ir}} - \sum_{r=1}^{s} \frac{A_{ir}}{B_{ir}} < -1 \tag{20}$$

and such that for $s' = s - 1$, (20) is not verified.

If s exists, two important points with respect to this partition can be demonstrated:

- $s > 1$ which implies that $\Theta(i)$ is never empty;

- s is unique; in other words, there is only one way to define a partition $\Theta(i) \subset T$, satisfying the conditions above.

If s does not exists, it means that (20) is never satisfied and that we are back to the case of section 2.3.2. We will now prove the two above points in sequence.

a) $s > 1$

It is sufficient to prove that there is always at least some value of s for which (20) is not verified. Now if we take s equal to 1, we find

$$\frac{A_{is}}{B_{is}} - \frac{A_{is}}{B_{is}} = 0$$

which always contradicts (20). The conclusion is that at least the cluster for which A_{is} is the largest will always receive a strictly positive membership from object i. This proves our first point.

b) s **is unique**

We will prove this by demonstrating that the left term of (20), identified as G_s, is an uniformly decreasing function, together with decreasing values of A_{is}. From a) above, we know that $G_1 > -1$. So, there will be at most one passage from $G_{s-1} \geq -1$ to $G_s < -1$.

Deleting for simplicity in (20) the index for object i, which is identical in all terms, G_s can be expressed as follows:

$$G_s = A_s \left(\frac{1}{B_s} + \frac{1}{B_{s-1}} + \ldots + \frac{1}{B_1} \right) - \frac{A_s}{B_s} - \frac{A_{s-1}}{B_{s-1}} - \ldots - \frac{A_1}{B_1}$$

which is equivalent to

$$G_s = (A_s - A_{s-1}) \left(\frac{1}{B_{s-1}} + \frac{1}{B_{s-2}} + \ldots + \frac{1}{B_1} \right) +$$
$$(A_{s-1} - A_{s-2}) \left(\frac{1}{B_{s-2}} + \frac{1}{B_{s-3}} + \ldots + \frac{1}{B_1} \right) + \ldots + (A_2 - A_1) \left(\frac{1}{B_1} \right)$$

or

$$G_s = (A_s - A_{s-1}) \left(\frac{1}{B_{s-1}} + \frac{1}{B_{s-2}} + \ldots + \frac{1}{B_1} \right) + G_{s-1} \tag{21}$$

Hence, as $A_s \leq A_{s-1}$ and as all B_r are strictly positive we show that

$$G_s \leq G_{s-1} \tag{22}$$

what was to be proven.

Hence if we define as $\Theta(i)$ the set of cluster indices for which (20) is not verified at object i, we can now formulate the correct solution for the memberships, taking into account constraint (2),

$$\text{for} \quad t \in \Theta(i)$$

$$u_{it} = \frac{\frac{1}{B_{it}}}{\sum\limits_{r \in \Theta(i)} \frac{1}{B_{ir}}} - \frac{1}{B_{it}} \left[\frac{\sum\limits_{r \in \Theta(i)} \frac{A_{ir}}{B_{ir}}}{\sum\limits_{r \in \Theta(i)} \frac{1}{B_{ir}}} - A_{it} \right] \geq 0 \tag{23}$$

$$\text{and for} \quad t \notin \Theta(i) \qquad u_{it} = 0$$

3 Algorithm

The corresponding algorithm becomes :

1. Assign some membership value to all objects with respect to each cluster and in accordance with the conditions (1) and (2);

2. calculate for each cluster the center vector according to (10) and the covariance matrix according to (5);

3. put all the cluster indices in the index set $\Theta(i)$ and evaluate the membership functions according to (14); if some of the membership functions are negative, put them equal to zero, eliminate their index from the index set $\Theta(i)$ and recalculate the other memberships according to (23); iterate as long as some memberships are strictly negative;

4. compare the memberships with those of the previous passage : if no value differs by more than a quantity ϵ, stop; otherwise go back to step 2.

Throughout the paper, we will call this the **fuzzy volume** algorithm.

4 Examples

Whatever the theoretical motivation, the ultimate test for any method is to confront it with data. But analyzing a new and unknown data set might be unsatisfactory when there is no consensus about its "true" structure.

Therefore, we will present two types of examples: some data sets from the literature, to illustrate the general good behavior of the **fuzzy volume** algorithm, in comparison with some other known algorithms, and some data that were especially designed to study the features for which the **fuzzy volume** algorithm was developed. To the first type belongs the Ruspini data and the stars data. For the second type we will present examples with very dissimilar hyperellipsoids with possibly different orientations.

The minimum **fuzzy volume** algorithm will be compared with two other algorithms: the **fuzzy** k-**means** method of Dunn [2] known to perform well with equal hyperspherical clusters but poorly in many other cases, and the **product of determinants** algorithm developed by Trauwaert, Kaufman and Rousseeuw [3], which is able to find dissimilar clusters (provided they are somewhat hyperellipsoidal in shape) but still bears some intrinsic limitations.

As all these algorithms are heuristic in nature, they are converging to some local optimum but not necessarily to the global optimum. Hence, in order to check whether an optimum is global or local and in order to enhance the chances to find the global optimum, each algorithm will be tried out from at least ten different starting positions. The frequency of arriving at the global optimum will be a supplementary measure of the relative quality of the methods.

4.1 Ruspini data

It is well known from the literature [5,6,7], but also from a graphical representation of these 75 two-dimensional observations, that the natural clustering calls for 4 clusters. Two of these clusters appear as fairly well separated, with 20 and 15 observations respectively. The

2 other clusters, with approximatively 20 observations each, have some bridges which can be considered as part of one or the other of the 2 clusters. This brings some uncertainty as to the right number of observations for each of these two clusters.

Applying the three above algorithms to this data set provides the results of Table 1, in which the global maximum is characterized as four near circular clusters with respectively 20, 23, 17 and 15 objects. The 12 different results are generated from different starting positions obtained by varying the initial arbitrary assignation of the membership values. Obviously **fuzzy k−means** has no problems in finding these natural clusters, but the other two algorithms perform nearly as well: only in one out of the 12 trials another partition was found in a different local optimum. This result may express the higher variability of these algorithms.

Table 1	natural clustering	alternative clusterings
Fuzzy k−means	12	0
Fuzzy product	11	1
Fuzzy volume	11	1

4.2   Two ellipsoidal clusters with different orientations

This data set consists of two groups of 25 two-dimensional objects. Each group has an approximately ellipsoidal shape with its main axis some six times larger than the small one. Moreover, the two ellipses have their main axes perpendicular to each other [3].

Applying the three algorithms to this data set produces the following results (Table 2) after ten trials.

Table 2	natural clustering	alternative clusterings
Fuzzy k−means	0	10
Fuzzy product	10	0
Fuzzy volume	10	0

The natural clusters have been correctly found by the last two algorithms in all instances (10 times). Only **fuzzy k−means** consistently failed to find these two ellipses; instead, it tried to identify two alternative clusters of more circular form.

4.3   The star group CYG OB1

This example consists of data on 47 stars, collected by C. Doom and reported by Kaufman and Rousseeuw [7]. Each star is characterized by its surface temperature and its light intensity, both on a logarithmic scale.

The plotted data [7] show two very dissimilar groups: one with only four stars and one with the remaining 43 stars.

Applying the three algorithms, the data is partitioned in two clusters as follows (Table 3):

Table 3	natural clustering (4/43)	alternative clusterings	
		(14/33)	(20/27)
Fuzzy k–means	0	0	10
Fuzzy product	0	10	0
Fuzzy volume	10	0	0

Here, not only **fuzzy k–means**, but also **fuzzy product** experiences difficulties to distinguish correctly two clusters of very unequal size and cardinality. Only **fuzzy volume** is completely successful in identifying the natural clusters, as was to be expected.

4.4   Artificial data set with two very unequal clusters

Up to now all algorithms have been used in their fuzzy mode. We will now present an example which justifies the use of the semi-fuzzy mode. The artificial data set consists of 64 four-dimensional objects, grouped in two neighbouring clusters of 56 and 8 objects respectively [8]. In our study we have omitted the **fuzzy k–means** algorithm, as it should be clear that it will not be able to provide any acceptable solution to this classification problem. The **fuzzy product** and **fuzzy volume** methods perform reasonably well, but on the basis of the preliminary knowledge we have about the data, none provide the expected natural clusters. This is because a typical fuzziness bias occurs: due to the many small memberships of the objects of the larger cluster to the smaller cluster, the center of the smaller cluster is shifted towards the larger. This results in the absorption, by the smaller cluster, of some objects of the larger cluster.

To avoid this fuzziness bias one can resort to a fuzziness factor strictly larger than zero. Table 4 shows, for different values of the fuzziness factor α, the frequency of obtaining the natural and alternative clusterings with the **total volume** method. It can be seen that with a fuzziness factor between 0.25 and 0.50, the **total volume** method succeeds in finding the natural clusters with a very high frequency, even in this difficult example.

Table 4
Clustering of two unequal clusters (8/56 objects)
with the semi-fuzzy total volume criterion.

clusterings obtained	fuzziness factor					
	0.00	0.33	0.50	0.67	0.8	0.95
8/56	-	10	9	7	4	-
13/51	10	-	-	1	-	-
30/34	-	-	-	2	2	-
others	-	-	1	-	4	10
total	10	10	10	10	10	10

5   Conclusions

In this paper an algorithm for fuzzy and semi-fuzzy clustering has been proposed, based on the minimization of the total (fuzzy) hypervolume. Applied to a series of very different examples, it performs almost as well as the classical algorithms when the clusters are not too different in shapes, sizes and cardinalities. However, in problems with clusters of very different shapes, sizes and cardinalities, the minimum hypervolume method appears to be the only algorithm able to reproduce the expected natural clusters, particularly in the semi-fuzzy mode.

References

[1] GUSTAFSON D.E., KESSEL W., *Fuzzy Clustering with a Fuzzy Covariance Matrix*, Proc. IEEE-CDC, Vol. 2 (K. S. Fu ed.), pp. 761-766, IEEE Press, Piscataway, New Jersey (1979).

[2] DUNN J.C., *A Fuzzy Relative of The ISODATA Process And Its Use In Detecting Compact Well-separated Clusters*, Journal of Cybernetics, 3, (1974), pp. 32-57.

[3] TRAUWAERT E., KAUFMAN L., ROUSSEEUW P., *Fuzzy Clustering Algorithms Based On The Maximum Likelihood Principle*, Fuzzy Sets and Systems, to appear (1991).

[4] BOCK H.H., *Fuzzy Clustering Procedures, Analyse des données et informatique*, Ed.: Amirchahy et Néel, pp. 205-218, Fontainebleau, (1979).

[5] ROUBENS M., *Pattern Classification Problems And Fuzzy Sets*, Fuzzy Sets And Systems, 1, (1978), pp. 239-253.

[6] TRAUWAERT E., *On The Meaning Of Dunn's Partition Coefficient For Fuzzy Clusters*, Fuzzy Sets and Systems, 25, (1988), pp. 217-242.

[7] KAUFMAN L., ROUSSEEUW P., *Finding Groups In Data*, John Wiley & Sons, NY, (1990).

[8] TRAUWAERT E., *Hybrid Clustering: Towards A Fuzzy Algorithm Without Bias*, Proceedings of the 3rd IFSA Congress, Ed.: J. Bezdek, Seattle, (1989).

Conceptual Data Systems

Patrick Scheich, Martin Skorsky, Frank Vogt,
Cornelia Wachter, and Rudolf Wille
Forschungsgruppe Begriffsanalyse, Fachbereich Mathematik
Technische Hochschule Darmstadt, Schloßgartenstraße 7, W–6100 Darmstadt

Abstract: Conceptual data systems shall serve with tools for exploring conceptually information which is stored in data bases. Ideas for conceptual data systems have been already developed in [3] by methods of formal concept analysis. This approach is extended to an elaborated system which allows a rich browsing through all available information. The basic tools are conceptual scales (cf. [2]) which preform the information so that always part of it can be visualized by nested line diagrams of concept lattices. The browsing procedure is performed with respect to a chosen sequence of conceptual scales by zooming back and forth in the given conceptual hierarchy. The implementation of a conceptual data system is based on a conceptual file which is structured according to requirements of effective browsing procedures and easy maintenance of the system.

1   Data Contexts and Concepts

Formal concept analysis has grown to a successful method of data analysis during the last decade. Its main advantage is that it clearly unfolds the inherent conceptual structure of a data context without loosing the original data or making unnecessary mathematical assumptions on their structure. This advantage combines however with the inconvenience that larger data contexts are difficult to explore. To diminish this problem we introduce the tool of a conceptual data system which allows a rich browsing through all available information. In this first section of the paper we recall the basic notions of formal concept analysis. By an example, we explain in Section 2 how to use a conceptual data system. The final section describes the structure and implementation of conceptual data systems.

Empirical data appear most often in the form of a table where the rows represent several objects and the columns represent attributes which have been evaluated for these objects. Each cell of the table contains the value of the corresponding object with respect to the corresponding attribute. In formal concept analysis this type of data is modelled by a *many–valued context* [*mehrwertiger Kontext*] which, in general, is a quadrupel (G, M, W, I) where G, M, and W are sets together with a ternary relation $I \subseteq G \times M \times W$ such that $(g, m, w_1) \in I$ and $(g, m, w_2) \in I$ imply $w_1 = w_2$. The set G is interpreted as a set of *objects* [Menge von *Gegenständen*], the set M as a set of *attributes* [Menge von *Merkmalen*]. Then $(g, m, w) \in I$ is read: *The object g has the value w for the attribute m.* Figure 1 shows part of a many–valued context which has been the result of a medical investigation at the *Children's Hospital* of the *McGill University* at *Montral.* The objects of this many–valued context are 111 children who have suffered from diabetes. For these children, 22 demographical, medical, and clinical attributes have been evaluated. This many–valued context will serve as the basic example in our paper for demonstrating the methods of formal concept analysis. An analysis of the data with respect to the aims of the medical investigation will not be given. For more detailed information about the data, see the preprint [1] of A. Ciampi, A. Schiffrin, J. Thiffault, H. Quintal, G. Weitzner, P. Poussier, and D. Lalla.

A many–valued context can be analysed by methods of formal concept analysis. For explaining this analysing procedure, we have to recall some of the basic notions. More

	sex	coma	age	symptom duration	glucose level	insulin requirement	pH level	15 more attributes →
C1	female	no	low	year	pathol.	low	pathol.	
C2	male	no	medium	month	pathol.	high	pathol.	
C3	male	no	high	year	pathol.	very low	normal	
C4	male	no	high	year	pathol.	high	pathol.	
C5	male	no	low	month	pathol.	low	normal	
C6	female	no	low	month	pathol.	very low	normal	
C7	male	no	medium	month	very p.	very low	pathol.	
C8	male	no	high	month	very p.	high	pathol.	
C9	male	no	high	month	pathol.	high	pathol.	
C10	female	no	low	month	pathol.	very low	normal	
C11	female	no	high	month	pathol.	very low	normal	
C12	female	yes	low	month	pathol.	high	normal	
C13	male	no	high	month	pathol.	low	normal	
C14	male	no	low	month	pathol.	very low	normal	
C15	male	no	medium	month	very p.	very low	normal	
C16	male	no	low	month	very p.	high	pathol.	
C17	male	no	medium	year	pathol.	low	pathol.	
C18	female	yes	medium	year	pathol.	high	pathol.	
C19	male	no	medium	year	pathol.	high	normal	
C20	male	no	medium	month	pathol.	high	pathol.	
C21	male	yes	low	month	pathol.	low	pathol.	
C22	male	no	medium	month	very p.	very low	pathol.	
C23	female	no	low	month	pathol.	very low	normal	
C24	female	no	medium	year	very p.	high	pathol.	
C25	female	no	high	none	pathol.	very low	normal	
C26	female	yes	high	month	pathol.	high	pathol.	
C27	male	yes	low	month	pathol.	high	pathol.	
C28	female	no	medium	month	pathol.	very low	normal	
C29	male	no	high	month	pathol.	high	pathol.	
C30	male	no	medium	none	pathol.	low	normal	
81 more children ↓								

Figure 1: A many–valued context about children suffering from diabetes

detailed information about formal concept analysis and the methods described in this section can be found in [2].

Formal concept analysis is based on the notion of a *(formal) context* [*(formaler) Kontext*] which is defined as a triple (G, M, I) where G and M are sets and I is a binary relation between G and M. The elements of G are called *objects* [*Gegenstände*], those of M *attributes* [*Merkmale*], and the incidence $(g, m) \in I$ is read: *The object g has the attribute m*. A *(formal) concept* [*(formaler) Begriff*] of the context (G, M, I) is defined as a pair (A, B)

where A and B are subsets of G and M, respectively, such that

$$A = \{g \in G | (g, m) \in I \text{ for all } m \in B\} \quad \text{and} \quad B = \{m \in M | (g, m) \in I \text{ for all } g \in A\} \,.$$

The sets A and B are called the *extent* [*Umfang*] and the *intent* [*Inhalt*] of the concept (A, B), respectively. The hierarchical *subconcept–superconcept* relation [*Unterbegriff–Oberbegriff–Relation*] is given by

$$(A_1, B_1) \leq (A_2, B_2) \quad :\Longleftrightarrow \quad A_1 \subseteq A_2 \quad (\Longleftrightarrow \quad B_1 \supseteq B_2)$$

for all concepts $(A_1, B_1), (A_2, B_2)$ of the formal context (G, M, I). The set $\mathcal{B}(G, M, I)$ of all concepts of (G, M, I) together with this order relation is a complete lattice called the *concept lattice* [*Begriffsverband*] of (G, M, I) and denoted by $\underline{\mathcal{B}}(G, M, I)$. Usually, contexts are represented by cross–tables and concept lattices by line diagrams (see Figure 2).

Many–valued contexts do not match the definition of a formal context. For unfolding the conceptual structure of a many–valued context (G, M, W, I) it has to be transformed into a formal context. This transformation is done by choosing a (*conceptual*) *scale* [(*begriff*"*–liche*) *Skala*] for each many–valued attribute m of M. A conceptual scale for an attribute m of M is a formal context $\mathcal{S}_m := (G_m, M_m, I_m)$ where the (potential) values of the attribute m are contained in the set G_m of objects of the scale $\mathcal{S}_m$. Figure 2 shows appropriate scales for some of the attributes of the many–valued context of Figure 1 together with their concept lattices. A scale gives a conceptual interpretation of the inherent structure of a many– valued attribute. Therefore, scales cannot be chosen automatically. Each choice of a scale requires an aim–oriented analysis of the nature of the attribute and is a first step of interpreting the data given by the many–valued context. After the scales have been chosen the *derived* formal context [*abgeleiteter* formaler Kontext] $(G, \bigcup_{m \in M} M_m, J)$ is created by defining

$$(g, n) \in J \quad :\Longleftrightarrow \quad (g, m, w) \in I \text{ and } (w, n) \in I_m \,.$$

This derived context contains all the information of the many– valued context (G, M, W, I) with respect to the view determined by the choice of the scales. The conceptual structure of this information can be visualized by a line diagram of the concept lattice of the derived context or of some parts of it. For that part of the derived context of the many–valued context of Figure 1 concerning the many–valued attributes *coma*, *pH level*, and *symptom duration*, the concept lattice is drawn in Figure 3 as a nested line diagram (cf. [5]). The object concepts are labelled only with the number of children belonging to them but not to proper subconcepts because the names of the children are supposed to be not of interest for the investigation.

2 How to Use the System

Conceptual data systems establish a shell which allows an interactive browsing through large nested line diagrams by presenting only parts of them and asking the user where he wants to have a more detailed insight. The advantage of this may become clear by the nested line diagram of Figure 3. This nested line diagram is not easy to read because the line diagrams of the inner level are rather small. It is clear that it would be almost impossible to draw a nested line diagram with, for instance, six or seven levels of nesting on one page or one

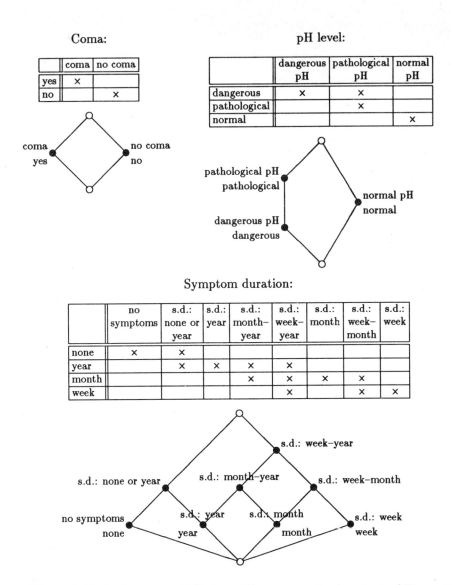

Figure 2: Scales for some attributes of the many–valued context of Figure 1

computer screen. An interactive browsing procedure may overcome this disadvantage by combining only so many scales on the screen that the resulting diagram is still readable. The browsing procedure and the ways of interaction are briefly described in the sequel.

Let us assume that we have a conceptual data system for the many– valued context of Figure 1. This system contains the data which are stored in a conceptual file and, furthermore, some software which enables the user to browse through the data (see Figure 7). The user has to decide which scales (i. e., attributes or some combinations of them) and which objects he wishes to consider, depending on the question he wants to analyse. The choice of scales and objects is transmitted to the conceptual data system via usual dialog

76

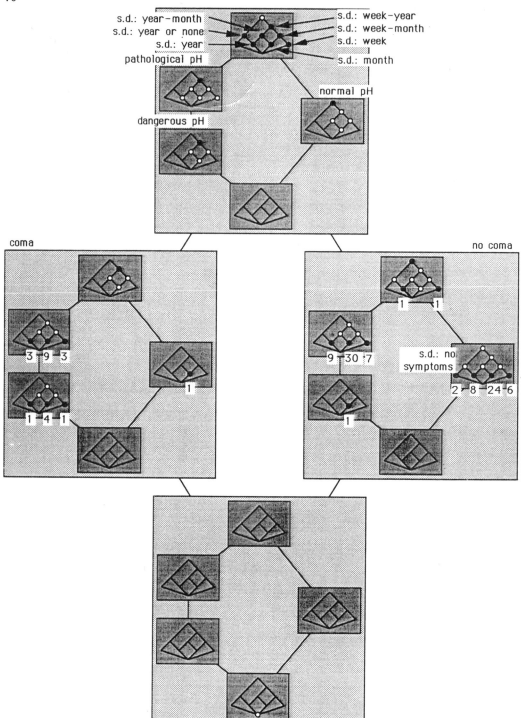

Figure 3: Nested line diagram of the concept lattice concerning the scales of Figure 2

components as they are common in interactive systems. For our example, we consider the following question which is important for the medical investigation: *Is a pathological pH level a good indicator for the danger that a child may fall into coma? How does the duration of the symptoms coincide with the interplay of pH level and coma?* A user who has this question might decide to consider the many–valued attributes *coma*, *pH level*, and *symptom duration* (in the listed sequence) together with all children. He transmits his specific choice to the system which *immediately* shows the nested line diagram of Figure 4 on the screen. In this case, the object concepts are again labelled with the number of children belonging to it but not to proper subconcepts. If the user wants more detailed information, the conceptual data system is able to label the object concepts with the names of the children instead.

The nested line diagram of Figure 4 allows a first interpretation with respect to the original question. In the right box labelled with *no coma*, it can be seen that there are 46 children who have a pathological pH level but have not been in coma. This may be interpreted that a pathological pH level is not a good indicator for the danger of coma. But the situation is much clearer for dangerous and normal pH levels which coincide with the presence and absence of coma, respectively.

After this first understanding, the user may decide to get more detailed information about those children who have the unexpected behaviour concerning the pH level and coma. For this he can choose a finer level of consideration by clicking with the mouse into the right box labelled with *no coma*. Then the diagram of Figure 4 is removed from the screen and (after a short moment) the nested line diagram of Figure 5 appears on the screen. The headline *no coma* indicates that the largest concept represented in this diagram is characterized by *no coma* (cf. Figure 3). In comparison to the previous nested line diagram (Figure 4) more details are available by the third scale which is drawn into the boxes representing the second scale. According to the posed question, the diagram suggests the interpretation that there is no obvious connection between the duration of the symptoms and the pH level because all durations occur and most children have symptoms for about one month independently from the pH level.

For comparing this interpretation with that about the children who have been in coma, the user may go back to the nested line diagram of Figure 4 by a simple command. Having this diagram on the screen, he may click into the left box labelled with *coma*. Then this box is refined to the nested line diagram shown in Figure 6. Again, almost all durations of symptoms occur having their maxima at the monthly duration. Together with the previous results, this yields the following interpretation: *The connection between the pH level and the presence or absence of coma seems to be complex. Therefore the pH level is not a good indicator for coma; furthermore, the duration of the symptoms does not contribute to a better understanding of the connection between pH level and coma.*

After this sequence of interactions is finished, the user can choose other scales and browse through the resulting nested line diagrams as described above. Within the interactive browsing procedure, the number of scales is not limited to three as in the example. Furthermore, it is possible to combine several many–valued attributes in one scale or to have different scales for one attribute. Therefore, the user can explore all available information by browsing through nested line diagrams with (theoretically) an arbitrary number of scales.

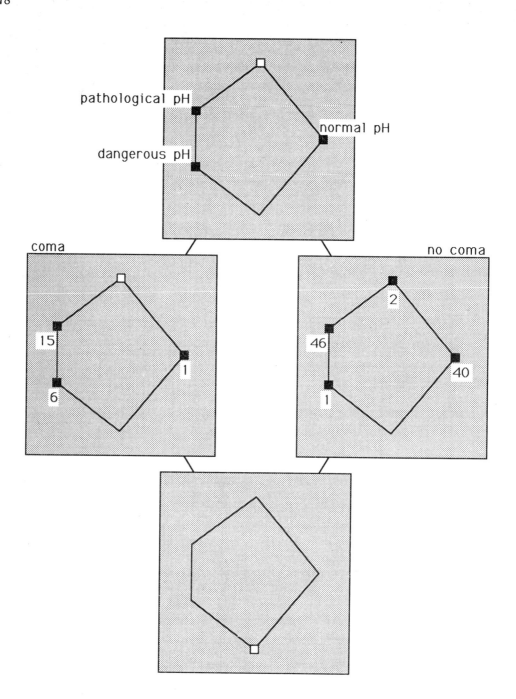

Figure 4: The nested line diagram first chosen via the conceptual data system

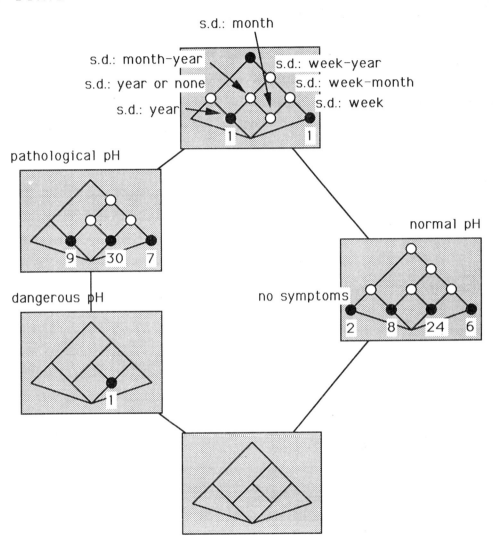

Figure 5: A finer view of the right box of the nested line diagram of Figure 4

coma

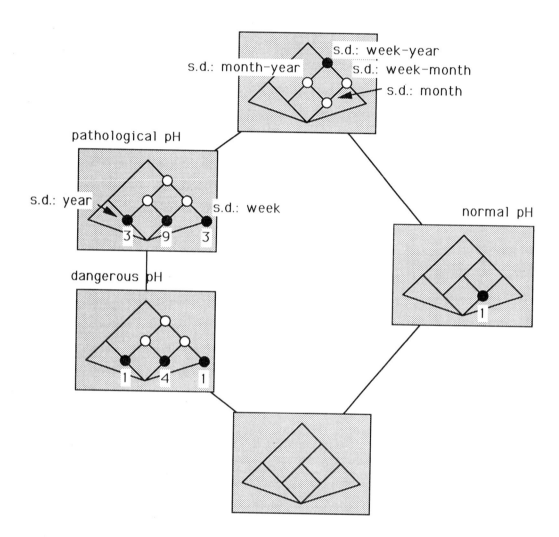

Figure 6: A finer view of the left box of the nested line diagram of Figure 4

3   The Structure of the System

A conceptual data system consists of data structures together with some software which allows the browsing described in the previous section as well as the upkeep and maintenance of the data. In this section we will omit a discussion of the software but briefly describe the structure of a conceptual file which is used to store the original data in the conceptual data system. The mathematical structure of a conceptual file has already been introduced in [3] whereas in this paper the implementation and the interplay of the different components of a conceptual file shall be outlined. Figure 7 indicates graphically what will be discussed in this section.

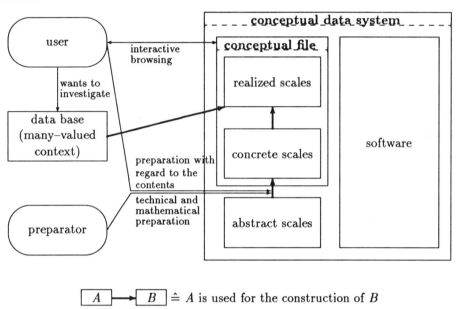

$\boxed{A} \longmapsto \boxed{B} \,\hat{=}\, A$ is used for the construction of B

Figure 7: The structure of a conceptual data system

The structure of a conceptual file is explained by describing the process of its construction. We assume that there is a user who wishes to investigate data available in the form of a many–valued context. Then this many–valued context has to be transformed into a conceptual file. For this transformation, the user is supported by the so–called *preparator* who knows the structure and implementation of a conceptual data system as well as the mathematical theory of formal concept analysis. It may occur that the user and the preparator are the same person (if the user knows enough of the theory) but the usual case will be that the user is a specialist in the area of the data whereas the preparator is a mathematician or computer scientist.

As first step of the transformation of the many–valued context into a conceptual file, one has to determine scales for the many–valued attributes, which is needed in general for analysing many–valued contexts by methods of formal concept analysis. For preparing a conceptual file for a conceptual data system, this determination of scales is splitted into several steps corresponding to the different aspects of scales which have to be considered.

Formally, a scale is a formal context (cf. Section 1) which is a pure mathematical structure

82

and has no specific meaning with respect to the data. Therefore, the same scales can be used with different meanings for different many–valued contexts. This leads to the notion of an *abstract scale* [*abstrakte Skala*] which combines a formal context together with a line diagram of its concept lattice and is stored into the conceptual data system independently from the specific data. An abstract scale contains the entire mathematical and geometrical information which is necessary for the computation and drawing of nested line diagrams. The advantage is that an abstract scale of a certain type has to be constructed only once and then can be used for several conceptual files or even several many–valued attributes of the same conceptual file. We assume that there is already a collection of abstract scales in the conceptual data system which is rich enough to transform the many–valued context. If this is not the case, the preparator can easily enlarge the collection by appropriate abstract scales. For the many–valued context of Figure 1, the collection should contain the abstract scales of Figure 8 which are represented by their line diagrams.

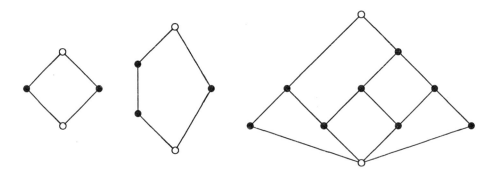

Figure 8: Some abstract scales

The abstract scales have to be interpreted with respect to the data available as many–valued context. This means that an abstract scale has to be chosen for each attribute of the many–valued context, and the abstract objects and attributes of this abstract scale have to be interpreted with respect to the given data. This is done by assigning the values of a many–valued attribute to the objects of the corresponding abstract scale and meaningful names to the attributes of the abstract scale. The resulting *concrete scale* [*konkrete Skala*], which gives a conceptual interpretation of the many–valued attribute, is stored in the conceptual file. More precisely, the name of the concrete scale is stored together with a reference to the underlying abstract scale and the assignment of the attribute values and names to the abstract objects and attributes, respectively. The notion of concrete scales allows to combine several many–valued attributes within one concrete scale and to store different concrete scales for the same many–valued attribute (cf. [3]). The construction of concrete scales requires mathematical and technical knowledge as well as some knowledge about the data and their background. Therefore, this step of preparing the conceptual file is a first step of interpreting the data which has to be done by the user and the preparator in cooperation.

The concrete scales which are shown in Figure 9 do not differ from the scales presented in Figure 2, concerning their shape and their meaning. The difference is that, under aspects of implementation, concrete scales consist only of a reference to the underlying abstract scale together with the assignments for the objects and attributes whereas scales in the sense of Section 1 are entire mathematical structures which have a certain interpretation.

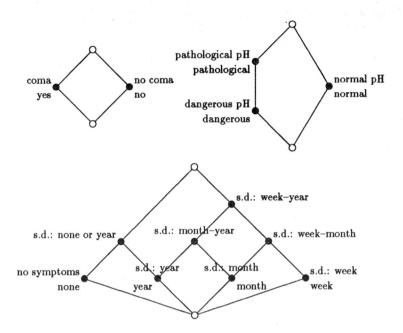

Figure 9: Concrete scales derived from the abstract scales of Figure 8

The concrete scales are used to derive a formal context from the many–valued context. In conceptual data systems, this formal context is not computed as an entire structure. Each concrete scale is used to derive a separate formal context from the corresponding part of the many–valued context. This construction of the so–called *realized scales [belegte Skalen]* can be done automatically by the conceptual data system by assigning to the objects of the concrete scales those objects of the many–valued context which have the considered object of the concrete scale as the value of the corresponding many–valued attribute. In Figure 10 this is demonstrated for the attribute *coma* of the many–valued context of Figure 1. The realized scale is added to the conceptual file by storing the assignment of the objects of the many–valued context together with a reference to the underlying concrete scale. After storing for each concrete scale the resulting realized scale, the preparation of the conceptual file is finished and the user can start to browse through the data.

The construction of the realized scales requires that the many– valued context has to be transmitted to the conceptual data system although it need not be stored in the conceptual file. This transmission would be most easy if the many–valued context exists in the form of a data base system. Then the conceptual data system could get the information which is necessary for constructing the realized scales automatically from the data base system. Also the update of the conceptual file would be easy in the case that the set of objects of the many–valued context (i. e., the data base) has been enlarged.

To summarize the advantages of using a conceptual data system, we start with the observation that a conceptual data system exempts the user from drawing a great number of large line diagrams by hand if he wants to investigate the conceptual structure of extensive data collections. The system can produce nested line diagrams from those few and small line diagrams which are stored in the abstract scales and can be drawn fast. Since the user can

84

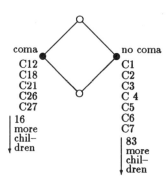

Figure 10: A realized scale derived from the many–valued context of Figure 1

arbitrarily combine the realized scales which are of interest to him the system can support him to analyse every reasonable question concerning the data. Furthermore, conceptual data systems allow to make conceptual knowledge available for many users because after the preparation of the conceptual file even a nonspecialist can easily browse through the data. This might be reasonable for information retrieval systems in public institutions (e. g., in libraries, documentation and information centers, etc.). A prototype of a conceptual data system for literature on interdisciplinary technology research is already designed as part of a more extensive conceptual knowledge system (see [4], [6]) and will be implemented in the near future.

Acknowledgement: This work was partly supported by the *"Zentrum für interdisziplinäre Technikforschung"* at the TH Darmstadt.

References

[1] A. Ciampi, A. Schiffrin, J. Thiffault, H. Quintal, G. Weitzner, P. Poussier, and D. Lalla. Cluster analysis of an insulin–dependent diabetic cohort towards the definition of clinical subtypes. Children's Hospital, McGill University, Montréal, Preprint.

[2] B. Ganter and R. Wille. Conceptual Scaling. In F. Roberts, editor, *Applications of combinatorics and graph theory to the biological and social sciences*, pages 139–167. Springer Verlag, New York, 1989.

[3] F. Vogt, C. Wachter, and R. Wille. Data analysis based on a conceptual file. In H. H. Bock and P. Ihm, editors, *Classification, data analysis, and knowledge organization*, pages 131–142, Berlin– Heidelberg, 1991. Springer–Verlag.

[4] C. Wachter and R. Wille. Formale Begriffsanalyse von Literaturdaten. FB4–Preprint 1443, TH Darmstadt, 1992.

[5] R. Wille. Lattices in data analysis: how to draw them with a computer. In I. Rival, editor, *Algorithms and order*, pages 33–58, Dordrecht–Boston, 1989. Kluwer.

[6] R. Wille. Concept lattices and conceptual knowledge systems. *Computers & Mathematics with Applications*, 23:493–515, 1992.

Conceptual Clustering via Convex-Ordinal Structures

Selma Strahringer and Rudolf Wille
Forschungsgruppe Begriffsanalyse,
Fachbereich Mathematik, Technische Hochschule Darmstadt,
Schloßgartenstr. 7, 6100 Darmstadt

Abstract: Conceptual clustering does not only follow formal principles for classifying objects, it uses also a human-comprehensible language for conceptual cluster descriptions. In doing so, an important question is how extensional and intensional arguments can be combined to support the aim which shall be fulfilled by specific classifications. This question is discussed for classifications based on ordinal data where the aim suggests clusters which are convex with respect to given order relations. The used conceptual language is taken from formal concept analysis and the chosen approach is based on the new theory of ordinal structures (see [SW92]). Especially, convex-ordinal structures and their concept lattices are used as tools for analysing ordinal data to find appropriate conceptual clusterings. For ordinal data there are elaborated methods to determine those convex-ordinal structures which are meaningful for the aim of the desired classification (see [St92]). The theoretical considerations are explained by examples.

1 Conceptual Clustering and Ordinal Convexity

Classifying objects is an important activity of human thinking which is basic for interpreting reality. There is a wide spectrum of methods to perform classifications and the interest is even increasing to develop further methods. Especially, there is a strong demand for mathematical methods of classification which can be implemented on computers. This has stimulated a rich development of numerical classification methods which are in extensive use today. But these methods are also criticized as in the following statement from [FL85]: *Despite the usefullness of numerical taxonomy techniques, any such method suffers from a major limitation, in that the resultant clusters may not be well characterized in some human-comprehensible conceptual language.* **Conceptual clustering** which was introduced in [Mi80], [MDS81] and [MS83] overcomes this limitation by incorporating a conceptual language based on attributes and attribute values. In conceptual clustering, the formation of clusters is not only guided by formal principles but also by the quality of conceptual cluster descriptions.

In this paper we want to consider a further important aspect, namely the aim which shall be fulfilled by specific classifications. With respect to this aspect, all formal methods of classification are limited because they can only take into account what of the aim can be formalized. But, in many cases, only part of the aim may be formally accessible. Let us discuss this by an example. In [Sch77], an investigation on dyslexia is described whose experimental part consists of 10 different tests. An aim could be to partition the 35 dyslexics of the investigation into appropriate learning groups. For discussing this aim, we concentrate to three representative tests concerning the ability of *remembering meaningless syllables*, of *memorizing audio-visually presented information*, and of *structuring word material*. In Figure 1, each of the 35 dyslexics is represented by a small circle to which the triple of its test scores is attached; the straight line segments indicate an order relation between

86

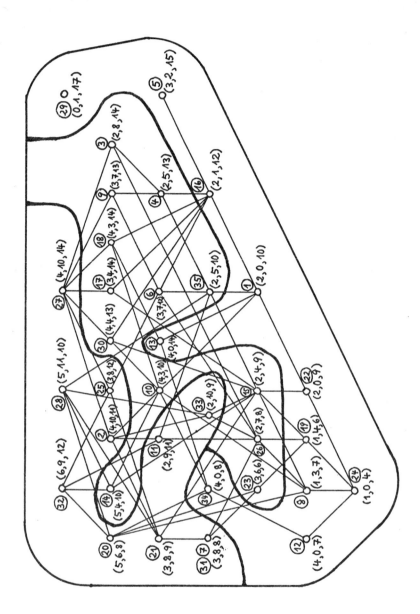

Figure 1: A clustering of 35 dyslexics ordered by their test scores

the dyslexics given by the component-wise order of their score triples. Certainly, the given formal information does not cover all aspects which should be considered for the formation of the learning groups. There could be many intensional arguments which are helpful for finding an appropriate partition (even close friendship between students could influence an adequate decision). Nevertheless, the formal data structure may support the decision. The question is: Which aspects of the aim can be formalized in the language of the given data structure?

In our example, we may deduce from the aim that the groups should be as homogeneous as possible. Since the test data are mainly ordinal in nature, homogeneity should be expressed in order-theoretic terms. Certainly, the homogeneous groups should fulfill as necessary condition that every group contains each student which is, with respect to the given ordering, between other students of this group (cf. [SW91]). In an ordered set $(P, \leq)$, an element x is said to be **between** the elements a and b if $a \leq x \leq b$ or $b \leq x \leq a$ is valid in $(P, \leq)$. A subset X of P is called **convex** if X contains each element of P which is between some pair of elements of X (cf. [Gr78], p. 17). Thus, the data representation as ordered set in Figure 1 supports the formation of homogeneous learning groups if it is assumed that homogeneity forces convexity with respect to the data ordering. Of course, the convexity condition reduces the groupings which have to be examined, but they are still very many. On the other hand, it seems difficult to deduce further formal principles from the aim which do not block appropriate solutions.

A reasonable way to approach a solution lies in an analysis of the presented data structure with regard to its content. The test scores show that the third ability of *structuring word material* does not seem to influence the other two abilities which are concerned with *memory*. Therefore a reasonable decision could be to concentrate first on a memory training which would lay emphasis on the first two scores in the presented triples. If the memory training is more general so that the specific differences between the first two abilities are less important, one could take the sum of the two scores as basis for the decision (despite the more ordinal nature of the data). The partition into three learning groups which is shown in Figure 1 is determined via those sums by the numerical intervals [1,5], [6,10], and [11,16]. Of course, the classes of the partition are convex with respect to the presented order relation. The chosen partition can be considered as a conceptual clustering if the occuring numerical expressions belong to the incorporated conceptual language. Even under the purely ordinal view, we have a conceptual description of the clusters: the bottom group consists of all students whose pairs of the first two scores are at most (4,0), (3,2), or (1,4), the top group constists of all students whose score pairs are at least (5,6), (3,8), or (2,9), and the description of the middle group are just the negation of the conjunction of the descriptions for the other two groups.

What we want to point out is the necessity to respect the aim of a classification. This has as consequence that the use of formal methods has to be coordinated with the aim. Thereby, one has the substantial problem that, in many cases, only parts of the aim can be formalized. Hence formal methods are not sufficient in these cases; they can only support decisions. Appropriate representations of data structures can help to analyse the extensional and intensional content of the data with respect to the aim. Such an analysis should particularly be concerned with the role of a conceptual language which allows human-comprehensible descriptions of clusters. The final decision for a specific clustering will then be a result of a rich spectrum of arguments integrating the different levels of considerations. In the following, we shall discuss further how convexity in ordinal data may contribute to conceptual clustering.

2 Concept Lattices of Ordinal Contexts

In this section we describe the conceptual language which shall be used for discussing conceptual clustering of ordinal data where the aim suggests convex clusters. As in Michalski's approach to conceptual clustering, attribute-value-pairs are also basic for our language which is already elaborated in *formal concept analysis* (cf.[SW92]). Our approach is based on the data model of a **(complete) ordinal context** $K := (G, M, (W_m, \sqsubseteq_m)_{m \in M}, I)$ which consists of sets G and M, a family $(W_m, \sqsubseteq_m)_{m \in M}$ of ordered sets, and a ternary relation $I \subseteq \bigcup_{m \in M} G \times \{m\} \times W_m$ such that for each $g \in G$ and $m \in M$ there is exactly one $w \in W_m$ with $(g, m, w) \in I$. The elements of G, M, and W_m are called **objects, attributes,** and **attribute values**, respectively; if $(g, m, w) \in I$ we say: *the object g has the value w for the attribute m.* An attribute m of K can be understood as a mapping from G into W_m; therefore we often write $m(g) = w$ for $(g, m, w) \in I$.

Ordinal contexts are mostly given by data tables as for example in Figure 2. The ordinal context described in Figure 2 is a result of an extensive investigation in developmental psychology (see [SKN92]). Its objects are 62 children from the age of 5 to 13 and its attributes are 9 general criteria of concept development and the attribute *age*; the attribute values are numbers which express the degree of fulfillment of the criteria. The investigation was performed with the aim to reconstruct the development of the concept of *work*. A final result of this reconstruction is a detailed description of developmental sequences which is empirically based on an appropriate conceptual clustering of the 62 children. As mathematical tools, the investigation has mainly used lattices of formal concepts which result from a set-theoretic formalization of concepts and conceptual hierarchies.

Let us briefly recall how the conceptual language is founded in formal concept analysis (see [Wi82]). It starts with the basic notion of a **formal context** as a triple (G, M, I) consisting of sets G and M linked by a binary relation I, i.e., $I \subseteq G \times M$. By incorporating words of the common language for certain set notions, a bridge between the formal model and its possible contents is established which is basic for understanding and interpreting. The elements of G and M are called **objects** and **attributes**, respectively, and $(g, m) \in I$ (or equivalently gIm) is read: *the object g has the attribute m.* The philosophical understanding of a concept as a unit of thoughts consisting of an extension and an intension now leads to the following formalization: a **formal concept** of a formal context (G, M, I) is a pair (A, B) for which $A \subseteq G$, $B \subseteq M$, $A = B^I := \{g \in G | gIm$ for all $m \in B\}$ and $B = A^I := \{m \in M | gIm$ for all $g \in A\}$; A and B are called the **extent** and the **intent** of the formal concept (A, B). The *subconcept-superconcept-relation* is captured by $(A_1, B_1) \leq (A_2, B_2) :\Leftrightarrow A_1 \subseteq A_2$ $(\Leftrightarrow B_1 \supseteq B_2)$. With this relation $\leq$, the formal concepts of a formal context (G, M, I) form a complete lattice, called the **concept lattice** of (G, M, I) and denoted by $\underline{\mathcal{B}}(G, M, I)$. The concept lattice of an ordinal context $K := (G, M, (W_m, \sqsubseteq_m)_{m \in M}, I)$ is in general defined as the concept lattice of its **derived** formal context $\widetilde{K} := (G, \bigcup_{m \in M} \{m\} \times W_m, \widetilde{I})$ where $g\widetilde{I}(m, w) :\Leftrightarrow m(g) \not\sqsupseteq w$ (see [SW92]).

No.	Age	Amount of motives 1	Quality of motives 2	Simple generalization 3	Variety of application 4	Generalization a 5a	b 5b	Differentiation of attributes 6	of objects 7	Differentiation 8	Integration 9
1	5	1	1	1	1	1	0	0	0	1	0
2	5	2	3	1	2	1	0	0	0	1	0
3	5	1	3	1	1	0	0	0	0	1	0
4	5	2	1	1	2	2	0	0	0	1	0
5	5	3	2	1	2	1	0	1	0	1	0
6	5	1	2	1	1	0	0	0	0	1	0
7	5	1	1	1	1	0	0	0	0	1	0
8	6	1	2	1	1	0	0	0	0	1	0
9	6	3	3	1	2	1	0	0	0	1	0
10	6	2	2	1	2	1	0	0	0	1	0
11	6	1	3	1	1	0	0	0	0	1	0
12	6	2	4	2	1	2	1	0	1	2	0
13	6	2	2	2	2	3	1	0	1	1	0
14	6	3	3	2	3	4	1	0	0	1	0
15	7	2	2	2	2	1	0	0	1	1	0
16	7	3	6	2	3	4	0	1	0	1	0
17	7	2	2	3	2	2	1	0	0	1	0
18	7	2	2	2	3	2	1	1	0	1	0
19	7	2	3	2	2	2	1	0	0	2	0
20	7	3	3	3	4	5	2	0	0	2	0
21	7	3	3	1	2	3	0	1	0	2	0
22	8	2	1	4	4	3	1	1	0	1	0
23	8	2	2	2	2	1	1	0	0	1	0
24	8	3	5	3	4	4	3	0	1	1	0
25	8	2	3	1	3	3	0	0	0	1	0
26	8	2	2	1	1	1	0	0	0	1	0
27	8	3	5	2	4	6	2	0	1	1	1
28	8	2	2	1	2	2	0	0	0	1	0
29	9	4	3	3	6	6	6	0	1	2	1
30	9	2	1	2	2	1	1	0	0	1	0
31	9	2	2	1	2	2	0	0	0	1	0
32	9	2	5	3	4	5	2	1	0	1	0
33	9	3	3	4	5	6	2	1	1	1	1
34	9	1	2	2	2	2	1	0	0	1	0
35	9	4	3	2	4	7	1	2	1	2	0
36	10	5	5	4	6	8	4	2	2	3	1
37	10	3	5	2	4	5	4	2	0	1	0
38	10	3	4	2	3	4	2	0	0	2	0
39	10	3	4	4	5	8	4	2	1	2	1
40	10	3	5	4	4	5	2	1	1	2	0
41	10	6	5	3	6	5	2	1	0	2	0
42	10	4	5	4	6	12	5	2	3	2	2
43	11	6	7	4	5	13	8	2	1	4	0
44	11	5	6	5	6	14	8	2	0	3	0
45	11	4	6	3	6	7	3	0	0	2	0
46	11	4	4	4	5	11	6	1	1	2	0
47	11	5	6	5	6	7	2	3	0	2	0
49	11	4	5	5	6	8	6	1	5	3	3
50	12	6	7	4	6	19	10	1	5	4	0
51	12	5	5	3	5	8	3	1	2	2	1
52	12	6	6	5	6	15	10	2	4	4	2
53	12	5	7	2	6	9	6	2	0	3	0
54	12	5	7	2	4	5	2	2	0	2	0
55	12	4	6	2	3	4	1	0	1	1	0
56	12	3	5	3	5	4	3	1	0	1	0
57	13	5	6	4	5	8	4	1	2	2	0
58	13	4	6	4	5	10	7	2	4	3	1
59	13	5	5	5	6	9	5	2	1	3	1
60	13	5	7	5	6	8	4	3	4	3	1
61	13	5	5	2	4	8	4	2	3	3	1
62	13	3	4	5	6	12	8	2	8	2	4
63	13	4	7	1	5	6	0	2	0	2	0

Figure 2: Evaluation of 62 children concerning their understanding of the concept *work* (No. 48 is missing)

Let us now come back to our example. In [SKN92], the analysis of developmental sequences is based on the criteria *quality of motives* (2), *generalization* (5b), and *structural differentiation* (8) which give the most differentiated view of changes and advances in development. The concept lattice of the ordinal context given by these attributes (see Figure 2) is represented in Figure 3 by its line diagram. The attribute orders are chosen in such a way that $g\tilde{I}(m,w)$ means: with respect to the criterium m, the child g has reached the degree w or higher. For instances, the circle on the left with the label 54, which represents the formal concept of the child 54, is connected upwards with the circle labelled with 2-7, which represents the formal concept of the degree 7 of criterium 2; this indicates that the child 54 has reached degree 7 in its quality of motives (and, following other ascending paths, degree 4 in generalization, but only degree 2 in structural differentiation). The black circles represent exactly the formal concepts of the 62 children where often several children constitute the same formal concept. The partition shown in **Figure 3** divides the circles, in particular the black circles, into seven classes which are convex with respect to the underlying order. The conceptual clustering of the 62 children described in this way is the result of a detailed argumentation integrating extensional and intensional views (see [SKN92]). Here we can only sketch this by a short description of the resulting levels of development (level 4 splits into two alternatives):

- *Level 1: Children mention only superficial, tautological or irrelevant motives, they hardly generalize or differentiate.*

- *Level 2: Children use simple types of motivational arguments, they hardly generalize and differentiate.*

- *Level 3: Children use simple types of motivational arguments, they begin to differentiate.*

- *Level 4a: Children reach higher quality of motives, they still generalize and differentiate on a low level.*

- *Level 4b: Children reach better quality of motives, they generalize on higher levels, but differentiate on a low level.*

- *Level 5: Children reach high quality of motives, they generalize on a high level and differentiate on a medium level.*

- *Level 6: Children reach extraordinary degrees for each of the criteria.*

The chosen clusters can be described purely in the formal language: Those objects are exactly in

- cluster 1 which have value < 3 for the attribute 2,

- cluster 2 which have value $= 3$ for the attribute 2, $= 1$ for the attribute 8,

- cluster 3 which have value $= 3$ for the attribute 2, < 3 for the attribute 5b, $= 2$ for the attribute 8,

- cluster 4a which have value > 3 for the attribute 2, < 3 for the attribute 5b,

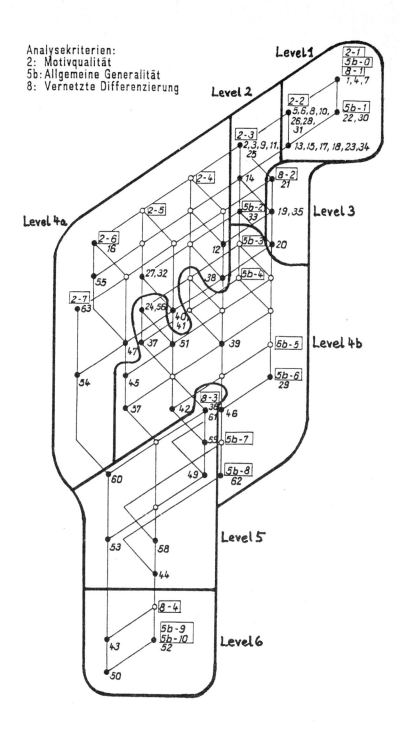

Figure 3: Concept lattice of the ordinal context of Figure 2 with clustering

- cluster 4b which have value ≥ 3 for the attribute 5b, < 3 for the attribute 8,

- cluster 5 which have value $= 3$ for the attribute 8,

- cluster 6 which have value $= 4$ for the attribute 8.

In spite of these formal descriptions, the clusters are not extents of the formal context which underlies the concept lattice of Figure 3. But if we extent the ordinal context to a suitable *convex-ordinal context*, each cluster can be characterized by a conjunction of derived attributes, i.e., it is an extent of the formal context derived from the convex-ordinal context; this follows from the fact that each class of the lattice partition in Figure 3 contains a greatest concept. A **convex-ordinal context** is an ordinal context $(G, M, (W_m, \sqsubseteq_m)_{m \in M}, I)$ in which, for each $m \in M$, there is an $m^d \in M$ such that $W_m = W_{m^d}$ and $v \sqsubseteq_m w \Leftrightarrow w \sqsubseteq_{m^d} v$ for all $v, w \in W_m$. An arbitrary ordinal context may be extended to a convex-ordinal context by adding for each attribute m a *dual attribute* m^d to the context. In the next section, we discuss in general the role of convex-ordinal contexts and structures for conceptual clustering.

3 Convex-Ordinal Structures and Contexts

The language of formal concept analysis yields conceptual descriptions for ordinally convex clusters if the ordinal data are formalized by a convex-ordinal context. Therefore convex-ordinal contexts may support the search for appropriate conceptual clusterings of ordinal data. The mathematical tools for this have been developed for convex-ordinal structures which have the same extent structures as the convex-ordinal contexts (see [St92]). A **convex-ordinal structure** is a relational structure $(S, (Q_n)_{n \in N})$ for which S is a set and $(Q_n)_{n \in N}$ is a family of quasi-orders (i.e., reflexive transitive binary relations) on S such that, for each $n \in N$, there is an $n^d \in N$ with $Q_{n^d} = Q_n^{-1}$ (Q_{n^d} is the dual quasi-order of Q_n). For each convex-ordinal context $\mathcal{K} := (G, M, (W_m, \sqsubseteq_m)_{m \in M}, I)$, one has the corresponding convex-ordinal structure $\underline{S}(\mathcal{K}) := (G, (\leq_m)_{m \in M})$ whose quasi-orders are defined by $g \leq_m h :\Leftrightarrow m(g) \sqsubseteq_m m(h)$ for $m \in M$. Conversely, each convex-ordinal structure corresponds in this way to a convex-ordinal context (see [SW92]).

Conceptual clustering via convex-ordinal structures shall be explained by an example too. The table in Figure 4 describes similarities between programming languages judged by experts (see [KL90]). The listed similarities give rise to a convex-ordinal context $\mathcal{K}$ shown in Figure 5. In this context

- $(g, \overset{m}{n}, >) \in I$ means: g is more similar to m than to n,

- $(g, \overset{m}{n}, =) \in I$ means: g is equally similar to m and to n,

- $(g, \overset{m}{n}, <) \in I$ means: g is less similar to m than to n.

The concept lattice of the derived context $\tilde{\mathcal{K}}$ is represented in Figure 6 by a nested line diagram (cf. [Wi89]). The extents of the 51 concepts are the possible clusters for conceptual clustering in the chosen setting of formal concept analysis. Of course, one could take the lattice of all extents as hierarchical clustering of the objects, but this might be too large for the considered aim in many cases. If we have the general aim in our example to understand

	M	A	Pa	F	C	L	Pr
Modula	10	9	9	7	6	2	1
Ada	9	10	9	7	7	3	0
Pascal	9	9	10	8	6	5	2
Fortran	7	7	8	10	4	4	3
Cobol	6	7	6	4	10	5	5
Lisp	2	3	5	4	5	10	8
Prolog	1	0	2	3	5	8	10

Figure 4: Similarities of programming languages judged by experts

	M A / A M	M Pa / Pa M	M F / F M	M C / C M	M L / L M	M Pr / Pr M	A Pa / Pa A	A F / F A	A C / C A	A L / L A	A Pr / Pr A	Pa F / F Pa	Pa C / C Pa	Pa L / L Pa	Pa Pr / Pr Pa	F C / C F	F L / L F	F Pr / Pr F	C L / L C	C Pr / Pr C	L Pr / Pr L
Modula	> <	> <	> <	> <	> <	> <	= =	> <	> <	> <	> <	> <	> <	> <	> <	> <	> <	> <	> <	> <	> <
Ada	< >	= =	> <	> <	> <	> <	> <	> <	> <	> <	> <	> <	> <	> <	> <	= =	> <	> <	> <	> <	> <
Pascal	= =	< >	> <	> <	> <	> <	< >	> <	> <	> <	> <	> <	> <	> <	> <	> <	> <	> <	> <	> <	> <
Fortran	= =	< >	< >	> <	> <	> <	< >	< >	> <	> <	> <	< >	> <	> <	> <	= =	> <	> <	> <	> <	> <
Cobol	< >	= =	> <	< >	> <	> <	> <	> <	< >	> <	> <	< >	> <	> <	> <	< >	> <	> <	> <	< >	= =
Lisp	< >	< >	< >	< >	< >	< >	< >	< >	< >	> <	= =	< >	< >	< >	< >	< >	< >	< >	< >	> <	> <
Prolog	> <	< >	< >	< >	< >	< >	< >	< >	< >	< >	< >	< >	< >	< >	< >	< >	< >	< >	< >	< >	> <

Figure 5: Convex-ordinal context derived from Figure 4

how experts see relationships between programming languages, then the lattice of extents would be an appropriate result from which further explorations could lead to satifactory answers concerning our aim. If we are interested in more simplifying views, our approach offers the use of *convex-ordinal scales* formed by suitable $\bigwedge$-subsemilattices of the lattice of all extents.

For a (quasi-)ordered set $(P, \leq)$, the corresponding **convex-ordinal scale** is the formal context $C_{(P,\leq)} := (P, \{\not\geq, \not\leq\} \times P, \square)$ for which $x \square (\not\geq, y) :\Leftrightarrow x \not\geq y$ and $x \square (\not\leq, y) :\Leftrightarrow x \not\leq y$ for $x, y \in P$. The extents of $C_{(P,\leq)}$ are exactly the convex subsets of $(P, \leq)$. The question is how to find appropriate convex-ordinal scales whose extents are extents of a given convex-ordinal context $\mathcal{K} := (G, M, (W_m, \sqsubseteq_m)_{m \in M}, I)$. Such convex-ordinal scales are given by *double-$\bigcup$-faithful* pairs (P_1, P_2) of subsets of M where (P_1, P_2) is said to be **double-$\bigcup$-faithful** if, for every $X_1 \subseteq P_1$ and $X_2 \subseteq P_2$, the unions

$$\bigcup_{m \in X_1} \{m\}^I \cup \bigcup_{n \in X_2} (G \setminus \{n\}^I) \text{ and } \bigcup_{m \in X_1} (G \setminus \{m\}^I) \cup \bigcup_{n \in X_2} \{n\}^I$$

are extents of the derived context $\tilde{\mathcal{K}}$ (see [St92]). The quasi-order belonging to (P_1, P_2) is defined on G by $g \leq_{(P_1,P_2)} h :\Leftrightarrow \{g\}^I \cap P_1 \supseteq \{h\}^I \cap P_1$ and $\{g\}^I \cap P_2 \subseteq \{h\}^I \cap P_2$. For our example, three double-$\bigcup$-faithful pairs (P_1, P_2) are shown in Figure 7 together mit their quasi-orders $\leq_{(P_1,P_2)}$ and the concept lattices of the convex-ordinal scales $C_{(G,\leq_{(P_1,P_2)})}$. Now, the question arises how to find appropriate double-$\bigcup$-faithful pairs of attribute sets for the convex-ordinal context $\mathcal{K}$. A suitable tool for this is the **double-hypergraph** $(\widetilde{M}_{irr}, K_U^d(\tilde{\mathcal{K}}))$ for which $\widetilde{M}_{irr}$ is the set of all irreducible attributes (m, w) of $\tilde{\mathcal{K}}$ (i.e., $\{(m, w)\}^I$ is a $\bigcap$-irreducible extent of $\tilde{\mathcal{K}}$) and $K_U^d(\tilde{\mathcal{K}})$ is the set of all minimal not double-$\bigcup$-faithful (unordered)

94

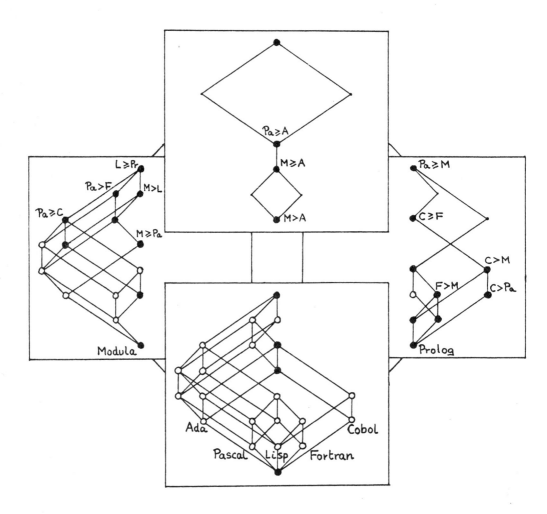

Figure 6: Concept lattice of the ordinal context of Figure 5

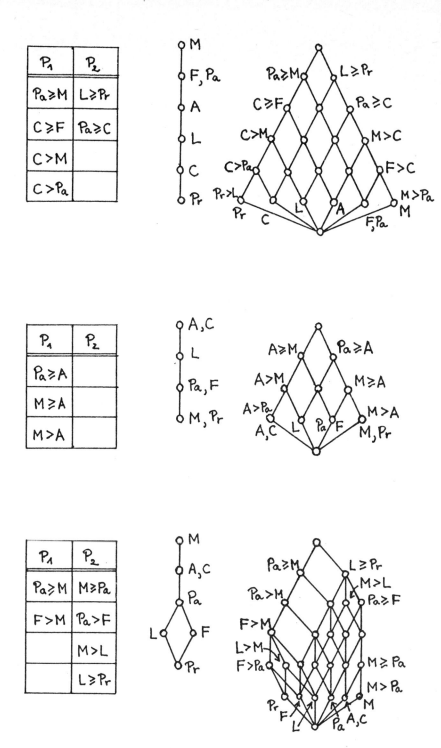

Figure 7: Double-∪-faithful pairs with their corresponding quasi-orders and concept lattices of their convex-ordinal scales

pairs of subsets of $\widetilde{M}_{irr}$. Then a pair (P_1, P_2) of subsets of $\widetilde{M}_{irr}$ is double-$\cup$-faithful if and only if there is no $\{Q_1, Q_2\}$ in $K_{\cup}^d(\widetilde{\mathcal{K}})$ with $Q_1 \subseteq P_1$ and $Q_2 \subseteq P_2$. Since the vertices (elements) of the double-hypergraph are attributes with a specific meaning, the choice of an appropriate convex-ordinal scale via a double-$\cup$-faithful pair of subsets can be made with regard to the content of the data and the specific aim of the desired conceptual clustering.

For our example the double-hypergraph is represented in Figure 8 by two diagrams: The thick line segments indicate the order on the set of irreducible attributes induced by the conceptual hierarchy of $\widetilde{\mathcal{K}}$; the thin line segments represent, in the first diagram, the minimal non-faithful unordered pairs consisting of a 2-element set and the empty set and, in the second diagram, those consisting of two 1-element sets. Therefore, the first diagram shows which subsets cannot be used for a component of a desired pair while the second diagram clarifies when two subsets cannot be combined to a pair. The three pairs of Figure 7 have been found via the double-hypergraph in Figure 8. They are specifically chosen to cover all irreducible attributes which guarantees that the lattice of all extents can be reconstructed by the extents of the corresponding convex-ordinal scales. This means in particular that the lattice of all extents can be represented by a nested line diagram where the levels are formed by the extent lattices of the three convex-ordinal scales. For such a representation three levels are necessary which follows from Proposition 4 in [St92]. By this Proposition, the smallest number of convex-ordinal scales yielding all extents equals the chromatic number of the double-hypergraph (which is 3 in our example). The nested representation of all extents based on convex-ordinal scales may support effectively the search for larger hierarchical clusterings where it is helpful to have only a small number of such scales.

Finally, the question remains how to determine the minimal not double-$\cup$-faithful (unordered) pairs of subsets of $\widetilde{M}_{irr}$. For this we introduce for each $(m, w) \in \widetilde{M}_{irr}$ an attribute $(m, w)^d$ with $g \tilde{I}^d (m, w)^d :\Leftrightarrow (g, (m, w)) \notin \tilde{I}$ and define $\widetilde{M}_{irr}^d := \{(m, w)^d | (m, w) \in \widetilde{M}_{irr}\}$; let us mention that $\{g \in G | g \tilde{I}^d (m, w)\}$ is always an extent of $\widetilde{\mathcal{K}}$ for $(m, w) \in \widetilde{M}_{irr}$ because $\mathcal{K}$ is a convex-ordinal context. For $P \subseteq \widetilde{M}_{irr}^d$, let $\bar{P} := \{(m, w) \in \widetilde{M}_{irr} | (m, w)^d \in P\}$. First we determine all extents $C_1, \ldots, C_k$ of $\widetilde{\mathcal{K}}$ whose complements are not extents of $\widetilde{\mathcal{K}}$; the indices are chosen so that $C_i \supseteq C_j$ implies $i \le j$. In Figure 6 such extents are represented by the non-black circles since complementation of extents is obtained by the 180° rotation of the diagram. Next we determine the subsets P_{ij} of $\widetilde{M}_{irr} \cup \widetilde{M}_{irr}^d$ which are minimal with $C_i = P_{ij}^{\tilde{I} \cup \tilde{I}^d}$. Then we define $dcrit(1) := \{\{P_1, \bar{P}_2\} | P_1 \subseteq \widetilde{M}_{irr}$ and $P_2 \subseteq \widetilde{M}_{irr}^d$ with $P_1 \cup P_2 = P_{1j}$ for some $j\}$ and recursively $dcrit(i) := \{\{P_1, \bar{P}_2\} | P_1 \subseteq \widetilde{M}_{irr}$ and $P_2 \subseteq \widetilde{M}_{irr}^d$ with $P_1 \cup P_2 = P_{ij}$ for some j so that $\{Q_1, Q_2\} \sqsubset \{P_1, \bar{P}_2\} \Rightarrow \{Q_1, Q_2\} \notin dcrit(l)$ for $l = 1, \ldots, i - 1\}$ where $\{Q_1, Q_2\} \sqsubset \{P_1, \bar{P}_2\} :\Leftrightarrow (Q_1 \subset P_1$ and $Q_2 \subseteq \bar{P}_2)$ or $(Q_1 \subseteq P_1$ and $Q_2 \subset \bar{P}_2)$. By Proposition 6 in [St92], the union of the sets $dcrit(i)$ with $i = 1, \ldots, k$ equals $K_{\cup}^d(\widetilde{\mathcal{K}})$.

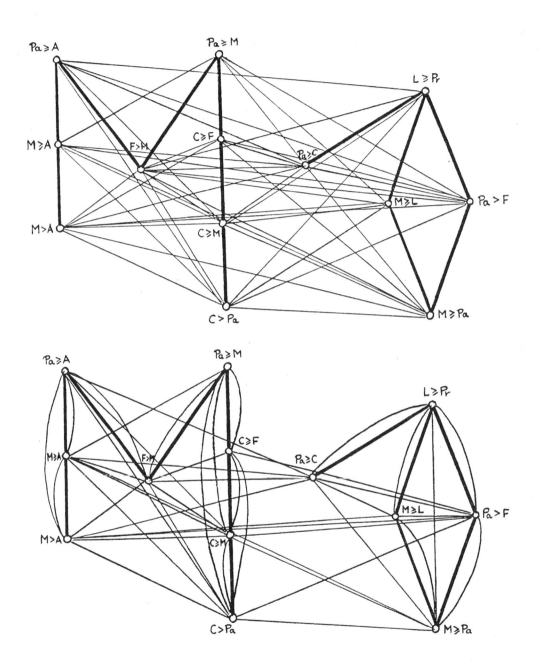

Figure 8: Double-hypergraph for the ordinal context of Figure 5

References

[FL85]   D. Fisher, P. Langley: Approaches to conceptual clustering. Technical Report 85-17, Dept. of Information and Computer Science, University of California, Irvine 1985.

[Gr78]   G. Grätzer: General lattice theory. Birkhäuser, Basel 1978.

[KL90]   W. Karbach, M. Linster: Wissensakquisition für Expertensysteme. Techniken, Modelle und Softwareerzeugnisse. Hanser, München/Wien 1990.

[Mi80]   R. S. Michalski: Knowledge acquisition through conceptual clustering: A theoretical framework and an algorithm for partitioning data into conjunctive concepts. International Journal of Policy Analysis and Information Systems, vol. 4, no. 3 (1980), 219-244.

[MDS81]   R. S. Michalski, E. Diday, R. E. Stepp: A recent advance in data analysis: Clustering objects into classes characterized by conjunctive concepts. In: L. N. Kanal, A. Rosenfeld (eds.): Progress in pattern recognition. North-Holland, Amsterdam 1981, 33-56.

[MS83]   R. S. Michalski, R. E. Stepp: Automated construction of classifications: Conceptual clustering versus numerical taxonomy. IEEE Transactions on Pattern Analysis and Machine Intelligence, vol. 5, no. 4 (1983), 396-410.

[Sch77]   W. Schneider: Taxonomie der Gedächtnisleistungen schwacher und normaler Rechtschreiber. In: H. Späth (ed.): Fallstudien Clusteranalyse. Oldenbourg, München/Wien (1977), 179-189.

[SKN92]   T. B. Seiler, B. Koböck, B. Niedermeier: Rekonstruktion der Entwicklung des Arbeitsbegriffs mit Mitteln der Formalen Begriffsanalyse. (in preparation)

[St92]   S. Strahringer: Dimensionality of ordinal structures. THD-Preprint Nr. 1452, TH Darmstadt 1992.

[SW91]   S. Strahringer, R. Wille: Convexity in ordinal data. In: H. H. Bock, P. Ihm (eds.): Classification, data analysis, and knowledge organization. Springer, Heidelberg 1991, 113-120.

[SW92]   S. Strahringer, R. Wille: Towards a structure theory for ordinal data. In: M. Schader (ed.): Analyzing and modeling data and knowledge. Springer, Heidelberg 1992, 129-139.

[Wi82]   R. Wille: Restructuring lattice theory: an approach based on hierarchies of concepts. In: I. Rival (ed.): Ordered sets. Reidel, Dordrecht-Boston 1982, 445-470.

[Wi89]   R. Wille: Lattices in data analysis: how to draw them with a computer. In: I. Rival (ed.): Algorithms and order. Kluwer, Dordrecht 1989, 33-58.

Diagrams of Similarity Lattices

Katja Lengnink

Forschungsgruppe Begriffsanalyse,
Fachbereich Mathematik, Technische Hochschule Darmstadt,
Schloßgartenstr. 7, 6100 Darmstadt

Abstract: Similarity is a main type of relation between objects in empirical data. For modelling similarity data the concept of an ordinal distance context is introduced. Based on the notions of formal concept analysis, graphical representations of ordinal distance contexts by so-called band diagrams of concept lattices are established. It is shown that hierarchies and pyramids can be understood as concept lattices of specific ordinal distance contexts. A useful reduction of similarity data to so-called threshold contexts and threshold lattices is discussed. The theoretical results are demonstrated by examples.

1 Concept Lattices of Similarity Data

A main activity of empirical research is the collection and interpretation of different kinds of data. The investigation of data is often concerned with similarity and dissimilarity, a basic type of relations on data objects. For analysing similarity data, the theory of hierarchical clustering yields frequently used methods which results in geometric representations by dendrograms (cf. [Boc74]). A more recent development is the theory of pyramidal clustering representing the resulting information in so-called indexed pyramids (see [BD85], [Did87], [BD89]). In [Len91], a lattice-theoretical method is established which represents similarity data by band diagrams of concept lattices. This approach is based on the theory of formal concept analysis (see [GW91], [Wil82], [Wil87]) and is explained in the sequel. In Section 2 this approach is compared with the above-mentioned methods of hierarchical and pyramidal clustering. In the final section we discuss a useful reduction of similarity data which yields so-called threshold lattices.

Let us first recall how a data matrix is mathematized in formal concept analysis by a **many-valued context** (see [Wil82], [GW89]). This is a structure $\mathcal{K} := (G, M, W, I)$ where G, M and W are (finite) sets and $I \subseteq G \times M \times W$ is a ternary relation such that $(g, m, w) \in I$ and $(g, m, v) \in I$ imply $w = v$. The elements of G, M and W are called objects, attributes and attribute values, respectively. $(g, m, w) \in I$ means that the object g has according to the attribute m the value w, we often write $m(g) = w$ instead of $(g, m, w) \in I$. A many-valued context may be described by a table where the rows and columns are indexed with the objects and attributes, the corresponding values are listed in the cells of the table.

	Social Studies	Latin	Chemistry	Sports
Stefan	1	1	1	2
Dieter	1	2	1	3
Xaver	1	2	2	2
Till	2	1	1	2

Figure 1: School grades of 4 students

Fig. 1 shows a data table which is used as a demonstrating example in [BD89] (the fifth row has been omitted and names for objects and attributes are added), it is also contained as part of a larger table of school grades from a class of a german Gymnasium in [GW86]. If we understand the table as a many-valued context the objects are students of the class, the attributes are school subjects and the values indicate the school grades which the students obtained for the different subjects. The example of Fig. 1 will be used for illustration throughout this paper.

For clustering objects it is common to compute distances from the given data matrix by some formula which, of course, have to be justified for the specific case. The resulting table can always be understood as an **ordinal distance context**. By this we mean a many-valued context $\mathcal{K} :=$ $(G, G, \mathbf{V}, J)$ with a set of objects G and a linearly ordered set $\mathbf{V} := (V, \leq)$ (cf. [SW92]). The elements of V are called distances or dissimilarity degrees. The equation $h(g) = v$ means that the distance between the object g and the object h is v. We assume that $h(g) = g(h)$ for all $g, h \in G$. In addition, we require $g(g) = 0$ for all $g \in G$. In [BD89] a distance measure $d : G \times G \rightarrow \mathbf{R}$ is used which is defined by

$$d(g, h) := \prod_{i=1}^{n} \frac{|\{m_i(g), m_i(h)\}|}{|m_i(G)|}$$

where, in the language of many-valued contexts, $M := \{m_1, \ldots, m_n\}$ and $m_i(G) := \{m_i(g) \mid g \in G\}$. For the example of Fig. 1 this yields the ordinal distance context shown in Fig. 2.

	Stefan	Dieter	Xaver	Till
Stefan	0	0.25	0.25	0.125
Dieter	0.25	0	0.25	0.5
Xaver	0.25	0.25	0	0.5
Till	0.125	0.5	0.5	0

Figure 2: Ordinal distance context resulting from Fig. 1 by $d(g, h) := \prod_{i=1}^{n} \frac{|\{m_i(g), m_i(h)\}|}{|m_i(G)|}$

In the sequel, we use this ordinal distance to discuss different representations of an ordinal distance context $\mathcal{K} := (G, G, \mathbf{V}, J)$ by diagrams. Following the paradigm of formal concept analysis, we first interpret the values of V to derive a formal context determining the conceptual structure of $\mathcal{K}$. This method is called **conceptual scaling** of many-valued contexts (see [GW89]). Since the structure on $\mathbf{V}$ is a linear order the interpretation of the values should be determined by this order. Therefore, the **derived context** $\tilde{\mathcal{K}} := (G, G \times V, \tilde{J})$ is defined by

$$g\tilde{J}(h, v) :\Leftrightarrow h(g) \leq v \text{ for every } g \in G \text{ and } (h, v) \in G \times V.$$

The binary relation $\tilde{J}$ is chosen in such a way that smaller distance values yield smaller extents of the formal concepts of $\tilde{\mathcal{K}}$. The concept lattice $\underline{\mathfrak{B}}(\tilde{\mathcal{K}})$ is considered to be the inherent conceptual structure of $\mathcal{K}$. The order-preserving map

$$\varphi : \underline{\mathfrak{B}}(\tilde{\mathcal{K}}) \rightarrow \mathbf{V} \text{ defined by } \varphi((A, B)) := max\{d(g, h) \mid g, h \in A\}$$

assigns to each concept a distance value indicating its inhomogeneity. The concept lattice $\underline{\mathfrak{B}}(\tilde{\mathcal{K}})$ can be represented by a band diagram in which the formal concepts are located in bands according to their distance values. The formal definition of a band diagram of a finite lattice $\mathbf{L}$ is the following:

Definition 1 Let $\mathbf{L}$ be a finite lattice, let $\mathbf{V} := (V, \leq)$ be a linearly ordered set, and let $\varphi : \mathbf{L} \rightarrow \mathbf{V}$ be an order-preserving map. A line diagram (Hasse diagram) of $\mathbf{L}$ is called **band diagram** **(Streifendiagramm)** of L **with respect to** φ if there are two positive real numbers $b, h \in \mathbf{R}^+$ such that

(i) there exist $|\varphi(L)|$ disjoint open intervals $(P_v)_{v \in \varphi(L)} \subseteq [0, h]$ with the property that $\overline{(\bigcup\limits_{v \in \varphi(L)} P_v)} =$ $[0, h]$ and $w \leq v$ implies $\sup P_w \leq \inf P_v$ and

(ii) a band $]0,b[\times P_v$ contains exactly those circles representing the elements $a \in L$ where $\varphi(a) = v$.

According to this definition we can prove the following proposition:

Proposition 1 *Let L be a finite lattice and let $\varphi : L \to V$ be an order-preserving map into a linearly ordered set $V := (V, \leq)$. Then L is representable by a band diagram with respect to φ.*

Proof Because φ is order-preserving, for every $v \in V$, the preimage $\varphi^{-1}(v)$ is a convex subset of L, i.e., $a \leq b \in \varphi^{-1}(v)$ implies $\{c \in L \mid a \leq c \leq b\} \subseteq \varphi^{-1}(v)$. Since $(\varphi^{-1}(v))_{v \in \varphi(L)}$ is a partition of L into $|\varphi(L)|$ convex subsets a band diagram can be constructed as follows:

(i) We consider arbitrary $b, h \in \mathbf{R}^+ \backslash \{0\}$ and partition the interval $[0, h]$ into $|\varphi(L)|$ intervals of the same length $h/|\varphi(L)|$: For $v \in \varphi(L)$, we choose $k_v := |\{w \in \varphi(L) \mid w \leq v\}|$ and define $P_v := \left] \frac{h(k_v - 1)}{|\varphi(L)|}, \frac{hk_v}{|\varphi(L)|} \right[$.

(ii) For every $v \in \varphi(L)$ we draw a line diagram of the ordered set $\varphi^{-1}(v)$ in the band $]0, b[\times P_v$. Since $\varphi(a) < \varphi(b)$ implies $b \nleq a$ the missing lines between neighbours can be inserted.

$\square$

For any finite ordinal distance context $\mathcal{K}$ the assumptions of Proposition 1 are fulfilled for $\underline{B}(\widetilde{\mathcal{K}})$ and φ. Hence the concept lattice $\underline{B}(\widetilde{\mathcal{K}})$ can be represented by a band diagram with respect to φ.

	$d(g,h) = 0$				$d(g,h) = 0.125$				$d(g,h) = 0.25$				$d(g,h) = 0.5$			
	S	D	X	T	S	D	X	T	S	D	X	T	S	D	X	T
S	x				x				x	x	x	x	x	x	x	x
D		x				x			x	x	x		x	x	x	x
X			x				x		x	x	x		x	x	x	x
T				x	x			x	x			x	x	x	x	x

Formal context derived from Fig. 2

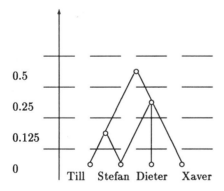

Figure 3: Band diagram of the concept lattice of the formal context above

The band diagram of the concept lattice of the ordinal distance context of Fig. 2 is shown in Fig. 3. As usual, every formal concept is represented by a circle, only the concept with the empty extent is omitted. The extent of a concept consists of all objects whose names are attached to circles reachable by descending paths from the circle of the concept. The dissimilarity degree of an extent A is given by $\varphi(A, A')$ which is the value of the band containing the circle of the concept (A, A'). The distance of two objects g and h can be seen in the band diagram as follows: We determine the lowest concept with g and h in its extent. Then $d(g, h)$ is the value of the band in which this concept is represented. Consequently, the ordinal distance context can be reconstructed completely from the band diagram.

2 Hierarchies and Pyramids as Concept Lattices

First we recall the definition of a hierarchy given in [Boc74].

Definition 2 Let G be a finite set. A **hierarchy** is a set $\mathcal{H}$ of non-empty subsets of G, ordered by inclusion, satisfying the following conditions:

(i) $G \in \mathcal{H}$,

(ii) $\{g\} \in \mathcal{H}$ for all $g \in G$,

(iii) $A, B \in \mathcal{H}$ implies $A \cap B \in \{A, B, \emptyset\}$.

The elements of $\mathcal{H}$ are called **clusters** of the hierarchy.

A pair $(\mathcal{H}, f)$ consisting of a hierarchy $\mathcal{H}$ and a function $f : \mathcal{H} \to \mathbf{R}^+$ is called an **indexed hierarchy** if $C \subset D$ always implies $f(C) < f(D)$ and if $f(C) = 0$ is equivalent to $|C| = 1$.

An indexed hierarchy can be represented by a dendrogram. It is proved (cf. [Boc74]) that every indexed hierarchy yields an **ultrametric** d which satisfies $d(g, k) \leq max\{d(g, h), d(h, k)\}$ for all $g, h, k \in G$. Conversely, if d is an ultrametric then the corresponding ordinal distance context can be represented by a dendrogram. Given an arbitrary ordinal distance context, the distance measure is not an ultrametric in general. In order to obtain a representation by a dendrogram it is common to modify the distance measure until an ultrametric is received. This can be done in various ways and the received ultrametric is not unique (cf. [Boc74]). Of course, the question of meaningfulness of such transformations have to be considered.

Let us apply two frequently used methods of computing ultrametrics to the distance measure given by the ordinal distance context of Fig. 2: the complete-linkage and the average-linkage method. The two ultrametric contexts and the corresponding dendrograms are shown in Fig. 4. To obtain an ultrametric it is necessary to modify the original ordinal distance context. This can lead to crucial changes as in our example where Stefan has the same ultrametric distance to Dieter and Xaver as Till. How can one justify such a change?

Next we recall the definition of a pyramid given in [BD89].

Definition 3 Let G be a finite set. A **pyramid** is a set $\mathcal{P}$ of non-empty subsets of G, ordered by inclusion, satisfying the following conditions:

(i) $G \in \mathcal{P}$,

(ii) $\{g\} \in \mathcal{P}$ for all $g \in G$,

(iii) $A, B \in \mathcal{P}$ implies $A \cap B \in \mathcal{P}$ or $A \cap B = \emptyset$,

(iv) there exists a linear order θ on G such that every element of $\mathcal{P}$ is an interval of (G, θ).

	S	D	X	T
S	0	0.5	0.5	0.125
D	0.5	0	0.25	0.5
X	0.5	0.25	0	0.5
T	0.125	0.5	0.5	0

	S	D	X	T
S	0	0.375	0.375	0.125
D	0.375	0	0.25	0.375
X	0.375	0.25	0	0.375
T	0.125	0.375	0.375	0

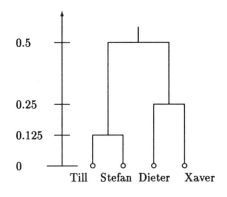

Complete-Linkage Method

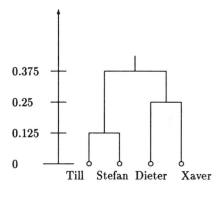

Average-Linkage Method

Figure 4: Hierarchical Clustering

The elements of $\mathcal{P}$ are called **clusters** of the pyramid.

A pair $(\mathcal{P}, f)$ consisting of a pyramid $\mathcal{P}$ and a function $f : \mathcal{P} \to \mathbf{R}^+$ is called an **indexed pyramid** if $A \subset B$ always implies $f(A) \leq f(B)$ and if $f(A) = 0$ is equivalent to $|A| = 1$. A **pyramidal index** d is a distance measure beween elements of G with the property that there exists a linear order θ on G which is compatible with d, i.e., $g\theta h\theta k$ always implies $d(g, k) \geq max\{d(g, h), d(h, k)\}$ for any $g, h, k \in G$.

In [Did87], it is proved that there is a bijection between indexed pyramids and pyramidal indices. The distance measure given by Fig. 2 is a pyramidal index with respect to the linear orders θ and ϕ determined by Till θ Stefan θ Dieter θ Xaver and Till ϕ Stefan ϕ Xaver ϕ Dieter. These linear orders give rise to graphic representations of indexed pyramids which are both shown in Fig. 4. The distances which can be read from the pyramidal representations are not those of Fig. 2. This is caused by the applied algorithm, presented in [BD89], which leads to special pyramids aggregating only two clusters to a new one.

In general, not every distance measure is a pyramidal index with respect to some linear order on the underlying set. Therefore, pyramidal clustering also leads to a modification of a given distance measure to a pyramidal index which corresponds to an indexed pyramid.

In [Did87], it is stated that every ultrametric is a pyramidal index. Hence the approach of pyramidal clustering can be seen as a generalization of hierarchical clustering. We can prove that the concept lattices of similarity data generalize hierarchies and indexed pyramids.

θ	S	D	X	T
S	0	0.25	0.5	0.125
D	0.25	0	0.25	0.5
X	0.5	0.25	0	1
T	0.125	0.5	1	0

ϕ	S	D	X	T
S	0	0.5	0.25	0.125
D	0.5	0	0.25	1
X	0.25	0.25	0	0.5
T	0.125	1	0.5	0

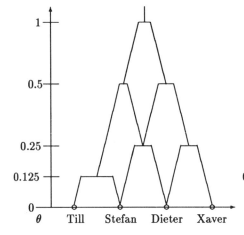 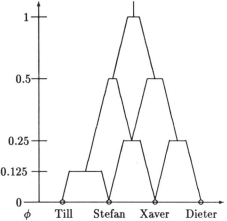

Figure 5: Pyramidal Clustering

Proposition 2 *Let $\mathcal{K} := (G, G, \mathbf{V}, J)$ be a finite ordinal distance context and let $\widetilde{\mathcal{K}} := (G, G \times V, \tilde{J})$ be its derived context with*

$$g\tilde{J}(h, v) :\Leftrightarrow h(g) \leq v \; (\Leftrightarrow d(g, h) \leq v) .$$

Furthermore, let $d : G \times G \to V$ be an ultrametric. Then the pair $(\underline{\mathcal{B}}(\widetilde{\mathcal{K}})\backslash\{(\emptyset, G)\}, \varphi)$ is isomorphic to the indexed hierarchy corresponding to d. In particular, the non-empty extents of $\widetilde{\mathcal{K}}$ are exactly the clusters given by d.

Proof For every extent A of $\widetilde{\mathcal{K}}$ there are $(h_i, v_i) \in G \times V$ $(i \in I)$ with $A = \{d(g, h_i) \leq v_i$ for all $i \in I\} = \bigcap_{i \in I}\{g \in G \mid d(g, h_i) \leq v_i\}$. If $A \neq \emptyset$ then, by the defining intersection property of hierarchies, there must be an $i \in I$ with $A = \{g \in G \mid d(g, h_i) \leq v_i\}$. Thus, the non-empty extents of $\widetilde{\mathcal{K}}$ are exactly the sets $\{g \in G \mid d(g, h) \leq v\}$ with $h \in G$ and $v \in V$. Hence the non-empty extents of $\widetilde{\mathcal{K}}$ are exacty the clusters given by d. By definition, for $(A, B) \in \underline{\mathcal{B}}(\widetilde{\mathcal{K}})$ with $A \neq \emptyset$, we have that $\varphi(A, B) := \{d(g, h) \mid g, h \in A\}$ equals the index value of the cluster A. Altogether yields that $(\underline{\mathcal{B}}(\widetilde{\mathcal{K}})\backslash\{(\emptyset, G)\}, \varphi)$ is isomorphic to the indexed hierarchy corresponding to d. $\quad\square$

Proposition 3 *Let $\mathcal{K} := (G, G, \mathbf{V}, J)$ be a finite ordinal distance context and let $\widetilde{\mathcal{K}} := (G, G \times V, \tilde{J})$ be its derived context with*

$$g\tilde{J}(h, v) :\Leftrightarrow h(g) \leq v \; (\Leftrightarrow d(g, h) \leq v) .$$

Furthermore, let $d : G \times G \rightarrow \mathbf{R}$ be a pyramidal index with respect to some linear order θ on G. Then the pair $(\underline{B}(\widetilde{K}) \backslash \{(\emptyset, G)\}, \varphi)$ is isomorphic to the indexed pyramid corresponding to d. In particular, the non-empty extents of $\widetilde{K}$ are exactly the clusters given by d.

Proof For every extent A of $\widetilde{K}$ there are $(h_i, v_i) \in G \times V$ $(i \in I)$ with $A = \{d(g, h_i) \leq v_i$ for all $i \in I\} = \bigcap_{i \in I}\{g \in G \mid d(g, h_i) \leq v_i\}$. In general, an attribute extent $\{g \in G \mid d(g, h) \leq v\}$ with $(h, v) \in G \times V$ is an interval of the linearly ordered set (G, θ) by the distance property of pyramidal indices. Such an interval is, by definition, a cluster given by the pyramidal index d. Since clusters given by d are defined to be the non-empty intersections of those intervals, the non-empty extents of $\widetilde{K}$ are exactly the clusters of the pyramidal index d. For $(A, B) \in \underline{B}(\widetilde{K})$ with $A \neq \emptyset$, we have $\varphi(A, B) = d(g, h)$ where g and h are the bounds of the interval A in (G, θ). By definition, $d(g, h)$ is the index value of the cluster A. Altogether yields that $(\underline{B}(\widetilde{K}) \backslash \{(\emptyset, G)\}, \varphi)$ is isomorphic to the indexed pyramid corresponding to d. $\qquad \square$

The understanding of hierarchies and pyramids as concept lattices of specific ordinal distance contexts opens a wider spectrum of methodological tools for investigations. It suggests in particular to study further types of distance measures which correspond to interesting classes of concept lattices. Of course, methods of transforming arbitrary distance measures to those of a specific type have always be analysed carefully so that they can be justified with respect to aims of concrete investigations.

3   Threshold Lattices

Representations of data by diagrams have the main purpose to give structural insights into the data and to support interpretations. Since for large data sets band diagrams may become very complex, it might be helpful to consider only interesting parts of the data and their representations. In this section we discuss a reduction of similarity data to so-called threshold contexts and their concept lattices.

For some investigations it is reasonable to analyse similarities between objects which are given by distances below a certain threshold. This leads to the following definition of a threshold context.

Definition 4 Let $\mathcal{K} := (G, G, \mathbf{V}, J)$ be an ordinal distance context. Then $\mathcal{K}_v := (G, G, J_v)$ where

$$gJ_v h :\Leftrightarrow h(g) \leq v$$

is called the **threshold context of the value** v.

For every ordinal distance context $\mathcal{K}$, the derived context $\widetilde{K}$ is isomorphic to the apposition of the threshold contexts which is denoted by $|_{v \in V} \mathcal{K}_v$. This means that the formal context $\widetilde{K}$ is received by listing the threshold contexts horizontally side by side. For each threshold context $\mathcal{K}_v$ the corresponding concept lattice $\underline{B}(\mathcal{K}_v)$ is called the **threshold lattice of the value** v. The incidence relation of every threshold context is reflexive and symmetric. Hence $(A, B) \in \underline{B}(\mathcal{K}_v)$ implies $(B, A) \in \underline{B}(\mathcal{K}_v)$. This yields an antiautomorphism on the threshold lattice described by

$$p : \underline{B}(\mathcal{K}_v) \rightarrow \underline{B}(\mathcal{K}_v) \text{ with } (A, B) \mapsto (B, A).$$

The mapping p is a so-called **polarity** on $\underline{B}(\mathcal{K}_v)$, i.e., an involutoric antiautomorphism. If the line diagram of a threshold lattice is arranged in such a way that the polarity is represented by a reflection with horizontal axis then the names of the attributes can be omitted.

In our example of Fig. 3 the derived context consists of the four threshold contexts $\mathcal{K}_0$, $\mathcal{K}_{0.125}$, $\mathcal{K}_{0.25}$ and $\mathcal{K}_{0.5}$ which are separated by double lines. The sequence of the corresponding threshold lattices is shown in Fig. 6.

Since $\widetilde{K}$ is isomorphic to the apposition $|_{v \in V} \mathcal{K}_v$ the extents of $\mathcal{K}_v$ are also extents of $\widetilde{K}$. This follows by a more general result which is proved in [GW91].

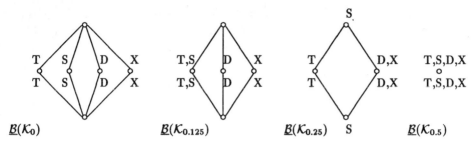

$$\underline{B}(\mathcal{K}_0) \qquad\qquad \underline{B}(\mathcal{K}_{0.125}) \qquad\qquad \underline{B}(\mathcal{K}_{0.25}) \quad S \qquad \underline{B}(\mathcal{K}_{0.5})$$

Figure 6: Sequence of threshold lattices belonging to the ordinal distance context of Fig. 2

Lemma 1 *Let $\mathcal{K}$ be an ordinal distance context, let $\widetilde{\mathcal{K}}$ be the formal context derived from $\mathcal{K}$, and let $\mathcal{K}_v$ be a threshold context of $\mathcal{K}$ corresponding to some value v. Then*

$$\eta : \underline{B}(\mathcal{K}_v) \to \underline{B}(\widetilde{\mathcal{K}}) \text{ with } (A, B) \mapsto (A, A^{\widetilde{J}})$$

is a meet-embedding.

The threshold lattices can be reconstructed within the band diagram of $\underline{B}(\widetilde{\mathcal{K}})$. We are able to locate the attribute concepts of $\mathcal{K}_v$ in the band diagram so that the threshold lattice $\underline{B}(\mathcal{K}_v)$ is isomorphic to the meet-subsemilattice generated by the located concepts. As usual, for $g \in G$ we write $\widetilde{\gamma}g$ for the concept $(\{g\}^{\widetilde{J}\widetilde{J}}, \{g\}^{\widetilde{J}})$.

Lemma 2 *For every $h \in G$ the attribute extent of h in $\mathcal{K}_v$ satisfies*

$$h^{J_v} = \{g \in G \mid gJ_vh\} = \bigcup\{A \mid \widetilde{\gamma}h \le (A, B) \in \underline{B}(\widetilde{\mathcal{K}}) \text{ and } \varphi(A, B) \le v\}.$$

Proof

"$\subseteq$" Let $g \in h^{J_v}$, i.e., gJ_vh. Since $(\{g, h\}^{J_vJ_v}, \{g, h\}^{J_v}) \in \underline{B}(\mathcal{K}_v)$
we have $(\{g, h\}^{J_vJ_v}, (\{g, h\}^{J_vJ_v})^{\widetilde{J}}) \in \underline{B}(\widetilde{\mathcal{K}})$. Obviously, $\widetilde{\gamma}h \le (\{g, h\}^{J_vJ_v}, (\{g, h\}^{J_vJ_v})^{\widetilde{J}})$
and moreover $\varphi(\{g, h\}^{J_vJ_v}, (\{g, h\}^{J_vJ_v})^{\widetilde{J}}) \le v$
because $\{g, h\}^{J_v} \supseteq \{g, h\}$ implies $\{g, h\}^{J_vJ_v} \subseteq \{g, h\}^{J_v}$.
Hence $h^{J_v} \subseteq \bigcup\{A \mid \widetilde{\gamma}h \le (A, B) \in \underline{B}(\widetilde{\mathcal{K}}) \text{ and } \varphi(A, B) \le v\}$.

"$\supseteq$" Let $g \in \bigcup\{A \mid \widetilde{\gamma}h \le (A, B) \in \underline{B}(\widetilde{\mathcal{K}}) \text{ and } \varphi(A, B) \le v\}$. Then there is a concept $(A, B) \in \underline{B}(\widetilde{\mathcal{K}})$
with $A \supseteq \{g, h\}$ and $\varphi(A, B) \le v$. Thus, gJ_vh and so $g \in h^{J_v}$.

$\square$

As a consequence of Lemma 1 and 2 we obtain the following result:

Proposition 4 *Let $\mathcal{M}_v := \{(A, B) \in \underline{B}(\widetilde{\mathcal{K}}) \mid (A, B) \text{ is maximal with } \varphi(A, B) \le v\}$. Then the following equation holds:*

$$(h^{J_v}, (h^{J_v})^{\widetilde{J}}) = \bigvee\{(A, B) \in \mathcal{M}_v \mid \widetilde{\gamma}h \le (A, B)\}.$$

The meet-subsemilattice of $\underline{B}(\widetilde{\mathcal{K}})$ generated by $\{(h^{J_v}, (h^{J_v})^{\widetilde{J}}) \mid h \in G\}$ is isomorphic to $\underline{B}(\mathcal{K}_v)$ as lattice.

Due to this theorem the threshold lattices can be reconstructed in the band diagram as follows:

(i) We fix a threshold v and mark the circles which represent the elements of $\mathcal{M}_v$ in the band diagram. This are all circles which are maximal with the property to be below or in the band which is indexed with v.

(ii) For an object concept $\tilde{\gamma}g$ we consider all marked concepts above $\tilde{\gamma}g$ and mark their join in the band diagram, too. This we do for all object concepts $\tilde{\gamma}g$.

(iii) Finally, we form all meets of the marked concepts whose circles represent the desired concept lattice $\underline{\mathcal{B}}(\mathcal{K}_v)$ in the band diagram.

References

[BD85] P. Bertrand and E. Diday. A visual representation of the compatibility between an order and a dissimilarity index: The pyramids. *Computed Statistics Quaterly*, 2:31–42, 1985.

[BD89] P. Brito and E. Diday. Pyramidal representation of symbolic objects. In *Knowledge, Data and Computer–Assisted Decisions — Proc. of the NATO Advanced Workshop on Data, Expert Knowledge and Decision, Hambourg*, Springer–Verlag, 1989.

[Boc74] H. H. Bock. *Automatische Klassifikation*. Vandenhoeck & Ruprecht, Göttingen, 1974.

[Did87] E. Diday. Orders and overlapping clusters by pyramids. Rapport de Recherche INRIA 730, INRIA, Rocquencourt, Le Chesnay, 1987.

[GW86] B. Ganter and R. Wille. Implikationen und Abhängigkeiten zwischen Merkmalen. In P. O. Degens, H.-J. Hermes and O. Opitz, editors, *Die Klassifikation und ihr Umfeld*, pages 171–185, Frankfurt, 1986. Indeks–Verlag.

[GW89] B. Ganter and R. Wille. Conceptual scaling. In F. Roberts, editor, *Applications of combinatorics and graph theory to the biological and social sciences*, pages 139–167. Springer Verlag, New York, 1989.

[GW91] B. Ganter and R. Wille. Skriptum zur Formalen Begriffsanalyse, 1991.

[Len91] K. Lengnink. Formale Begriffsanalyse von Ähnlichkeitsdaten, 1991. Staatsexamensarbeit.

[SW92] S. Strahringer and R. Wille. Towards a structure theory for ordinal data. In M. Schader, editor, *Analyzing and Modeling Data and Knowledge*, pages 129–139, Heidelberg, 1992. Springer–Verlag.

[Wil82] R. Wille. Restructuring lattice theory: an approach based on hierarchies of concepts. In I. Rival, editor, *Ordered Sets*, pages 445–470, Dordrecht–Boston, 1982. Reidel.

[Wil87] R. Wille. Bedeutungen von Begriffsverbänden. In B. Ganter, R. Wille and K. E. Wolff, editors, *Beiträge zur Begriffsanalyse*, pages 161–211. B. I.–Wissenschaftsverlag, Mannheim, 1987.

Approximate Galois Lattices of Formal Concepts

I. Van Mechelen

Department of Psychology, University of Leuven
Tiensestraat 102, B–3000 Leuven, Belgium

Abstract: The Galois lattice approach has been shown to provide a rich framework for the study of monothetic formal concepts within the context of an object by attribute correspondence. Yet, the approach is hampered by two problems: First, even for data sets of moderate size the Galois lattice is usually very complex. Second, small modifications of the correspondence under study (e.g., due to error) may lead to considerable modifications of the lattice. This paper presents a procedure to construct an approximate Galois lattice with limited order and length for a given binary correspondence. The procedure, which is based on Boolean matrix theory, makes use of an algorithm for Boolean factor analysis. Goodness–of–recovery results from a simulation study suggest that it allows one to retrieve a true correspondence from error–perturbed data.

1 Introduction and Problem

Quite a number of categorization problems start from a class of objects G, a class of attributes M, and a binary correspondence I, which indicates for each object $g \in G$ and each attribute $m \in M$ whether m applies to g (denoted by gIm), or not. Within such a *context* (G, M, I), one may consider pairs of an object set $A \subseteq G$ and an attribute set $B \subseteq M$ that are such that:

$$\forall a \in A, b \in B : aIb$$

and that

$$\forall a \in G \setminus A, \exists b \in B : \neg aIb$$

$$\forall b \in M \setminus B, \exists a \in A : \neg aIb$$

Such pairs (A, B) are maximal rectangles within the binary object by attribute incidence matrix H_I associated with the correspondence I (provided appropriate permutations of the rows and columns of the matrix in question). The set of all maximal rectangles of the correspondence under study will further be denoted by $L(G, M, I)$. Special maximal rectangles (A, B) are the one with $A = G$ and the one with $B = M$. An important property is that if for two maximal rectangles (A_1, B_1) and (A_2, B_2) it is true that $A_1 \subseteq A_2$, then $B_2 \subseteq B_1$, and vice versa.

A natural partial order may be defined on $L(G, M, I)$:

$$(A_1, B_1) \leq (A_2, B_2) \Leftrightarrow A_1 \subseteq A_2 (\Leftrightarrow B_2 \subseteq B_1)$$

It may further be shown that $L(G, M, I), \leq$ is a lattice, called the *Galois lattice* of the context (G, M, I).

The Galois framework constitutes a rich mathematical framework for the formalization of the study of *monothetic concepts* (i.e., concepts defined by a conjunctive combination of features or attributes). A maximal rectangle (A, B) may indeed be considered a *formal concept* of the context (G, M, I), A being the concept's *extension* and B its *intension* (Wille, 1982). The reciprocal relation between the partial orders of extensions and intensions further is a straightforward formalization of the duality between extension and intension observed by the authors of the Logique of Port–Royal and by G. W. Leibniz.

The Galois lattice can also be considered a tool to solve clustering problems. A major advantage of the Galois approach to clustering is that it leads to clusters of objects that are directly characterized in terms of an attribute pattern. Moreover, the object clusters are embedded in a comprehensive *taxonomy*, which simultaneously highlights the structure of the attributes.

Yet, despite its many advantages, the Galois approach is also hampered by two major problems: (1) Even for data sets of moderate size, the number of formal concepts is usually very high, and the lattice structure very complicated. (2) Small changes in the context (G, M, I) may lead to considerable changes in the list of formal concepts and in the lattice structure; this is especially troublesome if the (G, M, I) data are error–perturbed (as in many practical applications). Several authors have studied each of these two problems, and have proposed (partial) solutions for them.

(1) As regards the complexity of the Galois lattice, Wille (1984) and Duquenne (1991a) have proposed procedures to obtain 'readable' graphic (Hasse diagram) representations of the Galois lattice. In particular, Wille (1984) has suggested to look for appropriate partitions of the attribute set M leading to a set of subcontexts with relatively simple lattices that can be jointly represented in a so–called 'graded line diagram'. Considerable theoretical-mathematical efforts further have been done to reduce lattices to a simplest possible structure from which the whole lattice can be reconstructed. As such, for distributive lattices Birkhoff (1940) has shown that a reconstruction is possible from the set of all meet–irreducible elements. For other lattices, solutions have been proposed by Wille (1977) with the notion of *scaffolding*, and by Duquenne (1991b) with the notion of *meet–core of meet–essential elements*. It must be noted, however, that these solutions, although satisfactory from a theoretical viewpoint, remain fairly complicated for practical applications.

(2) The problem of dealing with noise or error in the data has been studied by Duquenne (1991a), Guénoche and Van Mechelen (1992), Luxenburger (1991), and Stöhr and Wille (1991). As such, Duquenne (1991a) and Guénoche and Van Mechelen (1992) have proposed procedures to retrieve dense (rather than homogeneous) rectangles. Luxenburger (1991) has introduced the notions of *partial implication* and *partial dependency*. Stöhr and Wille (1991) rely on external (so–called 'tolerance') information on the objects which is used to retrieve more robust as well as more simple lattice structures.

In this paper we propose a simple method that, similar to the Stöhr–Wille approach, deals with both the complexity problem and the problem of error–perturbed data. Unlike the Stöhr–Wille approach, it does not rely on external information on the objects. The method does not intend to replace the procedures listed above, but may be complementarily used in practical lattice applications.

2   Method

The method to be proposed here makes use of Boolean matrix theory (Kim, 1982), a framework that is closely related to lattice theory. Let β_0 denote the Boolean algebra of two elements $(0, 1)$, with the operations $+, \times$, and ^{C} (the latter denoting *complement*, i.e.,

$0^C = 1$ and $1^C = 0$). The set of all n–element row vectors over β_0 will be denoted by V_n, the set of all n–element column vectors by V^n, and the set of all $n \times m$ Boolean matrices by B_{nm}. The operations $+, \times$, and ^{C} on β_0 can be naturally extended to V_n, V^n, and B_{nm}. $V_n, +$ and $V^n, +$ are called Boolean vector spaces. A natural partial order further may be defined on V_n (V^n), in that for $\mathbf{v}$ and $\mathbf{w} \in V_n$ (V^n), $\mathbf{v} \leq \mathbf{w}$, iff $\forall i = 1 \ldots n : v(i) \leq w(i)$.

If $A \in B_{nm}$ is a Boolean matrix, one may consider within $V_m, +$ the span of all row vectors of A, denoted by $R(A)$, and similarly within $V^n, +$ the span of all column vectors of A, denoted by $C(A)$.

The following is a first important result:

THEOREM 1 (Kim, 1982)
$R(A), \leq$ and $C(A), \leq$ are anti–isomorphic lattices.

A second important result involves the concept of rank of a Boolean matrix. The situation here differs from that of matrices over a field, in that, for example, for Boolean matrices row rank does not need to equal column rank. For Boolean matrices one may further also consider a third rank concept: the *Schein rank* (Kim, 1982). To introduce it, we first present Theorem 2:

THEOREM 2
Let $A \in B_{nm}$. The following are equivalent:
(1) k is the smallest number for which $\exists P \in B_{nk}$ and $Q \in B_{km}$ such that $A = P \times Q$.
(2) k is the smallest number for which holds that A can be covered by k rectangles
(3) k is the smallest number such that $R(A)$ (resp. $C(A)$) is contained in a space spanned by k vectors.

DEFINITION 1
For any Boolean matrix the number k that satisfies the properties of Theorem 2 is called the Schein rank of that matrix.

In general, the Schein rank of a matrix is less than or equal to the minimum of its row rank and column rank. It is interesting to note that if $R(A)$ (and, hence, $C(A)$) is distributive, then the row rank, column rank and Schein rank of A are equal.

Given a Boolean matrix $A \in B_{nm}$, a practical data–analytic problem is to retrieve, for a low value of k, a Boolean matrix $\hat{A}$ with Schein rank k, such that $\hat{A}$ is as close as possible to A, closeness being defined in terms of the number of discrepancies between A and $\hat{A}$. This essentially is the problem of Boolean factor analysis, which looks for an approximate decomposition of a given matrix $A \in B_{nm}$ as the product $P \times Q$ of matrices $P \in B_{nk}$ and $Q \in B_{km}$. An algorithm for Boolean factor analysis has been proposed by Mickey, Mundle, and Engelman (1983), and an improved version of it has been developed by De Boeck and Rosenberg (1988) for their hierarchical classes analysis. De Boeck and Rosenberg (1988) also did a simulation study with 'true' matrices T to which error was systematically added, resulting in error–perturbed data matrices D. De Boeck and Rosenberg found that, for values of k at and beyond the Schein rank of T, their algorithm had an almost perfect goodness of recovery, in that $\hat{D}$ was usually very close to T.

These results on Boolean matrices now can be connected to the Galois framework through the following theorem, originally formulated in a different form by Zaretski (1963):

THEOREM 3

Let (G, M, I) be a context and denote the associated incidence matrix by H_I. It holds that $R((H_I)^C), \leq$ is isomorphic to $L(G, M, I), \leq$.

Proof: Define a mapping f from $R((H_I)^C)$ to $L(G, M, I)$, such that for any $\mathbf{v} \in R((H_I)^C)$, $f(\mathbf{v}) = (B', B) \in L(G, M, I)$, with $B = \{m \in M \mid \mathbf{v}(m) = 0\}$ and $B' = \{g \in G \mid \forall m \in B : gIm\}$. It is clear that f is an isotone bijection and, hence, a lattice isomorphism. QED

COROLLARY 1

If for a context (G, M, I) with incidence matrix H_I the Schein rank of $(H_I)^C$ equals k, then $L(G, M, I), \leq$ has order less or equal to 2^k and length less or equal to k.

Proof: This is a direct consequence of Theorem 3 and the definition of Schein rank (e.g., Property (3) of Theorem 2). QED

The practical importance of Corollary 1 is that it shows that a Boolean factor analysis of the complement of an incidence matrix H_I leads to an approximate Galois lattice of restricted order and length. In this respect the simulation results of De Boeck and Rosenberg also can be further interpreted in that they show that their algorithm allows one to approximately recover from an error–perturbed context (G, M, I_E) the underlying true correspondence I.

From an algorithmic point of view, it may be useful to point to the fact that for a rank k Boolean factor–analytic decomposition $P \times Q$ of the complement of an incidence matrix H_I, the rows of Q span a vector space that contains $R(\hat{H}_I^C)$. Via the isomorphism of Theorem 3, every concept of the approximate lattice then corresponds to a sum of rows of Q, and, as such to a subset of $\{1, \ldots, k\}$. This implies that in order to enumerate all concepts of the approximate lattice, it suffices to inspect the power set of $\{1, \ldots, k\}$. The latter can easily be done through an adjusted version of Ganter's (1984) algorithm.

3   Concluding Remarks

The proposed strategy deals with the two problems of Galois lattices that have been outlined above: (1) It results in a relatively simple lattice structure, and (2) it allows the researcher to retrieve a true correspondence from an error–perturbed context. A limitation of the strategy is that the simplicity of the lattice may still not be very satisfactory if values of k larger than 5 are needed. Furthermore, as regards the problem of error–perturbed data, it must be noted that the goodness–of–recovery results of De Boeck and Rosenberg's simulation study were satisfactory in terms of recovery of the true *correspondence*. Yet, even for a close approximation of the true correspondence, fairly large discrepancies with respect to the true *lattice* are possible.

As regards the approximate nature of Galois lattices resulting from our procedure, up to now in this paper we have interpreted the data as error–perturbed, and the resulting lattice as an intended recovery of an underlying true structure. This implies that discrepancies between the original context and the reconstructed context are (largely) attributed to errors in the data. An alternative interpretation is, however, also possible: The data may be considered to be the truth, whereas the resulting lattice is interpreted as an incorrect, but useful, approximate representation of them.

112

Acknowledgement

The author thanks Hans-J. Bandelt, Paul De Boeck, and Jean-Paul Doignon for their helpful comments on a previous version of this paper.

References

BIRKHOFF, G. (1940), *Lattice Theory*, American Mathematical Society, Providence.

DE BOECK, P., and ROSENBERG, S. (1988), Hierarchical classes: Model and data analysis, *Psychometrika, 53, 361–381.*

DUQUENNE, V. (1991a), The core of finite lattices, *Discrete Mathematics, 88, 133–147.*

DUQUENNE, V. (1991b), *GLAD: General Lattice Analysis & Design, A FORTRAN Program for a Glad User*, CAMS-EHESS, Paris.

GANTER, B. (1984), *Two basic algorithms in concept analysis (Preprint 831)*, Technische Hochschule, Darmstadt.

GUENOCHE, A., and VAN MECHELEN, I. (1992), Galois approach to the induction of concepts, in I. Van Mechelen, J. Hampton, R. Michalski, and P. Theuns (eds.), *Categories and concepts: Theoretical views and inductive data analysis*, Academic Press, London, 287–308.

KIM, K.H. (1982), *Boolean Matrix Theory*, Marcel Dekker, New York.

LUXENBURGER, M. (1991), Implications partielles et dépendances partielles, *Mathématiques, Informatique et Sciences Humaines, 113, 35–55.*

MICKEY, M.R., MUNDLE, P., and ENGELMAN, L. (1983), Boolean factor analysis, in W.J. Dixon (ed.), *BMDP Statistical Software*, University of California Press, Berkeley, 538–545, 692.

STOEHR, B., and WILLE, R. (1991), *Formal Concept Analysis of Data with Tolerances (Preprint 1401)*, Technische Hochschule, Darmstadt.

WILLE, R. (1977), Aspects of finite lattices, in M. Aigner (ed.), *Higher Combinatorics*, D. Reidel Publishing Company, Dordrecht, 79–100.

WILLE, R. (1982), Restructuring lattice theory: An approach based on hierarchies of concepts, in O. Rival (Ed.), *Ordered Sets*, Reidel, Boston, 445–470.

WILLE, R. (1984), Line diagrams of hierarchical concept systems, *International Classification, 11, 77–86.*

ZARETSKI, K.A. (1963), The semigroup of binary relations (Russian), *Mat. Sbornik, 61, 291–305.*

Representation of Data by Pseudoline Arrangements

Wolfgang Kollewe

Forschungsgruppe Begriffsanalyse, Fachbereich Mathematik, Technische Hochschule Darmstadt,

6100 Darmstadt, Schloßgartenstraße 7

Abstract: Formal contexts are under certain conditions representable by oriented pseudoline arrangements. Oriented pseudoline arrangements respect the hierarchical order of concept lattices. The formal context is reconstructable out of an oriented pseudoline arrangement. The representation of oriented pseudoline arrangements are quite easy to understand. In general there is no sufficient condition to decide if a context is representable (pictorial) or not. There is an infinite class of minor–minimal non–pictorial contexts. This tells us that excluding a finite number of examples is never sufficient to guarantee that a formal context is pictorial. Oriented pseudoline arrangements are isomorphic to oriented matroids of rank 3. We can therefore use an algorithmic approach which is generating oriented matroids of rank 3 and decide wether a context is representable by an oriented pseudoline arrangement.

1   Introduction

A frequently used method of data analysis is to describe structures of data with graphic representations. In Formal Concept Analysis data are usually represented by line diagrams of concept lattices. This method of representation unfolds the hierarchical order of the concepts which is inherent in the data context. This paper presents a geometrical method of representation which also shows the concepts and their hierarchical order.

In his article *'Testdaten als Merkmalsvektoren'* H. Feger (1989) describes a geometrical visualization of the structures of data. The main idea of this method is to decompose the space by hyperplanes into a positive and a negative halfspace. An object is located in a positve halfspace if it has the attribute which corresponds to this halfspace. The dimension of the space in which the data are represented depends on the complexity of the data structure. This article discusses conditions for formal contexts which guarantee the representability of contexts by line arrangements in the plane. We restrict our considerations to the plane because it is most common to use the euclidean plane for the visualization of data structures.

2   Basic Notions

A formal context $\mathbf{K} := (G, M, I)$ consists of two sets G and M together with a binary relation I between G and M. The elements of G and M are called objects and attributes, respectively, and gIm is read: the object g has the attribute m. A (formal) concept of the context (G, M, I) is defined as a pair (A, B) with $A \subseteq G$, $B \subseteq M$, $A = \{g \in G | gIm$ for all $m \in B\}$, and $B = \{m \in M | gIm$ for all $a \in A\}$; A and B are called the extent and the intent of the concept(A, B), respectively.

A finite set $\mathcal{L} := \{l_1, l_2, \ldots, l_n\}$ of n oriented lines in general position in the euclidean plane induces a cell decomposition of the plane. For a k–element subset $\mathcal{T} := \{t_1, t_2, \ldots, t_k\}$ of cells of dimension 2 (called topes), we define an incidence relation $I^* \subset \mathcal{T} \times \mathcal{L}$ as follows:

t_i and l_j are incident if and only if t_i lies on the positive side of l_j.

We call a formal context (G, M, I) represented by an oriented line arrangement $\mathcal{L}$ (with topes $\mathcal{T}$) if and only if there are bijections

$$p : G \to \mathcal{T} \quad \text{and} \quad q : M \to \mathcal{L} \quad \text{such that} \quad gIm \quad \text{if and only if} \quad p(g)I^*q(m).$$

In his paper 'Teilungen der Ebene durch Geraden oder topologische Geraden' (1956) G. Ringel has shown that there is an arrangement with nine pseudolines which is not strechable, i.e. there is no isomorphic arrangement of lines. In other words, every line arrangement gives rise to a pseudoline arrangement but not vice versa. This is the reason why we consider pseudoline arrangements more generally.

Definition 1 *An arrangement of pseudolines in the real projective plane P is any family of simply closed curves in P such that every two curves have precisely one point in common, at which they cross each other.*

In the following we will consider only simple arrangements, in which there are no three lines which have a common point. We call such an arrangement *oriented* if every pseudoline has an orientation.

Oriented matroids of rank 3 and oriented pseudoline arrangements can be identified as shown by Cordovil (1982). For a given (uniform) rank 3 oriented matroid with $n + 1$ elements, we consider its (simple) pseudoline representation with $n + 1$ oriented pseudolines $\mathcal{L} := \{l_0, l_1, \ldots, l_n\}$. When linking the projective plane with the euclidean plane, l_0 is interpreted as the line of infinity.

3 Examples

The first example represents repertory grid data of a patient with anorexia nervosa. In this grid test the client has determined the objects and the attributes. The standard evaluation of such grids uses the application of the Principal Component Analysis as suggested by Slater (1977) and Gabriel (1981). This leads via a projection from a high–dimensional space onto a plane to the biplot visualization which represents the persons and the pairs of constructs as points in the plane. The original data are not identifiable in this kind of representation. In Formal Concept Analysis it is important that the original data are identifiable within the derived hierarchy of concepts, so that every discovered connection can always be analysed and rated on the basis of the data which are responsible for this fact; for conceptual grid evaluation see Spangenberg and Wolff (1988). Oriented pseudoline arrangements fulfil this basic idea of data evaluation: the formal context is reconstructable out of an oriented pseudoline arrangement.

If an object has an attribute then the object lies on the positive side of the corresponding pseudoline. For instance, in the context of fig.1 one can see that the BROTHER–IN–LAW has the attribute 'talkative'. In the pseudoline arrangement of fig.2 the tope which is representing the BROTHER–IN–LAW is lying on the right side of the pseudoline which is labelled

	touchy	withdrawn	self-confident	dutiful	hearty	difficult	attentive	easily offended	not irascible	fearful	talkative	superficial	sensitive	ambitious
SELF	×	×	×		×	×	×		×	×			×	×
IDEAL	×		×	×	×		×		×				×	×
FATHER	×	×		×	×	×	×	×	×	×		×	×	×
MOTHER	×	×		×	×	×		×	×	×		×	×	×
SISTER	×	×		×	×	×	×		×	×			×	×
BROTHER-IN-LAW			×	×	×		×				×	×		×

Figure 1: Formal context of a repertory grid

with 'talkative'. We may thus determine for every object all the attributes which are incident with the corresponding object.

Furthermore it is possible to find all concepts of the formal context in this representation. In our example the client has characterized his family as 'withdrawn', 'difficult', 'fearful', 'sensitive', 'not irascible', 'touchy', 'hearty' and 'ambitious', which are the attributes forming the intent of the concept 'family'. The extent of this concept is: MOTHER, FATHER, SISTER, and SELF. To determine a concept in an oriented pseudoline arrangement we have to select a set of objects and then to look for all pseudolines whose positive halfspaces contain the selected set of objects. Finally we have to collect all the objects which are located in the intersection of those halfspaces.

FATHER, SISTER, SELF, IDEAL, and BROTHER–IN–LAW have the attributes 'attentive', 'hearty', and 'ambitious'. This yields another concept of the context. The intersection of the extents of the two described concepts gives rise to a new concept. The extent of this context is: FATHER, SISTER, and SELF. The intent of this concept is: 'withdrawn', 'difficult', 'fearful', 'sensitive', 'not irascible', 'touchy', 'hearty', 'ambitious', and 'attentive'. This is a subconcept of the first two ones. The hierarchical relation subconcept–superconcept, which is modelled as follows

$$(A_1, B_1) \le (A_2, B_2) : \iff A_1 \subseteq A_2 \quad (\iff B_1 \supseteq B_2),$$

is reflected in the representation by an oriented pseudoline arrangement.

In a line diagram of a concept lattice every circle represents exactly one concept. Thus, it is for instance easy to count all concepts of the formal context. In an oriented pseudoline arrangement it is more difficult to determine all concepts. To find all concepts of a formal context in an oriented pseudoline arrangement, we have to form the necessary intersections of positive halfspaces as described above.

A line diagram of a concept lattice is not an euclidean representation, therefore the euclidean distance between two circles does not say anything about the connection of the

116

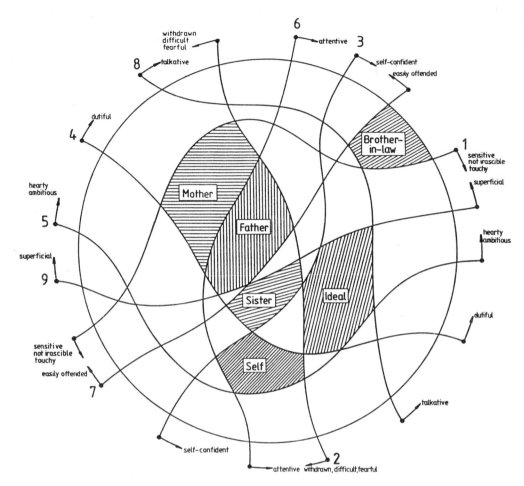

Figure 2: Oriented pseudoline arrangement of the context in figure 1

corresponding concepts. The hierarchical order is determined only by the line segments between the circles. For oriented pseudoline arrangements the distance between two topes gives some information about their intensional connection. If two objects differ only in one attribute their topes have one edge in common, i.e. the concepts 'father' and 'mother'. The topes of objects which differ in two attributes have either one vertex in common or one tope lying between them, i.e. the concepts 'father' and 'sister'. Topes of objects which differ in more than two attributes have neither an edge nor a vertex in common. The fact that the distance between two topes of objects which differ in many attributes is larger than for topes of objects which differ only in one or two attributes reflects our natural feeling and makes an oriented pseudoline arrangement useful.

The second example shown in fig.3 represents also a repertory grid test. In this case we have a many-valued context (G, M, W, I); W consists of attribute values. In our context every attribute is formed by two adjectives, i.e. lively and calm. The patient has the opportunity to answer to the questions with grades 1 up to 6. The answers 1, 2, or 3 belong

	lively/calm	thoughtful/egoistic	straightforward/problematical	superficial/empathic	listen attentively/impatient	deliberate/inconsiderate	honest/dishonest	sensitive/not sensitive
SELF	1	3	5	5	2	1	2	1
IDEAL	2	4	2	5	1	2	1	2
RALF	4	2	3	5	2	3	1	1
MOTHER	1	1	4	2	6	3	3	1
FATHER	5	3	2	4	3	1	3	3
EX-FRIEND R.	3	5	5	4	3	3	4	3
SISTER	4	3	2	4	2	2	2	2
EX-FRIEND W.	2	4	1	2	5	5	4	5
LODGER	2	3	2	4	2	2	2	2

Figure 3: Many–valued context of a repertory grid test

to the first adjective of the attribute, i.e. for the attribute lively 1 means that the judged person is very lively and 3 means the person is lively only to a small degree. The attribute values 4, 5, and 6 belong to the second adjective of the attribute.

There is no automatic way to transform a many–valued context into a one–valued one. The evaluator has to choose conceptual scales which yield the significant concepts for the evaluation. In the present example the structure of the answers leads to a partition into two classes. This results in the structure of the dichotomic scale. The concept lattice of dichotomic scale consists of a two element antichain with a common lower and upper bound. The derived context of the many–valued context has now 16 attributes as it is shown in fig.4; for basic facts of conceptual scaling the reader is referred to B. Ganter and R. Wille (1989), W. Kollewe (1989) and the references given there.

Every oriented pseudoline decomposes the plane into a positive and a negative halfspace. This can be understood as the formation of a two–element antichain. Therfore we need only one pseudoline for the representation of dichotomic attributes. The derived context of our example with 16 attributes is representable by just 8 lines. This reduction of pseudolines makes the arrangement analysable in a clearer and better way for the evaluator.

4 Oriented Matroids and Pseudoline Arrangements

An oriented pseudoline arrangement is describable by a set of hyperline sequences. This compact form of description stores the whole information of an oriented pseudoline arrangement. The way to get a set of hyperline sequences of an oriented pseudoline arrangement is the following. We pick a pseudoline different of l_0. We follow this line starting at pseudoline l_0 and we write down the directions of the other pseudolines as signs as we meet them when we go along our chosen pseudoline. For instance, in fig.2 line 1 meets first line 0, which runs

	lively	calm	thoughtful	egoistic	straightforward	problematical	superficial	empathic	listen attentively	impatient	deliberate	inconsiderate	honest	dishonest	sensitive	not sensitive
SELF	×		×	×				×	×		×		×		×	
IDEAL	×			×	×			×	×		×		×		×	
RALF		×	×		×			×	×		×		×		×	
MOTHER	×		×			×	×		×	×			×		×	
FATHER		×	×		×			×	×		×		×		×	
EX-FRIEND R.	×			×		×		×	×		×			×	×	
SISTER		×	×		×			×	×		×		×		×	
EX-FRIEND W.	×			×	×		×			×		×		×		×
LODGER	×		×			×			×	×		×		×		×

Figure 4: Derived context of the many-valued context of figure 3

counterclockwise, therefore we write 0. Then line 1 meets line 9, which crosses line 1 from the right to the left; in this case we write -9. We do the same for line 0 starting at line 1. The result for our first example is described in the scheme below:

```
0 :  1  -7   3   6  -2   8   4   5   9
1 :  0  -9  -5  -4   2  -6  -3   7  -8
2 :  0   8  -1  -6   7   9  -3   4   5
3 :  0  -5  -6  -4   2  -9  -7   1   8
4 :  0   8  -5  -2   3  -7  -9   6   1
5 :  0   8   4  -2   6   3  -7  -9   1
6 :  0  -5   3  -7  -9  -4   2   1   8
7 :  0   8  -1   3  -2   9   4   6   5
8 :  0  -4  -5  -9   1   7   3   6  -2
9 :  0   8   3  -2  -7   4   6   5   1
```

This abstract notion still makes sense for oriented matroids and, moreover, it is even a possible approach in defining oriented matroids since the Grassmann Plücker sign conditions are encoded in this structure. In the uniform case it turns out that the 3-term Grassmann Plücker relation is sufficient to check the oriented matroid property (J. Bokowski, A. Guedes de Oliviera and J. Richter-Gebert (1991)).

We write $E = \{1, \ldots, n\}$ for the finite set of n points, and we use the notation $\Lambda(n, d)$ for the set of all tuples $\lambda = (\lambda_1, \ldots, \lambda_d) \in E^d$, with $1 \leq \lambda_1 < \lambda_2 < \ldots < \lambda_d \leq n$. To each oriented hyperline $l = (l_1, \ldots, l_{d-2}) \in \Lambda(n, d-2)$ we assign a *normalized hyperline sequence*

$$h_l = (l_1, \ldots, l_{d-2} | s^l_{d-1} l_{d-1}, \ldots, s^l_n l_n)$$

with $\{l_1, \ldots, l_d\} = E$, $l_{d-1} < l_d, \ldots, l_n$ and $s^l_i \in \{-1, +1\}$, $d - 1 \leq i \leq n$ $(s^l_{d-1} = 1)$; finally, we assign corresponding bracket signs

$$sign[l_1, \ldots, l_{d-2}, s^l_j l_j, s^l_k l_k] := 1$$

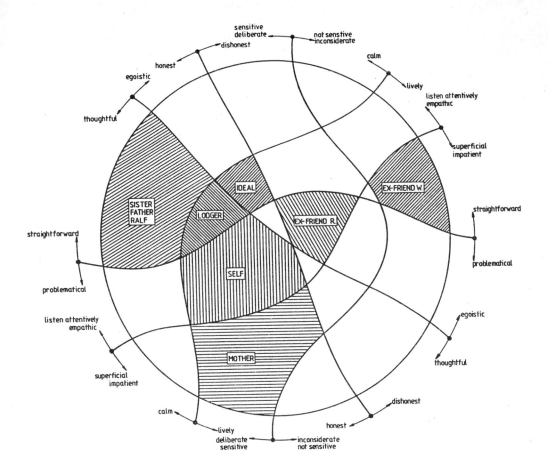

Figure 5: Oriented pseudoline arrangement of the derived context figure 4

for all $j, k : d - 1 \le j < k \le n$ such that for all bases $b = [b_1, \dots, b_d] \in \Lambda(n, d)$ and for all permutations $\pi : \{1, \dots, d\} \to \{1, \dots, d\}$

$$sign[b_1, \dots, b_d] = sign(\pi)sign[b_{\pi(1)}, \dots, b_{\pi(d)}],$$
$$-sign[b_1, \dots, b_j, \dots, b_d] = sign[b_1, \dots, -b_j, \dots, b_d].$$

The latter conditions tell us that the $\binom{d}{2}$ definitions of signs for a fixed bracket $b \in \Lambda(n, d)$ must compatible.

Definition 2 (Oriented Matroid in Terms of Hyperline Sequences) *Set E together with a set of hyperline sequences with compatible bracket signs is a uniform matroid of n points in rank d.*

5 Representability of Contexts

The problem of representability reads in our notion as follows: Given a context **K**, find an oriented matroid of rank 3 and a subset $\mathcal{T} = \{t_1, t_2, \ldots, t_k\}$ of its set of topes such that the triple $(\mathcal{T}, \mathcal{L}, I^*)$ is isomorphic to the given context. Such a context is called pictorial.

Now some necessary conditions are given for a context to be pictorial. Remember that for an oriented matroid χ with $(n+1)$ elements rank 3 means that the total number of topes is given by

$$t(\chi) = 1 + \tfrac{n(n+1)}{2}.$$

A context which has a larger number of objects than $1 + \tfrac{n(n+1)}{2}$, where n is the number of attributes, is not pictorial. A context which has eight pairwise different objects for at least one 3–element subset of attributes, is also not pictorial. It will be seen that in general these necessary conditions for a formal context to be pictorial are not sufficient. We have found some contexts which fulfil the above–mentioned conditions but which are nevertheless not pictorial.

Definition 3 *Given a context (G,M,I). A non–pictorial context is called minor– minimal if and only if all subcontexts are pictorial. We call a context reoriented if the relation I is replaced by I^c for some $N \subseteq M$.*

Proposition 1 *The property of a context to be pictorial is invariant when forming any reoriented context.*

Proposition 2 *The following contexts are minor–minimal non–pictorial. For these sizes, these are the only non–pictorial contexts up to reorientation.*

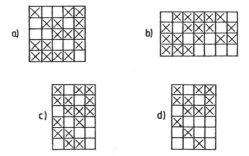

Furthermore, it turns out that there is an infinite class of minor–minimal non–pictorial contexts. This tell us that excluding a finite number of examples is never sufficient to guarantee that a context is pictorial. For further information see J. Bokowski and W. Kollewe (1992).

Since there is no simple criterium to decide for a formal context if it is pictorial or not, only an algorithmic approach may be used for this decision. The problem of finding an oriented matroid admissible to a given context uses the same basic algorithm of J. Bokowski which was applied for embedding combinatorial manifolds in real vector spaces when using

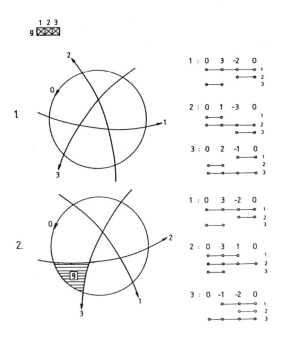

Figure 6:

oriented matroid theory. A transition to the context case was implemented by A. Guedes de Oliveira.

The main idea of this algorithmic construction is starting with a suitable minor of the oriented matroid we are looking for and appropriately extending it afterwards. In the context case we start with just one basis: an oriented matroid with 3 points of rank 3. In the extreme case we have to construct all possible extensions of the oriented matroid from which we started. But the context very often passes some properties to substructures which can be used for reducing the class of minors. Thus the tree of all extensions may substantially be cut. For further information see J. Bokowski and B. Sturmfels (1989), J.Bokowski and W. Kollewe (1992) and the literature given there.

There is still the question, how can be checked if there is a tope of the oriented pseudoline arrangement for every object of the formal context in which this object can be embedded. This can be done in terms of combinatorial intervals within the hyperline configuration.

Proposition 3 *An object g of a given context* **K** *can be embedded into a rank 3 oriented matroid with* $(n+1)$ *points if and only if there exists a hyperline* l_i *such that the intersection of all halfspaces* $h_j^{(i)}$ *is not empty for the intent of the row of the object g is not empty.*

The main idea of the proof can already be seen in the following example described by fig.6. An object g which has the three attributes 1, 2, and 3 shall be embedded into a pseudoline arrangement with 4 lines where line 0 is the line of infinity. We see in the first case that the intersection of the intervals for every hyperline is empty. The second solution shows intersections which are not empty for the hyperlines 2 and 3. This means that there is a tope where the object is embeddable.

References

BOKOWSKI, J., GUEDES DE OLIVEIRA, A., and RICHTER-GEBERT, J. (1991), Algebraic varieties characterizing matroids and oriented matroids, *Advances in Mathematics*, *87, 160–185.*

BOKOWSKI, J., and KOLLEWE, W. (1992), On representing contexts in line arrangements, *ORDER, (to appear).*

BOKOWSKI, J., and STURMFELS, B. (1989), Computational synthetic geometry, *Lecture Notes in Mathematics, 1355.*

CORDOVIL, R. (1983), Oriented matroids of rank three and arrangements of pseudolines, *Annals of Discrete Math., 17, 219–223.*

FEGER, H. (1989), Testdaten als Merkmalsvektoren, *Beitrag zur Festschrift für Karl Josef Klauer, 'Wissenschaft und Verantwortung'*, Hogrefe, Göttingen.

FOLKMAN, J., and LAWRENCE, J. (1978), Oriented matroids, *J. Combinatorial Theory*, B 25, 199–236.

GABRIEL, K. R. (1981), Biplot display of multivariate matrices of data and diagnosis, in: V. Barnett (ed.), *Interpreting multivariate data*, Wiley, Chichester, 147–173.

GANTER, B., and WILLE, R. (1989), Conceptual scaling, in: F. Roberts (ed.), *Applications of combinatorics and graph theory to the biological and social sciences*, Springer Verlag, New York, 139–167

GANTER, B., and WILLE, R. (1992), Formale Begriffsanalyse, Manuskript 75p.

GRÜNBAUM, B. (1972), *Arrangements and spreads*, American Math. Soc., Regional Conf. Ser. 10, Rhode Island.

KOLLEWE, W. (1989), Evaluation of a survey with methods of formal concept analysis, in: O. Opitz (ed.), *Conceptual and numerical analysis of data*, Springer Verlag, Berlin–Heidelberg, 123–134.

RINGEL, G. (1956), Teilungen der Ebene durch Geraden und topologische Geraden, *Math. Zeitschrift, 64, 79–102.*

SLATER, P. (1977), *The measurement of interpersonal space by grid technique*, Vol. I and II, Wiley, New York, 1977.

SPANGENBERG, N. and WOLFF, K.E. (1988), Conceptual grid evaluation, in: H.H. Bock (ed.), *Classification and related methods of data analysis*, North–Holland, Amsterdam, 577–580

WILLE, R. (1984), Liniendiagramme hierarchischer Begriffssysteme, in: H. H. Bock (Hrsg.), *Anwendungen der Klassifikation: Datenanalyse und numerische Klassifikation*, INDEKS–Verlag, Frankfurt, 32–51.

WILLE, R. (1987), Bedeutungen von Begriffsverbänden, in: B. Ganter, R. Wille, E. Wolff (Hrsg.), *Beiträge zur Begriffsanalyse*, B.I.–Wissenschaftsverlag, Mannheim–Wien–Zürich, 161–211

A Relational Approach to Split Decomposition

H.–J. Bandelt
Mathematisches Seminar der Universität
Bundesstr. 55, D–2000 Hamburg 13, GERMANY

A.W.M. Dress
Fakultät für Mathematik der Universität
Postfach 8640, D–4800 Bielefeld 1, GERMANY

Abstract: Given a set of objects, a split A, B (or distinctive feature) partitions the set into two complementary parts A and B. To a system of splits we associate the relation between pairs of objects that opposes a pair u, v to a pair x, y exactly when the pairs u, v and x, y are separated by at least one split from the system. Conversely, for a quaternary relation opposing pairs of objects, we consider those splits A, B for which each pair from A is opposed to each pair from B. This sets up an adjoint situation between systems of splits and relations between pairs. We characterize the closed systems and open relations. Systems of weakly compatible splits (arising in the decomposition theory of metrics) are always closed, and the corresponding relations can be characterized by a 6–point condition. Particular instances are described by 5–point conditions. A concluding example from biology illustrates this relational approach.

1   Introduction

"The classification of objects (e.g., countries, events, animals, books) plays an important role in the organization of knowledge" (Sattath and Tversky, 1987). Such a classification is typically achieved by partitioning the objects into several groups so that in each group any two objects are somehow more closely related to each other than to any third object outside this group. This task would be easy if any two objects which appear to be closely related to a third one would also be closely related to each other — a condition which typically fails for almost all real data. Cluster theory has been designed as a scientific discipline to deal precisely with this sort of failure. In other words, "the raison d'être" of cluster theory is the notorious intransitivity of similarity (Dress and von Haeseler, 1990).

To construct reasonable groupings of objects in spite of this intransitivity, one of the standard approaches taken in cluster theory is to depart from either all triplets or all quartets of distinct objects, since for three or four objects the data immediately suggest "local" groupings.The problem then is to aggregate this local information in a consistent way to obtain a global solution of the given classification problem. In this note we present a rather general framework for aggregating quartets into global distinctive features ("splits"). In contrast to previous approaches (cf. Bandelt and Dress, 1986) we relax the requirement that necessarily tree structures be obtained, by admitting alternative groupings within the quartets. This is motivated by the (additive) decomposition theory for distance matrices developed in Bandelt and Dress (1992 a) and its fruitful application to phylogenetic analyses, paradigmatically worked out in Dopazo et al. (1992) and Bandelt and Dress (1992 b). The "split decomposition" of a given distance matrix results in a (fairly small) number of weighted splits that can be represented by graphs built up from cyclic cells and cut edges in a grid–like or tree–like fashion.

To be more specific, recall that the deletion of any edge in a labelled tree gives rise to two components, thus determining a split $\{A, B\}$ (i.e. a partition into two parts A and B) of the set of objects (i.e. labels). The resulting splits have the characteristic property that they are pairwise compatible in the sense that the partition refining two tree splits $\{A_1, B_1\}$ and $\{A_2, B_2\}$ consists of three parts, i.e. at least one of the intersections $A_1 \cap A_2$, $A_1 \cap B_2$, $B_1 \cap A_2$, $B_1 \cap B_2$ is empty. For the more general systems of splits considered in Bandelt and Dress (1992 a) one relaxes pairwise compatibility to triplewise weak compatibility: for any three splits $\{A_i, B_i\}$ $(i = 1, 2, 3)$ at least one of the intersections $A_1 \cap A_2 \cap A_3$, $A_1 \cap B_2 \cap B_3$, $B_1 \cap A_2 \cap B_3$, $B_1 \cap B_2 \cap A_3$ is empty. In Propositions 5 and 3 below we will see that systems of compatible resp. weakly compatible splits are determined by their restrictions to all subsets consisting of at most four objects: for every quartet (u, v, x, y) of objects one would thus record whether there is a given split $\{A, B\}$ that separates u, v from x, y (so that A contains one pair and B the other). Moreover, the quaternary relations arising this way can easily be characterized. It turns out that even more general systems of splits, dubbed conformal (in analogy to a corresponding notion for hypergraphs), can be recovered from their associated quaternary separation relations; see Proposition 2 below.

Phylogeny reconstruction is one of the potential applications of our relational approach. There are several ways to obtain quaternary relations from molecular data. For every quartet (u, v, x, y) of (DNA or protein) sequences one has three possible groupings: u, v versus x, y, or u, x versus v, y, or u, y versus v, x. We propose to exclude only the most improbable grouping(s) (according to some criterion such as maximum parsimony or maximum likelihood), leaving us with typically two alternative groupings for each quartet. The splits aggregated (in a strict consensus fashion) from all these partial quartet splits are necessarily weakly compatible. Amenable to further analysis, these splits often give much better insight into the given data then just the application of standard tree approximation algorithms.

2   An adjoint situation

For a given set X one has an adjoint situation $S \mapsto \|_S$ and $\| \mapsto S_\|$ between systems S of ordered pairs (A, B) of nonempty sets $A, B \subseteq X$ and quaternary relations $\|$ on X, where

(*) $uv \parallel_S xy$ if and only if there is some $(A, B) \in S$ with $u, v \in A$ and $x, y \in B$,

(**) $(A, B) \in S_\|$ if and only if $uv \parallel xy$ for all $u, v \in A$ and $x, y \in B$.

Indeed, both assignments preserve inclusions, and moreover, each relation $\|$ includes $\|_{S_\|}$ and each system S is contained in $S_{\|_S}$. Those members $\|$ and S, respectively, for which the latter inclusions are actually equalities are called *open* and *closed*, respectively. The adjoint situation then naturally restricts to an (order) isomorphism between closed systems S of pairs of subsets and open quaternary relations $\|$ on X.

The open relations and closed systems are easily characterized: a relation $\|$ is open if and only if $uv \parallel xy$ implies $uu \parallel xy$, $uv \parallel xx$, $vu \parallel xy$, $uv \parallel yx$; whereas a system S is closed if and only if, for $\emptyset \neq A_0 \subseteq A$ and $\emptyset \neq B_0 \subseteq (B_1 \cap B_2) \cup (B_2 \cap B_3) \cup (B_3 \cap B_1)$ with $A, B_i \subseteq X$, we have $(A_0, B_0) \in S$ whenever $(A, B_i) \in S$ $(i = 1, 2, 3)$, and similarly $(B_0, A_0) \in S$ whenever $(B_i, A) \in S$ $(i = 1, 2, 3)$. To see that this requirement suffices to guarantee that S is closed, just observe that for each subset A one obtains a graph (possibly with loops) on X where vertices w and x are adjacent exactly when $(A, \{w, x\}) \in S$. It is clear that the subsets B with $(A, B) \in S$ comprise all complete subgraphs if and only if the first half of the above requirement is fulfilled; indeed, this is essentially what Gilmore's condition guarantees in the context of hypergraphs; cf. Berge (1989).

We are only interested in the symmetric case where one considers systems S of partial splits: a *partial split* $\{A, B\}$ is an unordered pair of disjoint nonempty subsets A, B of X. Then, $\{A, B\}$ being identified with (A, B) and (B, A), the canonical isomorphism between closed systems and open relations restricts in the following way:

Proposition 1. There is a one–to–one correspondence, set up by (∗) and (∗∗), between quaternary relations $\parallel$ on a set X satisfying the two axioms

(R1) $uv \parallel xy$ implies $vu \parallel xy$, $xy \parallel uv$, and $uu \parallel xy$ $(u, v, x, y \in X)$,

(R2) non $uu \parallel uu$ for every $u \in X$

and closed systems S of partial splits, i.e., systems S of subset pairs such that

(S1) $(A, B) \in S$ implies $(B, A) \in S$, $A \cap B = \emptyset$, and $A, B \neq \emptyset$,

(S2) $(A, B_i) \in S$ for $i = 1, 2, 3$ with $\emptyset \neq A_0 \subseteq A$ and
$\emptyset \neq B_0 \subseteq (B_1 \cap B_2) \cup (B_2 \cap B_3) \cup (B_3 \cap B_1)$ implies $(A_0, B_0) \in S$

hold.

Evidently, closed systems S of partial splits are determined by their maximal members. Particular interest attaches to the case where every maximal partial split $\{A, B\}$ constitutes a *split*, that is, $A \cup B = X$ holds. To every system S of splits one associates the system S_0 of those partial splits $\{A_0, B_0\}$ for which there is $\{A, B\} \in S$ with $A_0 \subseteq A$ and $B_0 \subseteq B$. Obviously, S_0 is closed if and only if S satisfies

(S3) for any $\{A_i, B_i\} \in S$ $(i = 1, 2, 3)$ with $A_1 \cap A_2 \cap A_3 \neq \emptyset$ there exists some $\{A, B\} \in S$ with $A_1 \cap A_2 \cap A_3 \subseteq A$ and $(B_1 \cap B_2) \cup (B_2 \cap B_3) \cup (B_3 \cap B_1) \subseteq B$.

Systems of splits fulfilling (S3) are henceforth called *conformal*. We will now show how the extensibility of partial splits to full splits translates into a 7–point condition for the associated relation $\parallel$. We make use of the following short–hand: given $x_1, \ldots, x_m, y_1, \ldots, y_n \in X$ with $m, n \geq 2$ we write $x_1 \ldots x_m \parallel y_1 \ldots y_n$ when $x_i x_j \parallel y_k y_l$ for all $i, j = 1, \ldots, m$ and $k, l = 1 \ldots, n$. Suppose $\{A, B\}$ is a maximal partial split with respect to $\parallel$ which is not full, i.e., there is some t in X outside $A \cup B$. Since neither $\{A \cup \{t\}, B\}$ nor $\{A, B \cup \{t\}\}$ qualifies as a split, there exist $u, v, w \in A$ and $x, y, z \in B$ such that neither $uv \parallel xt$ nor $tw \parallel yz$ holds. As $uvw \parallel xyz$ we thus arrive at the following consequence of Proposition 1.

Proposition 2. Quaternary relations $\parallel$ on a set X that satisfy (R1), (R2), and

(R3) $uvw \parallel xyz$ implies $uv \parallel xt$ or $tw \parallel yz$ (for $t, u, v, w, x, y, z \in X$)

correspond to conformal systems of splits.

3 Weakly compatible splits

A system S is certainly conformal (S3) if the stronger requirement

(S4) for any $\{A_i, B_i\} \in S$ $(i = 1, 2, 3)$ at least one of the intersections
$A_1 \cap A_2 \cap A_3$, $A_1 \cap B_2 \cap B_3$, $B_1 \cap A_2 \cap B_3$, $B_1 \cap B_2 \cap A_3$ is empty

holds, that is, in (S3) one can always choose one of $\{A_i, B_i\}$ $(i = 1, 2, 3)$ as the required split $\{A, B\}$. Such a system S is said to consist of *weakly compatible* splits. An equivalent description, used in Bandelt and Dress (1992 a), is in terms of a forbidden configuration: there are no three splits $\{A_i, B_i\} \in S$ $(i = 1, 2, 3)$ together with four distinct points $x_0, x_1, x_2, x_3 \in X$ such that $A_i \cap \{x_0, x_1, x_2, x_3\} = \{x_0, x_i\}$. Weak compatibility is thus a 4–point condition and can therefore be expressed by the corresponding quaternary relations.

Proposition 3. Quaternary relations $\parallel$ on a set X that satisfy the three axioms (R1),

(R4) at most two of the instances $uv \parallel xy$, $ux \parallel vy$, $uy \parallel vx$ hold for $u, v, x, y \in X$,

(R5) $uv \parallel xyz$ implies $uv \parallel xyzt$ or $tuv \parallel xyz$ $(t, u, v, x, y, z \in X)$

correspond to systems of weakly compatible splits of X.

Proof. First observe that (R1) and (R4) yield (R2). We claim that (R4) and (R5) imply (R3). Indeed, suppose $uvw \parallel xyz$ holds but neither $uv \parallel xt$ nor $tw \parallel yz$. Then, according to (R5), $uvw \parallel xy$ and $uvw \parallel xz$ extend to $tuvw \parallel xy$ and $tuvw \parallel xz$, respectively, since non $uv \parallel xt$. Further, $tuw \parallel xy$ and $tuw \parallel xz$ necessarily extend to $tuwz \parallel xy$ and $tuwy \parallel xz$, respectively, because non $tw \parallel yz$. Moreover, $uvw \parallel yz$ can only extend to $uvw \parallel yzt$, and $uv \parallel yzt$ finally leads to $uvx \parallel yzt$. Summarizing, for u, x, y, z we have $ux \parallel yz$, $uy \parallel xz$ as well as $uz \parallel xy$, contrary to (R4).

Then, as weak compatibility is expressed by (R4), Proposition 2 concludes the proof.

Systems S of weakly compatible splits are considerably restricted in their sizes: if X has n elements, then S can contain at most $\binom{n}{2}$ distinct splits (see Corollary 5 of Bandelt and Dress, 1992 a). Hence for a relation $\parallel$ on X satisfying (R1) and (R4) there are at most $\binom{n}{2}$ associated splits (in $S_\parallel$).

The following example shows that one cannot dispense with a 6–point axiom for $\parallel$ in the preceding proposition. Consider the relation $\parallel$ on the 6–point set $\{u, v, w, x, y, z\}$ defined by

$$uv \parallel wxy, \quad uv \parallel wxz, \quad wy \parallel uvxz, \quad xy \parallel uvwz, \quad wz \parallel uvxy, \quad xz \parallel uvwy,$$

so that (R1) is fulfilled. Then (R5) is violated, although every restriction of $\parallel$ to a proper subset satisfies all requirements in Proposition 3.

A natural example of a quaternary relation $\parallel$ that satisfies (R1) and (R2) but not (R4) is given by the separation relation in the Euclidean plane: let $uv \parallel xy$ exactly when the line segment between u and v does not intersect the one from x to y. Indeed, if x is an interior point of a triangle with vertices u, v, w, then all three instances $uv \parallel wx$, $uw \parallel vx$, $ux \parallel vw$ hold. Note that this relation restricted to a finite subset X of the Euclidean plane in general position satisfies (R4) if and only if X is the set of vertices of a convex polygon. In this case (R5) is satisfied as well, and the splits are induced by straight lines that separate X into two nonempty parts; see Section 3 of Bandelt and Dress (1992 a).

Particular systems S of weakly compatible splits can be characterized by simple 5–point conditions. Say that S is *closed under intersection–unions* (of incompatible splits) if S satisfies

(S5) $\{A, B\}, \{C, D\} \in S$ with $A \cap C$, $A \cap D$, $B \cap C$, $B \cap D$, all nonempty implies $\{A \cap C, B \cup D\} \in S$.

Proposition 4. Quaternary relations $\parallel$ on a set X that satisfy (R1), (R4) as well as

(R6) $uv \parallel xy$ implies $uv \parallel xt$ or $tu \parallel xy$ $(t, u, v, x, y \in X)$,.

(R7) $uv \parallel xy$ and $uv \parallel yz$ imply $uv \parallel xz$ $(u, v, x, y, z \in X)$

also satisfy (R5) and correspond to systems of weakly compatible splits of X closed under intersection–unions.

Proof. Assume $uv \parallel xyz$ and non $tu \parallel yz$. Then from $uv \parallel yz$ we infer $uv \parallel yt$ and $uv \parallel zt$ using (R6) (and (R1)). This together with $uv \parallel xy$ yields $uv \parallel xt$ by (R7), thus establishing (R5). It is easy to see that (R7) amounts to the intersection–union closure property for splits.

Finally, a system S of splits consists of *compatible* splits if

(S6) for $\{A, B\}, \{C, D\} \in S$ at least one of $A \cap C$, $A \cap D$, $B \cap C$, $B \cap D$ is empty.

Proposition 5. Quaternary relations $\parallel$ on a set X that satisfy (R1), (R6), and

(R8) $uv \parallel xy$ implies non $ux \parallel vy$ and non $uy \parallel vx$ for $u, v, x, y \in X$

correspond to systems of compatible splits of X.

This proposition is obtained from Proposition 4 since (R7) is a consequence of (R6) and (R8). Proposition 5 restates a result from Bandelt and Dress (1986), formulated for relations defined for quartets of distinct points in a finite set X. In the finite case this characterizes the partially labelled trees, where the labels form a partition of X, so that only (interior) vertices of degree larger than two may be without label. The splits are then in one–to–one correspondence with the edges, and $uv \parallel xy$ means that the path from u (i.e., more precisely, the vertex labelled with the block containing u) to v does not intersect the path from x to y.

4   Application to sequence space

Given an alphabet Λ with at least two distinct letters, the n–th power Λ^n consists of all sequences of length n with letters from Λ. The similarity $s(y, z)$ of two sequences $y = y_1 \ldots y_n$ and $z = z_1 \ldots z_n$ is the number of matches, whereas the distance $d(y, z)$ is the number of mismatches:

$$s(y, z) = \#\{i \mid y_i = z_i\}, \ d(y, z) = n - s(y, z).$$

The (*distance*) d–*relation* $\parallel_d$ is defined by $(u, v, x, y \in \Lambda^n)$:

$$uv \parallel_d xy \text{ if and only if } \ s(u, v) + s(x, y)$$
$$> \min \ \big(s(u, x) + s(v, y), \ s(u, y) + s(v, x) \big).$$

The associated splits are exactly the *d–splits* (in the sense of Bandelt and Dress, 1992 a), i.e., the splits $\{A, B\}$ having positive isolation index (*d–index*, for short)

$$\alpha_{A,B}^d := \tfrac{1}{2} \cdot \min_{\substack{u, v \in A, \\ x, y \in B}} \big(s(u, v) + s(x, y)$$
$$- \min \big(s(u, v) + s(x, y), s(u, x) + s(v, y), s(u, y) + s(v, x) \big) \big).$$

This notion of split and index is fundamental for a canonical decomposition theory of distance functions. Nevertheless, there are other ways to define a relation $\parallel$ for quartets of sequences that are meaningful in some applications. For instance, we define the (*parsimony*) *p–relation* $\|_p$ by

$$uv \parallel_p xy \quad \text{if and only if} \quad \#\{i \mid u_i = v_i \text{ and } x_i = y_i\}$$
$$> \min\left(\#\{i \mid u_i = x_i \text{ and } v_i = y_i\}, \#\{i \mid u_i = y_i \text{ and } v_i = x_i\}\right).$$

The splits $\{A, B\}$ with respect to $\|_p$ are called *p–splits* with (positive) isolation index (*p–index*, for short)

$$\alpha_{A,B}^p = \min_{\substack{u,v \in A, \\ x,y \in B}} \Big(\#\{i \mid u_i = v_i, \ x_i = y_i\}$$
$$- \min(\#\{i \mid u_i = v_i, \ x_i = y_i\}, (\#\{i \mid u_i = x_i, \ v_i = y_i\},$$
$$(\#\{i \mid u_i = y_i, \ v_i = x_i\})\Big).$$

Simple examples show that neither relation need satisfy (R5), though (R4) is trivially fulfilled. The d– and p–relations are particular instances of relations based on scoring functions σ that assign to each configuration $\{\{\lambda_1, \lambda_2\}, \{\mu_1, \mu_2\}\}$ of letters some real number $\sigma(\{\lambda_1, \lambda_2\}, \{\mu_1, \mu_2\})$. Given such a scoring function σ we put for $u, v, x, y \in \Lambda^n$

$$uv \parallel_\sigma xy \quad \text{exactly when the sum } \sum_{i=1}^n \sigma(\{u_i, v_i\}, \{x_i, y_i\})$$
is larger than the minimum of the two sums
$$\sum_{i=1}^n \sigma(\{u_i, x_i\}, \{v_i, y_i\}) \text{ and } \sum_{i=1}^n \sigma(\{u_i, y_i\}, \{v_i, x_i\}).$$

In particular, the following scoring function leads to the d–relation (and corresponding isolation indices):

$$\sigma_d(\{\lambda_1, \lambda_2\}, \{\mu_1, \mu_2\}) = \begin{cases} 1 & \text{if} & \lambda_1 = \lambda_2 \neq \mu_1 = \mu_2, \\ \frac{1}{2} & \text{if} & \mu_1 \neq \lambda_1 = \lambda_2 \neq \mu_2 \neq \mu_1 \quad \text{or} \\ & & \lambda_1 \neq \mu_1 = \mu_2 \neq \lambda_2 \neq \lambda_1, \\ 0 & \text{otherwise,} \end{cases}$$

whereas the p–relation (with p–indices) is obtained for the scores

$$\sigma_p(\{\lambda_1, \lambda_2\}, \{\mu_1, \mu_2\}) = \begin{cases} 1 & \text{if } \lambda_1 = \lambda_2 \neq \mu_1 = \mu_2, \\ 0 & \text{otherwise.} \end{cases}$$

Note that instead of σ_d and σ_p one can equally use the modified scoring functions σ_d' and σ_p' in the definitions of the d– and p–relations:

$$\sigma_d'(\{\lambda_1, \lambda_2\}, \{\mu_1, \mu_2\}) = \begin{cases} 1 & \text{if } \lambda_1 = \lambda_2 \text{ and } \mu_1 = \mu_2, \\ 0 & \text{if } \lambda_1 \neq \lambda_2 \text{ and } \mu_1 \neq \mu_2, \\ \frac{1}{2} & \text{otherwise;} \end{cases}$$

$$\sigma_p'(\{\lambda_1, \lambda_2\}, \{\mu_1, \mu_2\}) = \begin{cases} 1 & \text{if } \lambda_1 = \lambda_2 \text{ and } \mu_1 = \mu_2, \\ 0 & \text{otherwise.} \end{cases}$$

For $u, v, x, y \in \Lambda^n$ we then obtain

$$\sum_{i=1}^n \sigma_d'\Big(\{u_i, v_i\}, \{x_i, y_i\}\Big) = \tfrac{1}{2}\big(s(u, v) + s(x, y)\big),$$
$$\sum_{i=1}^n \sigma_p'\Big(\{u_i, v_i\}, \{x_i, y_i\}\Big) = \#\{i \mid u_i = v_i \text{ and } x_i = y_i\}.$$

When a Markov model is assumed for the substitution of letters, one can also consider the log–likelihood $\sigma_l(\{\lambda_1, \lambda_2\}, \{\mu_1, \mu_2\})$ for having the pairs λ_1, λ_2 and μ_1, μ_2 evolved along non–crossing paths in a tree. These three types of scores (and their weighted versions) are of considerable interest in biology for estimating phylogenetic trees. Further scoring functions are described in Table 1 of Li et al. (1987).

Here is an illustrative example. We re–analyse the mitochondrial DNA data from Human, Chimpanzee, Gorilla, Orangutan, and Gibbon, sequenced by Brown et al. (1982). The data consist of five homologous fragments of length 896 bp. The variable sites are also shown in Fig. 11.10 of Nei (1987). The 90 informative sites are reproduced in Table 1. From the frequencies of the polytypic patterns (counted in Table 1, or more conveniently, read off from Table 11.2 of Nei, 1987) one readily determines the parsimony relation $\|_p$. It turns out that the relation enjoys the properties of Proposition 3 in this particular case, viz., it is obtained from nine splits: five are trivial splits that separate one species from the other four; the nontrivial splits (supported by the 90 sites given in Table 1) and their isolation indices are described in Table 2. Human is grouped with the African apes at a high isolation index (24). Further, the Human–Chimpanzee grouping is favoured over the Chimpanzee–Gorilla grouping (with index 7 versus 4), although its significance may be disputable. This is at odds with the result from maximum parsimony, which supports the latter grouping; cf. Fig. 11.11 of Nei (1987).

```
              5    10   15   20   25   30   35   40   45
Human       GTCAT CATCCTTCTTTTTTTAGCAATTTCCTCACCTTCTCCGTCA
Chimpanzee  ATTAC CATTCCTTTTTTCCC CGCATTCTCCCTGCCTTCTT CATTA
Gorilla     GTTGT TATTACCTCCCTTTCAACAACCCCTTTGTTTCACC TATCG
Orangutan   ACCACTCCCACCCTTCCTCCTAAGACTCACACACTCAACTCGCCA
Gibbon      ACCGCCCCCATCCCCTCCCTCAAGT CCTAT CCATCCAATC TACTG

              50   55   60   65   70   75   80   85   90
Human       CGCTCTCGCCGCTCTCACTCCCCTT ATTTTCTTGTCCGGTGACCG
Chimpanzee  TGTCTCCGCCGTTCCCATTTTCCTC ATTTT CTTACTCAGTGACCG
Gorilla     CGTCCCCGTCATTCCAACTTTCCTT ATTCTTTCGC CTAGT GATTA
Orangutan   CACCT CTATTACCTTAGTCCCTACC GCCTA GCCATT TTCAC ACTAA
Gibbon      TACTT TTACTACCTTAGCCCCTACAGCCCAGCCAA ACGAC ACTAA
```

Table 1. Nucleotide sequences of mitochondrial DNA from humans and apes; data from Brown et al. (1982). Here only informative sites are presented.

isolation index	split	
24	{Orangutan, Gibbon},	{Human, Chimp., Gorilla}
7	{Human, Chimp.},	{Gorilla, Orangutan, Gibbon}
4	{Chimp., Gorilla},	{Human, Orangutan, Gibbon}
1	{Gorilla, Orangutan},	{Human, Chimp., Gibbon}

Table 2. The parsimony splits and indices for the sequences of Table 1.

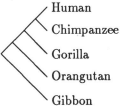

Fig. 1. Tree supported by the p–splits with isolation indices larger than 4.

In contrast to the principle of maximum parsimony (i.e., selecting the trees with minimum number of mutational changes) our criterion for establishing parsimony splits is local

130

and relaxed: we reject the least parsimonious solution(s) for every quartet of sequences (in defining $\|_p\|$ and then aggregate the medium and most parsimonious quartet trees in a strict consensus way (to obtain the splits). We expect that this approach is somewhat less sensitive to systematic and sampling errors than the standard parsimony and distance methods. At least, for the above mt DNA data, the tree built from the splits with highest indices (see Fig. 1) agrees with the maximum likelihood estimate (based on the same data) performed by Hasegawa and Kishino (1989) and is also in accord with several analyses of DNA–DNA hybridization data as well as nuclear genes; cf. Kishino and Hasegawa (1989). This tree receives strong support from comparisons of mitochondrial COII gene sequences (of length 684 bp), determined by Ruvolo et al. (1991). For instance, the observed numbers of nucleotide differences between Human, Chimpanzee, Gorilla, and Siamang perfectly fit the tree shown in Fig. 2, with a fairly long internal branch.

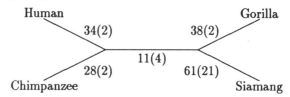

Fig. 2. Tree representation for observed numbers of nucleotide differences and transversional differences (in brackets), respectively, between four primates; data from Fig. 1 of Ruvolo et al. (1991).

References

BANDELT, H.–J. and DRESS, A. (1986), Reconstructing the shape of a tree from observed dissimilarity data, Advances Applied Mathematics, 7, 309–343. BANDELT, H.–J. and DRESS, A. (1992 a), A canonical decomposition theory for metrics on a finite set, Advances Mathematics, to appear.

BANDELT, H.–J. and DRESS, A.W.M. (1992 b), Split decomposition: a new and useful approach to phylogenetic analysis of distance data, Molecular Phylogenetics Evolution, to appear.

BERGE, C. (1989), Hypergraphs, North–Holland.

BROWN, W.M., PRAGER, E.M., WANG, A., and WILSON, A.C. (1982), Mitochondrial DNA sequences of primates: tempo and mode of evolution, J. Molecular Evolution, 18, 225–239.

DAPAZO, J., DRESS, A., and VON HAESELER, A. (1990), Split decomposition: a new technique to analyse viral evolution, Preprint 90–037, Universität Bielefeld.

DRESS, A. and VON HAESELER, eds. (1990), Trees and hierarchical structures, Lecture Notes Biomathematics, 84, Springer–Verlag.

HASEGAWA, M. and KISHINO, H. (1989), Confidence limits on the maximum likelihood estimate of the hominoid tree from mitochondrial–DNA sequences, Evolution, 43, 672–677.

KISHINO, H. and HASEGAWA, M. (1989), Evaluation of the maximum likelihood estimate of the evolutionary tree topologies from DNA sequence data, and the branching order in Hominoidea, J. Molecular Evolution, 29, 170–179.

LI, W.-H., WOLFE, K.H., SOURDIS, J., and SHARP, P.M. (1987), Reconstruction of phylogenetic trees and estimation of divergence times under nonconstant rates of evolution, Cold Spring Harbor Symposia Quantitative Biology, Vol. III, 847–856.

NEI, M. (1987), Molecular evolutionary genetics, Columbia University Press.

RUVOLO, M., DISOTELL, T.R., ALLARD, M.W., BROWN, W.M., and HONEYCUTT, R.L. (1991), Resolution of the African hominoid trichotomy by use of a mitochondrial gene sequence, Proc. National Academy Sciences USA, 88, 1570–1574.

SATTATH, S. and TVERSKY, A. (1987), On the relation between common and distinctive feature models, Psychological Review, 94, 16–22.

Some New Useful Representations of Dissimilarities in Mathematical Classification

Frank Critchley
Department of Statistics
University of Warwick,
CV4 7AL Coventry, GB

Bernard Van Cutsem
Laboratoire Modélisation et Calcul
Université Joseph Fourier, Institut National Polytechnique de Grenoble,
Centre National de la Recherche Scientifique
B.P.53X
38041 GRENOBLE Cedex, F

Abstract : Some recent results of Frank Critchley and Bernard Van Cutsem (1991) give new ways of representing dissimilarities defined on a general set and having values in an ordered set. These representations generalise and unify the well known representations of the classical bijections concerning ultrametrics on a finite set obtained by Benzécri (1965), Hartigan (1967), Jardine, Jardine and Sibson (1967) and Johnson (1967), their extensions to bijections between dissimilarities on a finite set and Numerical Stratified Clusterings discussed by Jardine and Sibson (1971), Janowitz (1978), Barthélémy, Leclerc and Monjardet (1984) and Critchley and Van Cutsem in a unpublished research report (1989). We first present our unifying and general result using equivalent definition of a function by its level map. Then we discuss in each case, how the classical results on dendrograms, prefilters and hierarchies can be obtained. We conclude by considering various possibilities opened by these results and looking at, for an example, the definition of set-valued ultrametrics.

1 Introduction

Some recent results of Frank Critchley and Bernard Van Cutsem (1991) give new ways of representing dissimilarities defined on a general set with values in an ordered set using various mathematical structures.

Let E denote a nonempty set and L a nonempty ordered set. Define $\mathcal{F} = L^E$ and $\mathcal{N} = \mathcal{P}(E)^L$. For every $f \in \mathcal{F}$, define $m_f \in \mathcal{N}$ by

$$m_f(l) = \{e \in E : f(e) \leq l\}.$$

and let u denote the map $\mathcal{F} \to \mathcal{N}$ defined by $u(f) = m_f$. In fact, u is a bijection between $\mathcal{F}$ and the set of level maps $\mathcal{M} = u(\mathcal{F}) \subseteq \mathcal{N}$ and there exists an inverse map $v : \mathcal{M} \to \mathcal{F}$. Moreover, u and v are dual order-isomorphisms between $\mathcal{F}$ and $\mathcal{M}$ when these sets are ordered pointwise.

Of course the relevant restrictions of u and v induce natural dual order-idsomorphisms between all pairs of corresponding subsets $\mathcal{F}$ and $\mathcal{M}$.

This very simple result implies many of the representations known in Classification Theory. Suppose that $E = S \times S$ where S is a finite set and consider dissimilarities or ultrametrics on S with values in $\mathbb{R}_+$. For instance, the following bijections are particular cases of the bijections u and v :

- bijections between ultrametrics on S and dendrograms (Hartigan (1967), Jardine, Jardine, Sibson (1967), Johnson (1967)),

- bijections between ultrametrics on S and indexed hierarchical classifications (Benzécri (1965), (1973)),

- bijections between dissimilarities on S and Numerical Stratified Clusterings (Jardine and Sibson (1971)),

In the case where dissimilarities on the finite set S take their values in an ordered set L, Janowitz (1978) has shown that there exists a bijection between the set of these dissimilarities and the set of residual functions from L to $\mathcal{P}(S \times S)$. This bijection is still a particular case of the result on level maps.

In the case where the set S is an arbitrary one (not necessarily finite), Critchley and Van Cutsem (1989) introduced bijections between the set of dissimilarities or ultrametrics with values in $\mathbb{R}_+$ and different sets of objects such as generalised dendrograms, prefilters and indexed regular generalised hierarchies. All these bijections are still particular cases of our result.

Hence this simple result allows us to have some representations of the set of dissimilarities or ultrametrics in a very general situation. It also gives the possibility of isolating precisely those conditions on L, and possibly E, for which the level maps can be variously characterised within $\mathcal{N}$. Specifically, we look at the case in which the ordered set L is chosen to be precisely $\mathcal{P}(S \times S)$. This brings to light weaker structures than the classical ones. In this way, we also give the definition and a few properties of set-valued hierarchies.

The main part of this communication is a summary of Critchley and Van Cutsem (1991).

2 Notations and prerequisites

We shall use intensively many notions and notations on ordered sets most of which can be found in Blyth and Janowitz (1972). We present here only the most important ones.

Let L be an ordered set. Introduce the following hypotheses and notations.

- Denote
$$[l) \equiv \{\tilde{l} \in L : \tilde{l} \geq l\} \quad \text{and} \quad]l) \equiv \{\tilde{l} \in L : \tilde{l} > l\}.$$

- LMIN : L has a minimum element, denoted 0.

- LTO : L is totally ordered.

- A *filter* is any nonempty subset $\tilde{L}$ of L such that $(l \in \tilde{L} \text{ and } \tilde{l} \geq l) \Rightarrow \tilde{l} \in \tilde{L}$, and a *principal filter* if there exists $l \in L$ such that $\tilde{L} = [l)$. In this case, we say that l generates the principal filter $[l)$.

- We say that an ordered set L is a *join semi-lattice* if it satisfies the condition :

  JSL : The set intersection of two principal filters of L is a principal filter.

  Then we define the *join* of l_1 and l_2, denoted $l_1 \vee l_2$, by $[l_1) \cap [l_2) = [l_1 \vee l_2)$.

  A join semi-lattice is called *join-complete* if every nonempty subset admits a least upper bound.

In addition, we shall use the following assumptions on L.

- **LUR** : L is upper regular, i.e.

$$\forall l \in L,]l) \quad \text{nonempty} \quad \Rightarrow \wedge]l) \quad \text{exists and is} \quad l.$$

- **LWF** : L is well filtered, i.e. every filter of L is either of the form $[l)$, or of the form $]l)$, for some $l \in L$.

- **LFILT** : L is filter complete. That is, every proper filter of L admits an infimum.

- Let L_1 and L_2 denote two nonempty ordered sets and $f : L_1 \to L_2$. The function f is said *isotone* if

$$\forall l_1 \in L_1, \forall l_2 \in L_2, l_1 \le l_2 \Rightarrow f(l_1) \le f(l_2)$$

and *antitone* if

$$\forall l_1 \in L_1, \forall l_2 \in L_2, l_1 \le l_2 \Rightarrow f(l_1) \ge f(l_2).$$

- The mapping $f : L_1 \to L_2$ is called *residual* if it is isotone and if the preimage of every principal filter of L_2 is a prefilter of L_1. More properties of residual maps and the definition of the conjugate notion of residuated map can be found in Blyth and Janowitz (1972).

- Let L_1 and L_2 denote two nonempty ordered sets and f a bijection $L_1 \to L_2$. The function f is called an *order-isomorphism* between L_1 and L_2 if f and its inverse f^{-1} are isotone and a *dual order-isomorphism* if f and its inverse f^{-1} are antitone.

We now consider the usual definitions of dissimilarities and ultrametrics in the general case. Let S denotes a nonempty set, L an ordered set with a minimum element 0, and f a function $S \times S \to L$. Then

- f is a *L-predissimilarity on S*, or simply a *predissimilarity*, if $\forall a \in S, f(a, a) = 0$, and a *dissimilarity* if also $\forall (a, b) \in S^2, f(a, b) = f(b, a)$.

- A predissimilarity is called a *preultrametric* if

$$\forall (a, b, c) \in S^3, \ (f(a, b) \le l \quad \text{and} \quad f(b, c) \le l) \Rightarrow (f(a, c) \le l).$$

An *ultrametric* is a symmetric preultrametric.

- A predissimilarity f is called *definite* if

$$f(a, b) = 0 \Rightarrow a = b.$$

We shall denote by $\mathcal{F}_\Delta$ the set of dissimilarities on S and by $\mathcal{F}_u$ the set of ultrametrics on S.

For any predissimilarity f, we shall denote $B_f(a, l)$ the closed ball centre $a \in S$ and radius $l \in L$ determined by f.

It will be convenient to identify binary relations on the set S with subsets of $S \times S$ and, as usual, to identify reflexive or symmetric or transitive binary relations with the relevant subsets of $S \times S$.

For the sake of homogeneity and to make easy the references, we use here the same numbers or labels as in Critchley and Van Cutsem (1991) to indicate the different properties listed in the definitions.

3 Level maps

We begin with our result which unifies and generalises the usual mathematical structures equivalent to dissimilarities. It is known that a function with real values can be defined by its level sets. We wish to extend this property to the functions $f : E \to L$ where E is any non empty set and L any nonempty ordered set.

Let $\mathcal{F}$ denote the set of all maps $f : E \to L$ and $\mathcal{N}$ the set of all maps $m : L \to \mathcal{P}(E)$. Note than $\mathcal{F}$ and $\mathcal{N}$ are ordered pointwise, i. e.

$$f \leq f' \Leftrightarrow \forall e \in E, f(e) \leq f'(e) \quad \text{and} \quad m \leq m' \quad \Leftrightarrow \quad \forall l \in E, \quad m(l) \leq m'(l),$$

where $\mathcal{P}(E)$ is ordered by inclusion.

To every function $f : E \to L$, we associate the, possibly empty, level sets of f setting

$$\forall l \in L, m_f(l) = \{e \in E : f(e) \leq l\}.$$

The function $u : f \to m_f$ associates to each $f \in \mathcal{F}$ a map in $\mathcal{N}$ which is called the level map of f. Let $\mathcal{M}$ denote the image set $u(\mathcal{F})$.

It is easy to verify that $m_f = m_g$ implies $f = g$ and so we can define the map $v : \mathcal{M} \to \mathcal{F}$ by $v(m) = f$ where $m = m_f$. This leads us to the following theorem.

Theorem 1 *The maps u and v establish a dual order-isomorphism between $\mathcal{F}$ and $\mathcal{M}$.* ∎

Characterisations of level maps are given by the following proposition.

Proposition 1 *Let $m \in \mathcal{N}$ and $L_{e,m} = \{e \in E : e \in m(l)\}$. Then the three following propositions are equivalent.*
    (1) m is level.
    (2)   (a) m is isotone,
            (b) $\forall e \in E, L_{e,m}$ has a minimum element.
    (3) $\forall e \in E, L_{e,m}$ is a principal filter of L. ∎

We shall now use this result to generalise the different bijections between sets of dissimilarities and mathematical structures such as dendrograms, prefilters and hierarchies.

4 Structures of type "dendrograms"

We give first the definitions of two new notions and then we show how they generalise the different known notions of dendrograms.

Definition 1 *A map $m \in \mathcal{N}$ is called an L-stratification on E, or simply a stratification, if it satisfies the following two conditions.*
    (S1) m is isotone.
    (S2) $\exists \bar{l} \in L$ such that $m(\bar{l}) = E$.
    (S3) $\forall l \in L,]l)$ nonempty $\Rightarrow \exists l' \in]l)$ such that $m(l') = m(l)$. ∎

The set of all L-stratifications is denoted $\mathcal{N}_S$.

Definition 2 *A map* $m \in \mathcal{N}$ *is called a generalised L-stratification on E, or simply a generalised stratification, if it satisfies the following two conditions.*
   ($GS1$) m is upper semi-continuous. That is

$$\forall l \in L, \;]l) \quad nonempty \quad \Rightarrow \quad m(l) = \cap\{m(l') : l' \in]l)\}.$$

   ($GS2$) m covers E. That is $\cup\{m(l) : l \in L\} = E$. ∎

The set of all generalised L-stratifications is denoted $\mathcal{N}_G$.

It is clear that every stratification is a generalised stratification. The following theorem resumes, with the notations and hypotheses previously introduced, the properties of stratifications and of generalised stratifications and the different inclusions or equalities between the sets $\mathcal{N}_S$, $\mathcal{N}_G$ and $\mathcal{M}$.

Theorem 2 (a) *(LUR and LWF)* $\Rightarrow \mathcal{M} = \mathcal{N}_G$.
   (b) $LTO \Rightarrow [\mathcal{M} = \mathcal{N}_G \Leftrightarrow (LUR \text{ and } LWF)]$.
   (c) $LTO \Rightarrow [\mathcal{M} = \mathcal{N}_S \Leftrightarrow \mathcal{M} = \mathcal{N}_S = \mathcal{N}_G \Leftrightarrow (LUR \text{ and } LWF \text{ and } \mathcal{N}_S = \mathcal{N}_G)]$.
   (d) *(LTO and LUR and LWF and E is finite)* $\Rightarrow [\mathcal{M} = \mathcal{N}_S = \mathcal{N}_G]$. ∎

Critchley and Van Cutsem (1991) give some examples showing that the four hypotheses of part (d) of this theorem are independent.

These definitions and theorem imply some classical results. We first recall the definitions. Dendrograms have been introduced in different papers by Hartigan (1967), by Jardine, Jardine and Sibson (1967) and by Johnson (1967). Numerically Stratified Clustering has been defined by Jardine and Sibson (1971) and Generalised Dendrograms have been defined by Critchley and Van Cutsem (1989).

Definition 3 *Let S be a finite nonempty set and $Eq(S)$ denote the set of all equivalence relations on S identified with the relevant subsets of $S \times S$. A dendrogram is a function $m : \mathbb{R}_+ \to Eq(S)$ satisfying the following properties.*
   ($D1$) *m is isotone.*
   ($D2$) $\exists \bar{l} \in \mathbb{R}_+$ *such that* $m(\bar{l}) = S \times S$.
   ($D3$) $\forall l \in \mathbb{R}_+, \; \exists \epsilon > 0$, *such that* $m(l) = m(l + \epsilon)$. ∎

Definition 4 *Let S be a finite nonempty set and $RS(S)$ denote the set of all reflexive and symmetric binary relations on S identified with the relevant subsets of $S \times S$. A Numerically Stratified Clustering (NSC) is a function $m : \mathbb{R}_+ \to RS(S)$ satisfying the following properties.*
   ($NSC1$) *m is isotone.*
   ($NSC2$) $\exists \bar{l} \in \mathbb{R}_+$ *such that* $m(\bar{l}) = S \times S$.
   ($NSC3$) $\forall l \in \mathbb{R}_+, \; \exists \epsilon > 0$, *such that* $m(l) = m(l + \epsilon)$. ∎

We note that the definitions of Dendrograms and of NSC are formally the same. Only the sets of values of m in $S \times S$ are different.

Definition 5 *Let S be a nonempty set. A Generalised Dendrogram (GD) is a function $m : \mathbb{R}_+ \to Eq(S)$ satisfying the following properties.*
   (GD1) $\forall l \in \mathbb{R}_+,\ m(l) = \cap\{m(l + \epsilon) : \epsilon > 0\}$.
   (GD2) $\cup\{m(l) : l \in \mathbb{R}_+\} = S \times S$. ∎

Theorem 1 implies, with $E = S \times S$, the following bijections and proves that they are moreover dual order-isomorphisms.

1) Let us suppose S is a finite set and $L = \mathbb{R}_+$. It is clear that dendrograms are L-stratifications on S and then, by part d of Theorem 2, are level maps. Thus the maps $u : \mathcal{F}_u \to \mathcal{M}$ and $v : \mathcal{M} \to \mathcal{F}_u$ defined by

$$\forall l \in \mathbb{R}_+,\ (u(f))(l) = \{(a,b) \in S^2 : f(a,b) \le l\}$$

and

$$\forall (a,b) \in S^2,\ (v(m))(a,b) = inf\{l \in \mathbb{R}_+ : (a,b) \in m(l)\}$$

are dual order-isomorphisms between $\mathcal{F}_u$ and the set of dendrograms on S.

2) Let us suppose S is a finite set and $L = \mathbb{R}_+$. It is clear that NSC are stratifications on S and then, by part d of Theorem 2, are level maps. So the maps $u : \mathcal{F}_\Delta \to \mathcal{M}$ and $v : \mathcal{M} \to \mathcal{F}_\Delta$ defined above in 1) are dual order-isomorphisms between $\mathcal{F}_\Delta$ and the set of NSC on S.

3) Let us suppose S is a nonempty set and $L = \mathbb{R}_+$. It is clear that Generalised Dendrograms are generalised stratifications on S and so, by part d of theorem 2, they are level maps. Thus the maps $u : \mathcal{F}_u \to \mathcal{M}$ and $v : \mathcal{M} \to \mathcal{F}_u$ defined as above in 1) are dual order-isomorphisms between $\mathcal{F}_u$ and the set of Generalised Dendrograms on S.

In these three cases, the restrictions of the considered bijections to definite ultrametrics or dissimilarities ($f(a,b) = 0 \Rightarrow a = b$) on one hand, and to definite dendrograms (for which $m(0)$ is the diagonal of $S \times S$), on the other hand, are still dual order-isomorphisms.

Janowitz (1978) has shown that there exists a dual order-isomorphism between the set of dissimilarities defined on a finite set S with values in an ordered set L verifying the properties LMIN and JSL, and the set of residual maps in $\mathcal{N}$. This was the first introduction of such dissimilarities. Janowitz wished to explore the exact properties of $\mathbb{R}_+$ that are indispensable for Johnson's theorem.

Let $\mathcal{M}_R$ denote the set of residual maps $m : L \to \mathcal{P}(E)$. It is clear that $\mathcal{M}_R \subseteq \mathcal{M}$ and it is easy to check that $\mathcal{M}_R$ is not empty if and only if L verifies LMIN.

The following theorem presents the conditions under which $\mathcal{M} = \mathcal{M}_R$ holds. A map $m \in \mathcal{M}$ is said to be *join complete* (with respect to E and L) if :
MJC : $\forall D \subseteq E$, $D\emptyset$, $\vee\{f_m(e) : e \in D\}$ exists,
in which case it is denoted by $l_m(D)$.

Theorem 3   (a) *If LMIN, then a level map is residual if and only if it is join complete.*
  (b) *Thus* $\mathcal{M} = \mathcal{M}_R$, *that is E and L are adapted, if and only if*
    (i) *LMIN*
and
    (ii) $\forall m \in \mathcal{M}$, *m is join complete.*
  (c) *In particular,* $\mathcal{M} = \mathcal{M}_R$ *if*
    either (1) *E is finite and L is a join semi-lattice with a minimum element,*
    or   (2) *L is a complete lattice.* ∎

Let $E = S \times S$ where S is a finite set and suppose LMIN and JSL. Then part c1 of Theorem 3 implies that $\mathcal{M} = \mathcal{M}_R$. So the maps u and v formally defined as above for dendrograms or NSC are, in this case, dual order-isomorphisms between $\mathcal{F}_\Delta$ and $\mathcal{M}_R$.

Critchley and Van Cutsem (1991), following Janowitz (1978), give some results and details on the links between the map v and the residuated map associated to the residual one m.

5   Structures of type "prefilters"

The notion of prefilter was introduced in Critchley and Van Cutsem (1989). The idea was to look at the precise structure that can be associated to the set of all balls $B_f(a, l)$ defined by a predissimilarity f. They viewed these balls as defining a map $\phi : S \times \mathbf{R}_+ \to \mathcal{P}(S)$, where S is a nonempty set and gave necessary and sufficient conditions for the map ϕ to define a predissimilarity on S. The map ϕ has been called a prefilter.

We first look at the correspondence between maps $m \in \mathcal{N}$ and prefilters. It is interesting to begin with the case $E = A \times B$ instead of $E = S \times S$ to make clear the different role of the sets A and B.

Suppose $E = A \times B$ where A and B are two nonempty sets and let L be a nonempty ordered set. Let $a \in A$. For every function $f \in \mathcal{F}$ and map $m \in \mathcal{N}$, we define the a-sections $f_a : B \to L$ and $m_a : L \to \mathcal{P}(B)$ setting

$$f_a(b) = f(a, b) \quad \text{and} \quad m_a(l) = \{b \in B : (a, b) \in m(l)\}.$$

Let $\mathcal{F}_A$ denote the set of all families of functions from B to L indexed by A. We define the map $\sigma_A : \mathcal{F} \to \mathcal{F}_A$ by $\sigma_A(f) = \{f_a\}_{a \in A}$. Clearly σ_A is a bijection which induces a natural order on $\mathcal{F}_A$ by

$$\{f_a\}_{a \in A} \le \{f'_a\}_{a \in A} \Leftrightarrow f \le f'.$$

Similarly, let $\mathcal{M}_A = \{m_a : a \in A\}$ denote the set of all families of level maps from L to $\mathcal{P}(E)$ indexed by A. The map $\zeta_A : \mathcal{M} \to \mathcal{M}_A$ defined by $\zeta_A(m) = \{m_a\}_{a \in A}$ is a bijection which induces a natural order on $\mathcal{M}_A$ by

$$\{m_a\}_{a \in A} \le \{m'_a\}_{a \in A} \Leftrightarrow m \le m'.$$

As $L_{b,m_a} = L_{(a,b),m}$, each m_a is level by Proposition 1.

Definition 6 *A function* $\phi : A \times L \to \mathcal{P}(B)$ *is called a foliation if* $\forall a \in A$, $\cup\{\phi(a, l) : l \in L\} = B$. *The foliation* ϕ *is called a level foliation if each* $\phi(a,.) : L \to \mathcal{P}(B)$ *is level.* ∎

Let Φ_A denote the set of all level foliations. Clearly, Φ_A and $\mathcal{M}_A$ can be identified via $\phi(a, l) = m_a(l)$.

Let $u_A : \mathcal{F}_A \to \mathcal{M}_A \equiv \Phi_A$ denote the map defined by $u_A(\{f_a\}_{a \in A}) = \{\tilde{u}_B(f_a)\}_{a \in A}$ where $\tilde{u}_B$ is the map u of Theorem 1 in the case where $E = B$. Similarly, let $v_A : \mathcal{M}_A \to \mathcal{F}_A$ denotes the map defined by $v_A(\{f_a\}_{a \in A}) = \{\tilde{v}_B(m_a)\}_{a \in A}$ where $\tilde{v}_B$ is the map v of Theorem 1 in the case where $E = B$. We then have the proposition

Proposition 2 *Suppose $E = A \times B$. Then the following diagrams commutes.*

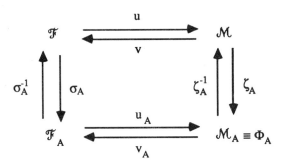

Moreover, the (compositions) of mappings shown establish natural
(a) order-isomorphisms between $\mathcal{F}$ and $\mathcal{F}_A$, and between $\mathcal{M}$ and $\mathcal{M}_A$.
(b) dual order-isomorphisms between the other four pairs of sets : $\mathcal{F}$ and $\mathcal{M}$, $\mathcal{F}$ and $\mathcal{M}_A$, $\mathcal{F}_A$ and $\mathcal{M}$, $\mathcal{F}_A$ and $\mathcal{M}_A$. ∎

The links with residual maps may be established in the following way.

Proposition 3 *Suppose $E = A \times B$ and LMIN. Then*
(a) $m \in M$ is residual $\Rightarrow \forall a \in A, m_a : L \to \mathcal{P}(B)$ is residual.
(b) $A \times B$ and L are adapted $\Rightarrow B$ and L are adapted. ∎

Critchley and Van Cutsem (1991) give some examples which show that the inverse implications of this proposition are not true.

We now come back to the case $E = S \times S$ and L verifies LMIN. We first recall the definition of a prefilter.

Definition 7 *Let S be a nonempty set . A prefilter on S is a function $\phi : S \times \mathbb{R}_+ \to \mathcal{P}(S)$ such that $\forall a \in S, \forall l \in \mathbb{R}_+$, verifies*
$(PF0)$ $a \in \phi(a, 0)$,
$(PF1)$ $\phi(a, l) = \cap\{\phi(a, l + \epsilon) : \epsilon > 0\}$,
$(PF2)$ $\cup\{\phi(a, l) : l \in \mathbb{R}_+\} = S$.
A prefilter is called definite if $\forall a \in S, \phi(a, 0) = \{a\}$, and symmetric if $b \in \phi(a, l)$ implies $a \in \phi(b, l)$. ∎

It is clear that, for a prefilter ϕ, the maps $\phi(a, .)$ are level. Thus by Proposition 2, the sets $\mathcal{F}_\Delta$ and $\mathcal{F}_u$ are dually order-isomorphic to the corresponding subsets of Φ_S. We give below characterisations of ultrametrics in terms of their corresponding reflexive level foliations (that is $m = \zeta_S^{-1}(\phi)$ is reflexive).

Theorem 4 *Suppose LMIN and $E = S \times S$. Let f be a L-predissimilarity on S and $\phi = (\zeta_S ou)f$ be the corresponding reflexive level foliation.*

(a) The following four propositions are equivalent.

    (1)   *f is an ultrametric on S.*

    (2)   *$\{b, c\} \subseteq \phi(a, l) \quad \Rightarrow \quad c \in \phi(b, l)$.*

    (3)   *for each $l \in L$, $a \overset{l}{\sim} b$ if $b \in \phi(a, l)$ defines an equivalence relation.*

    (4)   *$b \in \phi(a, l) \Rightarrow \phi(a, l) = \phi(b, l)$.*

(b) Suppose now that LTO holds. We define the following hierarchical condition on $Range(\phi)$:

    $(HIER)$ $\forall H \in Range(\phi), \forall H' \in Range(\phi), H \cap H' \in \{\emptyset, H, H'\}$.

Then the following seven propositions are equivalent :

    (1) *to* (4) *above and*

    (5)   *(i) HIER and (ii) $\phi(a, l) \subseteq \phi(a', l') \quad \Rightarrow \quad \phi(a', l') = \phi(a, l')$.*

    (6)   *(i) HIER and (ii) $\phi(a, l) \subset \phi(a', l') \quad \Rightarrow \quad l < l'$.*

    (7)   *(i) HIER and (ii) $b \in \phi(a, l) \quad \Leftrightarrow \quad a \in \phi(b, l)$.* ■

It is clear that this theorem summarises a lot of useful characterisations of ultrametricity often well known when S is finite and $L = \mathbb{R}_+$. Critchley and Van Cutsem (1991) reported many similar results in the case either LTO holds or f is symmetric or definite.

6 Structures of type "hierarchies"

We first give the definition of what Critchley and Van Cutsem (1991) called Benzécri Structures, which is the natural generalisation of classical hierarchical structures to the case where S is an arbitrary set and where the arbitrary ordered set L admitting a minimum element replaces $\mathbb{R}_+$. First we give the definition.

Definition 8 *Let S denote a nonempty set and L an ordered set admitting a minimum element. A Benzécri structure is a pair $(\mathcal{H}, g)$ where $\mathcal{H}$ is a collection of subsets of S and g is a function from $\mathcal{H}$ to L which satisfy the following conditions.*

    $H1$ $\forall (H, H') \in \mathcal{H}^2, H \cap H' \in \{\emptyset, H, H'\}$.

    $GH2$ $\forall a \in S, S = \cup\{H \in \mathcal{H} : a \in H\}$.

    $I1$ $\forall (H, H') \in \mathcal{H}^2, H \subset H' \Rightarrow g(H) < g(H')$.

    $R1$ $\forall (a, b) \in S^2, H_{ab} = \cap\{H \in \mathcal{H} : a \in H, b \in H\} \in \mathcal{H}$.

    $R2$ $\forall a \in S, \forall l \in L, H_a(l) = \{b \in S : g(H_{ab}) \le l\} \in \mathcal{H}$.

    $R3$ $\forall a \in S, \forall l \in L, g(H_a(l)) \le l$.

    $R4$ $\forall a \in S, g(H_{aa}) = 0$.

A Benzécri structure is called definite if $\forall a \in S, g(H_{aa}) = \{a\}$. ■

Whereas axioms $R1$ to $R3$ of a Benzécri structure are redundant when S is finite, Critchley and Van Cutsem (1991) give examples showing that they are all necessary in the case where S is not a finite set.

In the existing literature of classification we can find such analogous structures, the definition of which are recalled hereafter. The definition of Indexed Hierarchy is in Benzécri (1965, 1973) that of Indexed Regular Generalised Hierarchy is analogous to that of Critchley and Van Cutsem (1989).

Definition 9 *Let S denote a nonempty finite set. An Indexed Hierarchy is a pair $(\mathcal{H}, g)$ where $\mathcal{H}$ is a collection of subsets of S and g is a function from $\mathcal{H}$ to $\mathbb{R}_+$ which satisfy the following conditions.*

$H0$ $\emptyset \notin \mathcal{H}$.

$H1$ $\forall (H, H') \in \mathcal{H}^2, H \cap H' \in \{\emptyset, H, H'\}$.

$H2$ $S \in \mathcal{H}$.

$H3$ $\forall a \in S, \{a\} \in \mathcal{H}$.

$I1$ $\forall (H, H') \in \mathcal{H}^2, H \subset H' \Rightarrow g(H) < g(H')$.

$I2$ $g(H) = 0 \Leftrightarrow \exists a \in S$ such that $H = \{a\}$. ∎

Definition 10 *Let S denote a nonempty finite set. An Indexed Regular Generalised Hierarchy is a pair $(\mathcal{H}, g)$ where $\mathcal{H}$ is a collection of subsets of S and g is a function from $\mathcal{H}$ to $\mathbb{R}_+$ which satisfy the following conditions.*

$H0$ $\emptyset \notin \mathcal{H}$.

$H1$ $\forall (H, H') \in \mathcal{H}^2, H \cap H' \in \{\emptyset, H, H'\}$.

$GH2$ $\forall a \in S, S = \cup \{H \in \mathcal{H} : a \in H\}$.

$GH3$ $\exists$ a partition Π of S the subsets of which belong to $\mathcal{H}$ such that $\forall A \in \Pi$,

     (i) $A \in \mathcal{H}$,

     (ii) $(H \subseteq A$ and $H \in \mathcal{H}) \Rightarrow H = A$.

$I1$ $\forall (H, H') \in \mathcal{H}^2, H \subset H' \Rightarrow g(H) < g(H')$.

$GI2$ $g(H) = 0 \Leftrightarrow H$ belongs to the terminal partition Π of S in $\mathcal{H}$.

RH $\forall (a, b) \in S^2, H_{ab} = \cap \{H \in \mathcal{H} : a \in H, b \in H\} \in \mathcal{H}$.

$RH1$ $\forall a \in S, \forall l \in L, H_a(l) = \{b \in S : g(H_{ab}) \leq l\} \in \mathcal{H}$.

$RH2$ $\forall a \in S, \forall l \in L, g(H_a(l)) \leq l$.

The following proposition shows that these two structures are particular cases of Benzécri Structures.

Proposition 4 *Let $\mathcal{H}$ be a collection of subsets of the nonempty set S and let g be a function from $\mathcal{H}$ to L. Then :*

(a) If $L = \mathbb{R}_+$, $(\mathcal{H}, g)$ is an Indexed Regular Generalised Hierarchy $\leftrightarrow$ $(\mathcal{H}, g)$ is a Benzécri structure.

(b) If $L = \mathbb{R}_+$ and S is finite, $(\mathcal{H}, g)$ is an Indexed Hierarchy $\leftrightarrow$ $(\mathcal{H}, g)$ is a definite Benzécri structure. ∎

Before giving the theorem establishing the bijection between the sets of ultrametrics on S and the set of Benzécri Structures in the general case, we need to define a condition linking the sets S and L.

LSU. L and S are ultrametrically compatible. That is, L has a minimum element 0 and $\forall$ ultrametric f on S, $\forall a \in S, \forall l \in L, \{\tilde{l} \in L : \exists \tilde{a}$ such that $\phi(a, l) = \phi(\tilde{a}, \tilde{l})\}$ has a minimum, where ϕ denotes $(\zeta_S \circ u)f$.

Critchley and Van Cutsem (1991) establish that any one of the four following conditions : (1) S is finite, LMIN and JSL, (2) L is a complete lattice, (3) LUR and LWF (4) LTO and LWF is sufficient to the condition LSU be verifed.

142

Let $\mathcal{B}$ denotes the set of Benzécri Structures on the nonempty set S and an ordered set L admitting a minimum element. We define an order on $\mathcal{B}$ as follows.

$$(\mathcal{H}, g) \leq (\mathcal{H}', g') \Leftrightarrow \forall a \in S, \forall l \in L, H_a(l) \subseteq H'_a(l).$$

Let $\Phi_S^U \equiv (\zeta_S \circ u)\mathcal{F}_u$ be the set of all reflexive level foliations that correspond to ultrametrics. Whenever LSU holds, we can define the map $\alpha : \Phi_S^U \to \mathcal{B}$ by $\alpha(\phi) = (\mathcal{H}, g)$ where $\mathcal{H} = \text{Range}(\phi)$ and $g : \mathcal{H} \to L$ satisfies

$$\forall a \in S, \forall l \in L : g(\phi(a, l)) = \min\{\tilde{l} \in L : \exists \tilde{a} \text{ such that } \phi(a, l) = \phi(\tilde{a}, \tilde{l})\}.$$

We define the map $\beta : \mathcal{B} \to \Phi_S^U$ by $\beta(\mathcal{H}, g) = \phi$ where ϕ is given by

$$\forall a \in S, \forall l \in L, \phi(a, l) = H_a(l).$$

These two maps α and β allow us to write the bijection theorem.

Theorem 5 *Let S denote a nonempty set and L an ordered set verifying LMIN, LTO and LSU. Then*
(a) α and β define natural order-isomorphisms between Φ_S^U and $\mathcal{B}$.
(b) $(\alpha \circ \zeta_S \circ u)$ and $(v \circ \zeta_S^{-1} \circ \beta)$ define a natural dual order-isomorphism between $\mathcal{F}_u$ and $\mathcal{B}$.
(c) The relevant restrictions of these maps define natural (dual)- order-isomorphisms between the definite members of $\mathcal{F}_u$, Φ_S^U and $\mathcal{B}$ ∎

Critchley and Van Cutsem (1991) give some further results proving the existence of subdominants in this general framework.

7 Set-valued dissimilarities

The different definitions of the structures of "dendrograms", "prefilters" or "Benzécri Structures" which have been defined in the previous sections used two kinds of objects :

- a collection of subsets of a set (S or $S \times S$ depending on the case),

- a map g defined on this collection of subsets with values in the ordered set L having a minimum element.

It is possible to isolate in each case the properties of these collections of subsets. As an example of what can be done, we look at the definition and an ultrametric triangle property of set-valued hierarchies.

Definition 11 *Let S denote a nonempty set and $\mathcal{H}$ a family of nonempty subsets of S. We say that $\mathcal{H}$ is a set-valued hierarchy if the following properties are satisfied.*
SVH1 $\forall a \in S, S = \cup\{H \in \mathcal{H} : a \in H\}$,
SVH2 $\forall (H, H') \in \mathcal{H}^2, H \cap H' \in \{\emptyset, H, H'\}$.
SVH3 $\forall (a, b) \in S^2, H_{ab} = \cap\{H \in \mathcal{H} : a \in H, b \in H\} \in \mathcal{H}$
A hierarchy $\mathcal{H}$ is called definite if $\forall a \in S, H_{aa} = \{a\}$. ∎

Theorem 6 *Let S denote a nonempty set and $\mathcal{H}$ a collection of nonempty subsets of S verifying the two hypotheses SVH1 and SVH3.*

Let us consider the four propositions

   (1) $\forall (H, H') \in \mathcal{H}^2$, $H \cap H' \in \{\emptyset, H, H'\}$.

   (2) $\forall (a, b, c) \in S^3$, *two of the three subsets H_{ab}, H_{ac}, H_{bc} are equal and contain the third one. (that is the property of an ultrametric triangle)*

   (3) $\forall (a, b, c) \in S^3, \forall A \subseteq S,$

$$(H_{ab} \subseteq A, H_{bc} \subseteq A) \Rightarrow (H_{ac} \subseteq A).$$

   (4) $\forall (a, c) \in S^2$, $H_{ac} = \cap\{H_{ab} \cup H_{bc} : b \in S\}$.

Then we have the following implications $1 \Leftrightarrow 2 \Rightarrow 3 \Leftrightarrow 4$. ∎

Analogous definitions of set-valued level maps and set-valued level foliations can be given. It is clear that these two notions can be linked by some "sectioning techniques".

These first results on set valued dissimilarities will be extended in a further publication.

8 Conclusions

The bijections between sets of dissimilarities and structures such as dendrograms and others play a fundamental and indispensable role in the theory, methods and algorithms which underpin practical applications of mathematical classification. The direct benefit of the present approach is that it brings a unity to the domain and provides a number of further utilisations. It also open a wide range of methodological advances : such as the ability to treat multi-attribute dissimilarities, the possibility to study the asymptotics of hierarchical cluster analysis, the introduction of asymmetry and the facility to extend to multi-way data.

The introduction of set valued dissimilarities also leads to the introduction of nonparametric methods in cluster analysis.

References

BARTHÉLÉMY, J.P., LECLERC, B., MONJARDET, B., (1984), Ensembles ordonnés et taxonomie mathématique, *Annals of Discrete Mathematics, 23, 523–548.*

BENZÉCRI, J.P., (1965), Problmes et méthodes de la taxinomie, *Rapport de Recherche de l'Université de Rennes 1.*

BENZÉCRI, J.P., ET AL. (1973), *L'analyse des données I. La taxinomie,* Dunod, Paris.

BLYTH, T.S., JANOWITZ, M.F., (1972), *Residuation Theory,* Pergamon Press, Oxford.

CRITCHLEY, F., AND VAN CUTSEM, B., (1989), Predissimilarities prefilters and ultrametrics on an arbitrary set, *Joint research report of University of Warwick (G.B.) and of Laboratoire Modélisation et Calcul, Université Joseph Fourier, Grenoble (F)..*

CRITCHLEY, F., AND VAN CUTSEM, B., (1991), An order-theoretic unification and generalisation of certain fundamental bijections in mathematical classification, *Statistique et Analyse des Données, to appear.*

HARTIGAN, J. A., (1967), Representations of similarity matrices by trees, *J. Am. Statist. Ass., 62, 1140-1158.*

144

JANOWITZ, M. F., (1978), An order theoretic model for cluster analysis, *SIAM J. Appl. math.*, *34, 55-72.*

JARDINE, C. J., JARDINE, N. AND SIBSON, R., (1967), The structure and construction of taxonomic hierarchies, *Mathematical Biosciences, 1, 171-179.*

JARDINE, N. AND SIBSON, R., (1971), *Mathematical Taxonomy*, John Wiley, London.

JOHNSON, S. C., (1967), Hierarchical clustering schemes, *Psychometrika, 32, 241-254.*

A Comparison of Two Methods for Global Optimization in Multidimensional Scaling

P.J.F. Groenen

Department of Data Theory, University of Leiden
P.O. Box 9555, 2300 RB Leiden, The Netherlands
e-mail: groenen@rulfsw.LeidenUniv.nl

Abstract: One of the problems of metric multidimensional scaling defined by Kruskal's stress function, is that only convergence to a local minimum is guaranteed. Depending on the dissimilarity data and the chosen dimensionality, many local minima may exist. Here we aim at finding a global minimum of the stress function. Several general methods for global optimization exist, but we shall focus on two. The first one is the tunneling method, proposed by Montalvo (1979) and Gomez and Levy (1982), which aims at finding a decreasing series of local minima. Groenen and Heiser (1991) adapted and applied tunneling to multidimensional scaling. The second method is the multi-level-single-linkage clustering developed by Timmer (1984), which is a stochastic method based on properties of multistart using multiple random starts, and single linkage clustering. An implementation of this method to multidimensional scaling is discussed. Though the two methods aim at finding a global minimum, neither is guaranteed to find it in practice. A comparison is presented of the performance of the two methods in finding the global minima for three data sets. It turns out that both methods arrive at the same candidate global minimum for the three examples.

Keywords: multidimensional scaling, global optimization, tunneling, multi-level-single-linkage clustering.

1 Introduction

For some time it is known how to perform least squares multidimensional scaling MDS by minimizing Kruskal's STRESS function (Kruskal, 1964). The purpose of multidimensional scaling is to represent n objects in a p dimensional space in such a way that the interobject distances $d_{ij}(\mathbf{X})$ match as the given dissimilarities δ_{ij} as closely as possible. The STRESS function to be minimized is

$$\sigma^2(\mathbf{X}) \;=\; \frac{1}{2}\sum_{i,j}^{n} w_{ij}(\delta_{ij} - d_{ij}(\mathbf{X}))^2 \tag{1}$$

where $\mathbf{X}$ is the $n \times p$ matrix with coordinates and w_{ij} are non-negative weights. The SMACOF algorithm of De Leeuw (1977) and De Leeuw and Heiser (1980) minimizes STRESS and is proved to converge to a stationary point (see also Mathar, 1989). Although convergence is attained, nothing can be said whether the stationary point found is a global minimum. The main characteristic of a local minimum $\mathbf{X}^*$ as opposed to a global minimum is that for a local minimum within a small distance of $\mathbf{X}^*$ there exists no other point with smaller STRESS value than $\sigma^2(\mathbf{X}^*)$. For a global minimum $\mathbf{X}^*$ no other point with smaller STRESS value than $\sigma^2(\mathbf{X}^*)$ exists. Note that a global minimum configuration need not be unique; there may well exist other configurations with the same global minimum STRESS values. Finding a global optimum is generally a difficult task, unless the function exhibits special analytical properties. For a recent overview of global optimization methods see Törn and Žilinskas (1989).

This paper focuses on two methods that search for a global minimum of the STRESS function, i.e. the tunneling method and multi-level-single-linkage clustering (MLSL). Since these methods are not well known outside the field of global optimization, we discuss them in the next sections. Furthermore, we present for each method modifications that are aimed at improving the performance, or are necessary to apply them to MDS. Then both methods are compared by some numerical examples.

Before we continue one special case is excluded, i.e. uni-dimensional scaling ($p = 1$). In this case the problem of local minima is extra severe. For moderately sized problems the global optimum of unidimensional scaling may be obtained by the dynamic programming approach of Hubert and Arabie (1986). In the remaining part of this paper we only discuss the case of $p > 1$.

2   The tunneling method

The basic idea of the tunneling method for global optimization is proposed by Montalvo (1979) and Gomez and Levy (1982). The method alternates two steps of which the first one is a local search to find a local minimum. The second is the tunneling step, where another configuration is sought with the same STRESS value as obtained from the previous local search step. If such a configuration is found, a different local minimum configuration must exist with equal or smaller STRESS value, hence a horizontal tunnel is dug between two configurations with equal STRESS.

The following analogy may clarify tunneling method. Suppose we wish to find the lowest spot in a selected area in the Alps. First we pour some water and see where it stops. This is the local search step. From this point we dig tunnels horizontally until we come out of the mountain, the so-called global search. Note that we do not have to bother about the landscape (mountains and local minima) located above the tunnel. Then we pour water again and dig tunnels again. If we stay underground for a long time while digging the tunnel, we conclude that the last spot was in fact the lowest place in the area. An important and attractive feature of the tunneling algorithm is that successive spots are always lower.

The tunneling step is the crucial part of the tunneling method. In this step we search for a configuration $\mathbf{X}$ with $\sigma(\mathbf{X})$ equal to the STRESS value of configuration $\mathbf{X}^*$ obtained in the previous local search. This can be done by minimizing a special function, the so called tunneling function. This function has to meet several characteristics, some of which are specific for MDS. The first one is that it should have zero points if and only if $\sigma(\mathbf{X})$ equals $\sigma(\mathbf{X}^*)$. Additionally, these zero points are restricted to be the lowest function value of the tunneling function. Further, the solution $\mathbf{X}^*$ must be excluded as a zero point. Finally, we also exclude trivial solutions such as rotations, translations and reflections of $\mathbf{X}^*$. We have to exclude them due to the invariance of the STRESS function under these transformations of $\mathbf{X}^*$ since it does not affect the interobject distance $d_{ij}(\mathbf{X})$.

Let us build the tunneling function $\tau(\mathbf{X})$ in steps as to meet the requirements imposed above. The first requirement immediately suggests

$$\tau(\mathbf{X}) = (\sigma(\mathbf{X}) - \sigma(\mathbf{X}^*))^2 \qquad (2)$$

which guarantees that $\tau(\mathbf{X}) > 0$ and takes care of the appropriate zero points. Joining this with the requirement that a zero point at $\mathbf{X}^*$ and a trivial transformation of $\mathbf{X}^*$ must be excluded suggests to divide (2) by the factor $\psi(\mathbf{X}) = \frac{1}{2} \sum_{i,j}^n w_{ij}(d_{ij}(\mathbf{X}^*) - d_{ij}(\mathbf{X}))^2$, which tends to zero as $\mathbf{X}$ comes close to $\mathbf{X}^*$. The division by this factor does not change the other zero points, but does suggest large values of $\tau(\mathbf{X})$ as $\mathbf{X}$ comes close to $\mathbf{X}^*$ and thus exclusion of a zero point at $\mathbf{X}^*$. In other words, the division by $\psi(\mathbf{X})$ is intended to erect a pole for $\mathbf{X}$ that have the same distances between objects as $\mathbf{X}^*$. We return to this aspect in a moment. Working with distances between objects clearly removes zero points due to rotations, translations and reflections of $\mathbf{X}^*$, a problem which occurs if we use Montalvo's definition of the pole. Furthermore, this factor equals zero if and only if for all pairs i, j the difference between $d_{ij}(\mathbf{X}^*)$ and $d_{ij}(\mathbf{X})$ is zero. The tunneling function takes the form

$$\tau(\mathbf{X}) = \frac{(\sigma(\mathbf{X}) - \sigma(\mathbf{X}^*))^2}{\psi(\mathbf{X})}. \tag{3}$$

Although this definition of $\tau(\mathbf{X})$ seems to fulfill all our requirements, two new problems emerge. The first one concerns the strength of the pole. It turns out that we cannot be sure a priori that as $\psi(\mathbf{X})$ goes to zero $\tau(\mathbf{X})$ becomes non-zero (Groenen and Heiser, 1991). One way to solve this is to take a root of the numerator of $\tau(\mathbf{X})$ that is sufficiently strong. For some $0 < \lambda < 1$ the tunneling function

$$\tau(\mathbf{X}) = \frac{(\sigma(\mathbf{X}) - \sigma(\mathbf{X}^*))^{2\lambda}}{\psi(\mathbf{X})} \tag{4}$$

must have a pole that is strong enough. This immediately introduces a second problem, called attraction of the horizon. Although Levy and Gomez (1985) are aware of this problem, no clear-cut solution is given. For $\mathbf{X}$ at the horizon, that is if $\|\mathbf{X}\| \gg \|\mathbf{X}^*\|$, $\psi(\mathbf{X})$ tends to large values. Due to taking the root of the numerator of $\tau(\mathbf{X})$ it can be proved that (4) tends to zero. The attraction of $\mathbf{X}$ to the horizon is an undesirable feature, which may be solved by

$$\tau(\mathbf{X}) = (\sigma(\mathbf{X}) - \sigma(\mathbf{X}^*))^{2\lambda} \left(1 + \frac{1}{\psi(\mathbf{X})}\right). \tag{5}$$

This definition of the tunneling function (which is different from the one proposed by Montalvo (1979) and Levy and Gomez (1985)) meets all the requirements set above. The minimization of (5) is done by majorization. For details we refer to Groenen and Heiser (1991). This minimization algorithm only generates local minima, much in the same way as SMACOF does. However, we easily know when $\tau(\mathbf{X})$ is in a local minimum, because we are searching for zero points of $\tau(\mathbf{X})$. An additional pole is erected at a point were $\tau(\mathbf{X})$ has a local minimum by changing the pole factor of (5) into $(\prod_i^m (1 + 1/\psi_i(\mathbf{X}))^{1/m}$ with m the number of poles. In this paper we use a small alteration of the minimization procedure, which accelerates the convergence at the end of the tunneling step.

3 Multi-level-single-linkage clustering

The multi-level-single-linkage clustering algorithm (MLSL) for global optimization is a stochastic method with attractive theoretical properties developed by Timmer (1984) and Rinnooy Kan and Timmer (1987). It may be regarded as an improvement of the well-known multistart method. The latter amounts to drawing a (large) sample of random points and start a local search procedure from each sample point. The lowest function value is the candidate global minimum. Clearly, in most cases multistart is wasteful, because the same local minimum may be found many times. A better performance could be obtained if we were able to start a local search procedure only once for each *region of attraction* that belongs to a unique local minimum. This ideal is approximated by clustering algorithms. Various cluster methods are known in the literature, but according to Timmer (1984) an adapted version of single linkage is in the present context the most appealing one. Clustering is a technique that assigns points in a space to clusters in order to find groups of similar points (Hartigan, 1975). In this context we regard each $q = n \times p$ configuration matrix $\mathbf{X}$ to be a point in $\mathcal{R}^q$. Each cluster (and therefore the region of attraction) is defined by a seed point (here a local minimum configuration) and all points that are within a critical distance r_k of each other. As r_k becomes smaller over the iterations, the clusters approximate the region of attraction better. The single linkage clustering algorithm is known for the sausage shaped clusters it finds, the so called chaining effect. For our purpose it is a useful effect, since the level sets L_y, given by $\{\mathbf{X} \in \mathcal{R}^q | \sigma(\mathbf{X}) \leq y\}$, may well have this form. The method is defined over a compact and convex subset S of $\mathcal{R}^q$ which is assumed to contain the global minimum. Here we choose S to be a rectangular parallelepiped. The MLSL algorithm differs from ordinary single linkage clustering in that it takes into account the function value of the sample points, hence the name multi-level-single-linkage. The algorithm can be summarized as follows

1. Initialise. Set iteration counter k to zero.

2. (Global phase) Draw N configurations from a uniform distribution over S. Evaluate the STRESS value for each configuration. Add the points to the initially empty set X. Set k equal to $k + 1$.

3. (Local phase) Apply the local search procedure to every sample point $\mathbf{X}_i$ of X except if $\mathbf{X}_i$ is within a small distance ε of the boundary of S, or if there exists another $\mathbf{X}_j$ with $\sigma(\mathbf{X}_j) < \sigma(\mathbf{X}_i)$ within critical distance r_k.

4. Decide by means of a stopping rule to stop or return to 2.

Note that it is not necessary to actually keep track of the clusters. The only information that is needed for the decision to start a local search procedure is derived from the distance between some point $\mathbf{A}$ and the other points in the sample and their function values. MLSL depends critically on how r_k is defined. This should be done in such a way that eventually all local minima are found, including the global one, but not infinite many local search procedures are needed. Timmer (1984) proved rigorously that these requirements are met by deriving the critical distance from volume measures and probabilities. A discussion on how r_k could be chosen in case of MDS is presented here.

Let $B_{\mathbf{A},r}$ be the set of points $\mathbf{X}$ within critical distance r of $\mathbf{A}$ with $\mathbf{A}, \mathbf{X} \in S$ and $A_{\mathbf{A},r}$ is the set of points that are in $B_{\mathbf{A},r}$ and with $\sigma(\mathbf{X}) < \sigma(\mathbf{A})$. In the following the critical distance in iteration k is denoted interchangeably by r or r_k. We want to make sure that with increasing k the number of expected points within distance r and with smaller function value increases, so that eventually no local searches are performed. On the other hand r should be able to become arbitrary small so that no local minimum is overlooked.

Suppose we have kN sample points from a uniform distribution where $\mathbf{A}$ is one of them. The probability that none of the other points are within distance r of $\mathbf{A}$ with smaller function value is

$$\left(1 - \frac{m(A_{\mathbf{A},r})}{m(S)}\right)^{kN-1}, \tag{6}$$

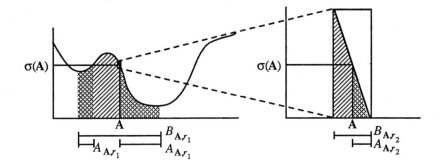

Figure 1: For decreasing r_k $m(A_{\mathbf{A},r})$ gets larger than $\beta m(B_{\mathbf{A},r})$ with $0 < \beta < \frac{1}{2}$.

where $m(S)$ is the Lebesque measure of S, a high-dimensional volume measure. We shall see that MLSL is defined in such a way that this probability tends to zero after a sufficient number of iterations. This implies that the probability of finding another point within distance r with lower function value goes to one, so that eventually no local search is started. However, given r we do not know a priori $m(A_{\mathbf{A},r})$. It turns out that a simple relationship exists between $m(A_{\mathbf{A},r})$ and $m(B_{\mathbf{A},r})$. It can be proven that as r decreases to 0, it holds that $m(A_{\mathbf{A},r}) \geq \beta m(B_{\mathbf{A},r})$ with $0 < \beta < \frac{1}{2}$ (see Timmer, 1984; Rinnooy Kan and Timmer, 1987). This can be seen intuitively by looking at Figure 1. For large r, $m(A_{\mathbf{A},r})$ can be larger than $\frac{1}{2}m(B_{\mathbf{A},r})$, as in the figure or smaller than $\frac{1}{2}m(B_{\mathbf{A},r})$. When zooming in, i.e. choosing r smaller and smaller, the surface of the function is approximated by a hyperplane. This holds for every distance measure that is symmetric around each of the q dimensions of $\mathbf{X}$ for $\mathbf{X}$ a non-stationary point. The points within distance r of $\mathbf{A}$ tend to be split in two equal parts, one with function values larger than $\sigma(\mathbf{A})$, the other half with smaller function values. It can be shown for STRESS that Timmer's theorem still holds, even though STRESS is not differentiable for every $\mathbf{X}$ in S. It allows us to express an upperbound of the probability (6) that no point is found in the sample set within distance r with smaller STRESS value:

$$\left(1 - \frac{m(A_{\mathbf{A},r})}{m(S)}\right)^{kN-1} \leq \left(1 - \frac{\beta m(B_{\mathbf{A},r})}{m(S)}\right)^{kN-1}. \tag{7}$$

The upperbound of this probability depends directly on $m(B_{\mathbf{A},r})$, which in turn depends directly on r for properly chosen distance measures. In the next section discusses how the distance measure is chosen for MDS.

Suppose that $B_{\mathbf{A},r}$ is entirely contained in S. Then the expected number of sample points within distance r_k is $kNm(B_{\mathbf{A},r})/m(S)$. By setting this equal to $\gamma \ln kN$, with γ a positive fixed constant, the density of points in $B_{\mathbf{A},r}$ gets higher with increasing number of iterations. As r becomes smaller, the number of points within distance r of $\mathbf{A}$ increases, and so does the number points within distance r with smaller STRESS values. Consequently, the probability that a local search is started from $\mathbf{A}$ goes to zero.

3.1 The curse of dimensionality

In MDS the size of $\mathbf{X}$ is dependent on the number of objects n and the number of coordinates p by which each object is represented. Even for moderately sized n and p the dimensionality of the domain of STRESS, q, becomes very large. This implies that almost every $\mathbf{X}$ is located closely to one of the boundaries of S, which we assume for the moment to be a hypercube. This may be seen by comparing the contents of a hypercube with the contents of a hyperball that fits exactly in the cube. For higher dimensions the difference between the two rises rapidly. This difference,

$$\left(2^q - \frac{\pi^{-q/2}}{\Gamma(1 + \frac{q}{2})}\right) r^q, \tag{8}$$

where $\Gamma(x)$ is the gamma function, rapidly tends to the volume of a hypercube $(2r)^q$ as q gets larger. Almost all sample points are expected to be located in one of the corners of S and thus close to one of its boundaries. Consequently, much of $B_{\mathbf{A},r}$ falls outside S, which implies that for relative large r a local search will be started at almost every sample point. To reduce the number of local searches, we would like $B_{\mathbf{A},r}$ to be always located entirely inside S to ensure that regardless of r the expected number of sample points within distance r of other sample points equals $\gamma \ln kN$. This may be obtained by moving $B_{\mathbf{A},r}$ to $B_{\mathbf{C},r}$ where $\mathbf{C}$ is as close as possible to $\mathbf{A}$ and $B_{\mathbf{C},r}$ is entirely contained in S. Note that only $B_{\mathbf{A},r}$ is moved, not the sample point $\mathbf{A}$ itself. Moving $B_{\mathbf{A},r}$ is only needed for points that are within distance r of the boundary. If $r < \varepsilon$ no local search is started for such points by definition, thus we may skip moving $B_{\mathbf{A},r}$ in this case too.

Timmer (1984) uses the Euclidean distance, so that $B_{\mathbf{A},r}$ defines a hypersphere. However, moving a hypersphere inside a hypercube may give the problem that $B_{\mathbf{C},r}$ does not contain $\mathbf{A}$ if $\mathbf{A}$ is too close to a corner of the hypercube. This situation applies particularly for relative large r and high dimensionality, conform (8). Therefore, we change the form of $B_{\mathbf{A},r}$ into a hypercube with center $\mathbf{A}$ and sides of length $2r$. Using moved hypercubes assures that irrespective of the dimensionality q, the number of expected points within distance r of $\mathbf{A}$ equals $\gamma \ln kN$.

Finally we are able to derive the critical distance r from

$$kN \frac{m(B_{\mathbf{A},r})}{m(S)} = \gamma \ln kN,$$

$$m(B_{\mathbf{A},r}) = m(S) \frac{\gamma \ln kN}{kN},$$

$$(2r)^q = m(S) \frac{\gamma \ln kN}{kN},$$

$$r = \frac{1}{2}\left(m(S)\frac{\gamma \ln kN}{kN}\right)^{1/q}. \tag{9}$$

Using moved hypercubes gives a larger reduction in the number of local searches in high dimensionality during first iterations without destroying the theoretical properties of MLSL.

The issue of invariance of STRESS under translation, rotation and reflection appears here too. Since the MLSL method searches for all local minima the trivial ones are also found. It can be solved by constraining the configuration matrix. Translation can be identified by fixing one point of the configuration matrix $\mathbf{X}$ at the origin, say the first row $\mathbf{x}_1'$. The rotation freedom is constrained by letting the second row $\mathbf{x}_2'$ vary along the first axis only and setting the remaining coordinates fixed to zero. For p larger than 2 we free the coordinates of the third row on the first two axes only and fixing the rest to zero, etc. Thus a configuration matrix $\mathbf{X}$ with 3 dimensions is constrained to be

$$\mathbf{X} = \begin{bmatrix} 0 & 0 & 0 \\ x_{21} & 0 & 0 \\ x_{31} & x_{32} & 0 \\ x_{41} & x_{42} & x_{43} \\ \vdots & \vdots & \vdots \\ x_{n1} & x_{n2} & x_{n3} \end{bmatrix}. \tag{10}$$

The reflection indeterminacy is resolved by restricting the first subdiagonal elements (x_{21}, x_{32} and x_{43} in the $p = 3$ dimensions example above) to be non-negative. The constraints reduce the number of free parameters q to $(n-1)p - \frac{1}{2}p(p-1)$.

Throughout the numerical experiments S is rectangular parallelepiped, which means that each of the q axes may have a different scale. It is determined so that the global minimum is inside S.

4 Numerical experiments

Three different examples are presented to compare different aspects of the two methods. The examples have a decreasing region of attraction to the global minimum, different q and one has a global minimum of zero. We present the results of MLSL with hypercubes, with moved hypercubes and with tunneling for each examples. Throughout the numerical experiments the dissimilarities are standardized to have sum of squares $\frac{1}{2}n(n-1)$ and the configurations are of fixed dimensionality $p = 2$.

The first example is the 4×4 constant dissimilarity matrix extensively discussed by De Leeuw (1988). He notices four types of stationary points, two of them in $p = 2$: a configuration with three points at the corners of an equilateral triangle and the remaining one at the centroid ('triangle') and a configuration with four points at the corners of a square ('square'). MLSL was started with the triangle configuration. 4300 random configurations were drawn with $N = 100$, which induced a total of 124 local searches. With the exception of the first local search, all resulted in the square configuration, albeit in the three different forms given in Figure 2. Only one triangle was found by MLSL due to the initial configuration;

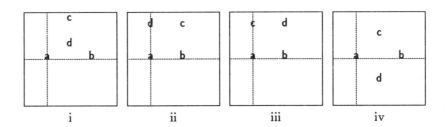

Figure 2: Four different local minima of the constant dissimilarity matrix found by MLSL.

the two others were not found. De Leeuw (1988) calls this an isolated stationary point. The square configuration has the least STRESS and a very large region of attraction, which suggests that it is the global minimum in two dimensions. Repeating the experiment, with moved hypercubes and the same sample size, decreased the number of local searches to 52 yielding the same local minima.

Tunneling was also started with configuration i of Figure 2 and with $\lambda = 1/3$. The start configuration of each tunneling step is a small perturbation of the previous local minimum. In 85 iterations the tunneling step yielded a configuration with the same STRESS as initial configuration and the subsequent local search yielded configuration ii of Figure 2. The next tunneling step needed 307 iterations to emerge from the tunnel, after which the local search reached configuration iii of Figure 2. When both of these configurations are inserted as poles in $\tau(\mathbf{X})$ then tunneling is able to find the remaining square configuration.

In the second example, the dissimilarity matrix is the Euclidean distance matrix of a configuration of an equally spaced grid of nine points. It shows what kind of local minima may be expected when the dissimilarities are Euclidean. 5800 configurations were obtained from the uniform distribution and 4070 local searches were started, which yielded 44 different local minima. Because of the structure of the data, only 6 different species of local minima were found as shown in Figure 3, each species having one or more configurations that differ only in labeling of the points, not in the form of configuration. Almost 72% of the local searches ended in the the global minimum indicating that it has a large region of attraction. With moved hypercubes, only 689 local were started from 5800 samples points, which led to 5 species of 22 local minima. Except for configuration iii, the same species were found. Tunneling was started from the worst fitting configuration (vi of Figure 3). After two tunneling steps it found the equally spaced grid of nine points.

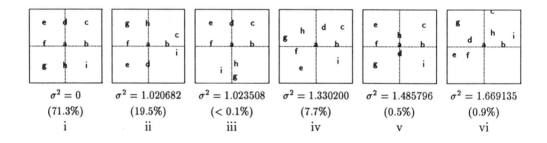

$\sigma^2 = 0$ $\sigma^2 = 1.020682$ $\sigma^2 = 1.023508$ $\sigma^2 = 1.330200$ $\sigma^2 = 1.485796$ $\sigma^2 = 1.669135$

(71.3%)   (19.5%)   ($< 0.1\%$)   (7.7%)   (0.5%)   (0.9%)

i   ii   iii   iv   v   vi

Figure 3: The species of local minimum configurations obtained by MLSL of the Euclidean distance matrix of configuration i. Also given is their STRESS and the percentage of random points whose local search ended in this configuration.

The final example concerns genetic dissimilarities of bacterial strains, which are reported in Mathar (1989) after experiments of Ihm (1986). Mathar (1991) noted that he obtained many different local minima using MDS on this data set, which makes it an interesting data set for the comparison of the two global optimization methods. Bacterial strains were investigated with respect to their phylogenetic similarity. DNA-sequences (chromosomes) of each two out of 17 different species have been brought together, heated and then annealed. If species are similar then a large portion of equal parts of the individual DNA-sequences are hanging together in some way. This can be measured and gives the so called "percent-binding", which is a well known term for biologists. The raw data were transformed to dissimilarities in the following way. If species i and j have a percent binding of $a\%$, say, then $\delta_{ij} = 1 - a/100$.

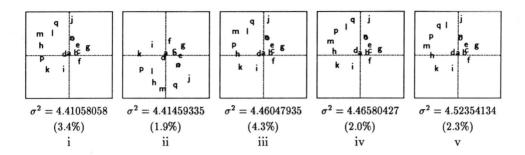

$\sigma^2 = 4.41058058$ $\sigma^2 = 4.41459335$ $\sigma^2 = 4.46047935$ $\sigma^2 = 4.46580427$ $\sigma^2 = 4.52354134$

(3.4%)   (1.9%)   (4.3%)   (2.0%)   (2.3%)

i   ii   iii   iv   v

Figure 4: The five best configurations of the genetic dissimilarity data obtained by MLSL, their STRESS and the percentage of random points whose local search ended in this configuration.

The initial configuration of MLSL was the classical scaling solution (Torgerson, 1958; Gower, 1966) with STRESS 20.04914295. A local search starting from this solution resulted in a local minimum with STRESS 4.46047935. Of 3200 random points that were drawn for MLSL, 3084 local searches were started that yielded 1098 different local minima. Although the convergence criterion of the local search procedure was strong (10^{-8}), we should not exclude that some of the 1098 minima may be regarded as the same. An histogram of the number of local minima with a certain STRESS value is presented in Figure 5. It shows that the majority of the local minima have rather high STRESS values. For this data set a global optimization method is clearly needed. In Figure 4 the five configurations with the lowest STRESS are given which represents the local minima of 14% of the local searches. The differences between the configurations are small. Due to the restrictions, configuration ii is more or less reflected along the second axis when compared to i, but the main difference is the positioning of objects h, l and m. Comparing i with iii, objects p and h are interchanged; i with iv, the difference is in objects p, h and m; i with v, the difference is in both the objects p, h, m and q, l. For this data set different local minima are formed by more or less interchanging the position of a few points. The experiment was repeated for moved hypercubes. With 6200 random points 29 local searches were needed, that resulted in 27 local minima, the best two being i and iii of Figure 4. Although only a very few local searches were started the assumed global minimum was found. The tunneling method was started with the worst local minimum found by MLSL and λ was set to 1/3. After a series of 15 decreasing local minima and at most 7 poles, the tunneling method found configuration i of Figure 4. In fact, the last five local minima of tunneling coincided with the five configurations of Figure 4.

The differences in the number of local searches of MLSL without moved hypercubes seems to depend on the dimensionality. In the first example relatively few local searches had to be performed as compared to the last two examples. Using moved hypercubes greatly reduced the number of local searches for all three examples. Tunneling was successful in all examples, even though the last example had a small region of attraction.

154

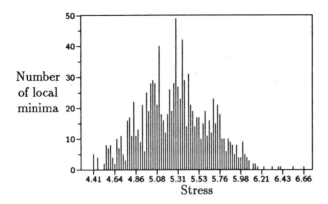

Figure 5: Histogram of the number of local minima found with a certain STRESS value of the genetic dissimilarity data.

5 Conclusions

Two methods for global optimization, i.e. tunneling and MLSL, were adapted to gain efficiency and to make them suitable for MDS. Both methods found the same candidate global minimum in the examples discussed. Each of the two methods for finding a global minimum in MDS has attractive and unattractive features, but none seems to be superior on all aspects. An advantage of the tunneling method is that it searches for a series of lower minima and disregards higher local minima. However, no guarantee can be given that the global minimum actually has been reached or that a longer tunneling step is needed, although the latter was not confirmed in the examples. MLSL has attractive theoretical properties for reaching the global minimum, apart from its algorithmic simplicity. However, this is a stochastic result which holds only for a large number of sample points. The high number of parameters in MDS constitutes a problem for both methods. The tunneling step may take a very long time as was noticed for the bacterial strains example. For high dimensionality MLSL reduces to plain multistart, loosing its efficiency even for moderately sized data sets. We found in the examples that using moved hypercubes greatly reduced the number of local searches for high dimensionality and resulted in the same candidate global minimum as ordinary MLSL. Note that MLSL has a storage problem which becomes extra severe for high dimensionality. An important aspect determining the succes of global optimization methods for STRESS is the structure of the dissimilarities, which may or may not have a large attraction to the global minimum configuration. Two examples showed that the structure of the dissimilarities can induce species of local minima that differ in labeling only, not in form.

For given dissimilarities it seems wise to use a few random start configurations to decide whether there are many different local minima, or not. If so, then both the tunneling method or MLSL may be used to search for the global minimum. If one wishes to compare all local minima then MLSL should be used, if one is interested only in the global minimum then tunneling can be considered.

References

DE LEEUW, J.(1977), Applications of convex analysis to multidimensional scaling. In: J. Barra et al. (Eds.) *Recent developments in statistics*, 133-145, Amsterdam: North-Holland.

DE LEEUW, J.(1988), Convergence of the majorization method for multidimensional scaling. *Journal of Classification*, *5*, 163-180.

DE LEEUW, J., HEISER, W.J.(1980), Multidimensional scaling with restrictions on the configuration. In: P.R. Krishnaiah (Ed.), *Multivariate analysis V*, 501-522, Amsterdam: North-Holland.

GOMEZ, S., LEVY, A.V. (1982), The tunneling method for solving the constrained global optimization problem with non-connected feasible regions. *Lecture notes in mathematics*, *909*, 34-47, Berlin: Springer-Verlag.

GOWER, J.C. (1966), Some distance properties of latent roots and vector methods used in multivariate analysis. *Biometrika*, *53*, 325-338.

GROENEN, P.J.F., HEISER, W.J. (1991), *An improved tunneling function for finding a decreasing series of local minima*. Internal report RR-91-06, Leiden: Department of Data Theory.

HARTIGAN, J.A. (1975), *Clustering algorithms*. New York: Wiley.

HUBERT, L.J., ARABIE, P. (1986), Unidimensional scaling and combinatorial optimization. In: J. De Leeuw, W.J. Heiser, J. Meulman and F. Critchley (Eds.), *Multidimensional Data Analysis*, 181-196, Leiden: DSWO Press.

IHM, P. (1986). *A problem on bacterial taxonomy*. Working paper, EURATOM: Ispra.

KRUSKAL, J.B. (1964), Nonmetric multidimensional scaling: a numerical method. *Psychometrika*, *29*, 115-129.

LEVY, A.V., GOMEZ, S. (1985), The tunneling method applied to global optimization. P.T. Boggs, R.H. Byrd, R.B. Schnabel (Ed.), *Numerical optimization 1984*, 213-244, Philadelphia: SIAM.

MATHAR, R. (1989), Algorithms in multidimensional scaling. In: O. Opitz (Ed.), *Conceptual and numerical analysis of data*, Berlin: Springer Verlag.

MATHAR, R. (1991), Personal communication.

MONTALVO, A. (1979), *Development of a new algorithm for the global minimization of functions*. Ph.D. thesis in Theoretical and applied Mechanics, Universidad Nacional Autonoma de Mexico.

RINNOOY KAN, A.H.G., TIMMER, G. T. (1987), Stochastic global optimization methods, Part I: clustering methods. *Mathematical Programming 39*, 27-56.

TIMMER, G.T. (1984), *Global optimization: a stochastic approach*. Ph.D. thesis, Rotterdam: Erasmus University.

TORGERSON, W.S. (1958), *Theory and methods of scaling*. New-York: Wiley.

TÖRN, A., ŽILINSKAS, A. (1989), Global optimization. G. Goos and J. Hartmanis (Eds.), *Lecture notes in computer science*, Vol. 350, Berlin: Springer-Verlag.

Directional Analysis of Three-Way Skew-Symmetric Matrices

Berrie Zielman

Department of Data Theory, University of Leiden

P.O. Box 9555, 2300 RB Leiden, The Netherlands

e-mail: zielman@rulfsw.LeidenUniv.nl

Abstract: Asymmetry is a problem in multidimensional scaling because these models predict symmetric values. A common way to solve this problem is to average the observations across the diagonal and then analyze the symmetrized component. In this paper a method for analyzing the departures from symmetry is extended to handle three-way tables. The departures from symmetry are skew-symmetric and this property is reflected in the model as a direction. An alternating least squares algorithm is presented for estimating the parameters of the model.

Keywords: multidimensional scaling, asymmetry, singular value decomposition, INDSCAL, three-way methods

1 Introduction

Tables where the rows and columns classify the same set of n objects occur frequently in the social sciences. The objects or stimuli can be anything; for example personality traits, soft drinks or occupational categories. In this paper we assume that the entries δ_{ij} ($i=1,...n$; $j=1,...n$) in the table Δ indicate a proximity relation between the objects. Multidimensional scaling (MDS) is a potential candidate for the analysis of proximity relations. Broadly defined MDS is a family of techniques that represent the objects as points in a low-dimensional space in such a way that the distances between the points correspond as closely as possible to the proximities. Usually the Euclidean distance function is computed, although other distance functions are possible. There may be a problem with distance analysis: the Euclidean distance is a symmetric function whereas the entries in the table can be asymmetric. A table is asymmetric if some of the elements in the upper-triangle differ from their corresponding elements in the lower-triangle.

The possibility of decomposing such a table into a symmetric component and a component indicating the departure from symmetry is a potential strategy for analysis. The decomposition of the table is:

$$\Delta = S + A, \tag{1}$$

where the matrix $S = \{s_{ij}\} = [\delta_{ij} + \delta_{ji}]/2$ is symmetric. The matrix $A = \{a_{ij}\}$ with elements $a_{ij} = [\delta_{ij}\text{-}\delta_{ji}]/2$ describes the departures from symmetry and is skew-symmetric: $a_{ij} = -a_{ji}$. The symmetric part can be analyzed by a symmetric model like MDS or cluster analysis; the matrix that contains the departures from symmetry can be studied by singular value decomposition (SVD). Gower (1977) and Constantine and Gower (1978) studied the singular value decomposition of a skew-symmetric matrix. The main result of their papers is that a display of the singular vectors should be interpreted in terms of areas and collinearities.

Suppose there are multiple tables $\mathbf{\Delta}_1, ..., \mathbf{\Delta}_m$ available, where the different tables correspond to subjects, experimental conditions, points in time, or other sources. The decomposition $\mathbf{\Delta} = \mathbf{S} + \mathbf{A}$ can be applied to each table $\mathbf{\Delta}_k$ separately and individual differences models, such as INDSCAL and IDIOSCAL (Carroll and Chang, 1971; Carroll, 1972) can be applied to the symmetric tables. An individual differences model describes relations between proximity-sources for the individuals by linear transformations of a common-space. In scalar product form the INDSCAL model for symmetric data can be written as:

$$\mathbf{S}_k = \mathbf{T}\mathbf{R}_k\mathbf{T}', \tag{2}$$

where $\mathbf{T}$ is an n by p matrix with coordinates, where p denotes the number of dimensions and $\mathbf{R}_k$ is a diagonal matrix with dimension weights, that indicate the relative importance or salience of the dimensions for replication k ($k=1,...,m$), where m is the number of sources. The IDIOSCAL model has the same structure, except for the matrix $\mathbf{R}_k$ which is now an p by p matrix of full rank.

No individual differences models seems to be available for skew-symmetric data. In this paper an individual differences model for three-way skew symmetric matrices is proposed where the common-space can be interpreted in terms of areas and collinearities, and the weights indicate the importance of a plane for a particular data source.

2 Singular value decomposition of skew-symmetric matrices

In this section we will discuss some properties of the singular value decomposition (SVD) of a skew-symmetric matrix. These properties are needed for interpreting diagrams derived from such analysis since the properties of the data displayed by these diagrams are different from a symmetric analysis. The departures from symmetry can be analyzed by computing the singular value decomposition of the skew-symmetric matrix $\mathbf{A}$. The SVD of any matrix $\mathbf{B}$ can be written as:

$$\mathbf{B} = \mathbf{K}\mathbf{\Phi}\mathbf{L}', \tag{3}$$

where $\mathbf{K}'\mathbf{K} = \mathbf{L}'\mathbf{L} = \mathbf{I}$ and $\mathbf{\Phi}$ is a diagonal matrix containing the singular values in descending order. The matrices $\mathbf{K}$ and $\mathbf{L}$ contain the left and right singular vectors. The SVD of a skew symmetric matrix can be written as:

$$\mathbf{A} = \mathbf{K}\mathbf{\Phi}\mathbf{J}\mathbf{K}', \tag{4}$$

where $\mathbf{J}$ is a block diagonal matrix with block matrices of order two by two, with one in above the diagonal, and minus one below the diagonal.

The matrix $\mathbf{K}$ containing the left singular vectors are rotations of the right singular vectors. The singular values of a skew symmetric matrix come in pairs: $\phi_1 = \phi_2$; $\phi_3 = \phi_4$; $\phi_{n-1} = \phi_n$. If n is odd the last singular value is zero. An alternative representation in terms of pairs of singular vectors that will be helpful for developing a three-way variant of the model is:

$$\mathbf{A} = \sum_{s=1}^{n/2} \phi_1(\mathbf{z}_s\mathbf{y}_s' - \mathbf{y}_s\mathbf{z}_s') \tag{5}$$

where the vector **z** denotes the sth column of **K** and the vector **y** denotes the $s+1$ column of **K**. This notation absorbs the matrix **J** into the difference of the two rank one matrices. From this last equation we see that the representation is invariant under scaling of pairs of axis by λ and $1/\lambda$, a result that can also be found in Gower (1977).

Gower has pointed out that diagrams obtained from an SVD by plotting subsequent pairs of dimensions found in the columns of **K** be interpreted in terms of areas and collinearities. A hypothetical example is provided in Figure 1 where four objects A, B, C, D are depicted. The skew-symmetry between A and B equals twice the area of the triangle with vertices OA and OB, this area is dark shaded. In this diagram skew-symmetry is modelled by a direction, hence the name directional plane. If we go counter clockwise, that is from B to A, the area is positive and if we go clockwise from A to B the area is negative, thus modelling skew-symmetry. The points B an D are collinear on a line with the origin giving zero area, this corresponds to symmetry between objects B and D.

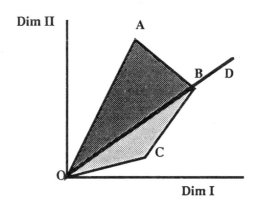

Figure 1: Interpretation of a directional plane in terms of areas

A special case of this model arises when all points are on a straight line, in this particular case the areas are proportional to the basis of the triangles. This gives a linear form of skew-symmetry that can be written as:

$$\mathbf{A} = \mathbf{e}\mathbf{y}' - \mathbf{y}\mathbf{e}', \qquad (6)$$

where one singular vector is constrained to a vector of ones. This linear model is also known under the name Thurstone Case-V model and is equivalent to the parameters to accommodate asymmetry of the Weeks and Bentler (1982) scaling model.

3 Directional planes representing individual difference

In multidimensional scaling, methods for representing individuals are very popular, in this section an INDSCAL representation will be developed for skew-symmetric data tables. The skew-symmetric generalization of the INDSCAL model that we wish to propose here is:

$$\mathbf{S}_k = \mathbf{TR}_k\mathbf{JT}',\tag{7}$$

where $\mathbf{T}$ is an n by p matrix with coordinates and $\mathbf{R}_k$ is a diagonal matrix with dimension weights coming in pairs, that indicate the relative importance or salience of the dimensions for replication k. The matrix $\mathbf{J}$ takes care of the skew-symmetry of the model and has the same block-diagonal structure as for the two-way model. Obviously the model is an INDSCAL generalization for skew-symmetric matrices, the only difference between INDSCAL and the skew-symmetric variant is the matrix $\mathbf{J}$.

As the sum of q pairs of dimensions the three-way model can be written as:

$$\mathbf{A}_k = \sum_{s=1}^{n/2} u_{ks}(\mathbf{z}_s\mathbf{y}_s' - \mathbf{y}_s\mathbf{z}_s')\tag{8}$$

where the matrix $\mathbf{J}$ is absorbed in the new notation, $\mathbf{z}$ and $\mathbf{y}$ are parameter vectors playing the same role as in the SVD, and the scalar u_{ks} denotes the regression weight indicating the relative importance or salience of the skew-symmetry modeled by the sth directional plane for the kth source.

4 Estimation of the three-way variant

The parameters of the model are estimated by Alternating Least Squares (ALS), the principle is that parameters are divided in subsets and that the parameters of one subset are improved keeping the other subsets fixed. Instead of estimating row and column parameters and the weights separately, as is usually done in ALS-algorithms, the present algorithm estimates in one substep the first dimension for rows and the second dimension for columns and in the second substep the first dimension for columns and the second dimension for rows. The model (1) can be rewritten by using the Kronecker product and Vec notation as

$$\mathrm{Vec}\mathbf{A}_k = \sum_{s=1}^{n/2} u_{ks}P(\mathbf{y}_s)\mathbf{z}_s\tag{9}$$

where Vec (.) transforms a matrix into a vector by stacking the columns of the matrix one underneath the other. The matrix $P(\mathbf{y}_s)$ is built up as follows:

$$P(\mathbf{y}_s) = (\mathbf{y}_s \otimes \mathbf{I}) - (\mathbf{I} \otimes \mathbf{y}_s),\tag{10}$$

where $\mathbf{I}$ is the identity matrix of order n; the symbol $\otimes$ denotes the Kronecker product of matrices; for example $A \otimes B = [a_{ij}\mathbf{B}]$. If the $\mathbf{y}_s$ vector is constrained to be a vector of unity's the three-way vector model of Zielman (1991) is obtained. Alternatively, with the design matrix being a function of the vector $\mathbf{z}_s$, the model can be written as:

$$\mathrm{Vec}\mathbf{A}_k = -\sum_{s=1}^{n/2} u_{ks}P(\mathbf{z}_s)\mathbf{y}_s\tag{11}$$

This notation shows that in an Alternating Least Squares (ALS) algorithm the parameters $\mathbf{z}_s$ and $\mathbf{y}_s$ can be found by the same operator, another advantage is that the importance weights automatically come in pairs.

The parameters are estimated by a paired dimension wise strategy, for each pair of dimensions each pair v of dimensions we subtract the contribution of the previous pairs of dimensions by computing the following quantities:

$$\underline{a}_{ijk} = a_{ijk} - \sum_{s<v}^{n/2} u_{ks}(\mathbf{z}_s\mathbf{y}'_s - \mathbf{y}_s\mathbf{z}'_s). \tag{12}$$

Collecting these quantities in the matrix $\mathbf{A}_k$, and keeping the other possible pairs of dimensions fixed, the following least squares loss function is minimized for each pair of dimensions:

$$L(u_{ks}; y_s; z_s) = \frac{1}{m}\sum_{k=1}^{m}(\mathrm{Vec}\underline{\mathbf{A}}_k - u_{ks}P(\mathbf{y}_s)\mathbf{z}_s)'\mathbf{M}_k(\mathrm{Vec}\underline{\mathbf{A}}_k - u_{ks}P(\mathbf{y}_s)\mathbf{z}_s), \tag{13}$$

where the matrix $\mathbf{M}_k$ is a diagonal matrix of order n^2 by n^2 with masses indicating the relative importance of the data values; if some of these masses are made zero the loss function value does not depend on the corresponding data elements; this provides us with a convenient tool for dealing with missing data. The update of the parameters $\mathbf{z}_s$ while keeping the other parameter sets fixed can be found by:

$$\mathbf{z}_s = [P'(\mathbf{y}_s)\sum_{k=1}^{m}\mathbf{M}_k u_{ks}^2 P(\mathbf{y}_s)]^{-}P'(\mathbf{y}_s)\sum_{k=1}^{m}\mathbf{M}_k\mathrm{Vec}\underline{\mathbf{A}}_k u_{ks}. \tag{14}$$

The matrix $[P'(\mathbf{y}_s)\sum_{k=1}^{m}\mathbf{M}_k u_{ks}^2 P(\mathbf{y}_s)]^{-}$ denotes the generalized inverse of the matrix $P'(\mathbf{y}_s)\sum_{k=1}^{m}\mathbf{M}_k u_{ks}^2 P(\mathbf{y}_s)$. This generalized inverse can be calculated as:

$$[P'(\mathbf{y}_s)\sum_{k=1}^{m}\mathbf{M}_k u_{ks}^2 P(\mathbf{y}_s) + \mathbf{y}_s(\mathbf{y}'_s\mathbf{y}_s)^{-1}\mathbf{y}'_s]^{-1} - \mathbf{y}_s(\mathbf{y}'_s\mathbf{y}_s)^{-1}\mathbf{y}' \tag{15}$$

where the vector $\mathbf{y}_s$ spans the null-space of the matrix. By using the same operator but with $\mathbf{y}_s$ playing now the role of $\mathbf{z}_s$ we find an improvement of the parameters $\mathbf{y}_s$.

$$\mathbf{y}_s = -[P'(\mathbf{z}_s)\sum_{k=1}^{m}\mathbf{M}_k u_{ks}^2 P(\mathbf{z}_s)]^{-}P'(\mathbf{z}_s)\sum_{k=1}^{m}\mathbf{M}_k\mathrm{Vec}\underline{\mathbf{A}}_k u_{ks}. \tag{16}$$

Assuming the inverse exists, the importance weights uks can be found by:

$$u_{ks} = [\mathbf{y}'P'(\mathbf{z}_s)\mathbf{M}_k P(\mathbf{z}_s)\mathbf{y}]^{-1}\mathbf{y}'P'(\mathbf{z}_s)\mathbf{M}_k\mathrm{Vec}\underline{\mathbf{A}}_k. \tag{17}$$

The algorithm alternates between subsets of parameters u_{ks}, $\mathbf{y}_s$, and $\mathbf{y}_s$ within each dimension loop by every sub-step the function value of the loss function is decreased and the algorithm converges to at least a local minimum. After improving all subsets of parameters corresponding to a particular pair of dimensions, we improve the subset of parameters corresponding to the next dimension. If all pairs of dimensions have been improved the algorithm starts another cycle by restarting at the first pair of dimensions. This proces is repeated untill convergence of the algorithm.

5 Conclusion

The present algorithm has been developed with an explicit decomposition of the data in mind, the skew-symmetry analysis and a symmetry analysis for three-way arrays can now be treated on an equal footing. After the analysis it can be desirable to compare the two analysis to come up with a single model for the complete data. How this can be done needs further study. Also, procedures for assessing the stability of the solution are not available, it might be desirable to develop them in future. Another line of future research is the development of an IDIOSCAL representation for skew-symmetric data.

References

Carroll, J. D., and Chang, J. J. (1970). Analysis of individual differences in multidimensional scaling via an N-way generalization of "Eckart-Young" decomposition. *Psychometrika, 35, 283-319.*

Carroll, J. D. (1972). Individual differences and multidimensional scaling. In: R. N. Shepard et al (Eds). *Multidimensional Scaling: Theory and Applications in the Behavioral Sciences.* Seminar press, New York, pp 115-155.

Constantine, A. G., and Gower, J. C. (1978). Graphical representation of asymmetric matrices. *Journal of the Royal Statistical Society(series C), 27, 297-304.*

Gower, J. C. (1977). The analysis of asymmetry and orthogonality. In: J. R. Barra, F. Brodeau, G. Romer, and B. van Cutsem (Eds.), *Recent developments in statistics.* North Holland, Amsterdam, 109-123.

Weeks, D. G., and Bentler, P. M. (1982). Restricted multidimensional scaling models for asymmetric proximities. *Psychometrika, 47, 201-208.*

Zielman, B. (1991). *Three-way scaling of asymmetric proximities. Unpublished master thesis.* Department of Data Theory, Leiden.

Clustering in Low-Dimensional Space

Willem J. Heiser

Department of Data Theory, University of Leiden,
P.O. Box 9555, 2300 RB Leiden, The Netherlands

Abstract: It is often asserted that clustering techniques and multidimensional scaling (MDS) have mutually exclusive roles in the analysis of a given set of data, the former being especially indicated when the configuration of points is high-dimensional and has local regions of high density, while the latter would be more useful when the points are low-dimensional and more evenly distributed. This view has been challenged by, amongst others, Critchley and Heiser (1988). They showed how a set of objects with ultrametric distances can be mapped as a series of points on a line. Taking one step further, it may be argued that clustering is a way to stabilize and robustify the multidimensional scaling task, the aim being to fit a low-dimensional distance model to groups of points, rather than to single points. Thus finding clusters is not an end in itself, but subordinate to some other task. This idea is discussed for the unfolding situation, in which we have to deal with single-peaked variables defined over a common set of points, and for the general MDS situation. A convergent least squares algorithm is described and illustrated. The approach leads to a useful decomposition of the badness-of-fit function into between and within components. For a fixed partitioning of the points a least squares version of Gower's (1989) canonical distance analysis is obtained.

1   Introduction

One can develop and use clustering methods to find natural groupings in observations of various kinds, but there is a large and fruitful area of research in which clustering is used to achieve something else. In psychometrics, for example, a long-standing tradition of exploratory factor analysis exists, where the overriding objective is to obtain invariance of factors with respect to random sampling from a prespecified domain of variables (Yates, 1987). Here one groups vectors into point clusters (factors) and (hyper)plane clusters (linear combinations of factors). Upon resampling of variables from the same domain - not merely resampling of individuals, i.e. obtaining fresh data, but actually using slightly different psychological tasks - it is expected to recover the same factors. Note that it would not be possible to speak or even to think of invariance unless there is an implicit or explicit allocation of variables into clusters - called factors in this context.

Another field of active development in clustering something else than plain observations is market segmentation. The idea of breaking down a possibly heterogeneous market into more homogeneous segments has led to a type of constrained multivariate regression in which different regression surfaces are sought in order to predict consumer preferences of *a priori* unknown groups of consumers for the same products. In this situation it is the set of regression coefficients that is clustered. The methodology is related to, but generally not equivalent with, the clusterwise regression method of Späth (1979, 1986). Examples are Kamakura's (1988) clustering procedure, some members of the family of methods for overlapping clusterwise multivariate linear regression of DeSarbo, Oliver & Rangaswamy (1989), the fuzzy clustering approach of Wedel & Steenkamp (1989), and Ogawa's (1987) segmentation procedure using logit estimation.

What these examples have in common is that variables are clustered, not the observation units. Why this circumstance leads to slightly different considerations will be explained in the next section. Then in section 3 *group differences unfolding* is defined, in which the variables of analysis are groups of judges, while the judgments are collected at an individual level. Section 4 looks at *cluster differences scaling*, a new technique to perform multidimensional scaling at a higher aggregation level than the one defined by the observational design. An illustrative example is presented in section 5, and the paper concludes with a discussion of some methodological issues connected with the concept of *a posteriori* creating a higher level of aggregation than the one specified in the design.

2 Clustering of variables versus clustering of objects

Why would not we use any similarity or dissimilarity coefficient for either rows (objects, observation units) or columns (variables, attributes) of a data matrix and subsequently apply any clustering method? One reason is that the data analytic situation is not symmetric:

- the data profile of an object *cannot* be changed meaningfully without considering analogous changes for the other objects;

- the object values of a variable *can* be changed without considering analogous changes for the other variables.

Thus rescaling, quantification, and transformation are operations that are carried out on variables, not on objects. While distance is the fundamental concept to express the relationships among objects or categories, inner product or correlation plays the same role for relationships among variables. In relation to the above aspects, observe that when we change the data profile of an object, we change its distance to the other objects; by contrast, when we rescale a variable, we do not change its angle (normalized inner product) with the other variables.

In the older psychometric literature, a frequently encountered objective is to cluster variables into hyperplanes, since such a constellation implies that they have a relatively simple factorial composition in terms of the vectors (called common factors) that are located in the intersection of hyperplanes. Translated into contemporary terminology, a hyperplane model allows us to draw simple path diagrams. It is clear that this rationale does not apply to objects.

When rescaling is called for, we might as well do it optimally, i.e. in such a way that the clustering objective maximally benefits from it. Instead of starting from a set of fixed similarity or dissimilarity coefficients, we could consider the following problem:

$$\min_{\{a_{jk}\},\{\mathbf{x}_k\}} \sum_k \sum_{j \in J_k} ||a_{jk}\mathbf{q}_j - \mathbf{x}_k||^2. \qquad (Problem\ 1)$$

Here $\mathbf{q}_j$ represents the data vector of variable j $(j = 1,...,m)$, the vector $\mathbf{x}_k$ is the representation of cluster k $(k = 1,...,K)$, the index set J_k provides the indices of the variables that are allocated to cluster k, and a_{jk} is a rescaling factor. The notation $||.||$ is used for the Euclidean norm. It is assumed that both $\mathbf{x}_k$ and $\mathbf{q}_j$ are normalized, i.e. they are in deviation from their mean, and have standard length $||\mathbf{x}_k||^2 = n$ and $||\mathbf{q}_j||^2 = n$, where n is the number of observations. Note that Problem 1 reduces to ordinary principal components analysis (PCA) when each variable is associated with each cluster, i.e. when $J_k = \{1, ..., m\}$, and if we require $x'_k x_l = 0$ for all k, l. For then we have the basic equivalence (Gifi, 1990, chap. 4):

$$\sum_k \sum_j \|a_{jk}q_j - x_k\|^2 = (K-1)nm + \sum_j \|q_j - \sum_k a_{jk}x_k\|^2,$$

where the second term on the right-hand side exhibits the well-known Eckart-Young (1936) form of PCA. When the J_k are mutually exclusive, a similar equivalence holds for Problem 1, since then

$$\sum_k \sum_{j \in J_k} \|a_{jk}q_j - x_k\|^2 = \sum_j \|q_j - \sum_{k \in J_j} a_{jk}x_k\|^2.$$

Here we no longer need to assume $x'_k x_l = 0$, since the inner summation $\sum_{k \in J_j}$ merely selects one of the x_k (the summation over k and $j \in J_k$ is replaced by the equivalent j and $k \in J_j$, with J_j the index set relating variable j to the clusters). Thus the cluster directions x_k can have any kind of angle with each other. In order to highlight the clustering aspect of Problem 1, we may replace the notation $\sum_k \sum_{j \in J_k}$ by ordinary double summation through the introduction of an indicator matrix G with $(0,1)$-elements g_{jk} indicating to which cluster(s) each variable is allocated. This allows us to write

$$\sum_k \sum_{j \in J_k} \|a_{jk}q_j - x_k\|^2 = \sum_j \sum_k g_{jk}\|a_{jk}q_j - x_k\|^2 = \sum_j \sum_k g_{jk}\|q_j - a_{jk}x_k\|^2, \qquad (1)$$

which brings the criterion in the form of an inner product of an indicator matrix and a squared Euclidean distance matrix, i.e. the usual mathematical programming form of K-means clustering (Gordon & Henderson, 1977; Selim & Ismail, 1984). The only difference with ordinary K-means is the presence of the rescaling coefficients a_{jk}, so Problem 1 with mutually exclusive index sets could be called *K-means with optimal rescaling*.

The squared distance form of the above criteria should not obscure the fact that they only depend upon the correlation between the variables. By minimizing over a_{jk} we obtain $a_{jk} = q'_j x_k / n$: the optimal rescaling factor equals the correlation between q_j and x_k. Substituting, it is not hard to show that the core of the problem is to maximize $\sum_j \sum_k g_{jk}(q'_j x_k / n)^2$, the sum of squared correlations between the variables and the clusters to which they have been allocated, a criterion originally suggested by Escoufier (1988).

Meulman and Verboon (1989, 1992) have studied Problem 1 independently, under the name *points-of-view analysis* (cf. Tucker & Messick, 1963), with the following additional requirements: (1) the role of q_j is played by Δ_j, a dissimilarity table obtained from one of m data sources; (2) instead of simple rescaling by some scalar quantity a_{jk}, the elements of Δ_j can be transformed either by some prechosen class of transformations, or be regarded as a function $\Delta(Z_j)$, where the columns of Z_j are transformed; (3) the role of x_k is played by a reparametrization in terms of the Euclidean distance matrix $D(X_k)$. These requirements lead to a separate distance model for each group of optimally aggregated dissimilarity matrices, with group membership that is *a priori* unknown. Verboon and Heiser (1990) used a similar approach for the analysis of multiple dynamic systems. Indeed, this wide variety of special cases illustrates the enormous potential of the idea.

Instead of fitting a separate distance model for each group of observations, we can also try to link several groups of observations in terms of a distance model; this forms the subject of the rest of the paper.

3 Group differences unfolding

One natural setting for unfolding is preference analysis; here the variables of analysis are *judges*, who evaluate a set of *options* (objects), for example by ranking them from very desirable (low rank number) to very undesirable (high rank number). The unfolding model (Coombs, 1964) is based on the idea that evaluative responses arise from choosing a different balance in a conflict between opposing tendencies. It follows that an arrangement of options must exist for which the response is single-dipped, with a judge-specific location of the dip (it is assumed that a low value indicates high preference). Suppose now that we want to model the situation by assuming a limited number of single-dipped response curves, rather than a single one for each judge. Such an assumption would appear to be especially indicated when the number of judges is large and number of options is small. Work by Van Blokland-Vogelesang et al. (1987), exclusively using observed and latent rank orders, is based on a similar assumption.

The multidimensional scaling (MDS) approach to unfolding (see for example Heiser, 1987) tries to model observed preferences with a response curve or surface that is linear in the Euclidean distance between p-dimensional points representing objects (object points) and a special kind of point, called the *ideal point*, representing the dip in the response (the point of maximal preference). A limited number of single-dipped response curves implies a limited number of ideal points compared to the number of judges, and thus a considerable reduction of the total number of parameters is obtained.

Let us call this method *group differences unfolding*, to distinguish it from ordinary unfolding, which is directed towards modelling the individual judges. One form of group differences unfolding can be obtained from a simple adaptation of Problem 1:

$$\min \quad \sigma(\mathbf{A}, \mathbf{G}, \mathbf{X}, \mathbf{Y}) = \sum_k \sum_j g_{jk} \|a_{jk}\delta_j - \mathbf{d}(\mathbf{x}_k, \mathbf{Y})\|^2. \qquad \text{(Problem 2)}$$

Here δ_j represents the data vector of judge j, and $\mathbf{d}(\mathbf{x}_k, \mathbf{Y})$ the vector of Euclidean distances $d(\mathbf{x}_k, \mathbf{y}_i)$ from the common ideal point $\mathbf{x}_k$ (of all judges for which column k of $\mathbf{G}$ is one) to object point $\mathbf{y}_i$, defined as $d(\mathbf{x}_k, \mathbf{y}_i) = \|\mathbf{x}_k - \mathbf{y}_i\|$, with $\mathbf{y}_i$ equal to (the transpose of) row i of $\mathbf{Y}$, an $n \times p$ matrix, where n is the number of objects and p the chosen dimensionality of the unfolding representation.

For fixed distances, Problem 2 reduces to K-means with optimal rescaling. Analogous to Meulman and Verboon (1992), this feature suggests an alternating least squares (ALS) algorithm, which alternates between optimization over $\mathbf{G}$ (with K-means reallocation based on the current residuals), over $\mathbf{A}$ (by finding the dominant principal component of each group J_k), and over $\mathbf{X}$ and $\mathbf{Y}$. The latter problem can be simplified, repeatedly using Huygens' Theorem (cf. Lebart et al., 1984, p. 126), by use of the orthogonal decomposition

$$\sigma(\mathbf{A}, \mathbf{G}, \mathbf{X}, \mathbf{Y}) = \sum_k \sum_{j \in J_k} ||a_{jk}\boldsymbol{\delta}_j - \underline{\boldsymbol{\delta}}_k||^2 + \sum_k n_k ||\underline{\boldsymbol{\delta}}_k - \mathbf{d}(\mathbf{x}_k, \mathbf{Y})||^2, \qquad (2)$$

with $\underline{\boldsymbol{\delta}}_k = (1/n_k) \sum_{j \in J_k} a_{jk}\boldsymbol{\delta}_j$.

Thus $\underline{\boldsymbol{\delta}}_k$ is the weighted average preference vector of group k, consisting of n_k judges, and both parts on the right-hand side of (2) have a useful role to play: the first part is a pooled within sum of squares that can be used to evaluate the heterogeneity left unaccounted for (influenced only by the number of groups K), and the second part is the approximation error, which is to be minimized by carrying out a weighted metric least squares unfolding analysis (Heiser, 1987).

Group differences unfolding appears to be a promising idea for a group of methods that traditionally suffers from an overload of parameters. De Soete and Heiser (1993) provide a latent class rationale of the method for continuous ratings, based on a mixture distribution with K multivariate normal densities, and propose a parametric bootstrap strategy for selecting the appropriate number of classes and the appropriate number of dimensions; they also illustrate this strategy on a Monte Carlo gauge (i.e., a test example with known stochastic properties), and illustrate the method on a set of party sympathy ratings from Members of Dutch Parliament.

4 Cluster differences scaling

If it is possible to replace a number of ideal points by a common point without having to specify *a priori* which points should be grouped together, then how about simultaneously partitioning the rows *and* columns of a symmetric dissimilarity table so that the groups are linked by low-dimensional Euclidean distances? In other words, what can we say about the ordinary MDS case? Because we are no longer considering observation vectors giving a column-wise structure to the cells of the table, as in the previous section, but arbitrary dissimilarities δ_{ij} (which could be summary statistics from some larger array of data), and because we want to keep the presentation as simple as possible, the rescaling factors are dropped. The indices $i = 1, ..., n$ and $j = 1, ..., n$ are used to indicate objects and object points, and $l = 1, ..., K$ serves as an alternate index for clusters. We consider

$$\min \quad \sigma(\mathbf{G}, \mathbf{X}) = \sum_k \sum_{l \neq k} \sum_{i \in J_k} \sum_{j \in J_l} [\delta_{ij} - d_{kl}(\mathbf{X})]^2. \qquad (Problem\ 3)$$

In this badness-of-fit function, $d_{kl}(\mathbf{X})$ denotes the Euclidean distance between row k and row l of the $K \times p$ configuration matrix $\mathbf{X}$. Note that the within cluster pairs $(i \in J_k, j \in J_k)$ are not included in $\sigma(\mathbf{G}, \mathbf{X})$. The index sets $J_k, k = 1, ..., K$, are explicitly included in the notation $\sigma(\mathbf{G}, \mathbf{X})$ as the indicator matrix $\mathbf{G}$, of order $n \times K$, with elements g_{ik}.

The method defined by Problem 3 will be called *cluster differences scaling* (CDS); it is a scaling method because the parameters $\mathbf{X}$ are (multidimensional) scale values, which account for the data through a distance function, and it is a cluster method because the set of objects is partitioned into disjoint groups. Each group of objects $i \in J_k$ is represented by a single common point $\mathbf{x}_k$; the number of clusters and the dimensionality can be chosen independently, provided $K \geq p + 1$.

A better understanding of Problem 3 can again be obtained from the orthogonal decomposition of the sum of squares involving all dissimilarities within pairs of distinct groups ($i \in J_k, j \in J_l$), and then adding over pairs of groups; this yields:

$$\sigma(\mathbf{G}, \mathbf{X}) = \sum_k \sum_{l \neq k} \sum_{i \in J_k} \sum_{j \in J_l} [\delta_{ij} - \underline{\delta}_{kl}]^2 + \sum_k \sum_{l \neq k} n_k n_l [\underline{\delta}_{kl} - d_{kl}(\mathbf{X})]^2, \tag{3}$$

with $\underline{\delta}_{kl} = (1/n_k n_l) \sum_{i \in J_k} \sum_{j \in J_l} \delta_{ij}$.

The quantities $\underline{\delta}_{kl}$ are the average dissimilarity between groups k and l, well-known from the *group average method* (Sokal & Michener, 1958), also called unweighted pair group method using arithmetic averages (UPGMA). They will be called Sokal-Michener dissimilarities here. The first part of the decomposition in (3) is the within-group dissimilarity variability that remains unaccounted for, while the second part is the between-group contribution to the badness-of-fit. Both look like being a generalization to multidimensional variation of the well-known ANOVA within-group sum of squares and between-group sum of squares, respectively.

For this decomposition to be truly a generalization, however, we need to add two qualifications. In the first place, the quantities $d_{kl}(\mathbf{X})$ provide a reparametrization intended to fit $\underline{\delta}_{kl}$ closely. But in standard ANOVA one fits the grand mean; therefore, the correspondence is only obtained when $d_{kl}(\mathbf{X})$ is constant for all (k, l), the case of a maximally uninformative MDS solution, in which $\mathbf{X}$ is a regular simplex in $K - 1$ dimensions. In the second place, when we substitute the one-dimensional quantities $\delta_{ij} = |z_i - z_j|$ to check if (3) yields the standard ANOVA decomposition, it turns out that the following equation must be satisfied for all k, l

$$(1/n_k n_l) \sum_{i \in J_k} \sum_{j \in J_l} |z_i - z_j| = |[\sum_{i \in J_k} z_i]/n_k - [\sum_{j \in J_l} z_j]/n_l| = |\beta_k - \beta_l|, \tag{4}$$

i.e. the Sokal-Michener dissimilarities should be equal to the positive difference between the means within each group (denoted by β_k, for $k = 1, ..., K$), and it is not hard to show that this is only the case when the point sets do not overlap. When the point sets do overlap, $\underline{\delta}_{kl}$ overestimates the difference between the means. From (4) it follows that

$$\sum_k \sum_{l \neq k} n_k n_l [\underline{\delta}_{kl}]^2 = 2n \sum_k n_k [\beta_k]^2, \tag{5}$$

i.e., the sum of squared Sokal-Michener dissimilarities among all pairs of groups is equal to the raw between group sum of squares (uncorrected for the mean), up to a factor $2n$. Of course, when we optimize over the partitioning in this case, the groups will never overlap (Fisher, 1958). But when the partitioning is fixed, as in Gower (1989), condition (4) needs to be checked.

From (4) it also becomes clear that when $\delta_{ij} = |z_i - z_j|$ it is not hard to find the group point $\mathbf{x}_k$: the second part of (3) will vanish when we choose $\mathbf{x}_k = \beta_k$. However, in the more general case of arbitrary dissimilarities the between part of (3) is minimized for a given partitioning by a weighted least squares MDS, with weights $w_{kl} = n_k n_l$, on the Sokal-

Michener dissimilarities. For a given configuration of points **X**, Problem 3 can be written more conveniently as finding the minimum of

$$\sigma(\mathbf{G}) = \sum_i \sum_k g_{ik} [\sum_{j \neq i} \sum_{l \neq k} g_{jl} (\delta_{ij} - d_{kl}(\mathbf{X}))^2]. \tag{6}$$

A significant and fortunate property of (6) is that the expression in square brackets does not depend on the the ith row of **G**, and therefore we can minimize $\sigma(\mathbf{G})$ in (6) row after row, by a K- means type of reallocation strategy. Alternating between weighted least squares MDS (for example with the SMACOF algorithm, as in Heiser & De Leeuw, 1977), and row-wise reallocations again gives a convergent process, which can be stopped when no reallocations occur during a single run along the rows of **G**.

From the fitted distances $d_{kl}(\mathbf{X})$, we have a p-dimensional representation of the clusters in **X**, but what can we say about the original objects? Well, we know their cluster membership from **G**, and we can find a Euclidean embedding $\mathbf{y}_i$ of any point i (one of the original points, or some other, supplementary point for which dissimilarities are available with respect to the original points) by solving the related problem

$$\min \quad \sigma(\mathbf{y}_i) = \sum_k \sum_{j \in J_k} [\delta_{ij} - d(\mathbf{y}_i, \mathbf{x}_k)]^2, \tag{7}$$

in which the $\mathbf{x}_k$ are kept fixed. In the same way as before, (7) can be split into a within and a between component, and the latter turns out to be a particular type of unfolding problem, weighted least squares external unfolding, which can be solved by a very simple iterative scheme (given in Heiser, 1987). The location found by the external unfolding represents the maximum improvement that can be obtained for point i by moving it away from its corresponding cluster point. The precise relationship with a constrained MDS in which we try to fit δ_{ij} with distances $d_{ij}(\mathbf{Y})$ under the condition $\mathbf{Y} = \mathbf{GX}$ awaits further study at another occasion.

5   Illustrative example

In order to illustrate cluster differences scaling, an empirical set of data will be employed. According to our source: "The particular example consists of twenty-four psychological tests given to 145 seventh and eighth grade school children in a suburb of Chicago, and is largely an outgrowth of the Spearman-Holzinger Unitary Trait Study [234]. The initial data were gathered by Holzinger and Swineford [245], while applications of these data by Holzinger and Harman [243], Kaiser [293], Neuhaus and Wrigley [379] and others have made this example a classic in factor analysis literature." (Harman, 1967, p. 124).

The correlations were converted into dissimilarities by the equation $\delta_{ij} = [2 - 2r_{ij}]^{1/2}$, which is the Euclidean distance between the endpoints of the unit-normalized vectors with angle $\arccos r_{ij}$. Most of the correlations are positive, as they almost always are in achievement testing, which implies that the test vectors occupy only a small portion of the surface of a hypersphere, and therefore two dimensions have been chosen for all analyses. The CDS method was implemented in an APL program, and initialized in a variety of ways, with four, five, and six clusters, from which a considerable number of local minima was identified. Most local minima differed only slightly, in particular yielding different allocations of individual points to (neighbouring) clusters; the cluster configurations were very similar most of the time. A typical result is displayed in Figure 1, a six- cluster solution with raw stress 44.813, and a between-cluster stress (second part of (3)) of a factor 100 smaller. Thus the

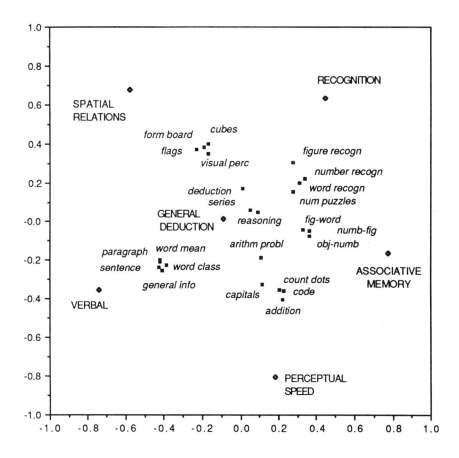

Figure 1: Six-cluster CDS solution for Holzinger/Swinford data.

Sokal-Michener dissimilarities are very well represented by the diamond-labelled points in Figure 1. The representation of the original points (the squares) was obtained by the unfolding procedure defined by (7). The allocation of the psychological tests to the six clusters is given below:

spatial relations	verbal	perceptual speed
visual perception	general information	addition
cubes	paragraph comprehension	code
paper form board	sentence completion	counting dots
flags	word classification	straight-curved capitals
word meaning		

recognition	*associative memory*	*general deduction*
word recognition	object-number	deduction
number recognition	number-figure	problem reasoning
figure recognition	figure-word	series completion
numerical puzzles		arithmetic problems

Here the labelling of the clusters corresponds to the psychological interpretation given by Harman (1967) to the so-called bi-factor pattern in his Table 7.6. This term indicates that each psychological test is connected to two factors: one general factor, and the other a specific group factor (in this case, *general deduction* is the general factor, and the other 5 clusters are the group factors). Harman's bi-factor solution is very similar to the present result, with two exceptions: 'numerical puzzles' is not allocated to the *recognition* factor, but to *general deduction*; and *all* tests are associated with general deduction in his analysis, while in ours only the four closest tests are allocated to it. This effect is a consequence of working with a distance model rather than a scalar product model, because the former lacks a fixed origin. But of course it is also true that the present analysis requires exclusive cluster membership, and therefore 'loading on two factors' is not at issue.

Although this example looks reasonable enough, showing that six clusters need not occupy five dimensions, but can be neatly arranged in two, there are reasons to be less than fully satisfied. Even from limited experience it is obvious that there are many local minima; the tight clustering of some of the psychological tests is spurious (within correlations are in the range .40 to .60); and the cluster points are rather far from the centroids of their members. The latter effect is not entirely unexpected, and it may be the expression of the bi-factor structure, in which each psychological test requires a mixture of *general deduction* and only one of the other five abilities, but we cannot exclude the operation of unresolved technical difficulties, like poorly chosen normalization factors.

6 Discussion

A major motivation of the work presented here - a preliminary version of which was given in Heiser (1991) - has been to contribute to a better methodological understanding of heterogeneity of variation. If one considers a homogeneous group of variables, it is sensible to replace them by one representative according to any reasonable definition of representativeness. It is less sensible to replace them by two or more representatives, because that stretches the concept of homogeneity beyond its meaning. If one is still tempted or urged to find two or more representatives, it is preferable to say that one is considering a heterogeneous group of variables.

Two concepts are strongly connected to the distinction between homogeneity and heterogeneity of variation: allocation and substitutability. Homogeneity implies substitutability, for although it recognizes differences, at the same time it regards them as irrelevant. In contrast, heterogeneity implies lack of substitutability, and that means that the variation cannot be reduced to a single kind. In that case allocation either on *a priori* or on *a posteriori* grounds – is unavoidable. Allocation is necessary to identify the different kinds of variation. Therefore there is room for considering clustering of variables into a number of homogeneous groups, and we have seen in section 2 how this can be done in view of the data - instead of on *a priori* grounds, as is done in PLS and LISREL modelling (Jöreskog &

Wold, 1982).

We have also seen that there is a rich variety of additional constraints that can be considered, giving us new forms of point-of-view analysis and unfolding that identify homogeneous groups of judges. Turning from the case of multiple variables to the MDS situation, there is again the default assumption of complete heterogeneity, which we might want to relax into *partial heterogeneity*: association into K homogeneous groups. When the groups are known in advance, we obtain versions of multivariate analysis of variance or discriminant analysis, depending upon the design of the study. In the latter case, the effective discriminant rule can be derived from (6): allocate object i to class k if

$$\sum_{j \in J_l} (\delta_{ij} - d_{kl}(\mathbf{X}))^2 \leq \sum_{j \in J_k} (\delta_{ij} - d_{kl}(\mathbf{X}))^2, \tag{8}$$

i.e., if the dissimilarities between object i and the objects that belong to class l are closer to the interclass distance than the dissimilarities between object i and the objects of class k. This rule allows non-linear discrimination, not only because of the possibly reduced dimensionality, but also because it focusses on *inter-class* dissimilarity. If the within cluster pairs ($i \in J_k, j \in J_k$) would be included in the badness-of-fit function, which corresponds to the more usual set-up, the intra-class dissimilarities would enter in the discriminant rule as well.

When the groups are not known in advance, cluster differences scaling in principle optimizes over all possible partitions. This process is comparable to searching for a few important contrasts between the means in an analysis of variance, instead of sticking to a global test between all of them. A considerable amount of computation is required here, but this seems to be offset by the benefits. Apart from yielding a more parsimoneous model, CDS creates a replication mode where initially we had none.

References

COOMBS, C.H. (1964), *A Theory of Data*, Wiley, New York.

CRITCHLEY, F., and HEISER, W.J. (1988), Hierarchical trees can be perfectly scaled in one dimension, *Journal of Classification*, 5, 5–20.

DESARBO, W.S., OLIVER, R.L., & A. RANGASWAMY (1989), A simulated annealing methodology for clusterwise linear regression, *Psychometrika*, 54, 707–736.

DE SOETE, G. & HEISER, W.J. (1993), A latent class unfolding model for analyzing single stimulus preference ratings, *Psychometrika*, 58, in press.

ESCOUFIER, Y. (1988), Beyond correspondence analysis. in: H.H. BOCK (eds.), *Classification and Related Methods of Data Analysis*, Elsevier (North-Holland), Amsterdam, 505–514.

FISHER, W.D. (1958), On grouping for maximum homogeneity, *J. Amer. Statist. Assoc.*, 53, 789–798.

GORDON, A.D. & J.T. HENDERSON (1977), Algorithm for Euclidean sum of squares classification, *Biometrics*, 33, 355–362.

GOWER, J.C. (1989), Generalised canonical analysis. in: R. COPPI & S. BOLASCO (eds.), *Multiway Data Analysis*, North-Holland, Amsterdam.

HARMAN, H.H. (1967), *Modern Factor Analysis*, (2nd ed.), University of Chicago Press, Chicago.

HEISER, W.J. (1987), Joint ordination of species and sites: the unfolding technique, in: P. LEGENDRE et al. (eds.), *Developments in Numerical Ecology*, Springer, New York, 189–221.

HEISER, W.J. (1991), Clustering of variables to optimise the fit of a low-dimensional Euclidean model, *Paper presented at the Third Conference of the International Federation of Classification Societies*, Edinburgh, Scotland.

HEISER, W.J. & J. DE LEEUW (1977), How to use Smacof-1. *Internal Report*, Department of Data Theory, University of Leiden, The Netherlands.

JÖRESKOG, K.G. & H. WOLD, eds. (1982), *Systems under indirect observation: causality, structure, prediction*, North-Holland, Amsterdam.

KAMAKURA, A.W. (1988), A least squares procedure for benefit segmentation with conjoint experiments, *Journal of Marketing Research*, 25, 157–167.

LEBART, L., MORINEAU, A. & K.M. WARWICK (1984), *Multivariate Descriptive Statistical Analysis: Correspondence Analysis and Related Techniques for Large Matrices*, Wiley, New York.

MEULMAN, J.J. & VERBOON, P. (1989), Nonlinear principal components analysis with distance restrictions on the component scores, *Paper presented at the Sixth European Meeting of the Psychometric Society*, Leuven, Belgium.

MEULMAN, J.J. & VERBOON, P. (1992), Points of view analysis revisited: fitting multidimensional structures to optimal distance components with cluster restrictions on the variables, *Psychometrika*, 57, in press.

OGAWA, K. (1987), An approach to simultaneous estimation and segmentation in conjoint analysis, *Marketing Science*, 6, 66–81.

SELIM, S.Z., & ISMAEL, M.A. (1984), K-means type algorithms: a generalized convergence theorem and characterization of local optimality, *IEEE Transactions on Pattern Analysis and Machine Intelligence*, 6, 81–87.

SOKAL, R.R. & MICHENER, C.D. (1958). A statistical method for evaluating systematic relationships, *Univ. Kansas Sci. Bull.*, 38, 1409–1438.

SPÄTH, H. (1979). Algorithm 39: Clusterwise linear regression, *Computing*, 22, 367–373.

SPÄTH, H. (1986). Clusterwise linear least squares versus least absolute deviations regression: a numerical comparison for a case study, in: W. GAUL & M. SCHADER (eds.), *Classification as a tool of research*, North-Holland, New York, 413–422.

TUCKER, L.R. & MESSICK, S. (1963), An individual differences model for multidimensional scaling, *Psychometrika*, 28, 333–367.

VAN BLOKLAND-VOGELESANG, R., VERBEEK, A., & EILERS, P. (1987), Iterative estimation of pattern and error parameters in a probabilistic unidimensional unfolding model, in: E.E. ROSKAM & R. SUCK (eds.), *Progress in Mathematical Psychology - 1*, North-Holland, Amsterdam, 393–418.

VERBOON, P. & HEISER, W.J. (1990), Some possibilities for the analysis of dynamic three-way data, *Internal Report RR-90-01*, University of Leiden: Dept. of Data Theory.

WEDEL, M. & STEENKAMP, J.B.E.M. (1989), Fuzzy clusterwise regression approach to benefit segmentation, *International Journal of Research in Marketing*, 6, 45–59.

YATES, A. (1987), *Multivariate Exploratory Data Analysis: A Perspective on Exploratory Factor Analysis*, SUNY Press, Albany, NY.

The Construction of Neighbour-Regions in Two Dimensions for Prediction with Multi-Level Categorical Variables

J. C. Gower

Department of Data Theory

University of Leiden

Abstract: In the multidimensional scaling of samples described by categorical variables, each l-level categorical variable is represented by a set of l points forming the vertices of a simplex. Any sample that has level i of a categorical variable will be nearer the corresponding vertex C_i than to any other vertex of the simplex, so defining convex neighbour-regions for each level. In r-dimensional approximations, predictions of levels are obtained by examining the intersections of the neighbour-regions with the r-dimensional space. The case $r = 2$ is of special practical importance and is the main concern of this paper; the methodology easily generalises. Examples are given of the forms taken by the neighbour-regions in planar sections of the $(l-1)$-dimensional simplex and an algorithm is proposed for their construction.

1 Introduction

1.1 Geometry

Gower (1992) has discussed the geometry of linear biplots (Gabriel, 1971), of non-linear biplots (Gower and Harding, 1988) and of generalised biplots (Gower, 1991). Linear and non-linear biplots refer only to quantitative variables but generalised biplots also admit categorical variables. The kth categorical variable is represented by a set of l_k category level points (CLP) which form an s-dimensional simplex that is contained in a subspace $\mathcal{M}$ of $\mathbb{R}_n$. Usually $s = l_k - 1$ but in special cases, such as with ordered categorical variables, we may have $s < l_k - 1$. Another subspace of $\mathbb{R}_n$ is an r-dimensional space $\mathcal{L}$ which contains the coordinates $\mathbf{Y}$ of n points representing samples. We shall assume that $\mathbf{e}'\mathbf{Y} = 0$, so that the centroid G of the sample-points is at the origin; normally $\mathbf{Y}$ will have been obtained from some form of multidimensional scaling. Under special conditions, G is also a point in $\mathcal{M}$ so that $\mathcal{L}$ and $\mathcal{M}$ are then not disjoint, even though both may be small sub-spaces of $\mathbb{R}_n$. However, Gower (1991) showed that usually $\mathcal{L}$ and $\mathcal{M}$ are disjoint, although the offset between the two spaces is likely to be small.

The whole of $\mathbb{R}_n$ may be partitioned into l_k n-dimensional regions, the qth of which contains all the points which are nearer Cq, the qth vertex of the simplex, than to any other vertex. These regions will be termed neighbour-regions and the qth neighbour-region for the kth categorical variable will be denoted by $\mathcal{Q}_q$ (for $q = 1, 2, ..., l_k$). Let $\mathcal{F}_{qt}$ be the $(n-1)$-dimensional flat that separates all points nearest C_q from all points nearest C_t. $\mathcal{F}_{qt}$ defines two trivial convex regions and $\mathcal{Q}_q$ is the intersection of all $\mathcal{F}_{qt}$ $(t \neq q)$ which, being the intersection of convex regions, is itself convex. It is easy to see that given the part $\mathcal{M}_q$, of $\mathcal{Q}_q$ that is within $\mathcal{M}$, then the whole of $\mathcal{Q}_q$ is found by extending orthogonally into the remaining $n - s$ dimensions of $\mathbb{R}_n$ (i.e. every $\mathbf{x} \in \mathcal{M}_q$ extends into the space normal to $\mathcal{M}_q$ at $\mathbf{x}$). When $s = l_k - 1$, the neighbour-regions $\mathcal{M}_q$ are convex cones, all with vertex

C at the circumcentre of the simplex. $\mathbb{R}_n$ may be partitioned into neighbour-regions $\mathcal{Q}_q$, ($q = 1, 2, ..., l_k$), in as many ways as there are categorical variables but attention will be focussed on the kth variable. Hence, in the following, the suffix k will be dropped, except when it is wished to emphasize reference to the kth variable. The neighbour-region $\mathcal{Q}_q$ intersects $\mathcal{L}$ and the qth category-level will be predicted for all sample points $\mathbf{Y}$ that lie in the intersection, which will therefore be termed a prediction-region. The prediction-regions must themselves be convex, because they are intersections of a linear sub-space with the convex regions $\mathcal{Q}_q$ ($q = 1, 2, ..., l$). This paper is concerned with the construction of the prediction-regions within $\mathcal{L}$.

The orthogonal projection, $\mathcal{M}^*$, of $\mathcal{L}$ onto $\mathcal{M}$ gives that part, often the whole, of $\mathcal{M}$ which may orthogonally extend into $\mathcal{L}$. Thus, the main problem in constructing the neighbour-regions $\mathcal{Q}_k$ is to find the intersections of $\mathcal{M}^*$ with the neighbour-regions $\mathcal{M}_k$ within $\mathcal{M}$. A secondary problem, the representation of these intersections in $\mathcal{L}$, depends on the relationships between the dimensions r, s and s^*, the dimensionality of $\mathcal{M}^*$. When $s^* < r$, an $(r - s^*)$-dimensional subspace of $\mathcal{L}$ projects into $\mathcal{N}^*$, a subspace of $\mathcal{N}(= \mathbb{R}_n - \mathcal{M})$, the space normal to $\mathcal{M}$. Then the representation of $\mathcal{M}^*$ in $\mathcal{L}$ has to be augmented by orthogonal extension into the whole of $\mathcal{L}$. This paper is mostly concerned with $r = 2$, the case of greatest practical importance, but also briefly examines $r = 1$. These two cases cover most of the variant situations and much of what follows generalises.

1.2 Algebra

The process of orthogonal extension requires the notion of what Gower (1992) termed back-projection. Back-projection of a point $\mathbf{x}$ in $\mathcal{M}$ onto $\mathcal{L}$ is defined to be the point $\mathbf{y} \in \mathcal{N} \cap \mathcal{L}$ that is closest to $\mathbf{x}$, where $\mathcal{N}$ is normal to $\mathcal{M}$ at $\mathbf{x}$. Writing $\mathbf{L}$ and $\mathbf{M}$ for matrices whose columns give orthonormal bases for $\mathcal{L}$ and $\mathcal{M}$, respectively, and $\mathbf{K}$ and $\mathbf{N}$ as their orthogonal complements, these conditions may be expressed as:

$$(i) \quad \mathbf{yMM'} = \mathbf{xMM'}$$
$$\text{and} \quad (ii) \quad \mathbf{yKK'} = \mathbf{0} \, .$$

Gower (1992) shows that provided $r \geq s$, the minimum of $(\mathbf{y} - \mathbf{x})(\mathbf{y} - \mathbf{x}')$ with respect to $\mathbf{y}$ subject to the constraints (i) and (ii) is given by:

$$\mathbf{y} = \mathbf{x}(\mathbf{I} + \mathbf{KK'M(M'LL'M)}^{-1}\mathbf{M'})\mathbf{LL'} \, . \tag{1}$$

An alternative expression is:

$$\mathbf{y} = \mathbf{x}(\mathbf{I} - \mathbf{K(K'NN'K)}^{-1}\mathbf{K'NN'}) \, . \tag{2}$$

When $\mathcal{M}$ contains the origin, G, then (2) simplifies to:

$$\mathbf{y} = \mathbf{xM(M'LL'M)}^{-1}\mathbf{M'LL'} \, . \tag{3}$$

In the following, the back-projection will be required of a point in $\mathcal{M}^*$ that is obtained by projecting $\mathcal{L}$ onto $\mathcal{M}$. Thus if $\mathbf{x} \in \mathcal{M}^*$ we have that $\mathbf{x} = \mathbf{vL'MM'} + \mathbf{q}$ for some vector $\mathbf{v}$. Substitution into (1) gives after a little manipulation:

$$\mathbf{y} = \mathbf{vL'M(M'LL'M)}^{-1}\mathbf{M'LL'} \, . \tag{4}$$

2 Examples

The geometry described in section 1.1 is difficult to visualise in its full generality. Thus in this section clarification is sought by studying a few simple special cases, which, as will be seen, point the way to a general solution to constructing prediction-regions in $\mathcal{L}$.

2.1 Three category-levels

Suppose the kth variable is Colour, with levels Blue or Green or Red, (say), in which case $l_k = 3$ and $s = 2$. Thus the CLPs form a triangle, the vertices C_1, C_2 and C_3 of which may be alternatively labelled as Blue, Green, and Red. These three CLPs determine the two-dimensional space $\mathcal{M}$. Within $\mathcal{M}$ the neighbour-regions $\mathcal{M}_1$, $\mathcal{M}_2$ and $\mathcal{M}_3$ are bounded by the perpendicular bisectors of the sides of the triangle and these are concurrent at the circumcentre (henceforth referred to as a c-centre) as is shown in Figure 1. Extending these boundaries by back-projection into the space $\mathcal{N}$ normal to $\mathcal{M}$ gives the partition of $\mathbb{R}_n$ into the neighbour-regions $\mathcal{Q}_1, \mathcal{Q}_2$ and $\mathcal{Q}_3$ that contain all the samples with the respective category-levels. The intersections of these neighbour-regions with $\mathcal{L}$ give the prediction-regions in $\mathcal{L}$ that predict the corresponding category-levels for the contained sample-points.

When $r \geq s$, equations (1), (2) and (3) allow the back-projections into $\mathcal{L}$ to be calculated for the CLPs. When $r = s = s^*$, as in Figure 1, it is necessary only to back-project $l_k + 1$ points, the l_k CLPs and their c-centre, which it is convenient to label with the name of the categorical variable itself. The prediction-regions are then easily constructed, because back-projection transforms mid-points into mid-points, so the boundaries of the prediction-regions can be constructed as shown in Figure 1. Samples which fall in the back-projected prediction-regions of $\mathcal{L}$ are predicted to have the corresponding labelled category-levels. Prediction-regions are not neighbour-regions for the back-projected CLPs and seem not necessarily to be neighbour-regions for any set of points in $\mathcal{L}$.

It would be impracticable to exhibit the prediction-regions for all categorical variables simultaneously but the prediction-CLPs obtained as the back-projections of the CLPs in $\mathcal{M}$ might suffice for doing most predictions by eye. Then all that would be seen in a plot of $\mathcal{L}$ are the points for Colour (the back-projected c-centre), Blue, Red and Green, and similarly for other categorical variables for which $r = s$.

When $r > s = s^*$, the only difficulty is that the back-projections define only an s-dimensional sub-space of $\mathcal{L}$. This is essentially the case of classical linear biplots, for which $s = 1$ (a linear coordinate axis) and, usually, $r = 2$. The geometry discussed by Gower (1992) for linear biplots requires only minor modification to handle categorical variables. With categories, linearity corresponds to an ordered categorical variable. The back-projections of the CLPs will give l collinear (in general s-dimensional) points in $\mathcal{L}$ and the prediction-regions can be completed by constructing the normals to the s-space within $\mathcal{L}$ at the mid-points between the back-projected CLPs to give the back-projection of $\mathcal{N}^*$ within $\mathcal{L}$ (i.e. the part of the orthogonal extension of the boundaries of the $\mathcal{M}_q$ that intersect $\mathcal{L}$). The geometry for the case $l = 4, s = s^* = 1$ and $r = 2$ is illustrated in Figure 2, where, for convenience, the mid-points, rather than the CLPs, have been back-projected immediately.

The case $s = 1$ is especially simple but when $1 < s < l_k - 1$, the problem of constructing the neighbour-regions of $\mathcal{M}$ is itself non-trivial. The tessellation algorithm of (Bowman,1980 and Sibson,1980) gives a solution for $s = 2, r \geq 2$, which requires only a back-projection of the tessellation onto $\mathcal{L}$.

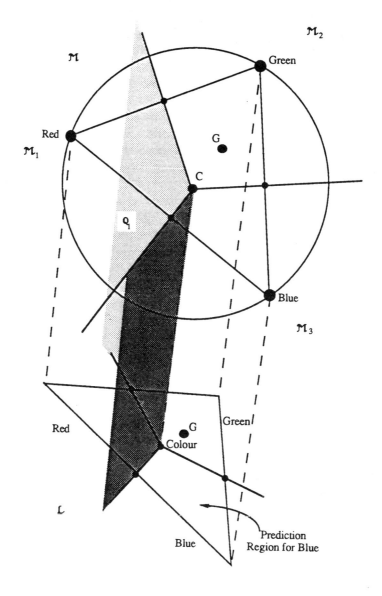

Figure 1: The case $r = s = s^* = 2$. For convenience $\mathcal{L}$ and $\mathcal{M}$ are shown as distinctly
disjoint spaces although they are usually close and the centroid G may be
common to both. The points for Red, Green and Blue in $\mathcal{L}$ are the back-
projections of the corresponding CLPs in $\mathcal{M}$, as are the mid-points which,
together with the c-centre, determine the neighbour-regions in $\mathcal{M}$ and the
prediction-regions in $\mathcal{L}$. The partition of $\mathcal{M}$ into neighbour-regions $\mathcal{M}_1$, $\mathcal{M}_2$
and $\mathcal{M}_3$ is indicated, as is the extension into $\mathcal{Q}_1$ to give the full neighbour-
region for Red; the prediction-region for Blue in $\mathcal{L}$ is labelled. The dashed
lines are orthogonal to $\mathcal{M}$.

The work of Devijver and Diekesei (1985) promises extensions to higher values of s but the problem of constructing the intersection of the tessellation in $\mathcal{M}$ with $\mathcal{M}^*$ is more difficult than in the unrestricted case ($s = l_k - 1$) discussed in the following, and remains to be solved. When $r < s$, the matrices in (1) and (2) that require inversion are of deficient rank

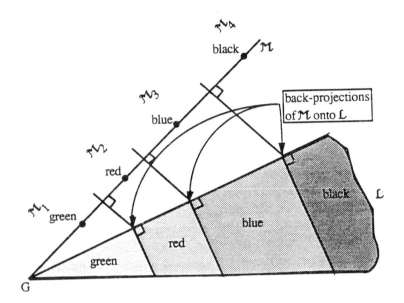

Figure 2: The prediction-regions for an ordered categorial variable with four levels (green, red, blue, black) in $\mathcal{M}$ of $s = 1$ dimension represented in $\mathcal{L}$ of $r = 2$ dimensions.

and the formulae are invalid. Normally $r = 2$, so this difficulty arises whenever $l_k > 3$, which is a common occurrence. First consider the case $s = 2$ and $r = s^* = 1$ illustrated in Figure 3. In this figure, the CLPs are the same as in Figure 1, so the neighbour-regions formed from back-projection of $\mathcal{M}$ are identical. For simplicity the triangle is not shown but the mid-points of its sides on the lines separating the neighbour-regions are retained.

Now, however, most points in $\mathcal{M}$ do not back-project into $\mathcal{L}$. Indeed, by definition, only points in $\mathcal{M}^*$ will back-project into $\mathcal{L}$. In Figure 3 two possibilities, $\mathcal{M}_1^*$ and $\mathcal{M}_2^*$, are shown, depending on the position, G_1 or G_2, of the projection of the centroid G within the triangle with the vertices Red, Green and Blue. When $\mathcal{M}_1^*$ passes through G_1, there are neighbour-regions for all three category-levels. When $\mathcal{M}_2^*$ passes through G_2, only the neighbour-regions for the levels Red and Green intersect with $\mathcal{M}^*$. Normally none of the CLPs will now back-project into $\mathcal{L}$ although, of course, there exist points in $\mathcal{L}$ that are nearest to the category-level points; however, the projection onto $\mathcal{M}_1^*$ of the CLP for Red falls into the Blue region showing that the prediction CLPs may fall into inconsistently labelled regions. It seems that for prediction, unlike for interpolation (see Gower, 1992), the projections of the CLPs onto $\mathcal{L}$ may be misleading. This example shows that when $r < s$ there are special considerations. When $r = 2$, this situation will have to be faced whenever there are categorical variables with four, or more, levels.

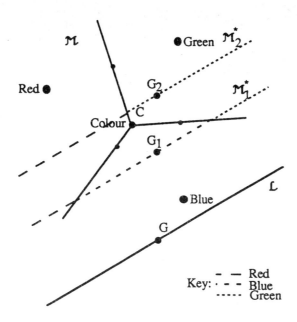

Figure 3: The case $s = 2$ and $r = s^* = 1$. The category-level points are as in Figure 1 but $\mathcal{L}$ is here one-dimensional. Two possibilities are shown: (i) $\mathcal{M}_1^*$ containing all three neighbour-regions and (ii) $\mathcal{M}_2^*$ where neighbour-regions exist only for red and green. The back-projections onto $\mathcal{L}$, having the same structures, are not exhibited.

2.2   Four Category-levels, Three Dimensions

With four levels, $\mathcal{M}$ is three dimensional, with CLPs denoted by the vertices C_1, C_2, C_3 and C_4 of a tetrahedron, and the neighbour regions in $\mathcal{M}$ are wedged-shaped, with plane boundaries whose edges are formed by the joins of the c-centre C with the c-centres of the triangles forming the four faces of the tetrahedron. For example, denoting the mid-point of C_iC_j by C_{ij} and the c-centre of $C_iC_jC_k$ by C_{ijk} then the region $\mathcal{M}_1$ has boundaries CC_{123}, CC_{124} and CC_{134} where, for example, the plane $CC_{123}C_{124}$ is normal to the edge C_1C_2 at C_{12}. First, to illustrate their deficiencies, we shall focus on the ordinary orthogonal projection of the tetrahedron onto $\mathcal{L}$ of two dimensions and see what becomes of the neighbour-regions; this is shown in Figure 4.

These regions, unlike the back-projections to be discussed shortly, overlap as is indicated in the lower part of the figure, so that, as well as samples falling into wrongly labelled regions, every region is of uncertain prediction. The overlapping regions are shown more clearly in Figure 5 which is the same as Figure 4 but with the constructional lines removed. In Figure 5, the bold lines are the projections of lines joining C to the six mid-points. To have overlapping regions may seem unsatisfactory but in fact this situation is better than with numerical information, where each point of $\mathcal{L}$ corresponds to all values that project from $\mathbb{R}_n$ into that point.

180

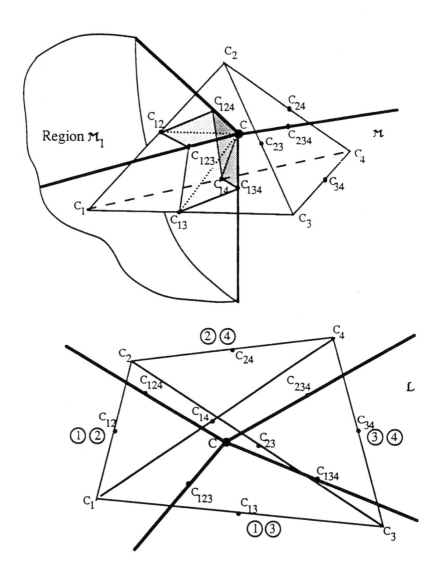

Figure 4: In M the four category-levels labelled C_1, C_2, C_3 and C_4 form a tetrahedron, c-centre C. The wedge-shaped neighbour-region M_1 is shown for C_1. The two-dimensional approximation in L shows the orthogonal projections of the neighbour-regions. The edges joining C to the c-centres of the four faces are shown bold.

With categorical information, each region of $\mathcal{L}$ is associated with only a subset of possible category-levels, as in Figure 5. However, as the number of levels increases, the over-

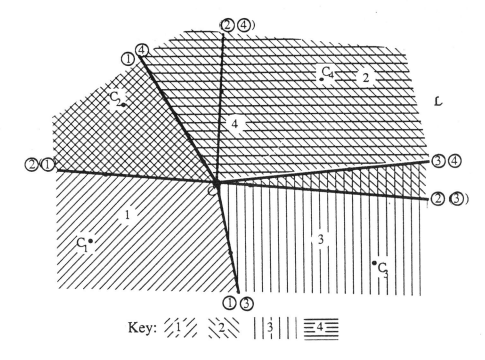

Key: (key symbols) 1, 2, 3, 4

Figure 5: Figure 4 with the constructional details omitted to clarify the degree of overlap of the regions assigning to the four category-levels. Regions shown are projections not back-projections. Bracketed region labels indicate an overlapping boundary.

lapping regions become increasingly difficult to describe. Proper prediction requires the back-projection of $\mathcal{M}^*$ onto $\mathcal{L}$. Because every point of $\mathbb{R}_n$ belongs to a unique neighbour-region, this must give non-overlapping prediction-regions. In terms of Figure 4, we require the intersection of the neighbour-regions of the tetrahedron with $\mathcal{M}^*$. Just as with Figure 3, the resulting configuration depends on the position and the orientation of the intersecting plane. There are three possible topologies depending on whether $\mathcal{M}^*$ intersects with one, two, or three of the lines joining C to the four c-centres C_{ijk}; four intersections are impossible, just as are three intersections in Figure 3. Intersections are at points on lines which separate the regions labelled $\mathcal{M}_i, \mathcal{M}_j$ and $\mathcal{M}_k$, so will be at the meeting-point of three edges. Figure 6(a) shows a slice with one intersection, being a slice through CC_{234} on the opposite side of C as is $\mathcal{M}_1$, while Figure 6(c) shows two intersections, arising from a slice on the other side of C, intersecting with all four regions. Figure 6(b) shows three intersections across the full range of $\mathcal{M}_1$. In these figures an intersection with a line joining C to a c-centre of a face is denoted by a triplet ijk and to a mid-point by a doublet ij. When an intersection with ijk occurs by extending a line or plane beyond a limiting point, such as C, then dotted lines are used, as with 234 in figures 6(b) and 6(c); such a point is termed a virtual point.

To understand virtual points better, consider the dotted line that joins 234 to 123 in Figure 6(b). This line separates regions M_2 and M_3, so that any point on it is equally close to C_2 and C_3 but their nearest vertex is at C_1. The line becomes undotted when it leaves M_1 and the nearest vertex is then given by C_2 or C_3.

Note that the neighbour-regions in one-dimensional approximations may be obtained by inspecting the intersection of a line with the different cases shown in Figure 6. In case (a) the situation is as in Figure 3, giving regions for either two or three levels; in cases (b) and (c), regions may be obtained that separate two, three or four levels.

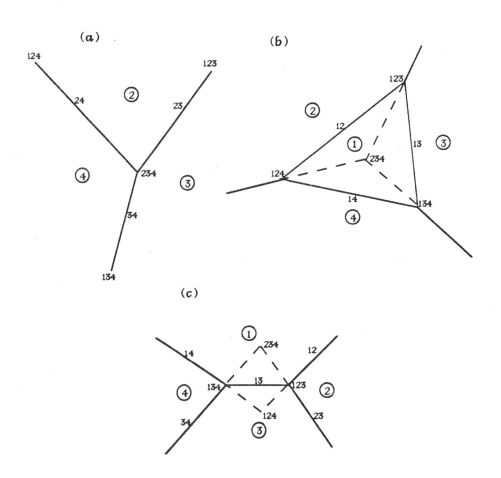

Figure 6: In (a) is shown the intersection with the neighbour-region M_1 of a tetrahedron, formed by a slice through the region that excludes C_1. In (b), the slice is through the space on the same side of C as is C_1 and cutting all three faces of M_1. In (c) the slide cuts two of the three edges of M_1.

2.3   Generalisations to Several Category-levels and Higher Dimensions

Next consider a two-dimensional section of a four-dimensional simplex, with vertices $C_i (i = 1, 2, \ldots, 5)$. Figure 7(a) is the analogue of Figure 6(b), being a three-dimensional intersection of the four-dimensional space, formed by slicing fully through $\mathcal{M}_1$. In this figure, the three-dimensional nature of the regions cannot be labelled convincingly in a two-dimensional drawing. The device has been used of attaching a label to each arm that is to be associated with the region delimited by that arm, the two arms on either side (as seen in two-dimensions) and the face of the tetrahedron bounding $\mathcal{M}_1$ that is common to all three of these arms. Figure 7(b) reduces the dimensionality to two by by slicing Figure 7(a) in its lower part towards the point marked 1245 but through $\mathcal{M}_1$ and avoiding $\mathcal{M}_5$. It follows that all triplets involving the suffix 5 are missing; this is shown in 7(b) by placing the points 235, 245 and 345 at the extremities of the arms, indicating that they are at infinity on the corresponding lines and that $\mathcal{M}_5$ may also be regarded as if it were at infinity. Figure 7(c) includes four virtual points (134, 135, 145, 345) that coincide; these represent four lines which intersect somewhere on 1345, so this common-point is so identified and indicates that 7(c) is derived from a slice of the original four-dimensional simplex through $\mathcal{M}_2$, on the opposite side of C as is CC_{1345}. Figures 7(d),7(e) and 7(f) show other possibilities.

The general pattern common to the variants shown in Figure 7 is now clear. All lines contain three triplets that share two suffices i, j (say). These lines are the intersections of $\mathcal{M}^*$, of dimension two, with the flat $\mathcal{F}_{ij}$ of dimension $l - 2$ that is normal to $C_i C_j$ at its mid-point C_{ij}. (Recall that $\mathcal{M}^*$ and $\mathcal{M}$ are not disjoint spaces.) $\mathcal{F}_{ij}$ contains all c-centres involving i and j, and, in particular, all c-centres of the form C_{ijk}. Indeed, these lie at the intersection of $\mathcal{F}_{ij}$, $\mathcal{F}_{ik}$ and $\mathcal{F}_{jk}$ so form an $(l - 3)$-dimensional space which meets the two-dimensional $\mathcal{M}^*$ in a point; it is such points that are labelled ijk in the figures and which mark the meeting of regions $\mathcal{M}_i$, $\mathcal{M}_j$ and $\mathcal{M}_k$. Thus three boundary lines lines must meet at each triplet and the non-virtual part of each boundary is terminated by a pair of triplets that share the suffix-pair that label the regions that are separated. When there is only one non-virtual point on a line, the non-virtual boundary is completed by constructing an arm starting at the non-virtual point and extending it away from the virtual points on the same boundary. In exceptional circumstances, two triplets ijk and ijl (virtual or non-virtual) adjacent on a line may coincide, in which case they become a point labelled $ijkl$ at the intersection of four regions, and so also coincide with ikl and jkl. Such special cases arise from a slice through a line contained in the space containing the c-centre C_{ijkl} and the lower-dimensional c-centres involving i, j, k and l; clearly, higher dimensional analogues inducing even more concurrent regions may exist. In another exceptional circumstance $\mathcal{M}^*$ may lie entirely within the $(l - 2)$-dimensional normal-space, in which case the order-three c-centres involving i and j meet $\mathcal{M}^*$ in lines and the order-four c-centres meet $\mathcal{M}^*$ in points; the regions $\mathcal{M}_i$ and $\mathcal{M}_j$ are then not separable for prediction. For example, in Figure 7(b) if the vertices of the triangle are labelled 1235, 1245 and 1345, then the central region is predicted as either $\mathcal{M}_1$ or $\mathcal{M}_5$.

An example of the extension to five dimensions is illustrated in Figure 8; a clear connection with Finite Geometries is noted. The general form of the cut of an $(l - 1)$-dimensional simplex by a two-dimensional plane $\mathcal{M}^*$ may now be stated:

(1) The prediction regions in $\mathcal{M}^*$ are convex and bounded by straight lines. Some, but not all, regions may be closed.

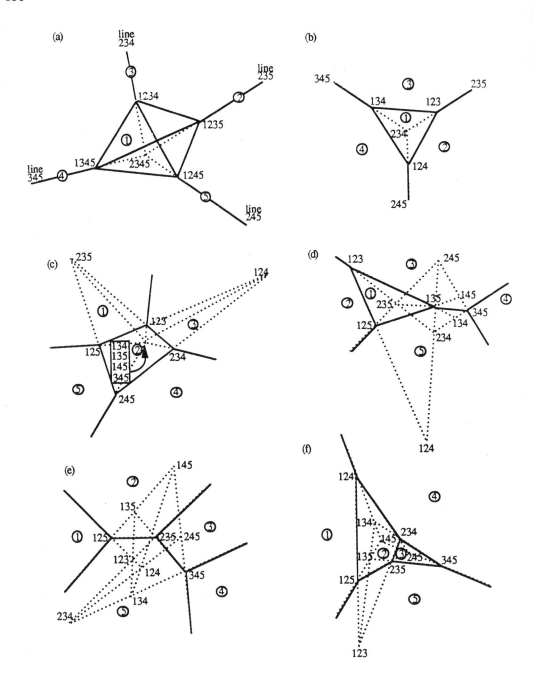

Figure 7: Additional topologies that may be obtained from two-dimensional slices of a four-dimensional simplex given by five category-level points.

(2) Each straight line contains $l-2$ triplets, most of which are virtual points, which share the two suffices labelling the regions separated by the lines.

(3) The non-virtual triplet-points form the vertices of the polygons bounding the prediction-regions. Boundaries are formed by joining all pairs of non-virtual points that share a pair of suffices.

(4) Special provision has to be made for boundaries that are outer arms containing only a single non-virtual point. These boundaries are at the end of a line and may be constructed by moving from the non-virtual point away from the remaining $l-3$ virtual points.

(5) All triplets, virtual or not, are at the meeting points of three lines and so are at the

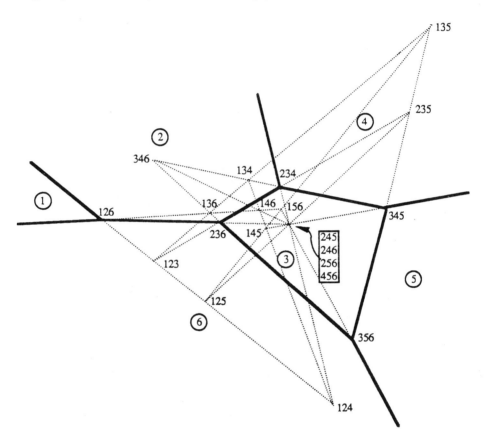

Figure 8: A two-dimensional slice of a five-dimensional simplex (six category-levels). All triplets sharing the same pair of suffices are collinear in sets of four. Similarly to Figure 7(c), the quadrilateral contains a common point that bears the four distinct triplets marked. This quadrilateral arises from a slice through the tetrahedron with c-centre C_{2456}, and CC_{2456} meets the plane shown at the common point.

join of three regions. For virtual points, these three regions are dominated by some fourth region.

(6) Closed regions are special, in the sense that the set of lines, one from each vertex of the region, that are not parts of the region's boundary, must be concurrent at an internal point of the region. For regions bounded by more than three lines, such concurrent points represent higher-order c-centres and the superposition of several sets of triplets.

(7) Exceptionally, higher-order c-centres may generate non-virtual points, in which case they form a vertex that is at the join of more than three prediction-regions.

3   Properties of Circumcentres

The above requires the c-centres of subsets of category-level coordinates given by the l vertices $C_1, C_2, \ldots, C_l$ of a simplex. In this paper, only the c-centres of sets of three points are required but for approximations in more than two dimensions, higher order c-centres are needed, so the following results are presented in their general form. Let the rows of $\mathbf{Z}$ give the coordinates of the simplex. Gower (1991) gives a general method for calculating these coordinates. With the extended matching coefficient (Gower, 1991), the distances between the CLPs are all equal and the simplex is regular and then the c-centre coincides with the centroid of the CLPs and are very easily constructed. With other choices of distance, such as the chi-squared distance of multiple correspondence analysis, the CLPs will form an irregular simplex. The squared-distances c_{hk}^2 between each pair of category-level points may be calculated from $\mathbf{Z}$ and formed into the $l \times l$ matrix $\mathbf{C} = \{-\frac{1}{2}c_{hk}^2\}$. Then Gower (1982) shows that the generalised c-centre of the coordinates $\mathbf{Z}$ is given by $\mathbf{s}'\mathbf{Z}$, where $\mathbf{s} = \mathbf{C}^{-1}\mathbf{e}/(\mathbf{e}'\mathbf{C}^{-1}\mathbf{e})$. The neighbour-regions remain convex cones but are more complicated than for the extended matching coefficient.

In the following we shall write $\mathbf{c}_i$ for the ith column of $\mathbf{C}^{-1}, c_i = \mathbf{e}'\mathbf{c}_i$ for the sum of the elements of the ith column, and c_{ii} for the ith diagonal element of $\mathbf{C}^{-1}$. To calculate the c-centre of a face of the simplex, say the face opposite C_1, first partition $\mathbf{C}$ as indicated in (5):

$$\mathbf{C}^{-1} = \begin{pmatrix} 0 & \mathbf{b}' \\ \mathbf{b} & \mathbf{B} \end{pmatrix}^{-1} = \begin{pmatrix} \beta & -\beta\mathbf{b}'\mathbf{B}^{-1} \\ -\beta\mathbf{B}^{-1}\mathbf{b} & \mathbf{B}^{-1} + \beta\mathbf{B}^{-1}\mathbf{b}\mathbf{b}'\mathbf{B}^{-1} \end{pmatrix} \tag{5}$$

where $\beta = -(\mathbf{b}'\mathbf{B}^{-1}\mathbf{b})^{-1}$. We require $\mathbf{t} = \mathbf{T}\mathbf{e}/(\mathbf{e}'\mathbf{T}\mathbf{e})$, the centering vector for the c-centre of the face opposite C_1, where $\mathbf{T} = \begin{pmatrix} 0 & \mathbf{0}' \\ \mathbf{0} & \mathbf{B}^{-1} \end{pmatrix}$. From (5) we have:

$$\mathbf{T} = \begin{pmatrix} 0 & \mathbf{0}' \\ \mathbf{0} & \mathbf{B}^{-1} \end{pmatrix} = \mathbf{C}^{-1} - (\mathbf{c}_1\mathbf{c}_1'/c_{11}) . \tag{6}$$

$$\text{Hence} \quad \mathbf{t} = \frac{\mathbf{C}^{-1}\mathbf{e} - \frac{c_1}{c_{11}}\mathbf{c}_1}{\mathbf{e}'\mathbf{C}^{-1}\mathbf{e} - \frac{c_1^2}{c_{11}}} . \tag{7}$$

Thus the centering-vector for the face opposite C_1 is obtained simply by deflating the inverse of $\mathbf{C}$ by its first row and column and then normalising to unit total; similarly for the other faces. Clearly, by replacing $\mathbf{C}^{-1}$ by $\mathbf{T}$ in (6) and (7), with corresponding changes in $\mathbf{c}_1$, these formulae may be used to obtain, by iterative deflation, the centering vectors for the c-centres of an ever-decreasing number of vertices.

Writing $\mathbf{T}_1$ for the matrix $\mathbf{T}$ of (5) and $\mathbf{t}_1$ for the vector (6), we may similarly define $\mathbf{T}_2$ from which may be obtained the centering-vector $\mathbf{t}_2$ for the face opposite C_2. Similarly, we may define $\mathbf{T}_{12}$ which gives the centering-vector $\mathbf{t}_{12}$ for the $l-2$ vertices excluding C_1 and C_2 from the full set. Repeated use of (5) shows that $\mathbf{s}$ has the form:

$$\mathbf{s} = \gamma_1\mathbf{t}_1 + \gamma_2\mathbf{t}_2 + \gamma_{12}\mathbf{t}_{12}$$

for suitable constants γ_1, γ_2 and γ_{12}. Writing $C_{1.12}$ for the c-centre corresponding to $\mathbf{t}_{12}$, this shows that the c-centres $C, C_{1.1}, C_{1.2}$ and $C_{1.12}$ are coplanar in two dimensions. Clearly this result remains valid for any similarly related set of c-centres, even when l refers to some subset of all of the vertices, provided that it does not contain C_1 or C_2.

The result of the previous paragraph may be obtained, and generalised, by direct geometric argument, as follows. $\mathcal{F}_{12}$ divides $\mathbb{R}_n$ into two parts, one containing all the points nearer C_1 than C_2 and the other, all the points nearer C_2 than C_1. Thus this normal has dimensionality $l-2$ and contains all c-centres sharing the pair of suffices 1,2 such as $C_{12}, C_{123}, C_{1234}, \ldots, C_{123\ldots l} = C$. Similarly, the space $\mathcal{F}_{123}$ that is normal to the plane defined by C_1, C_2 and C_3 at C_{123} has dimensionality $l-3$ and contains all c-centres sharing the three suffices such as $C_{123}, C_{1234}, \ldots, C_{123\ldots l} = C$. Proceeding in this way we eventually find an $l-(l-2)$ plane that contains all c-centres that share $l-2$ suffices. For the $l-2$ suffices 1.12, there are only the four possibilities: $C, C_{1.1}, C_{1.2}$ and $C_{1.12}$, as found above.

A further result which may be verified algebraically from the results of this section but which follows more directly from geometrical considerations, is that the lines $C_\mathbf{m}C_{\mathbf{m}.i}$ and $C_{\mathbf{m}.i}C_{\mathbf{m}.ij}$ are orthogonal for all sets of subscripts $\mathbf{m}$ containing i and j. This follows from considering the $m-1$ vertices C_k, for all $k \in \mathbf{m}.i$, which lie in a space $\mathbb{R}_{m-2}$. Their c-centre is at $C_{\mathbf{m}.i}$ and the corresponding c-sphere has radius $R_{\mathbf{m}.i}$, say. $\mathbb{R}_{m-2}$ also contains $C_{\mathbf{m}.ij}$ for all $i \neq j \in \mathbf{m}$. Now $C_{\mathbf{m}.i}$ is equidistant ($R_{\mathbf{m}.i}$) from all C_k. $C_\mathbf{m}$ also is equidistant from all C_k which lie on the c-sphere with radius $R_\mathbf{m}$. It follows that $R_\mathbf{m}^2 = R_{\mathbf{m}.i}^2 + (C_\mathbf{m}C_{\mathbf{m}.i})^2$ and that $C_\mathbf{m}$ lies on the normal to $\mathbb{R}_{m-2}$ at $C_{\mathbf{m}.i}$. Hence, $C_\mathbf{m}C_{\mathbf{m}.i}$ is normal to all subspaces of $\mathbb{R}_{m-2}$, including the lines $C_{\mathbf{m}.i}C_{\mathbf{m}.ij}$ for all $i \neq j \in \mathbf{m}$. Also $C_\mathbf{m}$ is further from C than is $C_{\mathbf{m}.i}$.

Consider vectors parallel to the lines $C_1C_{12}, C_{12}C_{123}, C_{123}C_{1234}, C_{1234}C_{12345}, \ldots$. The remark at the end of the previous paragraph shows that every vector of the list is orthogonal to all those which occur to its left. Thus these vectors form an orthogonal basis for the space of the simplex, but with special reference to $\mathcal{M}_1$. Clearly many other similar bases may be constructed; the rule is that in the course of constructing the list, when the next higher level c-centre is chosen, it must introduce a previously unused suffix.

4 Algorithmic Considerations

The remarks made at the end of section 2.3 form the basis of an algorithm for computing the neighbour-regions in $\mathcal{L}^*$ but there are several details that remain to be considered. In this section, first a general outline of an algorithm is given and then some of the outstanding details in its implementation are briefly discussed. The algorithm requires an extension to the concept of joining two points. When P is a non-virtual point and Q is virtual, we require to draw a line in the direction from Q to P but starting at P; when both points are non-virtual joining has its conventional meaning; when both points are virtual, joining is defined to have a null effect.

Algorithm

(1) For each triplet ijk, form the space $\mathcal{F}_{ijk}$ that is normal to $C_iC_jC_k$ at C_{ijk}. Evaluate $\mathcal{M}^* \cap \mathcal{F}_{ijk}$. This is, usually, a single point, which carries the label ijk. Determine whether or not ijk is a virtual point.

(2) For all pairs of (ijk) and (lmn) join those that share two suffices.

(3) Finally back-project the polygonal tessellation onto $\mathcal{L}$ and label the regions with the category-level labels.

(4) When $s^* < r$, complete the prediction-regions in $\mathcal{L}$ by orthogonal extension.

Remarks

Step (1). By results given in section 3, the coordinates $\mathbf{c}$, say, of C_{ijk} are easily found and a pair of orthogonal vectors in the plane $C_iC_jC_k$ are given by the directions C_iC_{ij} and $C_{ij}C_{ijk}$. Suppose these two vectors are placed in the columns of a matrix $\mathbf{P} = (\mathbf{p}_1, \mathbf{p}_2)$. The matrix $\mathbf{M}^*$ is given as described in section 1.1 as the orthogonal projection of $\mathcal{L}$ onto $\mathcal{M}$. Because $\mathcal{F}_{ijk}$ generally does not contain the origin G, the row-coordinate $\mathbf{x}$ of any point in $\mathcal{N}_{ijk}$ is given by $\mathbf{x} = \mathbf{c} + \mathbf{w}$, where the normality with $C_iC_jC_k$ requires that $\mathbf{wP} = 0$. Also, because $\mathbf{x} \in \mathcal{M}^*$ we have that $\mathbf{x} = \mathbf{vM}^{*'}$ for some row-vector $\mathbf{v}$. Post-multiplying by $\mathbf{P}$ gives $\mathbf{v}(\mathbf{M}^{*'}\mathbf{P}) = \mathbf{cP}$, so determining $\mathbf{v}$ and giving:

$$\mathbf{x} = \mathbf{cP}(\mathbf{M}^{*'}\mathbf{P})^{-1}\mathbf{M}^{*'} .$$

This is the formula for determining the vertex ijk. If $\mathbf{M}^{*'}\mathbf{P}$ is singular, then $\mathbf{v}$ is not unique and, rather than determining a single point, ijk will be a line, or even a plane, bounded by intersections with the spaces of higher order c-centres of the form $ijkl\ldots$. This will occur whenever $\mathcal{M}^*$ happens to contain an edge or lie in the surface of one of the convex cones $\mathcal{M}_i$. Such solutions are pathological and are not explored further here. Nevertheless, a fully robust algorithm would need to take them into consideration.

We must distinguish three kinds of vertex: those not represented, those that are virtual points and those that are non-virtual points. When the equations $\mathbf{v}(\mathbf{M}^{*'}\mathbf{P}) = \mathbf{cP}$ are not consistent, there is no solution for ijk, so determining a "point at infinity" arising from a neighbour-region that is not represented in $\mathcal{M}^*$- see Figure 7(b) and its discussion in section 2. By definition ijk is equidistant r_{ijk}, say, from C_i, C_j and C_k. It is a virtual point if another vertex, C_l, can be found such that the distance, r_l, from ijk to C_l is less than r_{ijk}. This is easily determined by computing all r_l, for $l \neq i, j, k$.

Step (2). Recall the extended definition of joining. Its effect is as follows:

(i) A pair of non-virtual points is joined in the conventional way

(ii) A non-virtual point joined to a virtual point generates an arm, one end of which is unjoined to further points

(iii) The join of any two virtual points is null and so has no visible effect. Note that the coordinates of the ends of arms computed in Step (2) are used in Step (3).

Step (3). Back-projection is given by the formulae (4) given at the beginning of section 2 operating on all non-virtual triplets obtained in step (2). A method for labelling the back-projected regions derives from noting that every line is labelled by a pair of suffices ij, indicating the boundary between Q_i and Q_j. Either the centroid or g-circumcentre (Gower, 1985) of all the vertices on lines that share the suffix i is suggested as a suitable position for placing the label for Q_i.

References

ANDERSON T. W. (1958), *An Introduction to Multivariate Statistical Analysis*, New York, John Wiley.

DEVIJVER P.A. and DIEKESEI M. (1985), Computing multidimensional Delauney tessellations, *Pattern Recognition Letters*, 1, 311-6.

ECKART C. and YOUNG G. (1936), The approximation of one matrix by another of lower rank, *Psychometrika*, 1, 211-8.

GABRIEL K. R. (1971), The biplot-graphic display of matrices with applications to principal components analysis, *Biometrika*, 58, 453-67.

GIFI A. (1990), *Non-linear Multivariate Analysis*, New York, J. Wiley and Son.

GOWER J. C. (1966), Some distance properties of latent root and vector methods used in multivariate analysis, *Biometrika*, 53, 325-38.

GOWER J. C. (1968), Adding a point to vector diagrams in multivariate analysis, *Biometrika*, 55, 582-5.

GOWER J.C. (1982), Euclidean distance geometry, *The Mathematical Scientist*, 7, 1-14.

GOWER J.C. (1985), Properties of Euclidean and non-Euclidean distance matrices, *Linear Algebra and its Applications*, 67, 81-97.

GOWER J. C. (1991), *Generalised biplots*, Research Report RR-91-02, Leiden, Department of Data Theory.

GOWER J.C. (1992), *Biplot Geometry*.

GOWER J. C. and HARDING S. (1988), Non-linear biplots. *Biometrika*, 73, 445-55.

GREENACRE M.J. (1984), *Theory and Applications of Correspondence Analysis*. London, Academic Press

SIBSON R. (1980), The Dirichlet tessellation as an aid to data analysis, *Scandinavian Journal of Statistics*, 7, 14-20.

TORGERSON W. S. (1955) *Theory and Methods of Scaling*. New York, John Wiley.

WATSON D. F. (1981) Computing the n-dimensional Delauney tessellation with applications to Voronoi polytopes, *The Computer Journal*, 24, 167-72.

Different Geometric Approaches to Correspondence Analysis of Multivariate Data

Michael J. Greenacre

Department of Statistics, University of South Africa
P.O. Box 392, Pretoria, 0001 South Africa

Abstract: Just as there are many algebraic generalisations of simple bivariate correspondence analysis to the multivariate case, so there are many possible generalisations of the geometric interpretation. A number of different approaches are presented and compared, including the geometry based on chi-squared distances between profiles (techniques illustrated: joint correspondence analysis and Procrustes analysis), the interpretation of scalar products between variables (biplot), and the barycentric property (homogeneity analysis and unfolding).

1 Introduction

Simple correspondence analysis applies to an $I \times J$ contingency table $\mathbf{N}$ with grand total n, or – to make notation and description simpler – to the *correspondence table* $\mathbf{P}$ which is the $I \times J$ table of the elements of $\mathbf{N}$ divided by the grand total: $p_{ij} = n_{ij}/n$. The row and column marginal totals of $\mathbf{P}$ are called the row masses r_i, $i = 1, \ldots, I$, and the column masses c_j, $j = 1, \ldots, J$, respectively. Row or column profiles of $\mathbf{P}$ (equivalently, of $\mathbf{N}$) are the rows or columns divided by their respective totals, e.g. the i-th row profile is the i-th row of the matrix $\mathbf{D}_r^{-1}\mathbf{P}$, and the j-th column profile is the j-th column of the matrix $\mathbf{PD}_c^{-1}$, where $\mathbf{D}_r$ and $\mathbf{D}_c$ are the diagonal matrices of row and column masses respectively. The technique is usually described as a generalisation of principal component analysis where each set of profiles defines a cloud of points in a multidimensional space, each point is weighted by its respective masses, and each space is structured by a weighted Euclidean metric called the chi-squared distance (see, for example, Greenacre (1984) for a detailed description). Apart from the weighting of the points and the weighting of the original dimensions of the space, reduction of dimensionality takes place in much the same way as in principal component analysis.

The geometry of simple correspondence analysis has been abundantly described, debated and criticised in the literature. Controversy has long raged over the convention of overlaying principal component type maps of the row profiles and of the column profiles in a joint map, which we call the *symmetric map*. Greenacre & Hastie (1987) took an "asymmetric" view of the geometry and considered the profiles of the rows, for example, as points within a J-pointed irregular simplex in $(J-1)$-dimensional space. The vertices of this simplex are the extreme unit profiles which effectively depict the column categories in the row space. A low-dimensional subspace, for example a plane, is identified which approximates the row profiles optimally in terms of weighted least-squares and both row profiles and column vertices are projected orthogonally onto the subspace to facilitate the interpretation of the data. This *asymmetric map* is a joint plot which is easier to justify than the symmetric map, and some distance interpretations are possible between the rows and columns, even in lower-dimensional maps. Greenacre (1989) further clarified the limits of the geometric interpretation of simple correspondence analysis, in a critical response to a series of articles by Carroll, Green & Schaffer (1986, 1987). These articles claimed to offer a joint display in which all interpoint distance comparisons were interpretable. In a rejoinder to Greenacre (1989)'s criticism, Carroll, Green & Schaffer remarked that, since their approach was essentially to think of simple correspondence analysis as a special case of multiple correspondence analysis, this criticism of their approach was equally applicable to multiple correspondence analysis. This is quite correct – multiple correspondence analysis is not a straightforward generalization of simple correspondence analysis and has hardly any of the geometric appeal of simple correspondence analysis, an aspect already discussed by Greenacre & Hastie (1987).

It has often been suggested in the literature that a correspondence analysis map is a biplot and this leads some users to interpret angles between row and column points in a joint map. The validity of this viewpoint may depend on what is meant by the term "biplot" and I personally use the definition originally proposed by Gabriel (1971), namely that it is a method which maps a rectangular data matrix in terms of row and column points in a low-dimensional space such that the scalar product between the i-th row point and j-th column point approximates the (i,j)-th element of the data matrix. Greenacre (1991b) pointed out that asymmetric maps were indeed biplots, but not the conventional symmetric maps, and showed how biplot axes could be defined and calibrated in profile units in order to read off approximate profile values from the map. In the light of a wider meaning being given recently to the biplot, the terminology *linear biplot* is preferred for Gabriel's "classical" biplot as defined above, to distinguish it from the *nonlinear* and *generalized* biplots recently proposed by Gower & Harding (1988) and Gower (1991).

The multiple case is far more complex than the simple two-way case. Just as there are many different ways of generalising the algebra (see, for example, Meyer 1991), so there appear to be many ways of generalising the geometry. The present article is a review of known geometric interpretations of the multiple form of correspondence analysis, and the description of some new methods of mapping multiway frequency data. In all these methods except the one described last, the data matrix of interest is the so-called *Burt matrix*, or super-matrix of all two-way cross-tabulations. Thus we are interested in the geometry of the variables themselves rather than that of the cases and we are only interested in mapping two-way associations amongst the variables and no higher-order associations. The example used throughout will be a four-variable data set from a survey of Danish home-owners reported by Madsen (1976), and also by Cox & Snell (1981). The original four-way table is reported in those articles, while our own Table 1 shows the Burt matrix.

2   Distance scaling approach

In multidimensional scaling the fundamental idea is to define a distance measure which justifiably measures dissimilarity between a set of objects for the particular purpose at hand, and then to try to represent the objects in a space of low dimensionality such that there is as close agreement as possible between the original exact distances and the approximate distances measured in the map. There are many different ways of defining the measure of agreement between the original (or "true") and mapped (or "fitted") distances, and many different algorithms for optimizing the agreement. In all cases, however, the quality of the map may be judged by a so-called "Shepard diagram" where true and fitted distances are plotted against one other in a scatterplot.

In simple correspondence analysis there are two sets of objects (rows and columns) and

	HOUSING				INFLUENCE			CONTACT		SATISFACTION		
	H1	H2	H3	H4	I1	I2	I3	C1	C2	S1	S2	S3
H1	400	0	0	0	140	172	88	219	181	99	101	200
H2	0	765	0	0	268	297	200	317	448	271	192	302
H3	0	0	239	0	95	84	60	82	157	64	79	96
H4	0	0	0	277	124	106	47	95	182	133	74	70
I1	140	268	95	124	627	0	0	234	393	282	170	175
I2	172	297	84	106	0	659	0	279	380	206	189	264
I3	88	200	60	47	0	0	395	200	195	79	87	229
C1	219	317	82	95	234	279	200	713	0	262	178	273
C2	181	448	157	182	393	380	195	0	968	305	268	395
S1	99	271	64	133	282	206	79	262	305	567	0	0
S2	101	192	79	74	170	189	87	178	268	0	446	0
S3	200	302	96	70	175	264	229	273	395	0	0	668

Table 1: Burt matrix for four-variable housing data

two types of distances: *within-variable distances*, i.e. row-to-row distances and column-to-column distances, and *between-variable distances*, i.e. row-to-column distances. Greenacre (1989) showed an example of the distance-scaling approach to simple correspondence analysis: firstly, exact within-variable chi-squared distances between profiles are compared to their mapped counterparts measured in a planar correspondence analysis map; and secondly, the (approximate) monotonically inverse relationship between true row profile elements, say, and mapped row-to-column distances is shown in the asymmetric map which maps the row profiles along with the column vertices.

In the multivariate case, the situation is much more complicated. There are now Q variables and a number of different ways of defining within-variable and between-variable distances. Performing a correspondence analysis on the Burt matrix as if it were a two-way table, which is one way of defining multiple correspondence analysis, is practically impossible to defend from a distance scaling viewpoint, since both within-variable and between-variable distances are inflated by the diagonal tables on the super-diagonal of the Burt matrix. Joint correspondence analysis (Greenacre 1988, 1991a) is an attempt to rectify the situation by ignoring the super-diagonal of the Burt matrix in the fitting procedure. In practice this means that the super-diagonal is replaced by interpolated values which are fitted exactly by the low-dimensional solution. The "true" within- and between-variable distances are still affected by terms due to the interpolated super-diagonal, but at least these terms are exactly fitted by the solution and have a substantive meaning (the situation is analogous to the estimation of communalities on the diagonal of a correlation matrix in factor analysis). The joint correspondence analysis of Table 1 in two-dimensions is shown in Figure 1.

A different generalisation to the multivariate case is to consider the complete set of $Q(Q-1)/2$ simple correspondence analyses of pairs of variables. In the case of the housing data, none of these analyses has dimensionality greater than two, so the full set can be laid out exactly in a type of draughtsman's plot of maps (Figure 2). The symmetric maps for each analysis are given in the same triangular format as the lower triangle of Table 1, but another useful possibility would be to give both forms of asymmetric map for each pair of variables in a square format on opposite sides of the diagonal, reserving the diagonal for a bar-chart of the marginal frequencies for each variable (cf. the histogram of each continuous

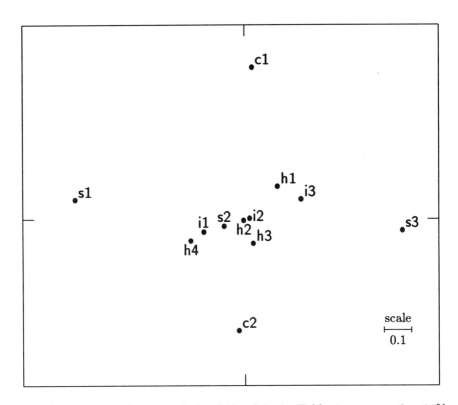

Figure 1: Joint correspondence analysis of the data in Table 1, representing 91% of the inertia

variable in the usual draughtsman's plot).

Figure 2 is thus a set of exact maps of all the two-way relationships, but each variable's set of categories appears $Q-1$ times, which complicates the inspection and interpretation of the maps. A synthesis of the maps may be obtained using generalised Procrustes analysis, by fitting each of the maps consecutively to an iteratively defined target map containing just one set of categories per variable. Only the rotation option in Procrustes analysis is required – the translation option is not necessary since each set of categories has its centroid at the origin; and the rescaling option is not applicable since the dispersion of the categories reflects how strong their association is. The Procrustes rotation should, however, be a weighted one, taking into account the category masses, which is an inherent aspect of correspondence analysis. The algorithm consists of obtaining an initial target configuration containing the four sets of categories, and then consecutively fitting each of the $Q(Q-1)/2$ maps to the target which is updated after each fitting. For example, with reference to Figure 2: use the I×H map as an initial target, fit the I points in the C×I map to the target and rotate the C points along with them; fit the C points in the S×C map and rotate the S points along with them – at this point there are two sets of C and I points which can each be averaged to obtain a target with one set of points for each variable; iterations can now commence, for example the C×H configuration is fitted to the target which is then updated, then the S×I map and is fitted, and so on.

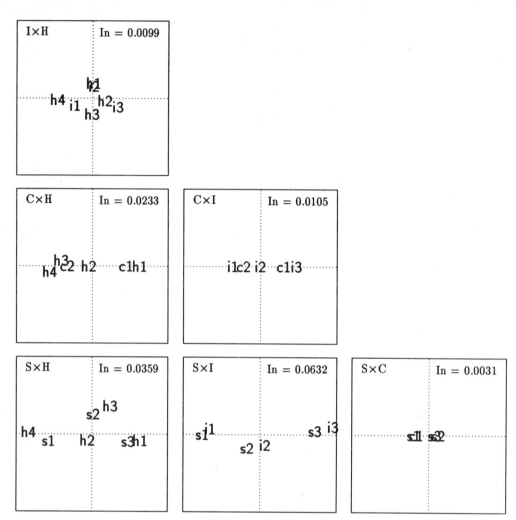

Figure 2: Simple correspondence analysis of all six pairwise cross-tabulations in lower triangle of Table 1, showing the inertia of each table. The scales of the maps are identical and each cross-tabulation is displayed exactly

At convergence, each correspondence analysis map will have been rotated to maximal agreement with a target configuration which is then interpreted as an average map of the bivariate associations. The average position of each category can be displayed along with its $Q-1$ individual positions which show its association with each of the other variables. There is an interesting decomposition of inertia in the resultant map (which in this example shows all the individual maps exactly). Each of the $Q(Q-1)/2$ pairwise inertias is represented twice by the individual category points, so that a measure of total inertia can be taken as the sum of all off-diagonal inertias of the Burt matrix, similar to joint correspondence analysis. By Huyghens' theorem, this total inertia is decomposed into an inertia concentrated into the average category points (a "between-groups" inertia) and an inertia of each sub-cloud of category points with respect to its respective average (a "within-groups" inertia). The

Procrustes analysis has achieved a minimum within-groups inertia, or equivalently a maximum between-groups inertia. In this particular example we would be performing such a decomposition exactly, but in general the full space of some of the cross-tabulations have dimensionality greater than two and the solutions would be projected onto a lower-dimensional subspace.

3 Linear biplot approach

To understand the link between correspondence analysis and the biplot, we recall the so-called "reconstitution formula" (see, for example, Greenacre 1984, p. 93) which expresses the proportions p_{ij} in the original data matrix in terms of the row and column masses and coordinates. One version of this formula is:

$$p_{ij} = r_i c_j (1 + \sum_k \sqrt{\lambda_k} \phi_{ik} \gamma_{jk}) + e_{ij} \tag{1}$$

where r_i and c_j are row and column masses, λ_k is the k-th principal inertia, and ϕ_{ik} and γ_{jk} are row and column standard coordinates, i.e. co-ordinates of the vertices in the respective row and column spaces. In the summation there are as many terms as there are dimensions in the solution and the residual error term e_{ij} absorbs that part of the data which is not accounted for by the solution. Formula (1) is now re-arranged so that the right hand side is in the form of a scalar product plus error:

$$\left(\frac{p_{ij}}{r_i} - c_j \right) = \sum_k f_{ik}(c_j \gamma_{jk}) + \tilde{e}_{ij} \tag{2}$$

where $f_{ik} = \sqrt{\lambda_k} \phi_{ik}$ is the principal coordinate of the i-th row profile on the k-th axis.

This shows that if we define the data to be the differences between the row profiles and their average profile, i.e. the lefthand side of (2), then the row profile points (with coordinates f_{ik}) and a rescaled version of the column vertex points (with coordinates $c_j \gamma_{jk}$) constitute a biplot. This is the correspondence analysis proposed by Gabriel & Odoroff (1990). Greenacre (1991b) showed that, for interpretation of a biplot, it is the *calibration* of the biplot axes that enables convenient reconstitution of the data in the map. Whether the biplot axes, defined by the directions of the vertex points, are rescaled or not is immaterial to this calibration. An example of a calibrated correspondence analysis biplot for a two-way table is given by Greenacre (1991b).

When we come to the multivariate case, the question of map quality becomes an issue. For a biplot to be successful, the data should be well-represented so that good data reconstitution is obtained by projecting points onto biplot axes. This immediately excludes the usual form of multiple correspondence analysis from being an acceptable biplot because the percentage of displayed inertia is always low. For example, in the multiple correspondence analysis of the Burt matrix of the housing data, the percentage of inertia displayed in the two-dimensional solution is only 37.8%. In the case of joint correspondence analysis, however, the percentage displayed in the two-dimensional solution is 91.0%. The biplot interpretation will thus be much more successful in the case of joint correspondence analysis, as illustrated in Figure 3. As an illustration we use the vertex position of category 1 of variable H to define a biplot axis, which has been calibrated in exactly the same manner as in simple correspondence analysis – for specific details see Greenacre (1991b). The estimated profile values of all the categories of the other three variables may now be read off the map by simply projecting the category points onto the biplot axis. For example, it appears that just over 30% of C1 (low contact) respondents fall into housing category H1 (tower blocks), while the corresponding figure for S3 (low satisfaction) respondents is just under 30%. This is why C1 and S3 are both on the side of the map which is high in the direction of H1. From the original data we can deduce that the actual profile values are $219/713 = .307$ and $200/668 = .299$ respectively. These values are thus very accurately represented in the display. The figure of 91% summarises the overall quality of reconstituting all the profile values from the map in this way, so we know that this biplot will be generally successful.

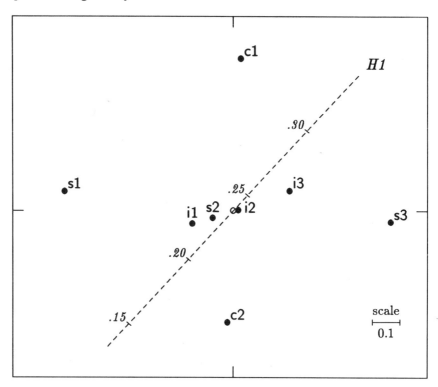

Figure 3: Joint correspondence analysis of the data in Table 1, showing calibrated biplot axis for category "H1"

4 Geometric relationship between cases and variables

In the above approaches, multiple correspondence analysis has come off rather poorly when we are concerned with a geometry of the variables which generalises the simple two-variable case. What, then, is the geometric rationale of multiple correspondence analysis? An argument by supporters of this method goes something like this. The difference in the argument is that the cases must be brought into the explanation – in fact, the analysis should rather be thought of as the correspondence analysis of the *super-indicator matrix* **Z**, the $N \times J$ matrix of zeros and ones which codes the categories of each of the N cases with resepct to the Q variables in dummy variable form, there being a J possible categories in total. Figure 4 shows the asymmetric map of the analysis of $\mathbf{Z}$ for the housing data, with the row points in standard coordinates and the column points (categories) in principal coordinates. Even though there are $N = 1681$ rows in $\mathbf{Z}$, there are only 72 unique ones, corresponding to the $4 \times 3 \times 2 \times 3 = 72$ cells in the four-way table. In Figure 4 we have coded each row with its respective value on the H (housing) variable, which has four categories. In this map the category point H1 is at the centroid (or weighted average, or barycentre) of all the row points labelled '1' (note that each of the '1' points accounts for a different number of respondents, not shown in the map, so that each of these points has different mass). This is known as the *barycentric property*, or the *centroid principle*. The next step in the justification would be to say that, if we had labelled the row points by the two categories of variable C, then the centroid of the '1' points would be the position of column point C1, which can be seen to be close to H1. H1 and C1 are thus close to one another, not because of any direct measure of distance between these two categories, but because of the 'H1' cases being in the same general area of the map as the 'C1' cases. A flaw in this argument is that no attention is paid to the validity of the display of the cases themselves, on which the justification depends. Is the geometry of the cases defensible, in other words how is distance between cases measured and how well are those distances displayed? On both counts the geometry of the cases is found to be wanting – first, one needs to justify the chi-squared distance between cases, where rare categories are emphasised, and second, the distances are usually very poorly displayed in a low-dimensional space.

Since homogeneity analysis is mathematically equivalent to multiple correspondence analysis, a comment is necessary about its geometric interpretation. Homogeneity analysis has the saving grace that it completely avoids the concept of a full space in which the points are supposed to lie. Each dimension is judged relative to a theoretical ideal of perfect homogeneity which the points might have attained in that dimension, and the performance of a solution on a dimension is measured by a "loss of homogeneity" (Gifi 1990). Imagine any random scatter of the 72 groups of row points of Figure 4 and imagine computing all the centroids of the 12 category points by successively identifying the subgroups of cases associated with each response category. If we were to connect each category point to the corresponding set of cases and measure the case-to-category distances, then the sum of squares of these distances would be the loss of homogeneity for the display (this exercise can be performed dimension by dimension and the losses for each dimension added, thanks to Pythagoras' theorem). For a random configuration of case points, the category points would tend towards the centre of the map and the loss would be high. In Figure 4, however, the loss is a minimum, in other words the sum of squared case-to-category distances is the least. Clearly, identification conditions are required for the case points in order to obtain a unique solution. In this case these conditions coincide with the normalisation of the row points to have sum-of-squares equal to 1 on each dimension, i.e. the same normalization as the standard coordinates in the correspondence analysis of $\mathbf{Z}$. This type of unfolding criterion between the case points and category points, where no distance scaling idea underlies the cases or the categories, is the only reasonable geometric rationale of multiple correspondence analysis, in my opinion. The success of such a criterion can be judged by counting how many smallest case-to-category distances correspond to ones in the indicator matrix.

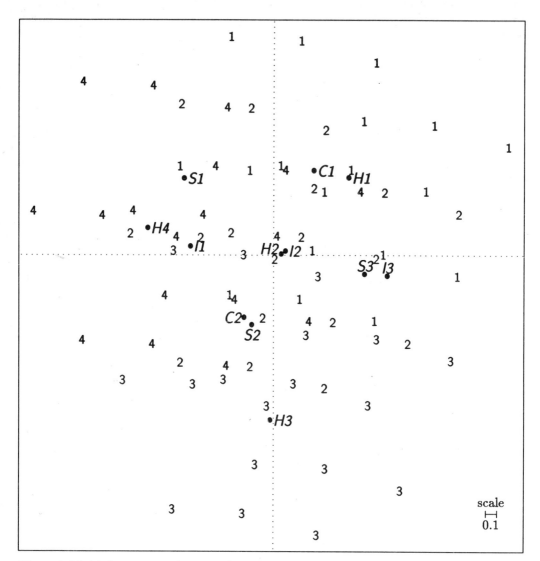

Figure 4: Multiple correspondence analysis of the indicator matrix corresponding to the data in Table 1. This is the asymmetric map with row points in standard coordinates and column points in principal coordinates

5 Conclusions

Whereas there are many ways of interpreting the same simple correspondence analysis map geometrically, generalisations to the multivariate case may have different interpretations. When interest is focussed on the association patterns between a set of categorical variables, joint correspondence analysis appears to be a natural generalisation of simple correspondence analysis. Procrustes analysis also needs to be investigated as an alternative way of constructing a map which is in some sense an average of all the two-way simple correspondence analyses. Joint correspondence analysis also generalises the linear biplot interpretation when it is of interest to reconstruct the profile elements in the map. The only valid geometric interpretation of multiple correspondence analysis appears to be the barycentric property which relates individual cases, or groups of cases, to their corresponding category points in an asymmetric map.

References

CARROLL, J.D., GREEN, P.E. & SCHAFFER, C. (1986), "Interpoint distance comparisons in correspondence analysis," *Journal of Marketing Research*, **23**, 271–280.

CARROLL, J.D., GREEN, P.E. & SCHAFFER, C. (1987), "Comparing interpoint distances in correspondence analysis: a clarification," *Journal of Marketing Research*, **24**, 445–450.

COX, D.R. & SNELL, E.J. (1981), *Applied Statistics: Principles and Examples*, London: Chapman and Hall.

GABRIEL, K.R. (1971), "The biplot graphical display of matrices with applications in principal component analysis," *Biometrika*, **58**, 453–467.

GABRIEL, K.R. (1981a), "Biplot display of multivariate matrices for inspection of data and diagnosis," in *Interpreting Multivariate Data*, ed. V. Barnett, Chichester: Wiley, 147–173.

GABRIEL, K.R. (1981b), "The complex correlational biplot," in *Theory Construction and Data Analysis in the Behavioral Sciences*, ed. S. Shye, San Francisco: Jossey-Bass, 350–370.

GABRIEL, K.R. & ODOROFF, C.L. (1990),"Biplots in biomedical research," *Statistics in Medecine*, **9**, 469–485.

GIFI, A. (1990), *Nonlinear Multivariate Analysis*, Chichester: Wiley.

GOWER, J. C. (1991), "Generalised biplots," Research report RR-91-02, Department of Data Theory, University of Leiden.

GOWER, J. C. & S. HARDING (1988), "Non-linear biplots," *Biometrika*, **73**, 445–455.

GREENACRE, M. J. (1984), *Theory and Applications of Correspondence Analysis*, London: Academic Press.

GREENACRE, M. J. (1988a), "Correspondence analysis on a personal computer," *Chemometrics and Intelligent Laboratory Systems*, **2**, 233–234.

GREENACRE, M. J. (1988b), "Correspondence analysis of multivariate categorical data by weighted least squares," *Biometrika*, **75**, 457–467.

GREENACRE, M. J. (1989), "The Carroll–Green–Schaffer scaling in correspondence analysis: a theoretical and empirical appraisal," *Journal of Marketing Research*, **26**, 358–365.

GREENACRE, M. J. (1990), "Some limitations of multiple correspondence analysis," *Computational Statistics Quarterly*, **3**, 249–256.

GREENACRE, M. J. (1991a), "Intepreting multiple correspondence analysis," *Applied Stochastic Models and Data Analysis*, **7**, 195–210.

GREENACRE, M. J. (1991b), "Biplots in correspondence analysis," Research Report CIT /91/R/07, Ecole des Mines, Sophia Antipolis, France, to be published in *Journal of Applied Statistics*.

GREENACRE, M. J. and T. J. HASTIE (1987), "The geometric interpretation of correspondence analysis," *Journal of the American Statistical Association*, **82**, 437–447.

MADSEN, M. (1976), "Statistical analysis of multiple contingency tables," *Scandinavian Journal of Statistics*, **3**, 97–106.

MEYER, R. (1992), "Canonical correlation analysis as a starting point for extensions of correspondence analysis," to appear in *Statistique et Analyse des Données*, June 1992.

Nonlinear Biplots for Nonlinear Mappings

Jacqueline J. Meulman and Willem J. Heiser

Department of Data Theory, University of Leiden

P.O. Box 9555, 2300 RB Leiden, The Netherlands

Abstract: The objective is the analysis of multivariate data, obtaining nonlinear transformations of the columns of the data matrix (the variables), nonlinear mappings for the rows (the objects), and in addition, a nonlinear biplot. The latter consists of nonlinear mappings of trajectories for the transformed variables in the space of the row-objects. Although ideas are applicable to the different techniques within the multivariate analysis framework, principal components analysis will be used as reference point.

Keywords: Principal Components Analysis, Principal Coordinates Analysis, Distance Approximation, External Unfolding, Joint Representation/Biplot

1 Introduction

Data are assumed to be available for n objects or individuals (the rows) and m variables (the columns) in the matrix $\mathbf{Z}$. The measurements on the objects for the m variables give the coordinates for each object in the observation space $\mathbf{Z}$; the columns in $\mathbf{Z}$ are assumed to be normalized to have means of zero and sum of squares of one. The starting point is classical principal components analysis, which can be viewed as a bilinear model (Kruskal, 1978). The aim is a joint representation of both the objects (in $\mathbf{X}$, of order $n \times p$) and the variables (in $\mathbf{A}$, of order $m \times p$) in low-dimensional p-space, with $p < m < n$, by minimizing a least squares loss function

$$\sigma(\mathbf{X}; \mathbf{A}) = \|\mathbf{Z} - \mathbf{X}\mathbf{A}'\|^2 = \sum_{j=1}^{m} \|\mathbf{z}_j - \mathbf{X}\mathbf{a}_j\|^2 . \tag{1}$$

A representation of objects and variables in the same space originates from Tucker (1960), and has become well-known as the biplot (Gabriel, 1971).

Principal components analysis in its original form (1) has been generalized to include nonlinear transformations of the variables (Kruskal & Shepard, 1974; Young, Takane & De Leeuw, 1978; Gifi, 1983, 1990; Winsberg & Ramsay, 1983). In the procedures proposed, the nonlinear transformation obtained is optimal in the sense that the loss according to the bilinear model will be as small as possible. Therefore, minimize

$$\sigma(\mathbf{Q}; \mathbf{X}; \mathbf{A}) = \|\mathbf{Q} - \mathbf{X}\mathbf{A}'\|^2 = \sum_{j=1}^{m} \|\mathbf{q}_j - \mathbf{X}\mathbf{a}_j\|^2 , \tag{2}$$

over $\mathbf{X}$, $\mathbf{A}$ and $\mathbf{Q}$, where columns of $\mathbf{Q}$ should satisfy $\mathbf{q}_j'\mathbf{q}_j = 1$ and $\mathbf{q}_j \in \Gamma_j$, where Γ_j indicates the set of admissible transformations of the given variable $\mathbf{z}_j$. The class of transformations may be defined differently for each variable $\mathbf{z}_j$, and includes nominal transformations (that preserve equal values in $\mathbf{z}_j$ by giving ties in $\mathbf{q}_j$), monotonic transformations (that maintain the order of the elements of $\mathbf{z}_j$ in $\mathbf{q}_j$), and linear transformations (that imply setting $\mathbf{q}_j = \mathbf{z}_j$, since it was required that $\mathbf{q}_j'\mathbf{q}_j = 1$).

2   PCA Viewed as a Multidimensional Scaling Technique

In contrast with the bilinear model (1), PCA can also be viewed as a particular multidimensional scaling problem. The Euclidean distances between the objects in the observation space $\mathbf{Z}$ are approximated by distances in a representation space $\mathbf{X}$ whose dimensionality, p, is again assumed to be (much) smaller than the dimensionality of the observation space. (Other MVA techniques have distance properties as well; see Meulman, 1986.) The representation space is a subspace of the observation space, and is obtained by a linear projection of the object points. This classical MDS mapping, based on projection, is known as Torgerson-Gower scaling or principal coordinates analysis (Torgerson, 1958; Gower, 1966). A squared distance between a pair of objects $\{i, k\}$ in $\mathbf{X}$ is defined by

$$d_{ik}^2(\mathbf{X}) \;=\; (\mathbf{e}_i - \mathbf{e}_k)'\mathbf{X}\mathbf{X}'(\mathbf{e}_i - \mathbf{e}_k)\,, \tag{3}$$

where $\mathbf{e}_i$ is the i-th column of the $n \times n$ identity matrix $\mathbf{I}$. The squared distance function, $D^2(\cdot)$, maps coordinates, $\mathbf{X}$, into squared distances in the matrix $D^2(\mathbf{X}) = \{d_{ik}^2(\mathbf{X})\}$, $i, k = 1,\ldots,n$. Distances in $\mathbf{Z}$ are defined analogously. The objective function that is minimized in classical MDS is a least squares loss function defined on the scalar products, and it can be written as

$$\sigma(\mathbf{X}) = \|\mathbf{Z}\mathbf{Z}' - \mathbf{X}\mathbf{X}'\|^2 \;=\; (1/4)\|\mathbf{J}(D^2(\mathbf{Z}) - D^2(\mathbf{X}))\mathbf{J}\|^2\,, \tag{4}$$

where $\mathbf{J}$ is a centering operator, $\mathbf{J} = \mathbf{I} - \mathbf{1}\mathbf{1}'/\mathbf{1}'\mathbf{1}$, with $\mathbf{1}$ an n-vector of all 1's.

In case the variables may be nonlinearly transformed, the columns in $\mathbf{Z}$ have to be replaced by optimal transformations in $\mathbf{Q}$ (see Meulman,1989), and the distances in $\mathbf{Q}$ are approximated in $\mathbf{X}$. Note that, although the transformation of the variables is nonlinear, the distance approximation would still involve a linear mapping from observation space into representation space.

3   Key Ideas of the Nonlinear Biplot

When loss function (4) is minimized, coordinates are obtained for the objects only. When the variables are to be represented as well, various possibilities exist. First of all, the (transformed) variables can be projected directly into the space, using multiple regression, in which the dimensions in $\mathbf{X}$ are the independent variables, and the $\mathbf{q}_j$ are the dependent variables. The coordinates are then obtained as $\mathbf{a}_j = (\mathbf{X}'\mathbf{X})^{-1}\mathbf{X}'\mathbf{q}_j$, and they will be equal to the rows of $\mathbf{A}$, found by minimizing (2) over $\mathbf{X}$ and $\mathbf{A}$ simultaneously, giving a linear biplot.

As an alternative to such a linear biplot, Gower & Harding (1988) considered to display objects and variables in a nonlinear biplot. The key ideas of a nonlinear biplot are as follows. First, regard each variable as a series of supplementary points (a trajectory) in the observation space. A supplementary point for variable j has coordinates in observation space that are all equal to zero, except for the j-th variable. So when $\mathbf{e}_j$ is the j-th column of the $m \times m$ identity matrix $\mathbf{I}$, the coordinates are given by $\mathbf{e}_{\rho_s} = \rho_s\mathbf{e}_j$, where $\min(\mathbf{q}_j) \leq \rho_s \leq \max(\mathbf{q}_j)$. Next, for each supplementary point, the distance is calculated to the n original points in observation space. The vector with squared distances between supplementary point $\mathbf{e}_{\rho_s}$ and the objects in $\mathbf{Q}$ is given by

$$d^2(\mathbf{e}_{\rho_s}, \mathbf{Q}) = \alpha + \mathbf{1}\mathbf{e}'_{\rho_s}\mathbf{e}_{\rho_s} - 2\mathbf{Q}\mathbf{e}_{\rho_s} = \alpha + \mathbf{1}\rho_s^2 - 2\mathbf{Q}\mathbf{e}_{\rho_s}, \tag{5}$$

with α an n-vector containing the diagonal elements of $\mathbf{Q}\mathbf{Q}'$. The k-th element of $d^2(\mathbf{e}_{\rho_s}, \mathbf{Q})$ gives the squared distance between the s-th supplementary point and the k-th object point in observation space, and will be written as $d^2(\mathbf{e}_{\rho_s}, \mathbf{q}_k^*)$, where $\mathbf{q}_k^*$ denotes the k-th row of $\mathbf{Q}$.

Mapping the trajectory for variable $\mathbf{q}_j$ involves the approximation of $d(\mathbf{e}_{\rho_s}, \mathbf{Q})$ by $d(\mathbf{y}_s, \mathbf{X})$, where $\mathbf{y}_s$ gives p-dimensional coordinates in the space of $\mathbf{X}$, for different values of ρ_s, $s = 1, \ldots, t$. Here t denotes a prechosen number, appropriate to cover the range of $\min(\mathbf{q}_j)$ to $\max(\mathbf{q}_j)$. Each supplementary point has to be mapped separately, using a coherent method. With the latter is meant that the mapping procedure to obtain $\mathbf{y}_s$ should be consistent with the method through which $\mathbf{X}$ itself was found.

4 Linear Mappings: Supplementary Points in a Principal Coordinates Analysis

The classic MDS approach finds the optimal object points by a linear projection; the principle of using a coherent method implies that the t supplementary points also should be projected from observation space into representation space. This can be done, for instance, by Gower's (1968) "add-a-point" procedure, which can be described as follows. To find the projection $\mathbf{y}_s$ of supplementary point s, define the vector $\mathbf{d}_s = \{d_{sk}\}$ whose k-th value is given by

$$d_{sk} = n^{-1}\sum_{i=1}^{n} d_{ik}^2(\mathbf{Q}) - 2n^{-2}\sum_{i=1}^{n}\sum_{k=1}^{n} d_{ik}^2(\mathbf{Q}) - d^2(\mathbf{e}_{\rho_s}, \mathbf{q}_k^*). \tag{6}$$

Then the supplementary point $\mathbf{y}_s$ in $\mathbf{X}$ is found as

$$\mathbf{y}_s = 1/2(\mathbf{X}'\mathbf{X})^{-1}\mathbf{X}'\mathbf{d}_s. \tag{7}$$

When $d^2(\mathbf{e}_{\rho_s}, \mathbf{q}_k^*)$ is a squared Euclidean distance (as in PCA), the supplementary points $\mathbf{y}_1, \ldots, \mathbf{y}_s, \ldots, \mathbf{y}_t$ for variable j will be located on a straight line through the origin. This straight line has exactly the same orientation as the vector $\mathbf{a}_j$ obtained from the regression $\mathbf{a}_j = (\mathbf{X}'\mathbf{X}) - \mathbf{1}\mathbf{X}'\mathbf{q}_j$. This can be seen from the following. The vector $\mathbf{d}_s = \{d_{sk}\}$ can be written as

$$\begin{aligned}\mathbf{d}_s &= n^{-1}D^2(\mathbf{Q})\mathbf{1} - 2n^{-2}(\mathbf{1}'D^2(\mathbf{Q})\mathbf{1})\mathbf{1} - d^2(\mathbf{e}_{\rho_s}, \mathbf{Q}) \\ &= n^{-1}(\alpha\mathbf{1}' + \mathbf{1}\alpha' - 2\mathbf{Q}\mathbf{Q}')\mathbf{1} - n^{-1}m\mathbf{1} - d^2(\mathbf{e}_{\rho_s}, \mathbf{Q}) \\ &= \alpha - d^2(\mathbf{e}_{\rho_s}, \mathbf{Q}).\end{aligned}$$

(Because $\mathbf{Q}$ is centered, the term $2\mathbf{Q}\mathbf{Q}'\mathbf{1}$ vanishes; because columns of $\mathbf{Q}$ have sum of squares equal to 1, $\alpha'\mathbf{1} = m$, and $\mathbf{1}'D^2(\mathbf{Q})\mathbf{1} = 2nm$). Since $\mathbf{e}_{\rho_s}$ has zero elements except for the j-th value, (3) can be written as

$$d^2(\mathbf{e}_{\rho_s}, \mathbf{Q}) = \alpha + \mathbf{1}\rho_s^2 - 2\mathbf{q}_j\rho_s,$$

and therefore

$$\mathbf{d}_s = 2\mathbf{q}_j\rho_s - \mathbf{1}\rho_s^2. \tag{8}$$

Substituting (8) in (7) gives .

$$y_s = 1/2(X'X)^{-1}X'(2q_j\rho_s - 1\rho_s^2) = (X'X)^{-1}X'q_j\rho_s \,. \tag{9}$$

(The term $X'1\rho_s^2$ vanishes, because X is centered). Since the optimal a_j in (2) are found as $a_j = (X'X)^{-1}X'q_j$, (9) reduces to $y_s = a_j\rho_s$, so the supplementary points are on a straight line through the origin, with the direction given by a_j.

There is an interesting relation between the trajectory for variable j and Gifi's (1990) approach to PCA that finds rank-one coordinates for each element of variable j. PCA in the Gifi system does not minimize $\sum_j \|q_j - Xa_j\|^2$, but starts with homogeneity analysis, also known as multiple correspondence analysis, in which each element of the variable j obtains coordinates in X. From there, the system applies proportionality restrictions on the set of coordinates for variable j (they have to be rank-one, i.e., on a straight line), and minimizes

$$\sigma(Q; A; X) = \sum_{j=1}^{m} \|q_j a_j' - X\|^2 \,, \tag{10}$$

where X and Q are centered, and X is normalized under the restriction that $X'X = I$. The a_j are dimension weights, and $q_{ij}a_j$ gives the coordinates for the i-th element of q_j in X. When $d^2(e_{\rho_s}, q_k^*)$ is a squared Euclidean distance, and ρ_s is chosen as $\rho_s = \{q_{ij}\}, i = 1, \ldots, n$, it follows that $y_s = q_{ij}a_j$. So when Gower's "add-a- point" procedure, with $\rho_s = \{q_{ij}\}$, is combined with classical MDS applied to $D(Q)$, the coordinates $y_1, \ldots, y_i, \ldots, y_k, \ldots, y_n$, giving the trajectory for variable j, are identical to the rank-one category coordinates $q_j a_j'$ in PCA according to Gifi (1990).

To conclude this section, two remarks should be made. First, (2) and (10) are only equivalent if X is normalized, $X'X = I$. If in PCA we wish distances in X to approximate distances in Q, X should be renormalized afterwards, giving $X'X = \Lambda^2$, with Λ^2 the diagonal matrix of eigenvalues, and a_j is then found as $a_j = \Lambda^{-2}X'q_j$. Secondly, the supplementary points in the classical procedure will only be on a straight line when Euclidean distances are used. For distance measures other than Euclidean, the projection of the supplementary points will give a nonlinear trajectory; hence the term nonlinear biplot in Gower and Harding (1988).

5 Nonlinear Mappings: Supplementary Points in a Principal Distance Analysis

In contrast with classical MDS and principal components analysis, multidimensional scaling techniques, as explored in Kruskal (1964), Guttman (1968), De Leeuw & Heiser (1980), Ramsay (1982), among others, give a nonlinear mapping of the objects: the representation space is not obtained by a projection. In the classical MDS approach the loss that is minimized is defined on the scalar products of the coordinates (in 4); in the MDS approach that will be discussed in the sequel, a least squares loss function is minimized that is defined on the distances. In the principal component analysis framework, a least squares loss function defined on the distances in Z and X is written as

$$\text{STRESS}(X) = \|D(Z) - D(X)\|^2 = \sum_{i=1}^{n}\sum_{k=1}^{n}(d_{ik}(Z) - d_{ik}(X))^2 \,. \tag{11}$$

When (11) is minimized over $\mathbf{X}$, we obtain a nonlinear mapping of the objects from the observation space $\mathbf{Z}$ in the representation space $\mathbf{X}$. Because the mapping is nonlinear, some distances in $\mathbf{X}$ may be larger than the corresponding distances in $\mathbf{Z}$, if this provides a minimum value for the loss measured in (11). Meulman (1986) reports that usually large distances in $\mathbf{Z}$ are approximated from above (i.e., they are larger in $\mathbf{X}$). Since the minimization of (11) has advantages over (4) with respect to distance fitting, it has been proposed to combine the nonlinear mapping of the objects with nonlinear transformation of the variables (De Leeuw & Meulman, 1986; Meulman, 1986). To attain this objective, we have to minimize a loss function that is a function of both the configuration $\mathbf{X}$ and the transformed variables in $\mathbf{Q}$; so, we minimize

$$\text{STRESS}(\mathbf{Q};\mathbf{X}) = \|D(\mathbf{Q}) - D(\mathbf{X})\|^2 , \tag{12}$$

over $\mathbf{X}$ and $\mathbf{Q}$, with $\mathbf{q}_j'\mathbf{q}_j = 1$ and $\mathbf{q}_j \in \Gamma_j$. The class of admissible transformations Γ_j again includes nominal, monotonic, and linear transformations. The corresponding technique has been called principal distance analysis in Meulman (1992).

When the representation space $\mathbf{X}$ is obtained by a nonlinear mapping, as in (12), the coherent method to fit the trajectory representing a variable should also use a nonlinear mapping, and not a projection as given by (7). To attain this objective, it is proposed to minimize

$$\text{STRESS}(\mathbf{y}_s) = \|d(\mathbf{e}_{\rho_s}, \mathbf{Q}) - d(\mathbf{y}_s, \mathbf{X})\|^2 \tag{13}$$

over $\mathbf{y}_s$ for given $\mathbf{Q}$ and $\mathbf{X}$. The loss function (13) represents the least squares external unfolding problem; it is called unfolding because it fits distances between two sets of points, $\mathbf{X}$ and $\mathbf{y}$, and it is called external because $\mathbf{X}$ is known and fixed. There is no closed-form solution for STRESS-based external unfolding as in (13), so the loss function has to be minimized iteratively.

The latter can be done using the SMACOF framework for unfolding (Heiser, 1981; 1987). The points on the j-th trajectory in $\mathbf{Q}$ will be mapped in $\mathbf{X}$ through the following procedure. For one point at the time, we start in some $\mathbf{y}_s$, and compute $\mathbf{b}_s = \{b_{sk}\}$ whose k-th value is given by

$$b_{sk} = \begin{cases} d(\mathbf{e}_{\rho_s}, \mathbf{q}_k^*)/d(\mathbf{y}_s, \mathbf{x}_k) \\ 0 \end{cases} \quad \text{if} \quad d(\mathbf{y}_s, \mathbf{x}_k) = 0 .$$

An update for the s-th point on the trajectory is given by

$$\tilde{\mathbf{y}}_s = n^{-1}\left[(\sum_{k=1}^{n} b_{sk})\mathbf{y}_s - \mathbf{X}'\mathbf{b}_s\right] . \tag{14}$$

Repeatedly computing (14), setting $\mathbf{y}_s = \tilde{\mathbf{y}}_s$ before each new step in the iterative scheme, gives a convergent series, until after a particular step $\|\mathbf{y}_s - \tilde{\mathbf{y}}_s\| \leq \epsilon$, with ϵ a preset small value. In contrast with the analysis in the previous section, where the projection of supplementary points using Euclidean distances resulted in a linear trajectory, the points $\mathbf{y}_1, \ldots, \mathbf{y}_s, \ldots, \mathbf{y}_t$, mapped in $\mathbf{X}$ using (14), will in general *not* be on a straight line, because the loss function (13) represents a nonlinear mapping. The complete process to obtain a nonlinear biplot is displayed in Figure 1.

206

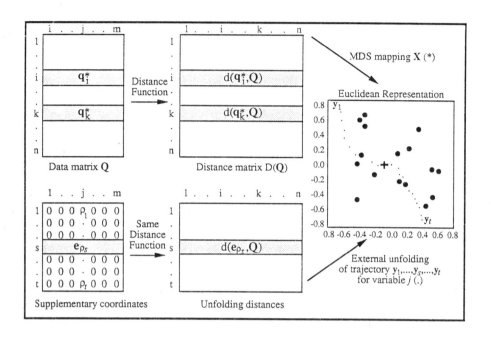

Figure 1: OVERVIEW OF THE PROCESS TO OBTAIN A NONLINEAR BIPLOT

6 Application of the Biplot using Nonlinear Mappings

The data analyzed to display the nonlinear biplot were collected in Leiden by Zeppa and Boon; they involve the relationship between 7 qualitative and 4 quantitative variables and the histologic diagnosis of cervical carcinoma and its precursor lesions (see Meulman, Zeppa, Boon & Rietveld, 1992). Data are available for 242 patients (50 cases each with mild dysplasia, moderate dysplasia, severe dysplasia, and carcinoma in situ, and 42 cases with invasive squamous cell carcinoma. The qualitative variables consist of subjective judgements of a sample of epithelial fragments, rated on a 4-point scale from normal to very abnormal. They concern Nuclear Shape, Nuclear Irregularity, Nucleus/Cytoplasm Ratio, Nucleus/Nucleolus Ratio, Nucleolar Irregularity, Chromatin Distribution, and Chromatin Pattern. In addition, 4 numerical features were established; these are Number of (abnormal) Cells per Fragment, Total Number of (abnormal) Cells, Absolute Number of Mitoses, and Number of Nucleoli.

The 4 numerical features were analyzed on an interval level. To obtain monotonic transformations for the 7 qualitative variables, spline functions were used (as in Winsberg & Ramsay, 1983; Ramsay, 1989). Using second degree monotonic spline functions, with one interior knot, fixes the number of parameters estimated for each qualitative variable to three.

The histologic diagnosis was entered into the analysis by using group variables: these are binary indicator variables for the 5 groups divided by the square root of the marginals. A linear mapping of the objects with nonlinear transformations of the variables would be equivalent to a PRINCALS analysis with mixed measurement levels (Gifi, 1990), and can be found in Meulman (1992).

For the principal distance analysis, which is the nonlinear mapping with nonlinear transformations, particular use is made of the SMACOF framework for restricted multidimensional scaling (De Leeuw & Heiser, 1980); the analysis is taken from Meulman (1992). The representation space **X** was obtained in two dimensions (with STRESS 0.0598), and is displayed in Figure 2; the points are labeled according to the histologic diagnosis. The five groups are clearly distinguished: the first dimension separates the groups 1, 2, and 3 from the groups 4 and 5, and the second dimension separates group 3 from the groups 1 and 2, and group 4 from group 5. The group points that are shown in Figure 2 are the centroids of the individual cases belonging to a particular group.

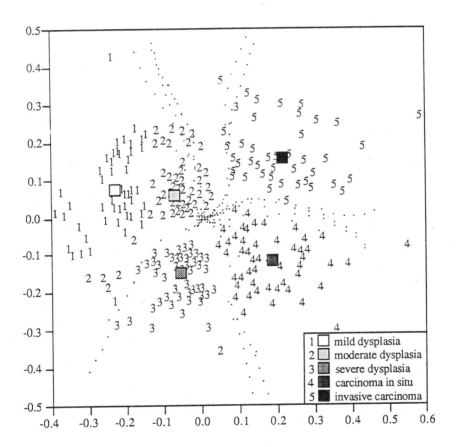

Figure 2: NONLINEAR BIPLOT FOR NONLINEAR MAPPINGIN CERVICAL CANCER EXAMPLE

Figure 2 also displays the low-dimensional trajectories of the variables, mapped into **X** by external unfolding using (13). We remark that they all go through the same point (when

$\rho_s = 0$ for all $j = 1, \ldots, m$), and that this point (with coordinates 0.0991 -0.057) is close to, but not coinciding with the centroid (0.0, 0.0) of all individual cases in the space $\mathbf{X}$. The trajectories have been highlighted in Figure 3, where the individual points have been omitted from the graphs, leaving only the group centroids. Each trajectory has a starting point and an endpoint; the latter represents the highest value, ρ_t, indicated by the symbol o.

The first row of three graphs displays trajectories that start in the upper left corner. Nuclear Shape and Nuclear Irregularity have their endpoint very close to the origin, from which we conclude that especially the small values of these variables are discriminating between groups. The trajectory of Chromatin Distribution passes the origin, and branches off towards the right-hand-side of the configuration, traversing the mid-area between groups 4 and 5. This particular trajectory is in fact the most nonlinear. The other trajectories are not very different from linear, but we obtain interesting information about the effective range of the variables.

Consider, for instance, the trajectories for Number of Cells per Fragment, Total Number of Abnormal Cells, and Number of Mitoses (the last row of graphs in Figure 3). These are very similar, and traverse the mid-area between groups 4 and 5, but their starting point is very close to the origin. The trajectories of Chromatin Pattern and Nucleus/Cytoplasm Ratio start in the group 2 area, and continue towards the bottom of the configuration, passing in between groups 3 and 4. Nucleolar Irregularity, Number of Nucleoli, and Nucleus/Nucleolus Ratio end in the group 5 area, but the latter starts close to the origin, while the first two start in the group 3 area. Also remarkable is the discontinuity that is sometimes displayed; consider, for example, the jump in the trajectory for Nucleolar Irregularity, in between the centroids for group 2 and 3.

The transformation of the qualitative variables is shown in Figure 4, with the optimal quantifications on the vertical axes versus the original scale values on the horizontal axes. If the nonlinear trajectories in Figure 3 are compared with the nonlinear transformations in Figure 4, we notice the following regularities. If the transformation is a concave function (Nuclear Shape, Nuclear Irregularity), the trajectory starts at the border of the configuration and ends near the origin. If the transformation is a convex function (Nucleus/Nucleolus Ratio), the trajectory starts near the origin, and extends towards the border. If the transformation is close to linear (Chromatin Pattern, Nucleus/Cytoplasm Ratio, Chromatin Distribution), both the starting point and the endpoint of the trajectories are remote from the origin; the shape of the trajectories, however, may be quite different.

The distance model that was fitted included fixed linear transformations of Number of Nucleoli, Number of Cells per Fragment, Total number of Cells, and Number of Mitoses. From the trajectories in low-dimensional space, we believe that the overall fit might be improved when for the latter three variables a monotonic transformation would be allowed. Since these trajectories start close to the origin, and end at the border, we conjecture that the transformations will be convex functions (compare Nucleus/Nucleolus Ratio).

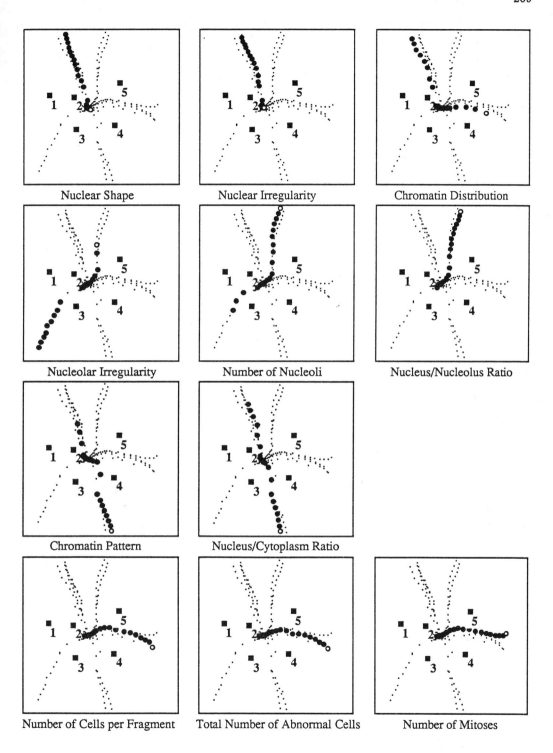

Nuclear Shape

Nuclear Irregularity

Chromatin Distribution

Nucleolar Irregularity

Number of Nucleoli

Nucleus/Nucleolus Ratio

Chromatin Pattern

Nucleus/Cytoplasm Ratio

Number of Cells per Fragment

Total Number of Abnormal Cells

Number of Mitoses

Figure 3: TRAJECTORIES IN **X**: HIGHLIGHTING EACH VARIABLE

210

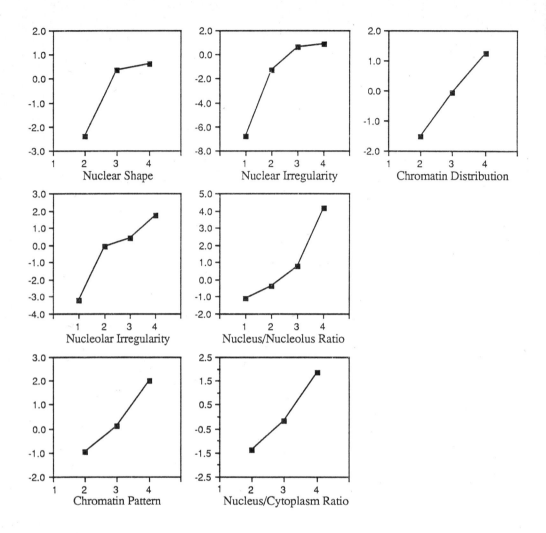

Figure 4: NONLINEAR TRANSFORMATIONS FOR VARIABLES IN CERVICAL CANCER EXAMPLE

7 A Nonlinear Biplot Using Chi-Squared Distances

In the previous sections, principal components analysis was taken as a reference point. In this section, we will go beyond ordinary Euclidean distance measures to be fitted, and focus on the chi-squared distance, operating in correspondence analysis. The chi-squared distance between two profiles i and k in the data matrix $\mathbf{F}$ (with positive entries) is defined as

$$d^2(\mathbf{f}_i, \mathbf{f}_k) = N \sum_j (f_{ij}/r_i - f_{kj}/r_k)^2/c_j ,$$

where r_i and r_k are the totals of row i and k, respectively, c_j is the total of column j, and $N = \sum_i \sum_j f_{ij} = \mathbf{c'1}$. The chi-squared distance is special, because it is invariant under scalar multiplication of the profiles. As a result, when the trajectory along one axis in high-dimensional space is considered, with $\rho_s = \{\min(\mathbf{f}_j), \ldots, \max(\mathbf{f}_j)\}$, the entire trajectory using the chi-squared metric accumulates into one point. This is rather unfortunate from the viewpoint of the nonlinear biplot. So, to make an interesting biplot, we have to consider other trajectories. Since profiles that are proportional to the column marginal $\mathbf{c} = \{c_j\}$ coincide with the centroid in high-dimensional space, we propose to define a supplementary row of coordinates as

$$N\mathbf{e}_{\rho_s} + (1 - \rho_s)\mathbf{c} \, .$$

If $\rho_1 = 0$, $N\mathbf{e}_{\rho_s} + (1 - \rho_s)\mathbf{c} = \mathbf{c}$, and the supplementary point will be located in the centroid. By varying ρ_s, we can now follow the rays extending from the centroid towards the corner points in high-dimensional space, and by using external unfolding again, these trajectories can be mapped into the low-dimensional space for the row points. The endpoint of the j-th trajectory is obtained for $\rho_1 = 1$, where $N\mathbf{e}_{\rho_s} + (1 - \rho_s)\mathbf{c} = N\mathbf{e}_j$.

If we apply classic scaling to the chi-squared distances between the row points in high-dimensional space, and project the trajectories for the columns into the low-dimensional space obtained for the row points, the result will become equivalent to an ordinary correspondence analysis, provided the right normalization is chosen. We obtain a joint display of row points and column trajectories, where the latter will be straight lines. If, instead, we approximate the chi-squared distances using a STRESS-based criterion, as in (11), and unfold the trajectories with the coherent method, as in (13), an alternative joint display will be obtained, with nonlinear trajectories extending from a common point.

8   Discussion

Although application of the nonlinear biplot is straightforward, a number of issues remain for further study. Why do some trajectories display discontinuities? Can we use information from the trajectory as a diagnostic for choosing a suitable nonlinear transformation of a particular variable (an axis of the high- dimensional space); for instance, if a trajectory is highly nonlinear, should we allow a completely nonlinear, i.e. nonmonotonic, transformation? If coordinates for a new object become available in low-dimensional space, can we predict the coordinates in high-dimensional space?

The classic biplot is a first order approximation; the nonlinear biplot is a genuine generalization, but since we no longer assume that the first order approximation will give a representation with an acceptable badness-of-fit, it also becomes a diagnostic: nonlinearity is assumed to indicate the discrepancy between the data and the representation in a linear biplot. Including nonlinear transformations of the axes in high-dimensional space, and mapping the new axes as possibly nonlinear trajectories, narrows the gap between data and representation even further. Other interesting trajectories than the axes may be mapped as well; we have shown that the search for other trajectories is imperative for chi-squared distances, used in correspondence analysis, but his pursuit may also be worthwhile or indispensable for other dissimilarity measures.

References

DE LEEUW, J., and HEISER, W.J. (1980). Multidimensional scaling with restrictions on the configuration. In P.R. Krishnaiah (ed.), *Multivariate analysis*, Vol. V (pp. 501-522). Amsterdam: North-Holland.

DE LEEUW, J., and MEULMAN, J.J. (1986). Principal component analysis and restricted multidimensional scaling. In W. Gaul and M. Schader (eds.), *Classification as a tool of research* (pp. 83-96). Amsterdam: North-Holland.

DE LEEUW, J. and HEISER, W.J. (1977). Convergence of correction-matrix algorithms for multidimensional scaling. In: J.C. Lingoes (ed.), *Geometric representations of relational data* (pp. 735-752). Ann Arbor, Michigan: Mathesis Press.

GABRIEL, K.R. (1971). The biplot-graphic display of matrices with applications to principal component analysis. *Biometrika*, 58, 453-467.

GIFI, A. (1983). *PRINCALS User's Guide*. Leiden: Department of Data Theory.

GIFI, A. (1990). *Nonlinear multivariate analysis* [first edition 1981, Department of Data Theory, The University of Leiden]. Chichester: Wiley.

GOWER, J.C. (1966). Some distance properties of latent roots and vector methods used in multivariate analysis. *Biometrika*, 53, 325-338.

GOWER, J.C. (1968) Adding a point to vector diagrams in multivariate analysis. *Biometrika*, 55, 582-585.

GOWER, J.C., and HARDING, S.A. (1988). Nonlinear biplots. *Biometrika*, 75, 445-455.

GUTTMAN, L. (1968). A general nonmetric technique for finding the smallest coordinate space for a configuration of points. *Psychometrika*, 33, 469-506.

HEISER, W.J. (1981). *Unfolding analysis of proximity data*. PhD thesis. Leiden: Department of Data Theory.

HEISER, W.J. (1987). Joint ordination of species and sites: the unfolding technique. In P. Legendre and L. Legendre (eds.), *Developments in numerical ecology* (pp. 189-221). New York: Springer.

KRUSKAL, J.B. (1964). Multidimensional scaling by optimizing goodness of fit to a non-metric hypothesis. *Psychometrika*, 29, 1-28.

KRUSKAL, J.B. (1978). Factor analysis and principal components analysis: bilinear methods. In W. H. Kruskal and J. M. Tanur (Eds.), *International Encyclopedia of Statistics*, (pp. 307- 330). New York: The Free Press.

KRUSKAL, J.B., and SHEPARD, R.N. (1974). A nonmetric variety of linear factor analysis. *Psychometrika*, 39, 123-157.

MEULMAN, J.J. (1986).*A distance approach to nonlinear multivariate analysis*. Leiden: DSWO Press.

MEULMAN, J.J. (1989). *Nonlinear principal coordinates analysis: minimizing the sum of squares of the smallest eigenvalues of a correlation matrix*. Research Report RR-89-05. Leiden: Department of Data Theory.

MEULMAN, J.J. (1992). The integration of multidimensional scaling and multivariate analysis with optimal transformations. *Psychometrika* (in press).

MEULMAN, J.J., ZEPPA, P., BOON, M.E., and RIETVELD, W.J. (1992). Prediction of various grades of cervical preneoplasia and neoplasia on plastic embedded cytobrush samples: discriminant analysis with qualitative and quantitative predictors. *Analytical and Quantitative Cytology and Histology*, 14, 60-72.

RAMSAY, J.O. (1982). *MULTISCALE II manual.* Department of Psychology, McGill University, Montréal.

RAMSAY, J.O. (1989). Monotone regression splines in action. *Statistical Science*, 4, 425-441.

TORGERSON, W.S. (1958). *Theory and methods of scaling.* New York: Wiley.

TUCKER, L. R (1960). Intra-individual and inter-individual multidimensionality. In: H. Gulliksen and S. Messick (eds.), *Psychological Scaling: Theory & Applications.* New York: Wiley.

WINSBERG, S., and RAMSAY, J.O. (1983). Monotone spline transformations for dimension reduction. *Psychometrika*, 48, 575-595.

YOUNG, F.W., TAKANE, Y., and DE LEEUW, J. (1978). The principal components of mixed measurement level multivariate data: An alternating least squares method with optimal scaling features. *Psychometrika*, 43, 279-281.

Gradient Filtering Projections for Recovering Structures in Multivariate Data

I.S. Yenyukov

Tverskaya Str. 11, Innofund, 103950 Moscow, GUS

Abstract: In this paper we introduce several sets of linear projections for low-dimensional projecting multivariate data witch are used for visual revealing structures in them. These projection sets are connected with the gradient filtering process defined by Fukunaga, Hosteller (1975). We define the following projection directions related to the filtering: directions with the extreme change of variance, directions with the extreme relocation of projected data points and the predominant gradient directions.

These projection sets coincide with the principal components set if the density of the data is ellipsoidally symmetric. Two first sets of projections approximate the Rao canonical discriminant projections if the density is a mixture of ellipsoidally symmetric densities with an equal within-covariance matrix.

Extraction of the projections is based on the solution of generalized eigenvector problems. We use "k-nearest neighbours" approch for estimating the density gradient.

Keywords: projection pursuit, gradient filtering, structure detecting, filtered estimates.

1 Introduction

It is well known that the computational realization of projection pursuit is quite a difficult problem (see Friedman (1987), for example) due to the local maximum problem in particular.

In this paper we design several new projection indexes for the projection pursuit technique (see Section 3). The corresponding computational procedures of obtaining the projections request only the solution of generalized eigenvector problems and avoid the local maximum problem (see Section 4).

These projections sets coincide with the principal components set if the density of the data is ellipsoidally symmetric. Two first sets of the projections approximate the Rao canonical discriminant projections if the density is a mixture of ellipsoidally symmetric densities with an equal within-covariance matrix (see Section 5). The gradient estimate is given in Section 6.

2 Gradient filtering process (GFP) and projections related to it

Let X be a p-variate random vector with the density $p(X)$. We define the gradient filtering as the transformation of the random vector X (see also Fukunaga, Hosteller (1975))

$$Y = X + a \nabla_X \ln p(X). \qquad (1)$$

For simplification of further expansions we are going to use $g(X)$ as $\nabla_X \ln p(X)$, i.e. $g(X) = \nabla_X \ln p(X)$.

It is possible to prove that under some conditions $EY = EX$ (see Fukunaga, Hosteller (1975)). So below we consider that $EX = 0$.

Now let U be a normalized p-variate vector, i.e. $\| U \| = 1$. We can define the following interesting projections related to GFP.

(i) Projections with extreme variance change. The corresponding projection index and extremum problem are

$$(U^t S_y U)/(U^t S_x U) \Longrightarrow \mathop{\text{extr}}_{U} , \qquad (2)$$

where S_y is the covariance matrix of Y, and S_x is the one of X.

(ii) Projections with extreme relocations of projected points. The corresponding projection index and extremum problem are

$$E(U^t(Y - X))^2 \Longrightarrow \mathop{\text{extr}}_{U} . \qquad (3)$$

(iii) Predominant gradient directions. The corresponding index and the extremum problem are

$$E((U^t g(X))^2/ \| g(X) \|^2) \Longrightarrow \mathop{\text{extr}}_{U} . \qquad (4)$$

In fact for revealing data structures the projections with both maximal and minimal values of these indexes (2), (3), (4) are of interest.

3 Computational procedure

The solution of problems (2), (3), (4) can be reduced to the solution of the following eigen-vector problems (or generalized eigenvectors)

(i) $$(S_y - \lambda S_x)U = 0 \qquad (5)$$
(ii) $$(V - \lambda I)U = 0, \qquad (6)$$
$$\text{where} \quad V = E(Y - X)(Y - X)^t \qquad (7)$$
(iii) $$(G - \lambda I)U = 0, \qquad (8)$$
$$\text{where} \quad G = E(\nabla_x \ln p(X)\nabla_x \ln p(X)^t)^2/ \| \nabla_x \ln p(X) \|^2 .$$

4 Filtered covariance matrices

Now we give the analytical expansions for the filtered covariance matrices S_y and V.

$$S_y = S_x - 2aI + a^2 E(g(X) g^t(X)) \qquad (9)$$
$$V = a^2 E(g(X) g^t(X)). \qquad (10)$$

Now let $p(X)$ be an ellipsoidally symmetric density

$$p(X) = f(X^t W^{-1} X).$$

Then covariance matrix $S_x = W$ and

$$g(X) = \nabla_x \ln p(X) = 2W^{-1}Xf'(X^tW^{-1}X)/p(X).$$

We have

$$
\begin{array}{rcl}
S_y &=& W - 2aI + c(a,f)W^{-1}, \\
V &=& c(a,f)W^{-1},
\end{array}
$$
(11)
(12)

where $c(a,f)$ is a positive constant dependent on the density f and the value a.

5 Why the projections are useful for revealing structures in data?

It is possible to prove the following statement: if the density $p(X)$ is ellipsoidally symmetric, then the eigenvector sets of the problems (5), (6), (8) coincide with the eigenvectors of matrix W, i.e. with the principal components set. Certainly the ordering of the vectors in the sets may be different. For problems (5), (6) the statement follows directly from (11), (12).

So we can see these sets of projections are some generalization of the principal components set.

Now let $p(X)$ be a mixture of ellipsoidally symmetric densities which have an equal within-covariance matrix W. At first we take the case when the densities of such a mixture have restricted supports and the latter are not overlapped. Then we have

$$
\begin{array}{rcl}
S_y &=& S - 2aI + t(a,p)W^{-1} \\
V &=& t(a,p)W^{-1},
\end{array}
$$
(13)
(14)

where $t(a,p)$ is a positive constant dependent on the density p and the value a.

Let us consider now that the covariance matrix $S = I$. We always can have it by suitable linear transformation. Then the Rao canonical discriminant vector set is defined as the eigenvector set of the problem

$$(S - lW)U = 0 \quad \text{or} \quad (I - lW)U = 0.$$

And it is easy to see that the eigenvectors of matrices (13), (14) coincide with the Rao canonical vector set.

In a more general case when overlapping is present, the expansions (13), (14) contain terms dependent on the overlapping level and their eigenvector systems are only some approximations of the canonical discriminant set.

6 Gradient estimating

For gradient estimating we use the "k-nearest neighbours mean-shift" estimator (Fukunaga, Hosteller (1975)). Let $X_1, \ldots, X_n$ be a random sample from p. Then the estimate of $g(X)$ in the point X is

$$\hat{g}(X) = b\sum(X_i - X)/k,$$

where k is the number of neighbours,

$b = (p+2)/d_k,$

d_k is the radius of the "k-nearest neighbours" sphere of the point X,

and the sum includes only points X_i belonging to the "k-nearest neighbours" sphere of the point X.

References

FRIEDMAN, J.H. (1987), Exploratory Projection Pursuit, *J. of Amer. Stat. Ass.*, 82, 249–266.

FUKUNAGA, K., HOSTELLER, L.D. (1975), The Estimation of the Gradient of a Density function, with Application in Pattern Recognition, *IEEE Tr. Information Theory*, 32–40.

YENYUKOV, I.S. (1988), Detecting Structures by Means of Projection Pursuit, in: *Proceedings of COMPSTAT-1988*, 48–58.

Classification with Set–Valued Decision Functions

Eugen Grycko
Department of Mathematics, University of Hagen
Lützowstr. 125, D 5800 Hagen

Abstract: An important variant of the classification problem can be formulated as follows: given a learning sample (supervised learning) a classifier shall be determined, whose probability of correct classification is a good approximation of the maximal (Bayesian) correct classification probability. If there is no parametric familiy available which encloses the class-specific distributions of the observable, then the statistician is forced to apply non-parametric procedures (e.g. density estimators).
In this paper a classical non-parametric classification model is approximated by a three-stage parametric Bayesian model. Basing on the approximating model classifiers are presented. The construction of these classifiers applies substantially set-valued decision functions – a concept which was consciously stated and thoroughly studied by O. Moeschlin 1982 and by H. Meister and O. Moeschlin 1986 and 1988. We conclude with the discussion of simulation results on the quality of the new family of classifiers.

1 Introduction

Let κ be an integer ≥ 2. Suppose we are given κ classes $K_1, \ldots, K_\kappa$ of objects. A measurement on an object belonging to an unknown class $K \in \{K_1, \ldots, K_\kappa\}$ is represented as a point x in a measurable space $(\mathbb{H}, \mathcal{H})$. The goal is to predict what class $K \in \{K_1, \ldots, K_\kappa\}$ the given object is in.

A well known possibility of describing the generation of measurements on objects is a two-stage Bayesian model:

(S1) The choice of an object generates an inobservable class $K \in \{K_1, \ldots, K_\kappa\}$
according to a probability distribution Q on $\{K_1, \ldots, K_\kappa\}$.

(S2) By the measurement a point (observable) $x \in \mathbb{H}$ is generated
according to the class-dependent probability measure P^K on $(\mathbb{H}, \mathcal{H})$.

Remark 1: If the probability measures Q and P^{K_j}, $j = 1, \ldots, \kappa$ are known, then an optimal (Bayesian) classification rule (classifier) $c_B : \mathbb{H} \longrightarrow \{K_1, \ldots, K_\kappa\}$ can easily be constructed (see L. Breiman et al. (1984), sec 1.5); the optimality of a classifier $c : \mathbb{H} \longrightarrow \{K_1, \ldots, K_\kappa\}$ means that the probability of correct classification

$$\tau(c) := \sum_{j=1}^{\kappa} P^{K_j}(c = K_j) \cdot Q(K_j) \tag{1}$$

is maximal.

A typical and interesting situation arises if the probability measures Q and $P^{K_1}, \ldots, P^{K_\kappa}$ are not known but the past experience is represented by a learning sample $(x^{(1)}, y^{(1)}), \ldots, (x^{(n)}, y^{(n)}) \in \mathbb{H} \times \{K_1, \ldots, K_\kappa\}$, each pair consisting of a measurement $x^{(i)}$ made on an object which is known to be a member of the class $y^{(i)} \in \{K_1, \ldots, K_\kappa\}$ for $i = 1, \ldots, \kappa$ respectively. We stress here on situations where the probability measures $P^{K_1}, \ldots, P^{K_\kappa}$ cannot be assumed to belong to a parametric family. In this case there are two important possibilities of constructing a classifier: non-parametric density estimation and the (k-th) nearest neighbor rule (see Breiman et al. (1984), p. 15).

In section 2 we present a Bayesian model for set-valued decision functions: Bayesian experiment for confidence estimation (BECE). In section 3 we apply the notion of BECE to construct a three-stage classification model which is shown to be consistent with the classical one ((S1) + S2)). In section 4 we give a learning algorithm for a classifier which applies our three-stage model. For the sake of comparison we report, in section 5, on a comparative simulation study concerning the quality of the new family of classifiers.

2 Bayesian Experiment for Confidence Estimation (BECE)

Let Θ be a Polish space and $\mathcal{T}$ its Borel-σ-algebra. Let P be a probability measure on $(\mathbb{H} \times \Theta, \mathcal{H} \otimes \mathcal{T})$. Let $P_{\mathbb{H}}$ and P_{Θ} denote the marginal distribution of P on $\mathbb{H}$ and on Θ respectively. The conditional distribution on $\mathbb{H}$ given $\vartheta \in \Theta$ under P will be denoted by $P(.|\vartheta)$; $(P(.|\vartheta))_{\vartheta \in \Theta}$ is a family of sample distributions. Let $g : \Theta \longrightarrow \mathbb{R}^k$ be a measurable function into the k-dimensional Euclidian space $\mathbb{R}^k$. We define the decision space:

$$\mathcal{C}(\mathbb{R}^k) := \{C \subset \mathbb{R}^k \,|\, C \neq \emptyset, C \text{ is compact and convex}\};$$

this decision space for set-valued decision functions was introduced by H. Meister and O. Moeschlin (1986) and (1988); $\mathcal{C}(\mathbb{R}^k)$ can be endowed with the Hausdorff metric which on its part induces the corresponding σ-algebra on $\mathcal{C}(\mathbb{R}^k)$. Suppose we are given a loss function $v : \mathbb{R}^k \times \mathcal{C}(\mathbb{R}^k) \longrightarrow \mathbb{R}_+$; intuitively, the expression $v(g(\vartheta), C)$ gives the loss of the set-valued decision $C \in \mathcal{C}(\mathbb{R}^k)$ in case where the true parameter value is ϑ.

Definition 1: The five-tupel $(\mathbb{H} \times \Theta, \mathcal{H} \otimes \mathcal{T}, P, g, v)$ is called BECE.

Let $P(.|x)$ denote the conditional distribution on Θ given $x \in \mathbb{H}$ under P (posterior distribution).

Definition 2: Let be $\alpha \in (0,1)$. A measurable map $Z : \mathbb{H} \longrightarrow \mathcal{C}(\mathbb{R}^k)$ is called α-correspondence, iff

$$P(g^{-1}(Z(x))|x) \geq \alpha \tag{2}$$

holds for every $x \in \mathbb{H}$. The set of all α-correspondences will be denoted by $\mathcal{Z}_\alpha$.

Remark 2: The inequality (2) is a classical confidence condition in Bayesian context.

Remark 3: In E. Grycko (1991) the following decision problem is considered:

For given $\alpha \in (0,1)$ find an (the) α-correspondence $\overline{Z} \in \mathcal{Z}_\alpha$ which minimizes the Bayes' risk; the Bayes' risk of a decision function $Z \in \mathcal{Z}_\alpha$ is given by

$$R(Z) := \int_{\mathbb{H} \times \Theta} v(g(\theta), Z(x)) dP(x, \theta)$$

Under mild conditions on the BECE existence and uniqueness theorems can be deduced for optimal α-correspondences; there are, moreover, numerical criteria available for calculating the optimal set-valued decisions (cf. E. Grycko (1991)).

3   A non-Classical three-Stage Classification Model

Recall the notation of sections 1 and 2. Let be $(\mathbb{H} \times \Theta, \mathcal{H} \otimes \mathcal{T}, P, g, v)$ a BECE. We assume the existence of probability measures $Q_{K_1}, \ldots, Q_{K_\kappa}$ on $(\Theta, \mathcal{T})$ satisfying the following two conditions

$$P_\Theta = \sum_{j=1}^{\kappa} Q(\{K_j\}) \cdot Q_{K_j} \tag{C1}$$

$$P^{K_j}(H) = \int_\Theta P(H|\vartheta) \, dQ_{K_j}(\vartheta) \tag{C2}$$

for all $H \in \mathcal{H}$ and $j = 1, \ldots, \kappa$.

Remark 4:   Condition (C1) states that the fitting of the BECE allows a convex combination of the a priori distribution P_Θ by the measures $Q_{K_1}, \ldots, Q_{K_\kappa}$. Condition (C2) states that the class-dependent distributions $P^{K_1}, \ldots, P^{K_\kappa}$ of the measurements can be represented by mixing the family $(P(.|\vartheta))_{\vartheta \in \Theta}$ of sample distributions.

Remark 5:   Condition (C2) is a mild one in the following approximative sense: if the family $(P(.|\vartheta))_{\vartheta \in \Theta}$ is Lebesgue-dominated and contains a translation and scale sub-family, then every Lebesgue-dominated probability measure can be approximated with respect to the variational distance by a measure with the representation (C2); this is an immidiate consequence of consistency results for kernel density estimators (cf. L. Devroye and L. Györfi (1985), Theorem 3.1).

Now we give a three-stage classification model:

(S1')   The choice of an object generates an inobservable class $K \in \{K_1, \ldots, K_\kappa\}$
        according to a probability distribution Q on $\{K_1, \ldots, K_\kappa\}$.

(S2')   During the measurement an inobservable parameter value $\vartheta \in \Theta$
        is generated according to the probability distribution Q_K

(S3')   The outcome of the measurement is a point $x \in \mathbb{H}$ which is
        generated according to the probability distribution $P(.|\vartheta)$

Remark 6: Let $\tilde{P}$ denote the probability distribution of the measurement $x \in \mathbb{H}$ taken on an object of unknown class assignment; by inspecting (S1) + (S2) of section 1 we obtain:

$$\tilde{P} = \sum_{j=1}^{\kappa} Q(K_j) \cdot P^{K_j};$$

it follows from (C2):

$$\tilde{P}(H) = \sum_{j=1}^{\kappa} Q(K_j) \cdot \int_\Theta P(H|\vartheta) \, dQ_{K_j}(\vartheta) \tag{3}$$

for $H \in \mathcal{H}$; condition (C1) implies:

$$\tilde{P}(H) = \int_\Theta P(H|\vartheta) \, dP_\Theta(\vartheta) = P_{\mathbb{H}}(H) \qquad (H \in \mathcal{H}) \tag{4}$$

we recognize that the measurement is described by the model (S1') + (S2') +(S3'), too.

4 Training the Classifier

Recall the notation of sections 1, 2 and 3.

Each pair $((P(.|\vartheta))_{\vartheta\in\Theta}, P_\Theta)$ consisting of a family $(P(.|\vartheta))_{\vartheta\in\Theta}$ of probability distributions on $(\mathbb{H}, \mathcal{H})$ and of a prior distribution P_Θ on $(\Theta, \mathcal{T})$ induces a probability distribution $P := P(.|\vartheta) \otimes P_\Theta$ on $(\mathbb{H} \times \Theta, \mathcal{H} \times \mathcal{T})$ which in turn induces its marginal distribution $P_\mathbb{H}$ on $(\mathbb{H}, \mathcal{H})$.

Suppose we are given a parametric family $((P_\gamma(.|\vartheta))_{\vartheta\in\Theta}, P_{\Theta,\gamma})_{\gamma\in\Gamma}$ of such pairs inducing the corresponding family $(P_{\mathbb{H},\gamma})_{\gamma\in\Gamma}$ of probability distributions on $(\mathbb{H}, \mathcal{H})$.

Let $(x^{(1)}, y^{(1)}), \ldots, (x^{(n)}, y^{(n)}) \in \mathbb{H} \times \{K_1, \ldots, K_\kappa\}$ be a learning sample; $x^{(i)}$ is the measurement on an object and $y^{(i)}$ its true class assignment, $i = 1, \ldots, n$. We choose a parameter value $\bar{\gamma} \in \Gamma$ such that the distribution $P_{\mathbb{H},\bar{\gamma}}$ explains well the data $x^{(1)}, \ldots, x^{(n)}$ in the sense of (4). Thus we have fitted a BECE $(\mathbb{H} \times \Theta, \mathcal{H} \otimes \mathcal{T}, P_{\bar{\gamma}}, g, v)$, where $P_{\bar{\gamma}} := P_{\bar{\gamma}}(.|\vartheta) \otimes P_{\Theta,\bar{\gamma}}$, to the learning sample.

For $\alpha \in (0,1)$ let us consider the optimal α-correspondence $\overline{Z}_\alpha \in \mathcal{Z}_\alpha$; cf. Remark 3. Take a sequence $0 < \alpha_1 < \ldots < \alpha_S < 1$ in the interval $(0,1)$ and calculate the convex compact sets $\overline{Z}_{\alpha_s}(x^{(i)})$ $(i = 1, \ldots, n; s = 1, \ldots, S)$. Consider the Bayesian (point-) estimator $\hat{g} : \mathbb{H} \longrightarrow \mathbb{R}^k$ for g. For $s = 1, \ldots, S$ define the discriminant functions

$$D_j^{(s)}(x) := \sum_{i=1}^{n} 1_{\overline{Z}_{\alpha_s}(x^{(i)})}(\hat{g}(x)) \cdot 1_{\{K_j\}}(y^{(i)}) \qquad (j = 1, \ldots, \kappa) \tag{5}$$

and the family of classifiers $c_s : \mathbb{H} \longrightarrow \{K_1, \ldots, K_\kappa\}$ $\quad s = 1, \ldots, S$:

$$c_s(x) := K_j \quad \text{iff} \quad D_j^{(s)}(x) = \max_{j'=1}^{\kappa} D_{j'}^{(s)}(x) \qquad (x \in \mathbb{H}).$$

For $s = 1, \ldots, S$ estimate the probability of correct classification by the classifier c_s:

$$\tau^*(c_s) = \frac{1}{n} \cdot \#\{ i \mid c_s(x^{(i)}) = y^{(i)} \}$$

and choose $s_0 \in \{1, \ldots, S\}$ such that c_{s_0} maximizes the estimated probability of correct classification.

Remark 7: The discriminant functions $D_j^{(s)}$ in (5) count the coverings of the estimating value $\hat{g}(x)$ by the optimal confidence sets $\overline{Z}_{\alpha_s}(x^{(i)})$ for the known class assignment $y^{(i)} = K_j$, $j = 1, \ldots, \kappa$.

Since there is no canonical choice for the confidence parameter α we let it vary over the interval $(0,1)$ and choose the "best" one, as far as it can be decided by evaluating the learning sample.

5 Simulative Evaluation

Let $\mathbb{H} := \Theta := \mathbb{R}^2$. Let S_2^+ denote the set of positive-definite symmetric 2×2-matrices. For $(a, T) \in \mathbb{R}^2 \times S_2^+$ let $N(a, T^{-1})$ denote the Gaussian distribution with mean vector a and covariance matrix T^{-1}. For $\gamma = (a, T) \in \mathbb{R}^2 \times S_2^+$ we put

$$P_\gamma(.|\vartheta) := N(\vartheta, T^{-1}) \qquad (\vartheta \in \Theta)$$

$$P_{\Theta,\gamma} := N(a, T^{-1})$$

and we define the sets

$$\Theta_j(\gamma) := \{\, \vartheta \in \Theta \mid k_{j-1} \le \langle \vartheta - a, T(\vartheta - a)\rangle < k_j(\gamma)\,\} \qquad (j = 1, 2, 3)$$

where $k_0(\gamma) = 0$, $k_3(\gamma) = \infty$ and $k_1(\gamma), k_2(\gamma) \in \mathbb{R}_+$ are chosen as to satisfy

$$Q(\{K_j\}) := P_{\Theta,\gamma}(\Theta_j(\gamma)) = \frac{1}{3} \tag{6}$$

for $j = 1, 2, 3, \gamma = (a, T) \in \mathbb{R}^2 \times S_2^+$, where Q is the uniform distribution on the set $\{K_1, K_2, K_3\}$ of classes. Let $Q_{K_j,\gamma}$ be the conditional probability of $P_{\Theta,\gamma}$ on $\Theta_j(\gamma)$, j=1,2,3. It is easy to see that the probability distributions $Q_{K_1,\gamma}$, $Q_{K_2,\gamma}$, $Q_{K_3,\gamma}$, Q and $P_{\Theta,\gamma}$ satisfy the condition (C1) of section 3. In view of condition (C2) we obtain the possible class-dependent distributions of the observations $x \in \mathbb{H}$:

$$P^{K_j,\gamma}(H) := \int_\Theta P_\gamma(H|\vartheta)\, dQ_{K_j,\gamma}(\vartheta) \tag{7}$$

for $H \in \mathcal{H}, j = 1, 2, 3$ and $\gamma \in \mathbb{R}^2 \times S_2^+$; the probability distributions $P^{K_j,\gamma}$ $(j = 2, 3)$ are obviously not Gaussian and even not unimodal and, therefore, the classical parametric discriminance analysis is not applicable.

By a straightforward calculation we have for the marginal distribution $P_{\mathbb{H},\gamma}$ of $P_\gamma := P_\gamma(.|\vartheta)\otimes P_{\Theta,\gamma}$ on $\mathbb{H}$:

$$P_{\mathbb{H},\gamma} = N(a, 2T^{-1}) \tag{8}$$

for $\gamma = (a, T) \in \mathbb{R}^2 \times S_2^+$; (8) is our clue for fitting the BECE to the learning sample $(x^{(1)}, y^{(1)})$, ..., $(x^{(n)}, y^{(n)}) \in \mathbb{H} \times \{K_1, K_2, K_3\}$ because, in view of Remark 6, $x^{(i)}$ is distributed according to $P_{\mathbb{H},\gamma}$.

We fix, for evaluation purposes, a parameter value $\gamma = (a, T) \in \mathbb{R}^2 \times S_2^+$ and generate a learning sample $(x^{(1)}, y^{(1)}), \ldots, (x^{(n)}, y^{(n)}) \in \mathbb{H} \times \{K_1, K_2, K_3\}$ according to (6) and (7); since our choice of γ is not known to the "statistician" who is training a classifier according to section 4, he, in view of (8) and of Remark 6, fits a BECE by estimating the parameter γ by $\overline{\gamma} = (\overline{a}, \overline{T})$, where

$$\overline{a} := \frac{1}{n}\sum_{i=1}^n x^{(i)} \qquad \overline{T}^{-1} := \frac{1}{2(n-1)}\sum_{i=1}^n (x^{(i)} - \overline{a})(x^{(i)} - \overline{a})',$$

obtaining $(\mathbb{H} \times \Theta, \mathcal{H} \otimes \mathcal{T}, P_{\overline{\gamma}}, g, v)$, where g is the identity map on Θ and $v(\theta, C)$ is the Lebesgue-content of the set-valued decision C.

Now the "statistician" is able to construct the Bayesian α-correspondence $\overline{Z}_\alpha$ for $\alpha \in (0,1)$ which assigns to each possible observation $x \in \mathbb{H}$ the highest probability density region with respect to the posterior distribution $P_{\overline{\gamma}}(.|x)$; cf. J.O. Berger (1985), sec. 4.3.2 and M.H. DeGroot (1970), sec. 9.9, Theorem 1.

The "statistician" constructs the classifier $c_{s_0} : \mathbb{H} \longrightarrow \{K_1, K_2, K_3\}$ as prescribed in section 4. The quality of the classifier is studied by estimating the probability $\tau(c_{s_0})$ of correct classification by the empirical rate $\hat{\tau}$ of correct classification obtained for a testing sample. The empirical rate $\hat{\tau}$ has to be viewed as a random number depending substantially on the (random) learning sample. In order to explore its distribution the above procedure (fixing the parameter value γ, generating a learning sample of size n according to (6) and (7), fitting the BECE, constructing the classifier and estimating its correct classification rate) is repeated 100 times leading to the empirical rates $\hat{\tau}_1, \ldots, \hat{\tau}_{100}$ of correct classification. Since the quality of this classification procedure is invariant under linear transformations and translations of $\mathbb{R}^2$ operating on the parameter values $\gamma = (a, T)$, we have restricted ourselves to the choice $\gamma = (0, I)$, I being the identity matrix. For orientation, we have calculated numerically the best attainable (Bayesian) correct classification probability $\tau(c_B) = 0.48$. The simulation was carried out for the learning sample size $n = 90$. In Diagram 1 the empirical distribution function of the sample $\hat{\tau}_1, \ldots, \hat{\tau}_{100}$ of rates is represented by a continuous line. For comparison, we have proceeded analogously with a classical classifier based on the kernel density estimation and we represent the empirical distribution function of the obtained rates of correct classification by the dotted line in Diagram 1.

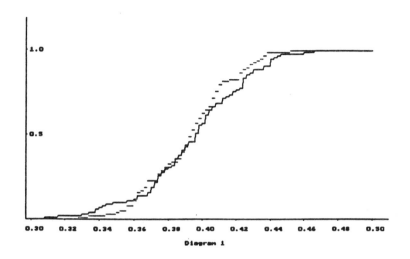

Diagram 1

Acknowledgment: The idea for the classification strategy presented in this paper arose during a discussion with Professor O. Moeschlin at the University of Hagen. The author would like to thank Professor O. Moeschlin for his encouragement and advice concerning this paper. The author would like to thank as well the referee whose suggestions led to an improvement of the text.

References

BERAN, R.J. and MILLAR, P.W. (1985), Asymptotic theory of confidence sets, *Proc. Berkeley Conf. Honor J. Neyman and J. Kiefer, vol.II, eds.: LeCam,L.M., Olshen,R.A.*

BERGER,J.O. (1985), *Statistical Decision Theory and Bayesian Analysis.* Springer-Verlag

BREIMAN,L., FRIEDMAN,J.H., OLSHEN,R.A., STONE,CH.J. (1984), *Classification and regression trees.* Wadsworth and Brooks.

DEGROOT,M.H. (1970), *Optimal statistical decisions.* McGraw-Hill

DEVROYE,L., GYÖRFI,L. (1985), *Nonparametric density estimation.* Willey and Sons

GRYCKO,E.(1991), *Zur Bayesschen Theorie mengenwertiger Entscheidungsfunktionen.* Doctoral dissertation, University of Hagen.

MEISTER, H. and MOESCHLIN, O. (1986), Conditions of Rao's covariance method type for set-valued estimators. *Proc. 4th Purdue Symp. Statist. Decision Theory and Related Topics, vol.2, 299-305.*

MEISTER, H. and MOESCHLIN, O. (1988), Unbiased set-valued estimators with minimal risk, *J. Math. Anal. Appl. 130, no.2, 426-438.*

MOESCHLIN,O. (1982), A Bayes concept of confidence estimation. *Trans. 9th Prague Conf. Inf. Theory, Statist. Decision Functios, Random Processes. Prague, pp. 91-94*

Canonical Discriminant Analysis: Comparison of Resampling Methods and Convex-Hull Approximation

C. Weihs

CIBA-GEIGY Ltd., Mathematical Applications, R-1008.Z2.22
CH-4002 Basel, Switzerland

Abstract: The predictive power of Canonical Discriminant Analysis is examined by means of variants of the delete-p%-jackknife, the bootstrap, and convex-hull approximation. Characteristics of predictive power are based upon estimated tolerance regions of the different classes of observations in that the coverage of the regions is investigated and one kind of misclassification rates is based upon these regions. Additionally, misclassification rates based upon densities are examined. Simulation data basically consist of 3 clusters of points in 2 dimensions with various distributional properties. The results indicate the superiority of those methods with the most varying samples, i.e. the delete-10%/50%-jackknife and the bootstrap, to set up prediction rules. Concerning misclassification rates convex-hull approximation of tolerance regions appeared to be competitive. Convex-hull approximation did not lead to a distinct improvement in case of non-normal data. For the characterization of predictive power, the misclassification rate based upon densities should be prefered.

Key Words: discriminant analysis, predictive power, delete-d-jackknife, bootstrap, balanced resampling, normal approximation, convex hull

1 Introduction

In order to characterize the predictive power of multivariate data-analytic methods like Principal Components Regression, Partial Least Squares analysis and Canonical Discriminant Analysis, it is customary to use resampling methods like the jackknife or the bootstrap (see e.g. Krzanowski and Radley (1989), Weihs and Schmidli (1991)). Predictive distributions are especially valuable in model determination, i.e. in the examination of model adequacy and in model selection. Also in this field a cross-validation viewpoint is argued for (Gelfand et al. (1991)).

In this paper we concentrate upon the examination of criteria for judging the predictive power of **Canonical Discriminant Analysis (CDA)**. Most of the analysis is based on ideas of Krzanowski and Radley (1989) concerning non-asymptotic tolerance regions in CDA, on the preliminary resampling methods discussed in Weihs and Schmidli (1991) and Weihs (1992), and various results on jackknife and bootstrap resampling in the recent literature (e.g. Shao (1989), Shao and Wu (1989), Davison et al. (1986)). As an alternative a convex-hull method is considered.

The **predictive power** of a model identified by CDA will be characterized by means of

- the estimated **coverage rate** of the simulated α%-**tolerance regions** compared with their nominal rate α,

- **misclassification rates** based upon prediction rules using the simulated $\alpha\%$-tolerance regions or the height of the corresponding densities.

The **resampling methods** compared are

- the classical delete-1-jackknife (Tukey (1958)),

- the more general delete-$p\%$-jackknife (Wu (1986), Shao (1989), Weihs and Schmidli (1991)),

- the bootstrap (e.g. Efron and Gong (1983)),

- a balanced delete-$p\%$-jackknife (Shao and Wu (1989)),

- a balanced bootstrap (Davison et al. (1986)).

For comparison, a **convex-hull method** is investigated.

The **simulation data** basically consist of 3 clusters of points in 2 dimensions. Distributional properties of the points in the clusters are varied.

In section 2 the resampling methods are motivated and discussed. In section 3 the estimation of the tolerance regions is described (including convex-hull approximation) and the characterizations of predictive power are introduced. In section 4 the simulation results are discussed. Section 5 provides a conclusion.

2 Resampling methods

Nonparametric estimation of tolerance (and confidence) regions is typically tackled by means of resampling methods like the jackknife and the bootstrap. In this section these methods will mainly be explained and motivated by the estimation of the expectation μ of a distribution F using a sample $x_1, x_2, \ldots, x_N$ (cp. Efron and Gong (1983)). The generalization to Canonical Discriminant Analysis is indicated in section 3.

One very obvious way to estimate μ and its accuracy is to compute sample mean and its variance:

$$\bar{x} = \frac{1}{N} \sum_{n=1,\ldots,N} x_n, \quad s^2 = \frac{1}{N(N-1)} \sum_{n=1,\ldots,N} (x_n - \bar{x})^2.$$

Unfortunately, the above formula for the accuracy does not, in any obvious way, extend to estimators other than $\bar{x}$, e.g. to the sample median. Two ways of making such an extension are the jackknife und the bootstrap.

The **classical jackknife** computes the sample mean by means of averaging the means of all data sets with 1 point deleted:

$$\bar{x} = \bar{x}_J = \frac{1}{N} \sum_{n=1,\ldots,N} \bar{x}_{(n)}, \quad \bar{x}_{(n)} = \frac{1}{N-1} \sum_{j \neq n} x_j.$$

The jackknife estimate of the accuracy of the sample mean is:

$$s_J^2 = \frac{N-1}{N} \sum_{n=1,\ldots,N} (\bar{x}_{(n)} - \bar{x}_J)^2, \quad \text{which is indeed equal to } s^2 \text{ !}$$

The **complete delete-d-jackknife** computes the sample mean by means of averaging the means of all data sets with d points deleted:

$$\bar{x} = \bar{x}_d = \frac{1}{K}\sum_S \bar{x}_{(S)}, \quad \bar{x}_{(S)} = \frac{1}{N-d}\sum_{(S)} x_j, \quad K = \binom{N}{d},$$

$\sum_S$ stands for summation over all subsets S with d points deleted,
$\sum_{(S)}$ stands for summation over all sample elements without the elements of subset S.

The delete-d-jackknife estimate of the accuracy of the sample mean is:

$$s_d^2 = \frac{N-d}{d \cdot K}\sum_S (\bar{x}_{(S)} - \bar{x}_d)^2, \text{ which is also equal to } s^2 \text{ !!}$$

Note that for half-sampling, i.e. $d = N/2$, the rescaling factor $(N - d)/d$ is 1. Also, the classical jackknife relates to the special case $d = 1$.

The delete-d-jackknife estimator has the disadvantage of dramatically increasing computational effort with increasing $d \leq N/2$ and fixed but fairly big N. The original estimator by Wu (1986), was thus modified in order to diminish the computational burden. One possible modification is the usage of **balanced subsampling** (Shao and Wu (1989)): Let $T = \{S_1, \ldots, S_t\}$ be a collection of t subsets of size $(N - d)$ satisfying the following two properties:
(1) Every object n, $1 \leq n \leq N$, appears with the same frequency in the subsets in T.
(2) Every pair (n, m), $1 \leq n, m \leq N$, appears together with the same frequency in T.

A special design satisfying these requirements is the **Plackett-Burman-design**, which is available for all N with $(N+1)$ a multiple of 4 (see Plackett and Burman (1946), Baumert et al. (1962) for all designs up to $N = 99$). The usage of Plackett-Burman-designs will be illustrated by means of an example. Let $N = 7$, then the Plackett-Burman-design looks as follows:

```
+  +  +  -  +  -  -
-  +  +  +  -  +  -
-  -  +  +  +  -  +
+  -  -  +  +  +  -
-  +  -  -  +  +  +
+  -  +  -  -  +  +
+  +  -  +  -  -  +
```

This design is cyclic, it can be generated from the first row only. Note that this way the number of $+$ and $-$ is the same in each row. Thus, if you take a '$-$' as an indicator for deletion, properties (1) and (2) are satisfied, and the design can be used for the balanced delete-3-jackknife.

For the **balanced delete-d-jackknife using a Plackett-Burman-design** the complete delete-d-jackknife formulas for $\bar{x}_d$ and s_d^2 apply, except that $K = t = N$, which is thus constant for different d. Since $K \geq t$ is necessary for (1), (2), Plackett-Burman-designs are minimal balanced subsampling schemes. The only disadvantage of the Plackett-Burman design is that it does not exist for all sample sizes N. Using a subsampling scheme with properties (1), (2) the equalities $\bar{x} = \bar{x}_d$, $s^2 = s_d^2$ are retained.

Another, perhaps more obvious, shortcut of the complete delete-d-jackknife is the usage of a random sample of size M of subsamples S with d elements of the original sample deleted (Shao (1989)). In this case one has to use $K = M$ in the above formulas. Shao recommends the usage of $M = N^{3/2}$ and calls the scheme **delete-d-jackknife-sampling**. Naturally, for this shortcut $\bar{x} \neq \bar{x}_d$, $s^2 \neq s^2_d$, in general.

The generalization of these estimators for a general parameter Θ to be estimated should be obvious. The (classical) delete-1-jackknife is known to give inconsistent variance estimates for nonsmooth estimators such as sample quantiles. For quantiles this deficiency can be rectified by using the complete delete-d-jackknife with $N^{1/2}/d \longrightarrow 0$ and $(N - d) \longrightarrow \infty$. The balanced delete-$d$-jackknife variance estimate shares the same asymptotic properties (Shao and Wu (1989)). Moreover, the delete-d-jackknife-sampling variance estimator retains the efficiency of the complete delete-d-jackknife if $N/M \longrightarrow 0$, i.e. if e.g. $M = N^\delta$, $\delta > 1$. Thus, the number of replicates M only has to be slightly larger than that required by the classical delete-1-jackknife variance estimator (Shao (1989)). Shao and Wu (1989) thus recommend the **delete-$p\%$-jackknife**, since $d = (p/100)N$ satisfies the above conditions for d. Balancing may be used if a corresponding design is at hand, and jackknife-sampling is an efficient shortcut at least otherwise. May be that the delete-$p\%$-jackknife is that near at hand, may be it was just luck, Weihs and Schmidli (1990, 1991) independently proposed the usage of $p\%$-crossvalidation for the determination of predictive power without any theoretical justification.

The **bootstrap** scheme estimates the sample mean and its variance in an apparently different way (see e.g. Efron and Gong (1983)). Let F_E be the empirical probability distribution of the data, putting probability mass $1/N$ on each x_n, and let $X_1^*, X_2^*, \ldots, X_N^*$ be random variables independently identically distributed with distribution F_E. In other words, each X_i^* is drawn independently with replacement and with equal probability from the original sample $x_1, x_2, \ldots, x_N$. Then $\overline{X}^* = \sum_{n=1,\ldots,N} X_n^*/N$ has variance $\text{var}_E(\overline{X}^*) = \sum_{n=1,\ldots,N}(x_n - \bar{x})^2/(N^2)$, var_E indicating variance under sampling from F_E. Obviously, $\text{var}_E(\overline{X}^*) = ((N-1)/N)s^2$. Thus, the bootstrap variance estimate is biased downward, but consistent. Note, that the jackknife estimates are rescaled to avoid such a bias.

Usually an explicit formula for the variance will not be available. Then, a Monte-Carlo simulation will be necessary. The bootstrap scheme estimates the unknown parameter Θ by means of averaging the parameter estimates of B bootstrap samples:

$$\Theta_B = \frac{1}{B} \sum_{b=1,\ldots,B} \Theta_b^*, \quad \Theta_b^* = \Theta_b^*(X_1^*, X_2^*, \ldots, X_N^*).$$

The bootstrap estimate of the accuracy of the parameter Θ is:

$$s_B^2(\Theta) = \frac{1}{B-1} \sum_{b=1,\ldots,B} (\Theta_b^* - \Theta_B)^2.$$

Note that this scheme essentially equals the delete-d-jackknife-resampling scheme with the exception that the jackknife uses sampling without replacement.

With the bootstrap the equalities $\bar{x}_B = \bar{x}$ and $s_B^2(\bar{x}) = s^2$ are not valid in general. **First order balancing** (property (1)) can be used to achieve $\bar{x}_B = \bar{x}$. First order balance can easily be accomplished for general N: Concatenate B copies of $x_1, x_2, \ldots, x_N$ in a string, randomly permute this string and then read off the B bootstrap samples as successive blocks of length N. Algorithms for second order balance (properties (1) and (2)) can be found in Graham et al. (1990).

3 Predictive power of Canonical Discriminant Analysis

Canonical Discriminant Analysis (CDA) is a well-known technique for the prediction of an a-priori group structure by means of multivariate quantitative data. Suppose that one has g groups with the reference set of observations consisting of n_i p-dimensional observations y_{ij}, $i = 1, \ldots, g$, $j = 1, \ldots, n_i$. Then CDA determines that linear combination $x_1 = a_1' y$ of the original variables y so that the ratio of the between-groups mean square and the within-groups mean square is maximized: $F_{a_1} = a_1' B a_1 / a_1' W a_1 = \max!$, where B is the between-groups covariance matrix $B = \sum_{i=1,\ldots,g} n_i (\bar{y}_i - \bar{y})(\bar{y}_i - \bar{y})' / (g-1)$, $\bar{y}_i$ the i-th group sample mean vector, $\bar{y} = \sum_{i=1,\ldots,g} n_i \bar{y}_i / n$, $n = \sum_{i=1,\ldots,g} n_i$, and W is the pooled within-groups covariance matrix $W = \sum_{i=1,\ldots,g} (n_i - 1) S_i / (n-g)$, S_i the sample covariance matrix of group i.

Having determined a_1, one can seek a second linear combination $x_2 = a_2 \bar{y}$ of the original variables with the next largest F-ratio, etc., imposing orthogonality constraints. The linear combinations x_1, x_2, etc. are called discriminant coordinates. This analysis is equivalent to an eigenanalysis of the matrix $W^{-1} B$. If the corresponding first 2 eigenvalues together constitute a sizeable portion of the sum of all eigenvalues, then a good visual impression of the relationships existing among the g groups of observations is obtained by plotting the first two discriminant coordinates, the so-called (2D-)discriminant space.

For **prediction** one traditionally assumes that the observations in each group are independent observations from a normal population whose mean depends on the group but whose dispersion matrix is constant across the groups. Since the discriminant coordinates are only linear combinations of the original variables, the **tolerance region** (of $(1 - \alpha)$ expectation) of the observations in group i can thus be constructed easily for any probability level. Assuming sample independence of the involved discriminant coordinates, each tolerance region turns out to be a hypersphere in s-dimensional discriminant space. In 2D these hyperspheres are simply circles drawn round each group mean. If group mean and dispersion matrix were known, the radius of the $100(1 - \alpha)\%$ tolerance circle would be $(\chi^2_{2,\alpha})^{1/2}$, where $\chi^2_{2,\alpha}$ is the value exceeded by $100\,\alpha\%$ of the χ^2-distribution on 2 degrees of freedom (see e.g. Krzanowski and Radley (1989)). If mean and dispersion have to be estimated, the radius is $((2(n_i - 1)/(n_i - 2))F_{2,n_i-2,\alpha})^{1/2}$, where $F_{k,l,\alpha}$ is the value exceeded by $100\alpha\%$ of the F-distribution on k and l degrees of freedom.

For prediction tolerance regions may be used in the following way:

Let z be the (2D-)discriminant coordinates of a new object to be classified into one of the g groups. Then, that group is chosen which has the smallest euclidean distance in discriminant space between z and its group mean. Obviously, this is equivalent to looking for that group with the smallest tolerance circle just including z. Moreover, one can show that this is also equivalent to looking for that group with the highest (2D normal) density in z (if all n_i are equal).

Using this prediction procedure, the overall performance of prediction, i.e. the **predictive power**, can be measured by the so-called **misclassification rate**, indicating that percentage of a test set with known group membership which was not correctly classified.

If one is not willing to assume normality of the original observations or equality of the population dispersion matrices of the groups, one may use the **jackknife and bootstrap** techniques indicated in section 2 **for estimating group means and their tolerance re-**

gions. Unfortunately, one cannot expect the discriminant coordinates resulting from the different choices of subsamples used by the jackknife and the bootstrap to be the same. This caused Krzanowski and Radley (1989) as well as Weihs and Schmidli (1990), again independently, to use **Procrustes transformations** (optimal rotation, translation and overall scaling) in order to achieve the best coincidence of the discriminant axes of the different subsamples. Assuming full coincidence, tolerance regions are then estimated assuming normality of the jackknife or bootstrap replicates in order to derive smooth regions easily (Davison et al. (1986)). Davison et al. have demonstrated the accuracy of this **normal approximation** for even quite complex situations. Unfortunately, using this normal approximation the resampling methods become parametric. Altogether, in the 2D-discriminant space this leads to $100(1-\alpha)\%$ tolerance ellipses $(x - \overline{x}_{iJ/B})'S_{iJ/B}^{-1}(x - \overline{x}_{iJ/B}) \leq C_{J/B}$ for group i, where $\overline{x}_{iJ}$, S_{iJ}, and $\overline{x}_{iB}$, S_{iB} are determined by means of the jackknife (J) and the bootstrap (B), respectively, and C_J is approximately equal to $(2(n_i - 1)/(n_i - 2))F_{2,n_i-2,\alpha}$ (n_i because of groupwise resampling), C_B is approximately equal to $(2(B-1)/(B-2))F_{2,B-2,\alpha}$.

There are (at least) **two ways to predict** a group for a new object. One way is to use the $100(1-\alpha)\%$ tolerance ellipsoids of which the size as well as the orientation will vary from group to group. To predict the group, one then looks for the 'smallest catching tolerance region', i.e. for the smallest Mahalanobis distance corresponding to $S_{iJ/B}^{-1}$ between the new object and the group means in discriminant space. A different way to predict is to look for the group with the highest (2D normal) density in the discriminant coordinates of the new object. Then, the criterion to prefer group i to group j is

$$2\log(f_i(z)/f_j(z)) = \log(\det(S_j)/\det(S_i)) - (z - \overline{x}_i)'S_i^{-1}(z - \overline{x}_i) + (z - \overline{x}_j)'S_j^{-1}(z - \overline{x}_j) > 0.$$

Note that this criterion differs from the Mahalanobis criterion by the term introducing the ratio of the determinants of the covariance matrices. Thus, the two procedures are not equivalent for different dispersion matrices for different groups.

A truely nonparametric method to approximate (convex) tolerance regions is **convex-hull** approximation. This again leads to two ways to predict a group for a new object. One way corresponds to looking for the smallest Mahalanobis distance, the other to looking for the highest density by generalized (2D) histogram approximation. In order to approximate $100(1-\alpha)\%$ tolerance regions, an algorithm of Hartigan (1987) was adopted, and convex hulls are determined using the Green and Silverman (1979) adaption J3 of an algorithm of Jarvis (1973). Convex-hull approximation suffers from the fact that the extension beyond the region of the actual sample is undefined. That is why new data outside of the 100% convex hulls of all g groups of the learning data are attached to that group with the nearest hull in Euclidean sense.

Another indicator for predictive power is the so-called **coverage rate** which is that percentage of elements in the test set which are known to belong to group i and are captured by the $100(1-\alpha)\%$ tolerance ellipsoids. Obviously, the nearer this rate to $100(1-\alpha)\%$ the better.

4 Simulations

Following Krzanowski and Radley (1989) resampling is performed groupwise, i.e. in one delete-$p\%$-jackknife- or bootstrap-replicate sampling is restricted to the objects of one group. Consider e.g. the delete-50%-jackknife. In one replication 50% of group i, say, is sampled randomly without replacement, and all objects of all other groups are also included in the sample. Note that this way in a bootstrap sample only objects of the subsampled group can appear more than once.

The following methods are compared with the standard 'one-shot' canonical discriminant analysis (**Fit**), which is not using any resampling to estimate tolerance regions of group populations:

- classical delete-1-jackknife (**Jack**),

- delete-10%-jackknife (**J10%**),

- delete-50%-jackknife (**J50%**),

- Placket-Burman-balanced jackknife (**Jbal**),

- bootstrap (**Boot**),

- 1st-order balanced bootstrap (**Bbal**), and

- convex-hull approximation (**Hull**).

The simulation data basically consist of 3 clusters of points in 2 dimensions. Variation of the distributional properties of the group populations leads to different simulations:

- all populations (bivariate) normal with equal standard deviations (**N1**),

- all populations (bivariate) normal, but with one standard deviation larger by factor 4 (**N4**),

- all populations (bivariate) normal, but with one standard deviation larger by factor 16 (**N16**),

- all populations (bivariate) χ^2-distributed with 3 degrees of freedom (**C3**),

- all populations (bivariate) χ^2-distributed, two with 3 degrees of freedom, one with 12 (**C12**).

The geometrical arrangement of the 3 groups of data in 2 dimensions corresponding to the different simulation data sets is indicated in Figure 1. Note the different degrees of overlapping of the 3 regions with the different data sets.

The **learning set** consists of 59 (**L59**) and 99 (**L99**) objects per group, respectively. The number of replications per group is set to the number of objects per group for both the delete-p%-jackknife and the bootstrap. Predictive power is tested by means of a **test set** with 500 objects per group.

In Table 1 the mean coverage rate (**cr**) of approximate 90%-tolerance ellipse, the Mahalanobis (**mm**) and the density (**dm**) misclassification rate are reported for all combinations of approximation methods and group populations. For the balanced versions of the jackknife and the bootstrap, only the smaller learning set L59 was examined. Also, some of the convex-hull simulations could not be finished because of computer time restrictions. The outcomes can be summarized as follows:

232

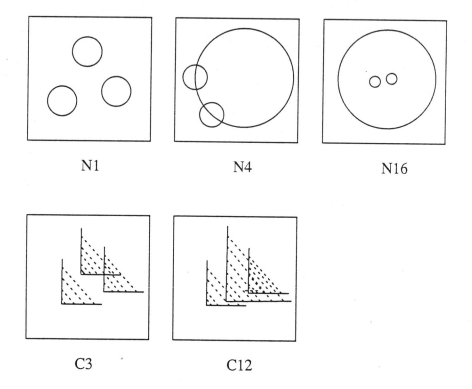

Figure 1: *Geometry of the 3 groups of data for the different simulations*

- The unbalanced versions of the bootstrap (Boot) and the delete-10%/50%-jackknife (J10%/J50%) appear to be best concerning both coverage and misclassification determined by density heights.

- Balanced sampling deteriorate coverage rates relative to the unbalanced versions.

- Misclassification rates determined by means of density heights are much lower than those determined by means of Mahalanobis distances, at least for (approximately) normal data.

- Convex-hull approximation (Hull) appears to be competitive concerning misclassification, but it always generates too low coverage rates.

- The classical delete-1-jackknife appears to be competitive concerning misclassification, but it generates bad and unequal coverage in all cases but the most simple case N1.

- The one-shot-method (fit) is competitive if all variances are equal (normal and χ^2 case), otherwise it tends to unequal coverage.

- Bad coverage rates do not automatically imply big misclassification.

- The larger the learning set the lower the misclassification.

Table 1: Predictive Power (in %)

meth\power	N1 cr	mm	dm	N4 cr	mm	dm	N16 cr	mm	dm
L59									
Fit	87.7	0	0	85.1 *	5.3	5.3	83.9 *	**17.7**	17.7
Jack	87.4	0	0	72.1 *	5.7	5.2	73.1 *	27.7	7.1
J10%	96.9	0	0	**90.3**	15.7	4.7	87.7	61.7	6.9
J50%	94.9	0	0	91.3	15.3	4.7	**88.2**	61.5	**6.6**
Jbal	55.1	0	0	51.6 *	6.5	5.8	53.3 *	26.5	15.5
Boot	98.0	0	0	86.8	12.7	5.4	84.1	62.3	9.1
Bbal	**90.5**	0	0	68.7	5.8	5.6	70.8 *	26.9	8.1
Hull	66.1	0	0	67.4	**4.5**	4.4	67.4	49.9	10.0
L99									
Fit	85.5	0	0	85.1 *	5.6	5.6	83.4 *	16.6	16.6
Jack	**88.7**	0	0	82.9 *	3.6	3.8	82.4 *	**8.3**	**1.6**
J10%	97.3	0	0	91.3	4.5	2.7	88.8	53.5	5.0
J50%	96.3	0	0	91.5	6.4	2.6	**89.7**	52.7	4.6
Boot	96.2	0	0	**90.9**	3.3	**1.8**	89.3	51.3	4.4
Hull	71.0	0	0	77.1	**2.5**	2.3	77.1	–	–

meth\power	C3 cr	mm	dm	C12 cr	mm	dm
L59						
Fit	**90.5**	**4.5**	**4.5**	88.1	24.7	24.7
Jack	70.9	6.6	6.2	66.9	27.9	23.5
J10%	88.1	6.7	6.7	**90.3**	**23.9**	22.5
J50%	88.9	5.5	5.5	91.8	**23.9**	22.7
Jbal	38.8	5.8	5.9	31.4	28.2	26.2
Boot	88.7	5.9	5.9	90.8	25.5	**22.5**
Bbal	68.3	7.1	6.7	63.6	26.5	24.0
Hull	73.0	5.0	5.6	75.3	29.5	27.3
L99						
Fit	87.9	4.5	4.5	83.5 *	23.5	23.5
Jack	73.4 *	3.4	3.7	59.6 *	22.7	21.5
J10%	90.7	4.6	4.9	**88.1**	**22.4**	21.2
J50%	**89.7**	4.1	4.3	85.6	22.5	**20.0**
Boot	89.2	4.5	4.5	85.8	22.5	20.1
Hull	77.1	**3.2**	**3.1**	76.7	27.3	–

* coverage rates of the 3 groups differ more than 25%

These results are also illustrated by means of 6 figures showing (approximate) 90%-tolerance ellipses or 100%-convex-hulls together with the test set:

- Figure 2 shows the close correspondence of the true dispersion of the test data and all the estimated tolerance ellipses of the delete-10%/50%-jackknife in the case of normality and equal group variance.

- Figure 3 indicates the problem of incorrect coverage of the one-shot method in the unequal group variance case.

- Figure 4 indicates the inadequacy (concerning coverage) of the classical delete-1-jack-knife.

- Figure 5 illustrates the inadequacy of 'normal' tolerance regions in the non-normal case.

- Figures 6, 7 indicate how unsatisfactory the coverage of convex-hulls can be.

5 Conclusion

The simulation results indicate the superiority of the delete-10%/50%-jackknife and the bootstrap to set up prediction rules using the approximated density height of the group populations in discriminant space. Moreover, concerning misclassification rates the results seem to indicate the competitiveness of convex-hull approximation of tolerance regions. Unexpectedly, convex-hull approximation did not lead to a distinct improvement in case of non-normal data.

Acknowledgement

I would like to thank H. Schmidli for many fruitful discussions and for programming, and W. Seewald for the suggestion of Plackett-Burman designs.

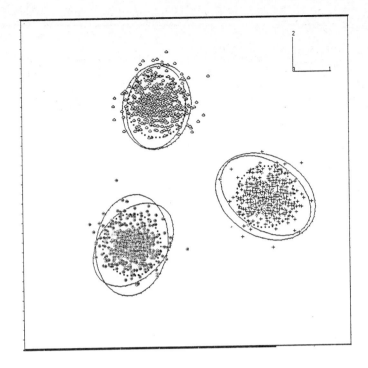

Figure 2: Delete-10%/50%-jackknife regions, all dispersions equal (N1), L59

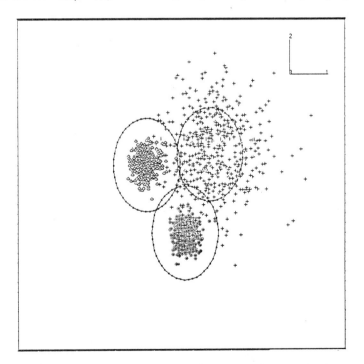

Figure 3: One-shot tolerance region, not all dispersions equal (N4), L59

Figure 4: Jackknife tolerance regions, not all group dispersions equal (N4), L99

Figure 5: Bootstrap tolerance regions in the χ^2 case (C3), L59

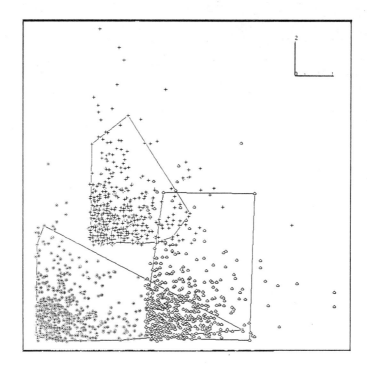

Figure 6: 100%-convex-hulls in the χ^2 case (C3), L59

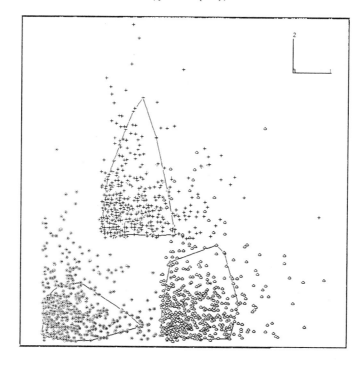

Figure 7: Convex-hull 90%-tolerance regions in the χ^2 case (C3), L59

References

BAUMERT, L.D., GOLUMB, S.W., HALL, M. (1962), Discovery of an Hadamard matrix of order 92, *Bulletin of the American Mathematical Society* 68, 237–238

DAVISON, A.C., HINKLEY, D.V., SCHECHTMAN, E. (1986), Efficient bootstrap simulation, *Biometrika* 73, 555–566

EFRON, B., GONG, G. (1983), A leisurely look at the bootstrap, the jackknife, and cross-validation, *The American Statistician* 37, 36–48

GELFAND, A.E., DEY, D.K., CHANG, H. (1991), Model determination using predictive distributions with implementations via sampling-based methods, paper presented at the 4th Valencia International Meeting on Bayesian Statistics, Peniscola, Spain, 15.-20. April, 1991

GRAHAM, R.L., HINKLEY, D.V., JOHN, P.W.M., SHI, S. (1990), Balanced design of bootstrap simulations, *Journal of the Royal Statistical Society*, Series B 52, 185–202

GREEN, P.J., SILVERMAN, B.W. (1979), Constructing the convex hull of a set of points in the plane, *The Computer Journal* 22, 262–266

HARTIGAN, J.A. (1987), Estimation of a convex contour in two dimensions, *Journal of the American Statistical Association* 82, 267–270

JARVIS, R.A. (1973), On the identification of the convex hull of a finite set of points in the plane, *Information Processing Letters* 2, 18–21

KRZANOWSKI, W.J., RADLEY, D. (1989), Nonparametric confidence and tolerance regions in canonical variate analysis, *Biometrics* 45, 1163–1173

PLACKETT, R.L., BURMAN, J.P. (1946), The design of optimum multifactorial experiments, *Biometrika* 33, 305–325

SHAO, J. (1989), The efficiency and consistiency of approximations to the jackknife variance estimators, *Journal of the American Statistical Association* 84, 114–119

SHAO, J., WU, C.F.J. (1989), A general theory for jackknife variance estimation, *The Annals of Statistics* 17, 1176–1197

TUKEY, J.W. (1958), Bias and confidence in not quite large samples (abstract), *The Annals of Mathematical Statistics* 29, 614

WEIHS, C., SCHMIDLI, H. (1990), OMEGA (Online Multivariate Exploratory Graphical Analysis): Routine searching for structure (with discussion); *Statistical Science* 5, 175–226

WEIHS, C., SCHMIDLI, H. (1991) : Multivariate Exploratory Data Analysis in Chemical Industry, *Mikrochimica Acta* II, 467–482

WEIHS, C. (1992), Vorhersagefähigkeit multivariater linearer Methoden, Simulation und Grafik (with discussion); in : H. Enke, J. Gölles, R. Haux, K.-D. Wernecke (Eds.), *Methoden und Werkzeuge für die exploratorische Datenanalyse in den Biowissenschaften*, G. Fischer, Stuttgart, 111–127

WU, C.F.J. (1986), Jackknife, bootstrap and other resampling methods in regression analysis (with discussion), *The Annals of Statistics* 14, 1261–1350

Nonparametric Prediction of Time Series on the Basis of Typical Course Patterns

P. Michels[1],

Faculty of Economics and Statistics, University of Konstanz, PO Box 5560
W-7750 Konstanz, Germany

Abstract: Long time series with high measurement frequencies (such as daily measurements) are often present in ecological studies. This holds true, for example, for pollution emissions in the air or for the run-off data of the Ruhr River analysed here. Parametric models, however, do not adequately deal with the frequently occurring nonlinear structures. The prediction techniques discussed in this article, which function on the basis of typical course patterns, are an alternative to parametric modelling. The predictions are averages of all time series data which have previous courses "similar" to the last known course. The methods differ only in the classification of those courses relevant to the forecast. The following three possibilities are presented for this: (i) The time series courses are divided into disjoined groups by means of nonhierarchical cluster analysis. (ii) Courses lying in a neighbourhood of the last known one form the forecast basis (kernel and nearest neighbour predictors). (iii) Such courses that after linear transformation lie near the last one and are positively correlated with it are included. The practical suitability of these methods will be tested on the basis of forecasts of the run-off of the Ruhr River. In this case the last method yields the best forecasts of the interesting peak data.

1 Introduction

In order to forecast time series with courses marked by trends and seasonal variations, these can be analysed and estimated. In this way meaningful predictions are obtained by means of extrapolation. However, it is much more difficult to make predictions when such regular structures are not discernible. This is often the case with many time series in ecological studies, such as daily averages of pollution emissions in the air or of the run-off of rivers. Parametric ARMA processes generally model the frequently underlying nonlinear relationships inadequately. As an alternative, techniques are proposed and discussed in this paper which function on the basis of typical course patterns that do not occur periodically. The forecasts done by means of these methods are averages of all time series data with previous courses lying near the last known course. The methods only differ in how the courses relevant to the forecast, which are "similar" to the last known one, are found.

Let $Y_t, t \in \mathbb{Z}$, be a real-valued strictly stationary stochastic process. Inference is to be made about this on the basis of a single realization $Y_1, \ldots, Y_T$ which is partitioned into the $T - p + 1$ courses $\mathbf{X}_t = (Y_{t-p+1}, \ldots, Y_t)'$, $t = p, \ldots, T$, of length p. If $Y_t, t \in \mathbb{Z}$, is markovian of the order p, then the future values of the time series only depend on the last p known data. In this case the best m-step prediction on the basis of the data $Y_1, \ldots, Y_T$ is given by the conditional expectation

$$\mu_m(\mathbf{X}_T) = E(Y_{T+m}|\mathbf{X}_T) \tag{1}$$

---

[1]Current address: American Express Int. Inc., PO Box 11 01 01, W-6000 Frankfurt 1, Germany

when a quadratic loss function is used. Since the term in (1) can be analytically determined only for sufficiently known processes, it must be estimated in the case of realistic time series data.

The forecast methods introduced in the next section are based on the classification of all time series courses $\mathbf{X}_t$, $t = p, \ldots, T$, in a few clusters. The m-step forecast is then simply given by the average of all data which, at distance m, follow courses belonging to the same cluster as the last known course. The kernel and nearest neighbour (NN) methods discussed in the third section do not require this rather rigid cluster partition and are therefore more flexible forecast instruments. The m-step kernel and NN predictors are weighted averages of all those observations that follow at distance m the courses $\mathbf{X}_t$ located in the neighbourhood of the last known course $\mathbf{X}_T$. The weights determined by the kernel function usually decrease with increasing distances between $\mathbf{X}_t$ and $\mathbf{X}_T$, and the size of the neighbourhood is specified by the so-called bandwidth. In the subsequent section a modification is proposed in which data with previous courses lying in the direct neighbourhood of the last known course are not the only values to go into the forecast. Further distant data are also included when the structure of their previous courses is similar to those of the last realizations $\mathbf{X}_T$. Finally, the methods discussed will be applied to a time series of the run-off of the Ruhr. The inclusion of more distant similar courses promises improved forecasts especially when there are only a few courses in the immediate neighbourhood of the last course $\mathbf{X}_T$.

2 Classification of time series courses by means of nonhierarchical cluster analysis

Cluster analysis is an instrument for the classification of M objects into K clusters. Applied to the classification of time series courses, this means dividing the $M = T - p + 1$ courses $\mathbf{X}_t$, $t = p, \ldots, T$, into the K clusters $C_1, \ldots, C_K$, which are represented by a typical course. The m-step predictor then has the form

$$\mu_m^P(\mathbf{X}_T) = \frac{\sum_{s=p}^{T-m} I_{C_l}(\mathbf{X}_s) Y_{s+m}}{\sum_{s=p}^{T-m} I_{C_l}(\mathbf{X}_s)} \quad \text{if } \mathbf{X}_T \epsilon C_l, \tag{2}$$

(with I_A as the indicator function of the set A) and corresponds to the arithmetic mean over all time series data Y_{s+m} with previous courses $\mathbf{X}_s$ belonging to the same cluster as $\mathbf{X}_T$. Collomb (1983) investigates the asymptotic properties of the predictor $\mu_m^P(\mathbf{X}_T)$, which, in analogy to the histogram, is referred to as a predictogram.

In order to calculate the predictogram, it is neccessary to find K representative states, such that each course $\mathbf{X}_t$, $t = p, \ldots, T$, can be assigned to exactly one of these states. With the K-medoid model, one clusters these M courses around K representative courses (called medoids) $\{\tilde{\mathbf{X}}_1, \ldots, \tilde{\mathbf{X}}_K\} \subset \{\mathbf{X}_p, \ldots, \mathbf{X}_T\}$, which are selected so that the mean distance

$$\frac{1}{M} \sum_{j=1}^{M} \min\{d(\tilde{\mathbf{X}}_i, \mathbf{X}_{j+p-1}) : i = 1, \ldots, K\} \tag{3}$$

between the courses belonging to a cluster and the appertaining medoids is minimized. In this case, d is a suitable distance measure. The term (3) should be minimized over all $\binom{M}{K}$

samples $\{\tilde{\mathbf{X}}_1, \ldots, \tilde{\mathbf{X}}_K\} \subset \{\mathbf{X}_p, \ldots, \mathbf{X}_T\}$, which leads to an enormous computational effort especially in the case of larger data sets. For this purpose a proposition of Kaufmann and Rousseeuw (1985) has been taken up. They recommend the algorithm CLARA (Clustering LARge Applications) for the clustering of large data sets. CLARA takes N samples of $n \ll M$ objects and carries out the above-mentioned complete K-medoid algorithm for each. Next, the objects not belonging to the sample are assigned to the cluster with the medoid to which they have the smallest Euclidean distance. The mean distance between all objects and their respective medoids can be calculated as a measure of the goodness of the cluster division. Finally, the sample is selected which leads to the minimal mean distance. For the practical application of this method, the length p of the courses, the number K of the clusters and the number n of the courses for which the complete K-medoid algorithm is carried out must be determined. The details of this are described by Michels and Heiler (1989). Predictograms can be calculated very quickly after the often lengthy classification algorithm has been done.

3 Kernel and Nearest Neighbour Methods

Because of its rigid cluster division, the predictogram often lacks the flexibility neccessary to produce good forecasts. For $p \geq 2$ very different courses are collected in a cluster leading to very heterogenous clusters. This causes the resulting forecast to be very biased, especially when the last known course differs too much from the typical course of the cluster to which it is assigned. When increasing the number of clusters, only a few data go into the forecast of unusual courses, which inflates the prediction variance. If the last course $\mathbf{X}_T$ is located on the periphery of the cluster C_l, then courses that lie far away from $\mathbf{X}_T$ in the same cluster C_l are relevant to the forecast. However, others that lie in another cluster near by $\mathbf{X}_T$ have no influence. In order to avoid these drawbacks, all courses in the neighbourhood of the last known course should be relevant to the forecast. Furthermore, it would be advantageous if the weight of an observation would depend on the nearness of its previous course to $\mathbf{X}_T$. This is performed by the kernel predictor

$$\mu_m^K(\mathbf{X}_T) = \frac{\sum_{s=p}^{T-m} K(\frac{\mathbf{X}_T - \mathbf{X}_s}{h}) Y_{s+m}}{\sum_{s=p}^{T-m} K(\frac{\mathbf{X}_T - \mathbf{X}_s}{h})}, \quad \frac{0}{0} := 0. \tag{4}$$

The analogon of (4) for $p = 1$ goes back to Nadaraya (1964) and Watson (1964). In numerous papers on the asymptotics of kernel estimates, many conditions have been given for the kernel function K. These are as a rule fulfilled when K has the properties of a bounded, unimodal density function that is symmetrical around zero and which tends rapidly towards zero if the absolute value of the argument tends to infinity. In addition to product kernels $K(\mathbf{u}) = \prod_{i=1}^p k_j(u_j)$, $\mathbf{u} = (u_1, \ldots, u_p)'$, (Euclidean) norm kernels of the form

$$K(\mathbf{u}) = k(\|\mathbf{u}\|), \quad \|\mathbf{u}\|^2 = \mathbf{u}'\mathbf{u}, \tag{5}$$

are especially suited for the construction of kernel functions defined on $\mathbb{R}^p$, where k and k_j are kernel functions defined on the real line. In the following a norm kernel with compact support $\{\|\mathbf{u}\| \leq 1\}$ is assumed. In this case the so-called bandwidth h is the radius of the neighbourhood of the last known course. An increase of h has the effect that the number of observations going into the forecast is tendentially increased and the forecast variance

Figure 1: 4 course patterns

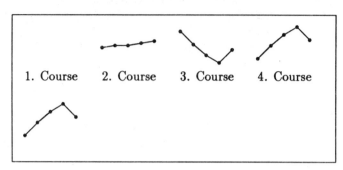

1. Course   2. Course   3. Course   4. Course

is thereby decreased. However, data which follow further distant courses are also included, which may lead to a larger bias. Conversely, a decrease in the bandwidth leads to a reduction of the bias and to an increase of the variance.

If only a few courses lie near the last known course, the forecast is an average of only few data which leads to a high variance. In order to avoid this, one can make sure that each forecast depends on a fixed number k of data points. For this purpose, in the equation (4) the fixed bandwidth h should be replaced by the distance $H_k(\mathbf{X}_T)$ between the last known course $\mathbf{X}_T$ and the k-th nearest course among the other courses $\mathbf{X}_s$. In this manner the k-nearest neighbour predictor is obtained.

In the mean time there exist a considerable number of papers on the asymptotics of kernel and NN-regression estimates. Here, let us refer simply to the monographs of Härdle (1990) and Michels (1992) and the literature cited therein.

4 Inclusion of distant similar courses

For the consistency (in L_2-norm) of the kernel predictors, the expected number of courses in a neighbourhood of the last known course that becomes smaller and smaller must increase above all bounds (cp. for this Michels, 1991, Robinson, 1983). The realization of the requirement for a neighbourhood of the last course $\mathbf{X}_T$ densely set with data vectors $\mathbf{X}_s$ becomes in practice more difficult for an increasing p. The inflation of the variance of the kernel predictor or bias of the NN-predictor in the case of an increase in dimension is often called the "curse of dimensionality".

In this paper a solution is proposed to the curse of dimensionality, which seems to be especially suited to time series data. The aim of this modification is to raise the number of observations going into the kernel estimation and to thereby reduce the variance in the forecast. In the case of the NN-forecast, the bias should be reduced by replacing less informative observations with those with a past that is "more informative" for the prediction of the time series. The following definitions have proven to be useful for the classification of time series courses: Let K be a kernel function with a compact support and $h > 0$ be a bandwidth (possibly dependent on $\mathbf{X}_t$).

Figure 2: 4 course patterns

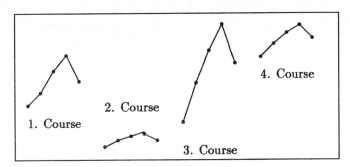

(i) The course $\mathbf{X}_s$ is said to be *directly similar* to the course $\mathbf{X}_t$ if $K((\mathbf{X}_t - \mathbf{X}_s)/h) > 0$ holds.

(ii) The course $\mathbf{X}_s$ is said to be *similar to course* $\mathbf{X}_t$ *after a shift of level* if $\theta_{0s} \in \mathbb{R}$ exists such that that $K((\mathbf{X}_t - (\theta_{0s}\mathbf{1}_p + \mathbf{X}_s))/h) > 0$ holds.

(iii) The course $\mathbf{X}_s$ is said to be *similar to course* $\mathbf{X}_t$ *after a linear transformation* if $(\theta_{0s}, \theta_{1s})' \in \mathbb{R} \times \mathbb{R}_+$ exist such that $K((\mathbf{X}_t - (\theta_{0s}\mathbf{1}_p + \theta_{1s}\mathbf{X}_s))/h) > 0$ holds.

Here, the notation $\mathbf{1}_p = (1, \ldots, 1)'$ was used. By means of the choice $h = H_k(\mathbf{X}_t)$, it is possible to include NN-forecasts as well. In the case of a fixed bandwidth h, the above similarity relations are reflexive und symmetrical, but not transitive.

Only observations with previous courses directly similar to the course $\mathbf{X}_T$ go into the forecast (4). In many cases, however, values Y_{s+m} with previous structures $\mathbf{X}_s$ that are similar to the course $\mathbf{X}_T$ after a shift of level can contain information relevant to the forecast. Although in figure 1 the Euclidean distance from course 4 to courses 2 and 3 is much smaller than that to course 1, the course patterns of 1 and 4 are very similar. Pattern 1, however, will not go into a usual kernel forecast with a kernel function defined on a compact support, although it apparently can be quite informative for the forecast of the time series after course 4.

In addition to similar courses on different levels, courses with similar passages, but with reaction patterns that are stronger or less strong, can also be informative. In figure 2, courses that are similar after a linear transformation are represented.

An intuitively obvious modification of the kernel or NN-predictors which, in addition to courses directly similar to $\mathbf{X}_T$, also includes those which are similar to $\mathbf{X}_T$ after linear transformation, is given by

$$\mu_m^{\theta}(\mathbf{X}_T) = \frac{\sum_{s=p}^{T-m} K(\frac{1}{h}(\mathbf{X}_T - (\theta_{0s}\mathbf{1}_p + \theta_{1s}\mathbf{X}_s)))(\theta_{0s} + \theta_{1s}Y_{s+m})}{\sum_{s=p}^{T-m} K(\frac{1}{h}(\mathbf{X}_T - (\theta_{0s}\mathbf{1}_p + \theta_{1s}\mathbf{X}_s)))}. \tag{6}$$

Of course, the successors Y_{s+m} must also be transformed according to the courses $\mathbf{X}_s$.

The values θ_{0s} and θ_{1s} should be determined in such a way that the course $\mathbf{X}_s$ lies as close as possible to the course $\mathbf{X}_T$ after linear transformation. For a kernel function that satisfies the condition $K(\lambda \mathbf{u}) \geq K(\mathbf{u})$ for all $\lambda \in [0,1]$, this is the case when θ_{0s} and θ_{1s} are selected so that $K((\mathbf{X}_T - (\theta_{0s}\mathbf{1}_p + \theta_{1s}\mathbf{X}_s))/h)$ is maximized. This condition generally leads to nonlinear optimization problems where the iterative solution for all $T - m - p + 1$ courses would be computationally expensive. However, if K is a Euclidean norm kernel (cp. (5)), then this maximization problem corresponds to the search for the minimum point of

$$\|\mathbf{X}_T - (\theta_{0s}\mathbf{1}_p + \theta_{1s}\mathbf{X}_s)\|^2, \; p \geq 2, \tag{7}$$

with respect to $\theta_s = (\theta_{0s}, \theta_{1s})' \in \mathbb{R} \times \mathbb{R}_+$. However, it is advisable to take into consideration the more general condition $0 \leq a \leq \theta_{1s} \leq b$ with $0 \leq a < 1 < b$ instead of the restriction $\theta_{1s} \geq 0$. The choice of $a > 0$ and $b < \infty$ makes it possible to assign a low weight to data when these follow courses with a dynamic which differs too much from that of $\mathbf{X}_T$. From the local Kuhn-Tucker conditions, one obtains for the minimum point from (7) under the restriction $0 \leq a \leq \theta_{1s} \leq b$ the equations

$$\theta_{0s}^* = \overline{X}_T - \theta_{1s}^* \overline{X}_s, \quad \theta_{1s}^* = \begin{cases} a & \text{if } s_{X_T,X_s} \leq a s_{X_s,X_s} \\ \dfrac{s_{X_T,X_s}}{s_{X_s,X_s}} & \text{if } a s_{X_s,X_s} < s_{X_T,X_s} < b s_{X_s,X_s} \\ b & \text{if } s_{X_T,X_s} \geq b s_{X_s,X_s}, \end{cases}$$

where $\overline{X}_t = \mathbf{1}'_p \mathbf{X}_t / p$ is the arithmetic mean of the components of $\mathbf{X}_t$ and $s_{X_s,X_t} = (\mathbf{X}'_s \mathbf{X}_t - p\overline{X}_s\overline{X}_t)/p$ is the empirical covariance between $\mathbf{X}_t$ and $\mathbf{X}_s$. The monograph of Michels (1992) is referred to for the explicit derivation of these results. Replacing in (6) θ_{0s} and θ_{1s} by θ_{0s}^* and θ_{1s}^* respectively allows the explicit determination of the predictor $\mu_m^\theta(\mathbf{X}_T)$. Since $\theta_{1,s}^* \neq \frac{s_{x,X_s}}{s_{X_s,X_s}}$ holds for all courses $\mathbf{X}_s$, for which $a \leq \frac{s_{x,X_s}}{s_{X_s,X_s}} \leq b$ is not satisfied, these are as a rule fitted worse and the ensuing data receive less weight in the average (6).

To calculate the kernel predictor, one looks for information not only in the neighbourhood of the last course. Structurally similar courses are also brought into play. This increases the number of the included observations and thereby decreases the prediction variance. The estimate (6) thus contributes to overcoming the problem of dimensionality.

5 Application to Data on the Run-off of the Ruhr

Yakowitz and Karlson (1987) report the successful modelling of the run-off of North American rivers with the help of nonparametric methods. In order to test the applicability of such methods for West German rivers, these methods are applied to the prediction of the run-off of the Ruhr. The analysed data are daily averages in $[m^3/sec]$ that were measured at Pegel Villigst over a period of five years from 01.01.1976 until 31.12.1980 — thus on a total of 1827 days. The course of this measurement series, which is partially illustrated in figure 3, shows neither a trend nor a clear seasonal figure. The peaks appear irregularly distributed over the year. Michels and Heiler (1989) compare the forecast characteristics of parametric ARMA-processes and nonparametric predictograms (see (2)) for the modelling of the run-off of the Ruhr. Only the predictogram with $p = 1$ leads to a slight improvement in comparison to the naive forecasts. The consequence that the future course should only depend on the level of the last known value is, however, interpretatively far from being satisfactory. One expects on the contrary that information about the rising or falling of the run-off should be by all means utilized in the forecast.

Figure 3: Daily averages of the run-off of the Ruhr in $[m^3/sec]$ at Pegel Villigst from 01.01.1980 until 31.12.1980.

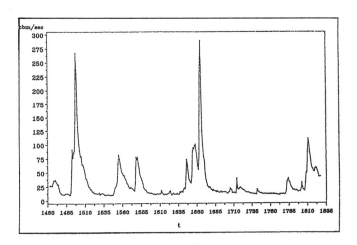

Because of the fixed bandwidth, kernel predictors are in principle totally unsuited to modelling this data set. The bandwidth which seems to be suitable for the majority of data (thus in the case of low run-off) is too small for the seldom, but interesting peaks. In the case of kernels with compact support, this causes that the forecast is generally not defined, since no further previous course lies in the neighbourhood of the last course. Because of this reason, only NN-predictors are considered here.

To illustrate the predictor characteristics, the 1-step predictors of an ARMA(1, 3)-model, those of the 10-NN-method with $p = 3$ and those of the modified 20-NN-method with $p = 4$ are plotted for the time period 1635 to 1685 (cp. figure 4). The corresponding Theil coefficients and success rates are listed on the bottom right in the plot. First, it is noticeable that the parametric ARMA-forecasts correspond almost exactly to the time-series delayed one day. Therefore ARMA-models do not satisfactorily explain the dynamic of the process. This is not true to that extent for the 10-NN-forecast. Of course, this method is also not able to predict the beginning of a high water period when the last known course contains only low values. Nevertheless, it deals with the development better if the last known time series value deviates from the "normal" data. A significant drawback of the NN-forecast lies in the considerable underestimation of the high water peak at observation number 1665. Since there are no 10 values of similarly high run-off in the past of the time series, this behaviour is quite typical of the NN-forecasts. The modified methods, which select possibly farther courses as a forecast basis and transform the following values correspondingly, yield considerably better forecasts for this peak value.

Figure 4: Run-off of the Ruhr: 1-step forecasts for the observations with numbers 1635-1685
a) ARMA(1,3)-model, b) 10-NN forecast, $p = 3$, c) modified 20-NN forecast

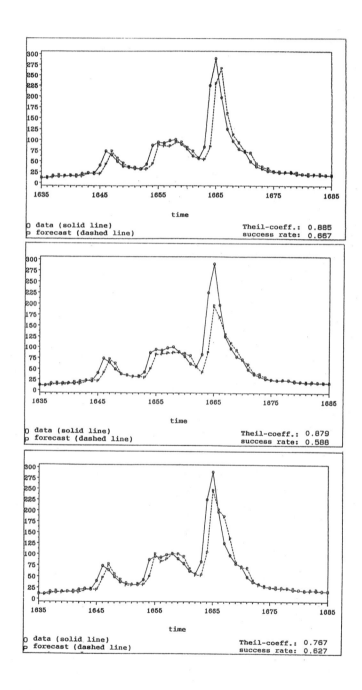

References

COLLOMB, G. (1983) From nonparametric regression to nonparametric prediction: Survey of the mean square error and original results on the predictogram. *Lecture Notes in Statistics, 16, 182-204.*

HÄRDLE, W. (1990) *Applied Nonparametric Regression*, Cambridge University Press, New York, etc.

KAUFMANN, L. and ROUSSEEUW, P.J. (1985) Clustering large data sets. *Report of the Dpt. of Mathematics and Informatics no. 85-13, Delft University of Technology.*

MICHELS, P. (1991) Moments of nonparametric kernel regression estimates for asymmetric kernel functions and *-mixing data. *Discussion paper No. 131/s, Faculty of Economics and Statistics, University of Konstanz.*

MICHELS, P. (1992) *Nichtparametrische Analyse und Prognose von Zeitreihen.* Arbeiten zur Angewandten Statistik, 36, Physica-Verlag, Heidelberg.

MICHELS, P. and HEILER, S. (1989) Die Wasserführung eines Flusses. Statistische Model-lierungsversuche, *Diskussion Paper No.118/s, Faculty of Economics and Statistics, University of Konstanz.*

NADARAYA, E.A. (1964) On estimating regression. *Theory of Probability and Application, 9, 141-142.*

ROBINSON, P.M. (1983) Nonparametric estimators for time series. *Journal of Time Series Analysis, 4, 185-207.*

WATSON, G.S. (1964) Smooth regression analysis. *Sankhyã, A, 26, 359-372.*

YAKOWITZ, S. und M. KARLSSON (1987) Nearest neighbor methods for time series, with application to rainfall-runoff prediction. *Stochastic Hydrology, 149-160.* D. Reidel, New York.

Moving Point Patterns: The Poisson Case

Dietmar Pfeifer, Hans–Peter Bäumer and Matthias Albrecht
Department of Mathematics and Computing Center, University of Oldenburg
P.O. Box 25 03, 2900 Oldenburg, Germany

Abstract: Motivated by time–dependent spatial random patterns occuring in marine or terrestrial ecosystems we investigate the limiting behaviour over time of certain Poisson point processes with possible movements of points according to a stochastic process. In particular, the possibilities of equilibrium, extinction or explosion of the system are discussed.

1 Introduction

A statistical analysis of random spatial patterns in marine or terrestrial ecosystems often requires the simultaneous consideration of time, space and migration. For instance, the spatial distribution of birds or geese in a certain observation area is varying over time due to incoming or outgoing flights and movements on the ground; similarly, the spatial distribution of sand worms in the wadden sea (like *arenicola marina*) depends on death, birth and migration of adults and larvae. Frequently the distributional patterns in space caused by such species are very much Poisson–like; hence it seems reasonable to study time–dependent point processes of this type and their long–time behaviour. Although the use of mathematical models in ecosystem theory is discussed controversely (see e.g. Wiegleb (1989) for similar problems in vegetation science) stochastic concepts in modelling of ecosystems are becoming more popular recently (see e.g. Richter and Söndgerath (1990)). A first simple model to study at least qualitative properties of systems as outlined above is a spatial birth–death process with migration (e.g., Brownian motion) whose time–marginal distributions are all Poisson point processes. Similar processes have been studied before (see e.g. Tsiatefas (1982), Volkova (1984), Madras (1989)), however with emphasis on different aspects of the model. Our approach resembles the presentation of Cox and Isham (1980), chapter 6.5 (iii); the Poisson property of the process here allows for easy calculations of possible limiting distributions over time. In particular, equilibrium, extinction, amd explosion of the system will be discussed.

2 The basic model

First we shall give a short account of the mathematical prerequisites in point process theory which are necessary to formulate our main results. For a more elementary treatment, see Pfeifer, Bäumer and Albrecht (1992) or the monographs of Daley and Vere–Jones (1988), Cressie (1991) or König and Schmidt (1992).

Although we shall for practical applications only consider Euclidean spaces such as $\mathbf{R}^1$, $\mathbf{R}^2$ or $\mathbf{R}^3$ it is easier from a theoretical point of view to start from general topological spaces such as locally compact Polish spaces $\mathcal{X}$ with corresponding σ–field $\mathcal{B}$ generated by the topology over $\mathcal{X}$ (cf. Kallenberg (1983)). Further, we may assume that the topology itself is induced by a metric ρ. Let $\mathcal{R} \subseteq \mathcal{B}$ denote the set of all relatively compact sets in $\mathcal{X}$. A measure μ on

$\mathcal{B}$ with finite values on $\mathcal{R}$ is called a *Radon–measure*. The set $\mathcal{M}$ of Radon measures over $\mathcal{X}$ can in a natural way be equipped with a σ–field $\mathbf{M}$ generated by the so–called *evolution mappings*

$$\tau_B : (\mathcal{M}, \mathbf{M}) \longrightarrow (\mathbf{R}^1, \mathcal{B}^1) : \mu \mapsto \mu(B)$$

where $\mathcal{B}^1$ denotes the Borel σ–field over $\mathbf{R}^1$. Any random variable ξ defined on some probability space with values in $(\mathcal{M}, \mathbf{M})$ is called a *random measure*, and if, in particular, the realizations of ξ are counting measures, i.e. $\xi(B) \in \mathbf{Z}^+ \cup \{\infty\}$ for all $B \in \mathcal{B}$, then ξ is called a *point process*. The realizations of point processes can be interpreted as random point configurations in the space $\mathcal{X}$; $\xi(B)$ here is the (random) number of points which fall into the set $B \in \mathcal{B}$. Sometimes it is convenient to consider point processes of the form

$$\xi = \sum_{k=1}^{N} \varepsilon_{X_k} \tag{1}$$

where ε_x denotes the Dirac measure concentrated in the point $x \in \mathcal{X}$, N is a $\mathbf{Z}^+$–valued random variable and the $\{X_k\}$ are random variables with values in $(\mathcal{X}, \mathcal{B})$; here $\xi(B) = \sum_{k=1}^{N} \varepsilon_{X_k}(B) = \sum_{k=1}^{N} \mathbb{1}_B(X_k)$ with $\mathbb{1}_B$ denoting the indicator variable of the event $B \in \mathcal{B}$, which makes the correspondance between measures in $\mathcal{M}$ and points in $\mathcal{X}$ more transparent. If, in particular, the random variables $\{X_k\}$ have ties with probabilities zero only, then ξ is called almost surely (a.s.) *simple*; i.e. there are no multiple counts of points in the pattern with positive probability.

A basic point process is the so–called *Poisson process* which is characterized by the following two properties:

1. There exists a measure μ on $\mathcal{B}$ (not necessarily Radon) such that for all $B \in \mathcal{B}$, the random variables $\xi(B)$ are $P(\lambda)$ Poisson–distributed with parameter $\lambda = \mu(B)$ with the convention that for $\mu(B) = 0$, $\xi(B) = 0$ a.s., and for $\mu(B) = \infty$, $\xi(B) = \infty$ a.s.

2. For any disjoint sequence of sets $\{B_n\}_{n \in \mathbf{N}}$, the sequence of random variables $\{\xi(B_n)\}_{n \in \mathbf{N}}$ is independent.

Especially by the latter property, point patterns realized by such a Poisson process are called *completely random*; a good illustrative example is perhaps the distributional pattern of raindrops on a walkway, or the spatial distribution of *arenicola marina* in the wadden sea.

Note that for Poisson processes, also $\mu(B) = E(\xi(B))$, $B \in \mathcal{B}$. In general,

$$E\xi(B) := E(\xi(B)), \quad B \in \mathcal{B},$$

always defines a measure on $\mathcal{B}$, called *intensity measure* of ξ; in the Poisson case, the intensity measure obviously defines the distribution of a Poisson process uniquely. Due to properties of the Poisson distribution, the superposition $\xi = \xi_1 + \xi_2$ of two independent Poisson point processes ξ_1 and ξ_2 again is a Poisson process with intensity measure $E\xi = E\xi_1 + E\xi_2$. Likewise it is possible to define a *p–thinning* of a Poisson process ξ; here a "point" of the process is retained independently of the other ones with probability $p \in [0, 1]$. The resulting process ζ is again Poisson with intensity measure $E\zeta = p \cdot E\xi$. This can be proved rigorously by the fact that a Poisson process ξ with *finite* intensity measure $\nu = E\xi$ can be represented in the form (1), where N is $P(\nu(\mathcal{X}))$–distributed and independent of the (also independent) random variables $\{X_k\}$ which follow the distribution $Q = \nu(\cdot)/\nu(\mathcal{X})$ provided $\nu(\mathcal{X}) > 0$

(otherwise there are a.s. no points realized). General Poisson point processes with σ–finite intensity measures can be constructed by superposition of independent Poisson processes with finite intensity measures concentrated on (at most countably many) disjoint subsets of $\mathcal{X}$. Note that, in particular, Poisson point processes are a.s. simple if the intensity measure is diffuse (i.e. all atoms have zero probabilities). For details, we refer the reader to Kallenberg (1983), Daley and Vere–Jones (1988), or König and Schmidt (1992).

Whereas for the study of Poisson point processes it is not absolutely necessary to use the full topological machinery as outlined in the beginning of this section it will become inevitable to do so when weak convergence of point processes is considered, as in the sequel of this paper. Some more notions will be needed to give simple sufficient conditions for such kinds of convergence.

A semiring $\mathcal{I} \subseteq \mathcal{R}$ is said to have the *DC–property* (i.e. being *dissecting* and *covering*), if for every set $B \in \mathcal{R}$ and every $\varepsilon > 0$ there exist finitely many sets $I_1, \ldots, I_n \in \mathcal{I}$ such that $B \subseteq \bigcup_{j=1}^{n} I_j$ and $\sup\{\rho(x,y) \mid x,y \in I_j\} < \varepsilon$, $1 \le j \le n$. (Such a set B is sometimes also called *pre–compact* or *totally bounded*.)

The following result on weak convergence to a simple point process is due to Kallenberg (1983), Theorem 4.7.

Theorem 1. *Let $\{\xi_n\}_{n \in \mathbb{N}}$ be a sequence of (not necessarily simple) point processes and ξ an a.s. simple point process such that $\mathcal{R}_\xi = \{B \in \mathcal{R} \mid \xi(\partial B) = 0 \text{ a.s.}\}$ contains a DC–semiring $\mathcal{I}$. Then the following two conditions are sufficient for weak convergence of $\{\xi_n\}$ to ξ, i.e. $P^{\xi_n} \overset{w}{\longrightarrow} P^{\xi}$:*

$$\lim_{n \to \infty} P\left(\xi_n\Big(\bigcup_{j=1}^{k} I_j\Big) = 0\right) = P\left(\xi\Big(\bigcup_{j=1}^{k} I_j\Big) = 0\right) \tag{2}$$

for all $k \in \mathbb{N}$, $I_1, \ldots, I_k \in \mathcal{I}$;

$$\limsup_{n \to \infty} E\xi_n(I) \le E\xi(I) < \infty \tag{3}$$

for all $I \in \mathcal{I}$.

For Euclidean spaces $(\mathcal{X}, \mathcal{B}) = (\mathbf{R}^d, \mathcal{B}^d)$ with finite dimension $d \in \mathbb{N}$ (and Borel σ–field $\mathcal{B}^d$) the semiring $\mathcal{I}$ of left–open, right–closed Intervals $I = \times_{j=1}^{d}(a_j, b_j]$ with $a_j < b_j$ will be a suitable DC–semiring fulfilling the conditions of Theorem 1 (Kallenberg (1983), p. 11 and Lemma 4.3).

In the particular case of *Poisson point processes* $\{\xi_n\}$, $\{\xi\}$ with finite total intensities $E\xi_n(\mathcal{X}) < \infty$, $E\xi(\mathcal{X}) < \infty$, conditions (2) and (3) in Theorem 1 simplify to the following simple condition:

$$\lim_{n \to \infty} E\xi_n(I) = E\xi(I), \quad I \in \mathcal{I}. \tag{4}$$

This is obvious since Poisson processes have independent increments, and the union of sets in (2) can w.l.o.g. be taken to be pairwise disjoint; condition (2) then simplifies to the consideration of the case $k = 1$ since by independence, $P\left(\xi_n\left(\bigcup_{j=1}^{k} I_j\right) = 0\right) = \prod_{j=1}^{k} P\left(\xi_n(I_j) = 0\right)$. But by (4),

$$\lim_{n \to \infty} P\left(\xi_n(I) = 0\right) = \lim_{n \to \infty} e^{-E\xi_n(I)} = e^{-E\xi(I)} = P\left(\xi(I) = 0\right)$$

for all $I \in \mathcal{I}$, hence (2) and (3) are satisfied.

Since in actual ecosystems, only a bounded (but possibly very large) number of objects can occur, the assumptions of finite total intensities in (4) are no real restriction for modeling purposes here.

3   Time–dependent Poisson point patterns

In this section we want to investigate the long–time behaviour of spatial Poisson point patterns in which objects are allowed to be newly created (by birth) or discarded (by death), and have the possibility of movements. For this purpose, we shall consider Poisson point processes $\{\xi_t\}_{t\geq 0}$ of the form (1), depending on the time t as

$$\xi_t = \sum_{k=1}^{N(t)} \mathbb{1}_{\{T_k > t\}} \varepsilon_{X_k(t)}, \quad t \geq 0, \tag{5}$$

where $\{N(t)\}_{t\geq 0}$ is an ordinary Poisson counting process on the line with finite and positive intensity $\lambda(t) = \Lambda'(t)$, where $\Lambda(t) = E(N(t))$ with $\Lambda(0) \geq 0$, $t \geq 0$ is some weakly increasing absolutely continuous function. $\{T_k\}_{k\in\mathbb{N}}$ is a family of (also from $\{N(t)\}$) independent and identically distributed life times with absolutely continuous cdf F with $F(0) = 0$ and density $f = F'$; $\{X_k(t) \mid t \geq 0\}_{k\in\mathbb{N}}$ is a family of (also from $\{N(t)\}$ and $\{T_k\}$) independent and identically distributed stochastic processes taking values in a locally compact Polish space $(\mathcal{X}, \mathcal{B})$ such as $(\mathbf{R}^d, \mathcal{B}^d)$.

The process $\{N(t) \mid t \geq 0\}$ here governs the creation of new particles whereas the counting processes $I_k(t) = \mathbb{1}_{\{T_k > t\}}$, $t \geq 0$, describe the life lengths T_k of each individual particle in the system. The movement of (alive) particles is governed by the processes $\{X_k(t) \mid t \geq 0\}$.

Note that if the processes $\{X_k(t) \mid t \geq 0\}$ are Markov processes then the process $\{\xi_t\}_{t\geq 0}$ also is a Markov process since for any strictly increasing non–negative sequence $\{t_n\}_{n\in\mathbb{N}}$ of time points the Poisson point processes $\{\xi_{t_n}\}_{n\in\mathbb{N}}$ are obtained successively by independent thinnings, superpositions and Markovian shifts of points. This follows from the Markov chain generation theorem as in Mathar and Pfeifer (1990), Lemma 3.2.2.

Let Q_t, $t \geq 0$, denote the distribution of $X_k(t)$, $k \in \mathbb{N}$. Then by Theorem 1, we obtain the following result concerning the long–time behaviour of the pattern process $\{\xi_t\}_{t\geq 0}$.

Theorem 2. *Let $\{\xi_t\}_{t\geq 0}$ be a family of Poisson point processes of type (5) with points located in a locally compact Polish space $(\mathcal{X}, \mathcal{B})$ such that $\mathcal{B}$ contains a suitable DC–semiring $\mathcal{I}$ of relatively compact subsets fulfilling the requirements of Theorem 1. Then if there exists some Radon measure μ over $\mathcal{B}$ such that*

$$\Lambda(t)(1 - F(t))Q_t(I) \longrightarrow \mu(I), \quad t \to \infty, \quad \text{for all } I \in \mathcal{I}, \tag{6}$$

we have weak convergence of ξ_t to some Poisson point process ξ with intensity measure μ, i.e. $P^{\xi_t} \xrightarrow{w} P^{\xi}$. In particular, if $T = \inf\{t > 0 \mid \xi_s(\mathcal{X}) = 0 \text{ for all } s \geq t\}$ denotes the time of (possible) extinction of the system, then

$$
\begin{aligned}
P(T \leq t) &= \exp\left(-\Lambda(t)(1 - F(t)) - \int_t^\infty \lambda(s)(1 - F(s))\, ds\right) \\
&= e^{-\mu(\mathcal{X})} \exp\left(-\int_t^\infty \Lambda(s)f(s)\, ds\right), \quad t \geq 0
\end{aligned}
\tag{7}
$$

with $e^{-\mu(\mathcal{X})} = 0$ for $\mu(\mathcal{X}) = \infty$. In the latter case, $T = \infty$ a.s., i.e. the system will a.s. not die out.

Proof. The first part follows immediately from Theorem 1, relation (4) and the fact that for the intensity measure of ξ_t, we have

$$E\xi_t(I) = E(N(t))E\Big(\mathbf{1}_{\{T_k>t\}}\varepsilon_{X_k(t)}(I)\Big) = \Lambda(t)(1 - F(t))Q_t(I)$$

for all $I \in \mathcal{I}$. For the second part, observe that by our assumptions,

$$\xi_t(\mathcal{X}) = \sum_{k=1}^{N(t)} \mathbf{1}_{\{T_k>t\}}, \ t \geq 0,$$

is a Markov birth–death process with birth and death rates $\beta_n(t)$ and $\delta_n(t)$, resp., given by

$$
\begin{aligned}
\beta_n(t) &= \lim_{h\downarrow 0} \frac{1}{h} P(\xi_{t+h}(\mathcal{X}) = n + 1 \mid \xi_t(\mathcal{X}) = n) = \lambda(t)(1 - F(t)), \quad n \in \mathbf{Z}^+, \\
\delta_n(t) &= \lim_{h\downarrow 0} \frac{1}{h} P(\xi_{t+h}(\mathcal{X}) = n - 1 \mid \xi_t(\mathcal{X}) = n) = \frac{nf(t)}{1 - F(t)}, \quad n \in \mathbf{N},
\end{aligned}
\tag{8}
$$

for $t \geq 0$. Note that the first part of (8) is due to the fact that the counting process $\{N(t)\}$ has birth rate $\lambda(t)$ at time t, with time–depending thinning by the (joint) survival probabilitites $1 - F(t)$, while for the second part, the death probability for a single individual alive particle in the time interval $[t, t+h]$ is approximately h times the hazard rate $f(t)/(1 - F(t))$. To prove (7), observe that by the independent increments property ii) of Poisson processes,

$$
\begin{aligned}
P(T \leq t) &= P\Big(\{\xi_t(\mathcal{X}) = 0\} \cap \bigcap_{s>t}\{\xi_s(\mathcal{X}) = 0\}\Big) \\
&= P\Big(\{\xi_t(\mathcal{X}) = 0\}\Big) P\Big(\bigcap_{s>t}\{\xi_s(\mathcal{X}) = 0\} \mid \xi_t(\mathcal{X}) = 0\Big)
\end{aligned}
$$

$$= P(\{\xi_t(\mathcal{X}) = 0\}) \lim_{h\downarrow 0} \prod_{k=0}^{\infty} (1 - h\beta_0(t + kh)) = \exp\big(-E\xi_t(\mathcal{X})\big)\exp\left(-\int_t^\infty \beta_0(s)\,ds\right)$$

which gives the first part of (7). The second equation follows by partial integration and (6). ∎

4 Applications

We shall give an example of a possible application in $(\mathcal{X}, \mathcal{B}) = (\mathbf{R}^d, \mathcal{B}^d)$ with finite dimension $d \in \mathbf{N}$. Suppose that the lifetime distributions are exponential with mean $\frac{1}{\tau}$, $\tau > 0$, and the movements of points governed by componentwise independent Brownian motions with zero mean and variance $\sigma^2 > 0$, and initial multivariate normal distribution with independent components of zero mean and variance $\sigma_0^2 \geq 0$. Let further the cumulative intensity of the Poisson birth process be given by

$$\Lambda(t) = \sqrt{t}^d e^{crt} + \Lambda(0), \quad t \geq 0,\tag{9}$$

with a parameter $c \geq 0$. Then if m_d denotes the d-dimensional Lebesgue measure, one of the following three cases will occur, depending on the choice of the parameter c:

Case 1, $c = 1$:

Here asymptotic equilibrium of the system will be achieved by weak convergence of $\{\xi_t\}$ to a Poisson point process ξ with mean measure given by

$$E\xi(B) = \frac{1}{\sqrt{2\pi\sigma^2}} m_d(B), \quad B \in \mathcal{B}^d. \tag{10}$$

This follows from Theorem 2 since for any bounded Interval I in $\mathbf{R}^d$ we have

$$Q_t(I) = \frac{1}{\sqrt{2\pi(\sigma_0^2 + \sigma^2 t)}^d} \int_I \exp\left(-\frac{1}{2(\sigma_0^2 + \sigma^2 t)}\sum_{k=1}^d x_k^2\right) dx_1\, dx_2 \ldots dx_d, \quad t > 0, \tag{11}$$

with

$$\sqrt{t}^d Q_t(I) \longrightarrow \frac{1}{\sqrt{2\pi\sigma^2}} m_d(I), \quad t \to \infty, \tag{12}$$

hence

$$\Lambda(t)(1 - F(t))Q_t(I) = \left(e^{(c-1)\tau t} + \frac{\Lambda(0)}{\sqrt{t}^d}\right)\sqrt{t}^d Q_t(I) \longrightarrow \frac{1}{\sqrt{2\pi\sigma^2}} m_d(I), \quad t \to \infty, \tag{13}$$

which proves (10).

Case 2, $c < 1$:

Here extinction of the system will eventually happen since $\mu(\mathcal{X}) = 0$ and by (7),

$$\begin{aligned}
P(T \le t) &= \exp\left(-\int_t^\infty \Lambda(s)f(s)\,ds\right) \\
&= e^{-\Lambda(0)e^{-\tau t}} \exp\left(-\int_t^\infty \sqrt{s}^d e^{-(1-c)\tau s}\,ds\right), \quad t \ge 0,
\end{aligned} \tag{14}$$

which is a proper cdf. Accordingly, similar as in (10) to (13), we have weak convergence of $\{\xi_t\}$ to the void point process, i.e. the Poisson process with zero intensity measure. The following figure shows the cdf of T for the parameter choices $d = 2$, $\Lambda(0) = 0$, $\tau = 1$ and c ranging from 0.2 to 0.8 with step 0.1.

254

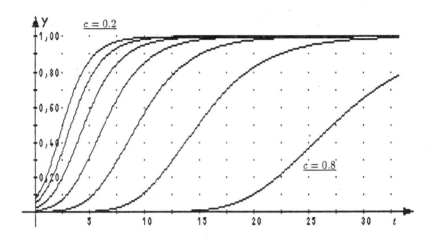

Fig. 1

Cumulative distribution function for extinction time T

Case 3, $c > 1$:

Here the system will eventually explode since the mean number of particles in any bounded set B at time $t > 0$ is $E\xi_t(B) \sim e^{(c-1)\tau t}m_d(B)/\sqrt{2\pi\sigma^2}$, which grows to ∞ for large t.

For birds, one might take $d = 2$, $c < 1$ and $\Lambda(0) = \sigma_0^2 = 0$; then $E\xi_t(\mathcal{X}) = t \cdot \exp\left(-(1-c)\tau t\right)$, $t \geq 0$, hence the system will start from zero individuals, growing up to a maximal average flock size of $1/[e(1-c)\tau]$ (achieved at time $t = 1/[(1-c)\tau]$), and then gradually decrease to zero again. For $c = 1$, the system would again start with zero individuals, but increase gradually to a stable average flock size within every bounded region of $\mathbf{R}^2$. In the case of worms, one might take $d = 2$, $c \leq 1$ and $\Lambda(0) > 0$, $\sigma_0^2 > 0$; then the system starts with a random configuration of an average positive number of individuals, and is either asymptotically stable over bounded regions (for $c = 1$) or dies out (for $c < 1$). Of course, similar results would hold true with other initial distributions $P^{X_k(0)}$ than normal distributions. For instance, in revitalization experiments in the wadden sea rectangular areas are covered with impermeable sheets such that after some time, no more individuals will be inside this area. This corresponds to a deterministic thinning procedure of a Poisson point process at the beginning ($t = 0$), hence one starts with an initial distribution for the location of individuals which is non–normal. However, the same analysis as before shows that in case $c = 1$, a random spatial pattern will develop over time which approaches a homogeneous Poisson process as before, which is in coincidence with observations made in field experiments.

Fig. 2 shows a simulation study of such a system for a starting homogeneous Poisson point process with total deletion of points within a rectangular area (taken from Pfeifer, Bäumer and Albrecht (1992)).

Software for visualizing moving point patterns is presently being developed by the authors and will be available upon request.

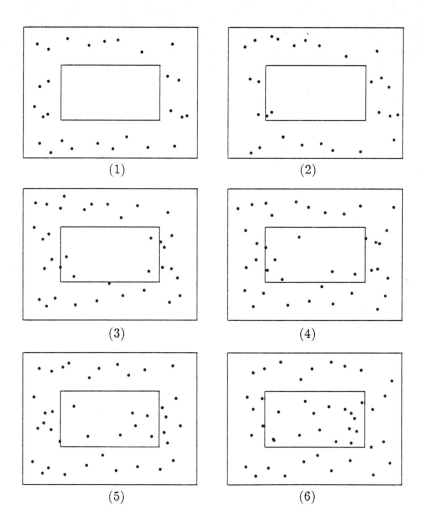

Fig. 2
Simulation of a revitalization experiment by Poisson point patterns

Acknowledgement

This work is in part supported by the *Bundesminister für Forschung und Technologie*, contract no. FZ 03F0023F, within the research project *Ökosystemforschung Niedersächsisches Wattenmeer — Pilotphase — : Beiträge der Angewandten Statistik zur Bearbeitung von Maßstabsfragen und zur Versuchsplanung für die Analyse räumlicher Strukturen und dynamischer Vorgänge im Watt.*

References

COX, D.R. and ISHAM, V. (1980), *Point Processes*. Chapman and Hall, London.

CRESSIE, N. (1991), *Statistics for Spatial Data*. Wiley, N.Y.

DALEY, D.J., and VERE–JONES, D. (1988), *An Introduction to the Theory of Point Processes*. Springer, N.Y.

MADRAS, N. (1989), Random walks with killing. *Probab. Th. Related Fields 80, no. 4, 581 – 600*.

KALLENBERG, O. (1983), *Random Measures*. Ac. Press, N.Y.

KÖNIG, D., SCHMIDT, V. (1992), *Zufällige Punktprozesse*. Teubner Skripten zur Mathematischen Stochastik, Teubner, Stuttgart.

MATHAR, R., and PFEIFER, D. (1990), *Stochastik für Informatiker*. Leitfäden und Monographien der Informatik. Teubner, Stuttgart.

PFEIFER, D., BÄUMER, H.-P., and ALBRECHT, M. (1992), Spatial point processes and their applications to biology and ecology. To appear in: *Modeling Geo–Biosphere Processes*.

RICHTER, O., and SÖNDGERATH, D. (1990), *Parameter Estimation in Ecology*. VCH Verlagsgesellschaft, Weinheim.

TSIATEFAS, G.N. (1982), Zur Bestimmung des Wanderungsprozesses [Determination of the migration process]. *Biometrical J. 24, no. 1, 87 – 92*.

VOLKOVA, E.I. (1984), Some asymptotic properties of branching processes with particle motion [in Russian]. *Dokl. Akad. Nauk SSSR 279, no. 2, 290 – 293*.

WIEGLEB, G. (1989), Explanation and prediction in vegetation science. *Vegetatio 83, 17 – 34*.

Part II

Information Retrieval,
Knowledge Processing
and
Software

Representations, Models and Abstractions in Probabilistic Information Retrieval

Norbert Fuhr

Informatik VI, Universität Dortmund, W-4600 Dortmund, Germany

Abstract: We show that most approaches in probabilistic information retrieval can be regarded as a combination of the three concepts representation, model and abstraction. First, documents and queries have to be represented in a certain form, e.g. as a sets of terms. Probabilistic models use certain assumptions about the distribution of the elements of the representation in relevant and nonrelevant documents in order to estimate the probability of relevance of a document w.r.t. a query. Older approaches based on query-specific relevance feedback are restricted to simple representations and models. Using abstractions from specific documents, terms and queries, more powerful approaches can be realized.

1 Introduction

The major problem in information retrieval (IR) is the search for items in a collection of documents which are relevant to the current request. This process bears an intrinsic uncertainty and vagueness: First, the information need cannot be represented uniquely by a formulation in the system's query language. Second, there is no clear procedure that decides whether a document is an answer to the information need or not. (IR systems with a Boolean query language are not an exception from the second point; they only shift all problems associated with uncertainty from the system to the user.) As the most successful approach for dealing with uncertainty and vagueness in IR, probabilistic models have been developed. In fact, the abstract models have been developed mostly in other disciplines (e.g. pattern matching and machine learning). Although probabilistic IR is theoretically well-founded, the application of these models to IR problems sometimes has been rather ad-hoc. In this paper, a new framework for the application of probabilistic models in IR is developed.

2 Probabilistic IR

We give a brief outline of the theoretical basis of probabilistic IR, along with the example of a simple probabilistic model which serves for the illustration of the concepts discussed in the following. (A more detailed presentation and a broader survey is given in Fuhr (1992).) First, we describe the event space of probabilistic IR. Let $\underline{D}$ denote the set of documents in the database and $\underline{Q}$ the (possibly infinite) set of queries submitted to the system. Then the event space is $\underline{Q} \times \underline{D}$. A single element of this event space is a query-document pair (q_k, d_m), where we assume that all these elements are equiprobable. Furthermore, let $\mathcal{R} = \{R, \bar{R}\}$ denote the set of possible relevance judgements (relevant / nonrelevant). Then with each element of the event space, there is associated a relevance judgement $r(q_k, d_m) \epsilon \mathcal{R}$. It is assumed that the relevance judgements for different documents w.r.t. the same query are independent of each other. Since an IR system can only have a limited understanding of documents and queries, it is based on representations of these objects, which we denote by d_m and q_k. As a consequence, several objects may have identical representations, and there may be different relevance judgements for query-document pairs sharing a single pair of representations (q_k, d_m). Now the task of a probabilistic IR system is the estimation of the probability of relevance $P(R|q_k, d_m)$. This is the probability that a random query-document pair with the representations (q_k, d_m) will be judged relevant. As retrieval output for a single query q_k, the documents of the database are ranked according to decreasing values of the probability of relevance $P(R|q_k, d_m)$. It can be shown theoretically that this strategy called "Probability Ranking Principle" yields optimum retrieval quality (Robertson (1977)). This finding is a major advantage of probabilistic IR in comparison to other IR approaches (e.g. Boolean and fuzzy retrieval, vector space model and cluster search) where no relationship between the theoretical model and retrieval quality can be derived. In order to illustrate the concepts discussed later in this paper, we now describe a simple probabilistic IR model, the binary independence retrieval (BIR) model from Robertson & Sparck Jones (1976). In this model, queries and documents are represented as sets of terms q_k^T and d_m^T, respectively. Let $T = \{t_1, \ldots, t_n\}$ denote the set of terms in the collection. Then the representation of a document can be mapped uniquely onto a binary vector $\vec{x} = (x_1, \ldots, x_n)$ with $x_i = 1$, if $t_i \in d_m^T$ and $x_i = 0$ otherwise. So the probability of relevance that we want to estimate now can be written as $P(R|q_k, \vec{x})$, the probability that a document represented by the binary vector $\vec{x}$ will be judged relevant w.r.t. a query represented by q_k. For the following derivation, we use odds instead of probabilities, where $O(y) = P(y)/P(\bar{y}) = P(y)/[1 - P(y)]$.

By applying Bayes' theorem, we can compute the odds of a document represented by a binary vector $\vec{x}$ being relevant to a query q_k as

$$O(R|q_k, \vec{x}) = \frac{P(R|q_k, \vec{x})}{P(\bar{R}|q_k, \vec{x})} = \frac{P(R|q_k)}{P(\bar{R}|q_k)} \cdot \frac{P(\vec{x}|R, q_k)}{P(\vec{x}|\bar{R}, q_k)}. \tag{1}$$

In order to arrive at a formula that is applicable for retrieval of documents, a so-called "linked dependence assumption" (Cooper (1991)) of the form

$$\frac{P(\vec{x}|R, q_k)}{P(\vec{x}|\bar{R}, q_k)} = \prod_{i=1}^{n} \frac{P(x_i|R, q_k)}{P(x_i|\bar{R}, q_k)} \tag{2}$$

is applied. This assumption says that the ratio between the probabilities of $\vec{x}$ occurring in the relevant and the nonrelevant documents is equal to the product of the corresponding ratios of the single terms. Now we can transform (1) into

$$O(R|q_k, \vec{x}) = O(R|q_k) \prod_{i=1}^{n} \frac{P(x_i|R, q_k)}{P(x_i|\bar{R}, q_k)} \tag{3}$$

Here $O(R|q_k)$ is the odds that an arbitrary document will be judged relevant w.r.t. q_k. $P(x_i|R, q_k)$ is the probability that term t_i occurs ($x_i = 1$) / does not occur ($x_i = 0$) in a random relevant document. $P(x_i|\bar{R}, q_k)$ is the corresponding probability for the nonrelevant documents. Now let $p_{ik} = P(x_i{=}1|R, q_k)$ denote the probability that t_i occurs in a random relevant document and $q_{ik} = P(x_i{=}1|\bar{R}, q_k)$ is the probability that t_i occurs in a random nonrelevant document. In addition, we assume that $p_{ik} = q_{ik}$ for all terms not occurring in the set q_k^T of query terms. With these notations and simplifications, we arrive at the formula

$$O(R|q_k, \vec{x}) = O(R|q_k) \prod_{t_i \in d_m^T \cap q_k^T} \frac{p_{ik}}{q_{ik}} \prod_{t_i \in q_k^T \setminus d_m^T} \frac{1 - p_{ik}}{1 - q_{ik}} \tag{4}$$

In order to apply the BIR model, we have to estimate the parameters p_{ik} and q_{ik} for the terms $t_i \epsilon q_k^T$. In principle, there are two methods for estimating parameters of probabilistic IR models[1]:

1. If no relevance information is available for this process, then plausible assumptions may yield reasonable approximations of the parameters to be estimated. In the case regarded here, we might assume that almost all documents in the database will be nonrelevant w.r.t. the current query, so q_{ik} is estimated by the proportion of documents in the collection containing term t_i. For the p_{ik}s, it is assumed that the value of these parameters is a constant for all terms, e.g. $p_{ik} = 0.5$. Successful results with this approach are described in Croft & Harper (1979).

2. Usually, relevance feedback methods are applied. For that, the IR system must first retrieve a number of documents by some other retrieval method (e.g. with the BIR model as described above). Now the user is asked to give relevance judgements for these documents. From this relevance feedback data, we can estimate the parameters p_{ik} and q_{ik} according to their definitions.

As the BIR model is a rather primitive probabilistic model, there are two directions of possible improvements: First, instead of the poor representation of the documents, more detailed information about the occurrence of a term in a document should be used. This line of improvement is described in the following section. Second, the linked dependence assumption underlying the BIR model does not hold in most applications; for this reason, one might seek for models based on more realistic assumptions (see section 4).

3 Representations

For the BIR model, we have assumed that documents as well as queries are sets of terms. However, the concept of a term has not been described any further. Here we will present different approaches for deriving terms from texts and for describing their occurrence within the text. In the following, we assume that only the representation of documents is changed, while queries are regarded as sets of terms in each case. In the BIR model, a document is mapped onto a **set of terms**. This assumption leads us to the representation by a vector $\vec{x}$ with binary-valued elements $x_i \in \{0, 1\}$. As a simple improvement, one can consider the within-document-frequency of the terms. So we have instead of a set of terms a **bag of terms** (a bag is a set with multiple occurrences of elements). Here the elements of the vector are nonnegative numbers, that is $x_i \in \{0, 1, 2, \ldots\}$. In this case, a variation of the BIR model can be applied that is based on formula 3 (possibly restricted to the set of terms from the query). However, due to the fact that instead of two parameters per term, several parameters of the form $P(x_i = j | R, q_k)$ and $P(x_i = j | \bar{R}, q_k)$ with $j = 0, 1, 2, \ldots$ have to be estimated, the small number of observations from relevance feedback yields only poor estimates. On the other hand, we will show in section 5 that this weakness can be overcome by applying appropriate abstractions. For an even more detailed description of the occurrence of a term in a text, the concept of **forms of occurrence** has been developed Fuhr (1989a). This concept comprises actually two aspects, namely the significance of the term w.r.t. the current text and the certainty of identification of the term in the text. Both aspects are described by the values of certain features. For the significance of a term, the within-document-frequency

---

[1]For a discussion of the different problems involved in estimating probabilistic parameters in IR, see Fuhr & Huether (1989).

as well as the location of the term in the document (i.e. a term occurring in the title of a document in general is more significant than a term occurring in the remainder of the text). For single words as terms, the certainty of identification can be described by using two different stemming algorithms. For this purpose, we consider as the "basic form" of a term the singular of nouns and the infinitive of verbs. In a second step, the basic form is reduced to the word stem (e.g. computers $\rightarrow$ computer $\rightarrow$ comput, computes $\rightarrow$ compute $\rightarrow$ comput). In retrieval, a document term matching a search term in its basic form (e.g. computers — computer) has a higher certainty of identification than another document term matching only the stem (e.g. computes — computer). The concept of forms of occurrence becomes even more important when noun phrases are considered as terms (in addition to single words). For phrases, the difficulty is to decide whether a phrase actually occurs in the document. As an example, the phrase 'indexing algorithm' occurs in a sentence starting " ... *The analysis of* indexing algorithms *...* ", whereas the phrase 'document indexing' may be located in a passage "... indexing *technique for retrieving* documents *...* ". In the first case, the phrase is identified correctly, while the second phrase obviously is not occurring in the text. However, the latter statement is only true from a syntactic point of view; regarding the semantics of the text, the phrase matches the text. For this reason, it may not be feasible to base the decision on a pure syntactic analysis; moreover, no perfect parsing systems for texts from broad subject fields are available. With the aspect of certainty of identification, the information about the occurrence of a phrase can be represented as a value of a feature of the form of occurrence, either as the word distance of the components of the phrase in the text or by a value characterizing the structure of the parse tree. Based on the set of features chosen for describing significance and certainty of identification, forms of occurrence can be defined as combinations of certain values of these features (see e.g. Fuhr et al. (1991)). Let $V = \{v_1, \ldots, v_L\}$ denote the set of features defined this way, then documents can be represented again by a vector $\vec{x}$ with elements $x_i \in V$ denoting the form of occurrence of the term t_i in the current document.

The representations discussed so far are based on terms occurring actually in the text. A different strategy for representing texts is based on a **controlled vocabulary**. In this case, only terms (so-called descriptors) from the controlled vocabulary may be used as search terms. For a document, a descriptor should not only be an element of the representation if it occurs in the document text, it also should be assigned if there is a semantic relationship between the text and the the descriptor. For this purpose, an indexing dictionary containing pairs of text terms and descriptors is required. If manually indexed documents are available, these pairs can be derived by computing association factors between text terms and descriptors (e.g. by estimating the conditional probability $P(s|t)$ that descriptor s will be assigned to a document, given that term t occurs in the text). Now the so-called "relevance description" (Biebricher et al. (1988)) of a descriptor w.r.t. a document contains information about the descriptor, the document, the relationships between the text terms and the descriptor, and the forms of occurrence of the text terms. Since the mapping of all this information onto a single value is not feasible, it is not surprising that rather different probabilistic models in combination with the new concept of abstraction have been developed within the context of the work based on this form of representation.

4 Models

The general goal of a probabilistic model is the estimation of a probability of the form $P(R|\vec{x})$ with $\vec{x} = (x_1, \ldots, x_n)^T$. Here and in the following, we will omit the restriction 'q_k' for reasons of simplicity as well as for a more general level of discussion. Furthermore, we will only regard the estimation process based on relevance feedback data. Since the number of possible values of $\vec{x}$ will be large in most cases, a direct estimation of the probabilities $P(R|\vec{x})$ is not possible. For this reason, a more analytic approach is required which considers the distribution of the values of single vector elements in relevant and nonrelevant query-document pairs. A probabilistic model assumes certain forms of relationship between these distributions and the distribution of the whole tuple in both classes. Below, we will describe several models that have been applied successfully in probabilistic IR. The BIR model introduced in section 2 uses **Bayesian inversion** in order to estimate the probabilities $P(\vec{x}|R)$ and $P(\vec{x}|\bar{R})$ instead of $P(R|\vec{x})$. Based on this technique, several models have been investigated in IR which differ only in the assumptions about the stochastic dependence or independence of the elements of $\vec{x}$. The strongest assumption for this purpose is the **linked dependence** assumption (eqn 2) underlying the BIR model. Instead of assuming single elements to be distributed independently in this form, one may use the weaker assumption of **pairwise dependencies**. This yields the following approximations for $r = R$ and $r = \bar{R}$: $P(\vec{x}|r) = \prod_{i=1}^n P(x_{m_i}|x_{m_{j(i)}}, r)$ with $0 \le j(i) < i$. Here $(m_1, \ldots, m_n)$ is a permutation of the numbers $1, \ldots, n$ and $j(i)$ is a function which maps i onto a number smaller than i and $P(x_i|x_{m_0}, r) = P(x_i|r)$. Now the problem is to find a good set of pairwise dependencies to be used in this formula. An optimum approximation of the probability $P(\vec{x})$ with pairwise dependencies can be achieved by the following method: For all possible pairs (x_i, x_j), first the stochastic dependence is computed by the measure

$$I(x_i, x_j) = \sum_{x_i, x_j} P(x_i, x_j) \log \frac{P(x_i, x_j)}{P(x_i)P(x_j)}.$$

These values are the edge weights of a graph with the x_i as nodes. Then the optimum set of pairwise dependencies is the maximum spanning tree of this graph. The whole procedure has to be performed separately for the relevant and the nonrelevant class (see Rijsbergen (1977)). In Lam & Yu (1982), **higher order dependencies** are considered by the Bahadur-Lazarsfeld-expansion (BLE), which regards all possible dependencies of term pairs, triplets etc. For its application, the BLE has to be truncated in order to limit the number of parameters to be estimated. However, in the truncated expression, negative probabilities might occur. Furthermore, the experimental results for the different forms of dependence described in Salton et al. (1983) show that this approach is not feasible as an extension of the BIR model, mainly because the amount of feedback data available for estimating the increased number of parameters is not sufficient. In contrast to the methods based on Bayesian inversion described above, the following methods aim at a direct estimation of the probability of relevance $P(R|\vec{x})$. **Probabilistic classification trees** are based on the probabilistic version of the learning algorithm ID3 which requires a description vector $\vec{x}$ with discrete-valued elements. Now ID3 constructs a probabilistic classification tree using a top-down approach: select an attribute (= a component of $\vec{x}$), divide the training set into subsets characterized by the possible values of the attribute, and repeat this procedure recursively with each subset until no further attribute can be selected. Each leaf l of the final tree represents a class of vectors $\vec{x}$, where each possible vector $\vec{x}$ belongs to exactly one

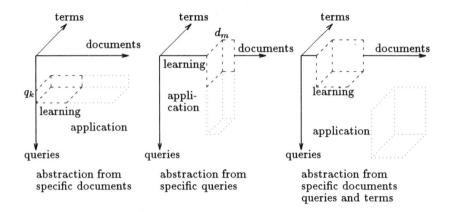

Figure 1: Learning strategies in IR

class l. Let l_j denote the class into which vector $\vec{x}$ falls, then the probability of relevance is estimated as $P(R|\vec{x}) = P(R|\vec{x} \in l_j)$ (see e.g. Fuhr et al. (1991)). While ID3 regards the distribution of an attribute within a specific subset of vectors, **regression methods** consider the distribution within the whole set. In the case of **logistic regression**, the probability of relevance is estimated by a function $P(R|\vec{x}) \approx \frac{\exp(\vec{b}^T \cdot \vec{x})}{1+\exp(\vec{b}^T \cdot \vec{x})}$, where $\vec{b}$ is a coefficient vector that is estimated based on the maximum likelihood method (see e.g. Fuhr & Pfeifer (1991)). **Least square polynomials** yield approximations of the form $P(R|\vec{x}) \approx \vec{b}^T \cdot \vec{x}$ (in the linear case), where $\vec{b}$ is a coefficient vector that minimizes the expectation of the squared error $(P(R|\vec{x}) - \vec{b}^T \cdot \vec{x})^2$ (see e.g. Fuhr (1989b)) .

5 Abstractions

The different probabilistic models described in the previous section all have in common that they can be regarded as a parameter learning process. Figure 1 shows three different learning approaches that are used in IR. The three axis indicate to what kinds of objects probabilistic parameters may relate to: documents, queries and terms. In each of the three approaches, we can distinguish a learning phase and an application phase: In the learning phase, we have relevance feedback data for a certain subset $Q_L \times D_L \times T_L$ of $Q \times D \times T$ from which we can derive probabilistic parameters. These parameters can be used in the application phase for the improvement of the descriptions of documents and queries. The BIR model is an example for the first type of learning, where a parameter relates to a certain term (or pairs, triplets in case of dependence models) and a specific query. Here we have relevance information from a set of documents D_L, and we can estimate parameters for the set of terms T_L occurring in these documents. In the application phase, we are restricted to the same query q_k and the set of terms T_L, but we can apply our model to all documents in D. The second type of learning is orthogonal to the first approach. Here a parameter relates to a certain term and a specific document. As an example, probabilistic indexing models (like e.g. the one described in Maron & Kuhns (1960)) collect relevance feedback data for a specific document d_m from a set of queries Q_L with the set of terms

T_L occurring in these queries. The parameters derived from this data can be used for the same document and the same set of terms T_L (occurring in queries) only, but for all queries submitted to the system. The major problem with this approach, however, is the fact that there are not enough relevance judgements for a single document in real databases, so it is almost impossible to estimate the parameters in this approach. The major drawback of these two learning approaches is their limited application range: in the first learning strategy, the relevance information collected for one query is worthless for any other query. In the same way, the probabilistic indexing approach restricts the use of relevance data to a single document. The third learning strategy depicted in figure 1 overcomes these restrictions. Here parameters are derived from a learning sample $Q_L \times D_L \times T_L$. Then these parameters can be applied to an arbitrary subset $Q_A \times D_A \times T_A$ of objects which may be totally different from the learning sample; especially, new queries as well as new documents and new terms can be considered. Now feedback data can be collected from all queries submitted to the IR system, thus increasing the size of the learning sample over time. This feature is the breakthrough for the application of the more sophisticated probabilistic models described in the previous section: Whereas in query-specific learning, only the feedback information from a few documents can be exploited for parameter estimation (and thus is only sufficient for rather simple models and representations), the third learning strategy extends the size of the learning sample drastically and thus allows for the estimation of a significantly higher number of parameters. In order to implement the third learning strategy, we must perform an abstraction from specific queries, documents and terms. Looking at the BIR model, one can see that this approach abstracts from specific documents by regarding features of documents instead of the documents itself. However, these features are related to specific terms (i.e. presence / absence of a term). A further abstraction from terms can only be achieved if we redefine the set of features, namely by regarding features of terms instead of the terms itself. For example, instead of a specific term, we can consider its inverse document frequency (the inverse of the number of documents in the collection in which the term occurs). Overall, now features of terms, documents and term-document pairs have to be defined. Examples for features of term-document pairs are the parameters relating to the representation of texts discussed in section 3 like e.g. the within-document-frequency or the form of occurrence of a term. A document-related feature is e.g. the length of the document text.

In order to reach our final goal — the third learning strategy — we have to go furtheron and abstract also from the specific query. For this purpose, appropriate features of queries (e.g. the number of query terms, their average or maximum inverse frequency) and query-document-pairs (e.g. the number of terms document and query have in common, the accumulated within-document frequencies of these common terms) have to be defined. So we finally have defined a feature vector $\vec{x} = \vec{x}(q_k, d_m, T)$ describing features of the triplet (query, document, terms). Now one of the probabilistic models can be applied in order to estimate the probability $P(R|\vec{x})$ that a query-document pair will be relevant, given that it has features $\vec{x}$. So we have seen that there are thre kinds of objects from which we can abstract — namely documents, queries and terms. In principle, any combination of abstraction from two of these types as well as only abstracting from a single type is possible. Not all of these possibilities are meaningful — especially no abstraction as well as abstraction over terms only. From the other cases, however, most have been investigated in different models:

- D: The BIR model abstracts from specific documents only.
- Q: The probabilistic indexing model described in Maron & Kuhns (1960) is an example for this strategy.

- (D, Q): A model abstracting from specific documents and queries, but with term-specific parameters is described in Wong & Yao (1990a).
- (D, Q, T): In Fuhr (1989b), this approach has been taken by using least square polynomials in combination with a text representation based on a controlled vocabulary.

6 Representations, models and abstractions

In the previous sections, we have discussed the three important elements of any probabilistic IR approach, namely representations, models and abstractions. We can regard these elements as three dimensions making up a space of possible approaches. If we would map publications on probabilistic IR onto the corresponding points in this space, then there would be two clusters: The first cluster is inspired by the BIR model and thus uses dependence or independence models in combination with sets or bags of terms, but abstracts only from specific documents. The second (smaller) cluster is based upon the work with the Darmstadt Indexing Approach (Biebricher et al. (1988), Fuhr (1989a), Fuhr et al. (1991)) and uses forms of occurrence or controlled vocabulary in combination with abstraction from specific terms, documents and queries. As models, mainly regression has been applied here, but the other types of models also have been investigated.

7 Outlook

The "approach space" described in the previous section suggests that further work in probabilistic IR can either fill the gaps in this space or extend one or more of the axis by developing completely new approaches. However, all the approaches developed so far are based more or less on the concept of batch retrieval. Even the feedback methods assume a procedure where feedback from the first retrieval run is used in order to improve the quality of the second run. Since an interactive system allows a larger variety of interactions than just query formulation and relevance feedback (see e.g. Croft & Thompson (1987)), these interactions should be incorporated in approaches for interactive probabilistic retrieval. As a consequence, the representations of interactive queries should become more complex. Furthermore, probabilistic IR systems should provide an appropriate set of commands and display functions. Finally, the scope of research has to be extended from the system-oriented perspective to a cognitive view of IR (see e.g. Pejtersen (1989)). Regarding the original task the user has to solve, the role that an IR system plays in this task has to be investigated, and systems that optimize their contribution within this task have to be developed.

References

BIEBRICHER, P.; FUHR, N.; KNORZ, G.; LUSTIG, G.; SCHWANTNER, M. (1988). The Automatic Indexing System AIR/PHYS - from Research to Application. In: Chiaramella, Y. (ed.): *11th International Conference on Research and Development in Information Retrieval*, pages 333–342. Presses Universitaires de Grenoble, Grenoble, France.

COOPER, W. (1991). Some Inconsistencies and Misnomers in Probabilistic IR. In: Bookstein, A.; Chiaramella, Y.; Salton, G.; Raghavan, V. (eds.): *Proceedings of the Fourteenth Annual International ACM/SIGIR Conference on Research and Development in Information Retrieval*, pages 57–61. ACM, New York.

CROFT, W.; HARPER, D. (1979). Using Probabilistic Models of Document Retrieval without Relevance Information. *Journal of Documentation 35*, pages 285–295.

CROFT, W. B.; THOMPSON, R. H. (1987). I3R: A New Approach to the Design of Document Retrieval Systems. *Journal of the American Society for Information Science 38(6)*, pages 389–404.

FUHR, N.; HÜTHER, H. (1989). Optimum Probability Estimation from Empirical Distributions. *Information Processing and Management 25(5)*, pages 493–507.

FUHR, N.; PFEIFER, U. (1991). Combining Model-Oriented and Description-Oriented Approaches for Probabilistic Indexing. In: Bookstein, A.; Chiaramella, Y.; Salton, G.; Raghavan, V. (eds.): *Proceedings of the Fourteenth Annual International ACM/SIGIR Conference on Research and Development in Information Retrieval*, pages 46–56. ACM, New York.

FUHR, N. (1989a). Models for Retrieval with Probabilistic Indexing. *Information Processing and Management 25(1)*, pages 55–72.

FUHR, N. (1989b). Optimum Polynomial Retrieval Functions Based on the Probability Ranking Principle. *ACM Transactions on Information Systems 7(3)*, pages 183–204.

FUHR, N. (1992). Probabilistic Models in Information Retrieval. *The Computer Journal 35*.

FUHR, N.; HARTMANN, S.; KNORZ, G.; LUSTIG, G.; SCHWANTNER, M.; TZERAS, K. (1991). AIR/X - a Rule-Based Multistage Indexing System for Large Subject Fields. In: *Proceedings of the RIAO'91, Barcelona, Spain, April 2-5, 1991*, pages 606–623.

LAM, K.; YU, C. (1982). A Clustered Search Algorithm Incorporating Arbitrary Term Dependencies. *ACM Transactions on Database Systems 7*.

MARON, M.; KUHNS, J. (1960). On Relevance, Probabilistic Indexing, and Information Retrieval. *Journal of the ACM 7*, pages 216–244.

PEJTERSEN, A. (1989). A Library System for Information Retrieval Based on a Cognitive Task Analysis and Supported by a Icon-Based Interface. In: Belkin, N.; van Rijsbergen, C. (eds.): *Proceedings of the Twelfth Annual International ACMSIGIR Conference on Research and Development in Information Retrieval*, pages 40–47. ACM, New York.

VAN RIJSBERGEN, C. (1977). A Theoretical Basis for the Use of Co-Occurrence Data in Information Retrieval. *Journal of Documentation 33*, pages 106–119.

ROBERTSON, S.; SPARCK JONES, K. (1976). Relevance Weighting of Search Terms. *Journal of the American Society for Information Science 27*, pages 129–146.

ROBERTSON, S. (1977). The Probability Ranking Principle in IR. *Journal of Documentation 33*, pages 294–304.

SALTON, G.; BUCKLEY, C.; YU, C. (1983). An Evaluation of Term Dependence Models in Information Retrieval. In: Salton, G.; Schneider, H.-J. (eds.): *Research and Development in Information Retrieval*, pages 151–173. Springer, Berlin et al.

WONG, S.; YAO, Y. (1990). A Generalized Binary Probabilistic Independence Model. *Journal of the American Society for Information Science 41(5)*, pages 324–329.

Fuzzy Graphs as a Basic Tool for Agglomerative Clustering and Information Retrieval

S. Miyamoto

Department of Information Science and Intelligent Systems
Faculty of Engineering, University of Tokushima
Tokushima 770, Japan

Abstract: The present paper reveals that fuzzy graphs are a basic tool for agglomerative hierarchical clustering. A new theorem is given that states equivalence between Wishart's mode analysis using kth nearest neighbor element (Wishart, 1968, 1969) and connected components of a fuzzy graph with membership values on vertices. This theorem is proved by generalizing a basic result in agglomerative clustering, which states the equivalence between the nearest neighbor method and the connected components of the standard fuzzy graph (Miyamoto, 1990). As a consequence, it is easily seen that Wishart's method can be replaced by the nearest neighbor method with a modified similarity. Second aim of this paper is to show how clusters are used in information retrieval with fuzziness in indices. A typical example is an input-output diagram that represents information retrieval process using fuzzy algebra (min-max algebra). A diagram representation for information retrieval with the ordinary algebra has been proposed and feedback has been introduced (Heaps, 1978). This system of the diagram has two drawbacks, one of which is that the system is unable to generalize the ordinary retrieval method of binary indexing, and the other is that feedback process fails to converge. The use of min-max algebra in fuzzy set theory solves these problems. The diagram with this algebra which corresponds to calculation of attainability of a fuzzy graph generalizes the binary retrieval method, and feedback always converges; the result shows a fuzzy equivalence relation, namely, a clustering. Application of Wishart's mode method to information retrieval using the above equivalence is also discussed.

1 Introduction

Theoretical basis of some methods of data analysis is different from the probabilistic basis of statistics. A typical example is found in automatic classification: some standard methods of agglomerative hierarchical clustering such as the nearest neighbor method does not use probability theory.

In this paper, we show that a part of fuzzy set theory originated by L.A.Zadeh (1965) is the basis of the nearest neighbor method and Wishart's mode method (1968,1969) which is a generalization of the nearest neighbor clustering.

Controversies often arise as to usefulness of fuzzy sets in statistical analysis and data analysis. The overall usefulness of fuzzy sets is not asserted here. Instead, a small part of the theory, that is, fuzzy graphs are shown to be a basic tool in agglomerative clustering. Indeed, fuzzy graphs, which are not inconsistent with the present theory of agglomerative clustering, provide a new algebraic approach that is complementary with the current approach.

The nearest neighbor method which is also called the single linkage method is known to have the most theoretically sound properties of all the agglomerative methods. In general, agglomerative linkage methods such as the nearest neighbor linkage, the furthest neighbor linkage, and the average linkage, are defined in terms of algorithms of pairwise link of clusters (Anderberg, 1973). This form of the algorithms are not convenient for proving mathematical properties of the linkage method. In the case of the nearest neighbor method, the property that the clusters are independent of the ordering of the objects is proved after the nearst neighbor algorithm has been transformed into the method of the minimum spanning trees (MST). In other words, the nearest neighbor method is sound because that is equivalent to generation of an MST of a weighted graph.

Discussion on fuzzy relations on a finite set and its transitive closure is closely related to the MST. Indeed, the transitive closure is proved to be a compact algebraic representation of a hierarchical classification that is equivalent to the nearest neighbor method (Miyamoto, 1990). The key concept of the proof of the equivalence is a fuzzy graph, which is a visualization of a fuzzy relation on a finite set.

Fuzzy relations are useful in describing information retrieval processes, in particular when information items are attached with weights of importance or interests (see, e.g., Bookstein, 1985). A process of information retrieval can be expressed as a block diagram in which a block is an index structure of a database whose input is an information request (a query) and output is a set of retrieved items in the database. When the index is weighted, the diagram calculation is carried out by the fuzzy algebra in which minimum is used for multiplication and maximum is used for addition.

Feedback in such a diagram implies a repeated application of an information search procedure. The feedback using the fuzzy algebra leads to the transitive closure of the relation of the block concerned. Namely, clusters of indices or information items are obtained from the fuzzy retrieval with feedback.

These two results show a methodological relationship between agglomerative clustering and information retrieval, by the link of fuzzy graphs.

Wishart's method is known to be a generalization of the nearest neighbor method, while its theoretical properties have been uninvestigated. The equivalence between the Wishart method and a fuzzy graph with membership values on the vertices is proved here. Application of the mode clustering to a diagram of information retrieval with feedback is considered, which deals with a prior weighting or preference on indices.

Throughout the present paper, the concept of the fuzzy graph is explicitly or implicitly used, as the visualization of fuzzy relations with the operation of the fuzzy algebra. Although the proof of a theorem herein looks cumbersome, the heart of the proof is simple: visualize the Wishart method as a fuzzy graph with membership values on the vertices. Composition of two fuzzy relations using the fuzzy algebra means calculation of the attainability grade of a pair of vertices on a fuzzy graph. In this way, the fuzzy graph is the main concept of this paper, even when this concept is not referred explicitly.

2 Symbols and preliminary results

2.1 Fuzzy relations and clustering

Let $X = \{x_1, x_2, ..., x_n\}$ be a finite set of objects to be classified, on which a measure of similarity $S(x_i, x_j)$ is defined. We assume, without loss of generality, that the similarity is normalized:

$$0 \leq S(x_i, x_j) \leq 1, \quad 1 \leq i, j \leq n$$

The measure S shows degree of similarity between x_i and x_j; $S(x_i, x_j) = 1$ means that the two objects x_i and x_j are most similar, whereas $S(x_i, x_j) = 0$ implies that x_i and x_j are not similar at all. Clustering means that a classification of X should be generated using S, so that similar objects are in the same class and two objects in different classes are not similar. In general, diagonal parts $S(x_i, x_i)$, $i = 1, ..., n$, are unnecessary for clustering. Nevertheless, we set $S(x_i, x_i) = 1$, $i = 1, ..., n$, to regard this measure as a fuzzy relation on X. Notice also that the symmetry $S(x_i, x_j) = S(x_j, x_i)$, $1 \leq i, j \leq n$, is assumed.

Fuzzy sets proposed by Zadeh (1965) is a generalization of the concept of ordinary sets. A fuzzy set a of X is characterized by the membership function $\mu_a(x)$, $x \in X$, in which the function μ_a takes its value in the unit interval $[0, 1]$. The meaning of μ_a is as follows. When $\mu_a(x) = 1$, x absolutely belongs to the fuzzy set a, whereas $\mu_a(x) = 0$ means that x does not belong to a at all. If $0 < \mu_a(x) < 1$, the belongingness of x to the set a is ambiguous. The value $\mu_a(x)$ is called the grade of membership.

A fuzzy relation on X is defined to be a fuzzy set on $X \times X$. Fuzzy relations are denoted by R, S, and so on. The grade of membership of R at $(x, y) \in X \times X$ is denoted by $R(x, y)$. In general, the value $R(x, y)$ implies degree of relatedness between x and y. The condition $R(x, x) = 1$, $\forall x \in X$, is called reflexive property of a fuzzy relation.

A simplified notation is sometimes used here for fuzzy sets and fuzzy relations on X. Since X is finite, we can identify a matrix and a fuzzy relation. Here the fuzzy relation R is identified with a matrix

$$R = (r_{ij}): \quad r_{ij} = R(x_i, x_j), \quad 1 \leq i, j \leq n$$

In the same way, a fuzzy set c whose grade of membership at x_i is $c_i = \mu_c(x_i)$ is identified with a vector $c = (c_1, c_2, ..., c_n)^T$ of the same symbol. (a^T means transpose of a.) Accordingly, matrix addition and multiplication are performed by the fuzzy algebra. Namely, for $R = (r_{ij})$ and $T = (t_{ij})$,

$$R + T = (\max[r_{ij}, t_{ij}])$$

and

$$RT = (\max_{1 \leq k \leq n} \min[r_{ik}, t_{kj}])$$

Furthermore, we use the symbol $\wedge$ for componentwise minimum operation:

$$R \wedge T = (\min[r_{ij}, t_{ij}])$$

In the case of clustering, most relations are symmetric, namely, $R(x, y) = R(y, x)$, for all $x, y \in X$. The above measure $S(x_i, x_j)$ may be interpreted as a fuzzy relation, since the measure is normalized.

A fuzzy relation R is called a fuzzy equivalence relation when R is reflexive, symmetric, and transitive. The transitivity means that for all $x, y, z \in X$,

$$R(x, y) \geq \min[R(x, z), R(z, y)]$$

holds. The significance of a fuzzy equivalence relation is understood after its α-cut is introduced. An α-cut is one of the most important operation which transforms a fuzzy set into an ordinary set using the parameter α ($\alpha \in [0, 1]$). Namely, an α-cut of R is given by

$$R_\alpha(x, y) = \begin{cases} 1 & (R(x, y) \geq \alpha) \\ 0 & (R(x, y) < \alpha) \end{cases}$$

Thus, a fuzzy equivalence relation means that every α-cut of the relation is an equivalence relation in the ordinary sense. Namely, for any $\alpha \in [0, 1]$, the following three conditions are satisfied:

(i) $R_\alpha(x,x) = 1, \quad x \in X$

(ii) $R_\alpha(x,y) = R_\alpha(x,y), \quad x,y \in X$

(iii) if $R_\alpha(x,z) = 1$ and $R_\alpha(z,y) = 1$, then $R_\alpha(x,y) = 1$

In other words, for each α, a classification of X is defined by a fuzzy equivalence relation. Note that a classification or equivalence classes mean a family of subsets of X in which the subsets are mutually disjoint and the union of all the subsets is equal to X. When α is increased, we have a finer classification; when α is decreased, a coarser classification is obtained. Thus, a fuzzy equivalence relation defines a hierarchical classification.

Since the fuzzy transitivity is a very strong condition, a fuzzy equivalence relation is not directly available in real applications. In agglomerative clustering, we start from an adequate definition of a similarity measure that is symmetric, and then some linkage method is applied. A similarity measure can be regarded as a fuzzy relation, and reflexivity is unimportant as stated before, but the transtivity is not satisfied. In the case of fuzzy relations, the transitive closure operation is used to define an equivalence relation. The transitive closure R^* is defined by

$$R^* = R + R^2 + ... + R^k + ...$$

where the fuzzy algebra is used. It is proved that if R is reflexive and symmetric, then

$$R^* = R^{n-1}$$

We omit the proof (See Miyamoto, 1990). ¿From now, we assume that the similarity and the fuzzy relation for clustering is denoted by the same symbol $S(x_i, x_j)$, $1 \leq i,j \leq n$.

When we consider a fuzzy graph GR associated with a fuzzy reflexive relation R, the set of vertices $V(GR)$ is the object set X with the membership value $R(x,y)$ on the edge (x,y). When a fuzzy relation defines a fuzzy graph, an α-cut of the graph has an edge between a pair of vertices if the corresponding value of the relation is not less than α. Namely, the edge set $E(GR_\alpha)$ is

$$E(GR_\alpha) = \{(x,y)|R(x,y) \geq \alpha, \quad x,y \in X\}$$

An alternative notation $GR = (X,R)$ for the same fuzzy graph is also used. In general, a fuzzy graph defined above is a directed graph. Nevertheless, when a fuzzy relation is symmetric, the corresponding fuzzy graph can be identified with an undirected graph, with the definition of an undirected edge $\{x_i, x_j\}$ with the grade $R(x_i, x_j) = R(x_j, x_i)$. In other words, An undirected edge is identified with the pair of directed edges (x_i, x_j) and (x_i, x_j) with the same value of the grade. As far as cluster analysis is concerned, we assume that fuzzy relations are symmetric, unless stated otherwise.

After introducing the fuzzy graph, it is evident that the transitive closure is a fuzzy equivalence, since the value $R^*(x,y)$ means the maximum value of the α when that α-cut is applied, x and y are in the same connected component of the graph defined by R^*_α.

Now, it is easily seen that the transitive closure generates hierarchical clusters as an agglomerative clustering procedure does. Indeed, it is well-known that the transitive closure is an equivalent statement of the nearest neighbor clustering. More precisely, the equivalence among the following three methods holds. (See, e.g., Miyamoto, 1990).

Given an arbitrary $\alpha \in [0,1]$, the following classifications (A–C) are equivalent.

(A) A classification generated by $[S^*]_\alpha$, where S^* is the transitive closure. Note that x_i and x_j are in the same class if ij entry of the matrix $[S^*]_\alpha$ is unity; they are in different classes if the corresponding entry is zero.

(B) A classification generated by connected components of the α-cut of the fuzzy graph GS obtained from S. (If two vertices are in the same connected component, then they are in the same class. ¿From now on, the set of vertices in a connected component is also called a *connected component*.)

(C) A classification generated at the level of similarity α using the nearest neighbor method of agglomerative clustering. (In other words, the classification when we stop the merging by the nearest neighbor clustering when the current similarity value for the pairwise merging becomes less than α for the first time.)

2.2 Diagram representation for information retrieval

Although information retrieval in general is not confined to search of documents, we use the terminology of document retrieval here to simplify the arguments. In considering problems in (document) information retrieval, we are concerned, at least, with two sets and a relation: $W = \{w_1, w_2, ..., w_m\}$ means a set of index terms, $D = \{d_1, d_2, ..., d_p\}$ means a set of documents, and U means a relation on $D \times W$ describing an index. Namely, if the term $w \in W$ is indexed to the document $d \in D$, then $U(d, w) = 1$; otherwise $U(d, w) = 0$. We assume that U is a fuzzy relation in general. The value $U(d, w)$ implies the weight with which w is indexed to d. Sometimes a fuzzy dictionary or a fuzzy thesaurus is used for finding a more adequate term or expanding a given term to its related terms. A fuzzy relation in a fuzzy thesaurus is denoted by F defined on $W \times W$. F may be symmetric or nonsymmetric. We call the relation F itself a fuzzy thesaurus, for simplicity. The way in which the grades of membership for U and F are specified is omitted here. Readers may see Miyamoto (1990).

Among various studies in theoretical information retrieval, a diagram representation by Heaps (1978) is considered here. A simple diagram of information retrieval is shown in Fig. 1, where the diagram is labeled by the index U, having the input q and the output r. The index U is $p \times m$ matrix (identified with the relation U on $D \times W$). The input $q = (q_1, q_2, ..., q_m)^T$ is m vector which is identified with the fuzzy set q in W (q_j is the grade of membership, usually written $\mu_q(w_j)$). In the same way, the output $r = (r_1, r_2, ..., r_p)^T$ means the fuzzy set r in D ($r_i = \mu_r(d_i)$).

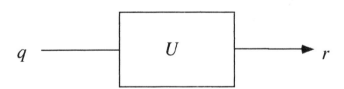

Figure 1: A simple diagram for information retrieval

The main characteristic in this diagram model is that a query and a response are expressed as vectors, this model is therefore called a vector model or vector processing model for information retrieval. Another feature is that queries are not confined to binary (0/1) valued vectors. Thus, this model shows a method of retrieval with weights in which q_j implies the weight on the term w_j.

The diagram in Fig. 1 enables a compact mathematical representation of an information retrieval process. Namely, given a query vector q, the retrieved document vector r is expressed as

$$r = Uq \tag{1}$$

If a query is expanded into its related terms using a thesaurus F, the expanded query is represented by Fq. Consequently, the response is

$$r = UFq \tag{2}$$

Figure 2 shows the diagram for this equation.

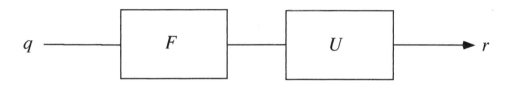

$q \longrightarrow \boxed{F} \longrightarrow \boxed{U} \longrightarrow r$

Figure 2: A diagram for information retrieval with a fuzzy thesaurus

Since Heaps considers the ordinary calculation for the retrieval method with weights, he uses the ordinary algebra of addition and multiplication. Now, we have the same expression (1) or (2) with two different calculation methods. One is the ordinary calculation method and the other is the fuzzy algebra of min-max calculation. We consider which is better.

Two drawbacks of the ordinary method of retrieval with weights are as follows, which are overcome by using the fuzzy algebra.

(a) The ordinary method is unable to extend the binary retrieval, which means that even if all the entries of q and U (and F) are 0/1 valued, the resulting response is not binary-valued. If the min-max calculation is used, it is easily seen that when q and U (and F) are binary, then the response is also 0/1 valued.

(b) The vector model without the fuzzy algebra lacks logical operations, and therefore is unable to reveal its relationship to the logical model of information retrieval commonly used.

Note that the logical model means that a query has the form of a logical expression of terms such as $(w_1 \ OR \ w_2) \ AND \ w_3$. Since a vector in the fuzzy algebraic model is the OR combination of the words w_j with their grades q_j $(j = 1, ..., m)$, the response r is also the OR combination of the documents d_i with the grades r_i, $j = 1, ..., n$. It is also clear that the vector model is unable to express the AND operation. To include AND in the vector model, the third algebraic operator such as the componentwise minimum $q_1 \wedge q_2$ is necessary. Thus, the method of fuzzy retrieval is advantageous over the ordinary retrieval method with weights.

Heaps considers a feedback retrieval and its diagram. Figure 3 shows a simple type of the feedback, in which the element U^T in the feedback loop means the inverse relation of U. As a matrix, U^T is the transpose of U.

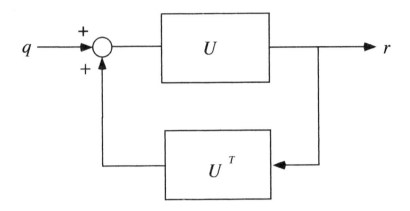

Figure 3: A feedback diagram for information retrieval

The ordinary algebraic calculation for the feedback leads to the infinite sum

$$r = Uq + (UU^T)Uq + ... + (UU^T)^n Uq + ...$$

which fails to converge in general. If we apply the fuzzy algebra, the above sum always converges to

$$r = (UU^T + I)^* Uq = U(U^T U + I)^* q$$

using the transitive closure. Interestingly enough, the above equation implies that the nearest neighbor clustering is carried out using the similarity UU^T between documents. Thus, feedback of retrieval is closely related to agglomerative clustering. More details can be found in Miyamoto (1990).

Let $G = U^T U + I$ and $W = UU^T + I$. Then, from the above equation,

$$r = W^* Uq = UG^* q$$

Here W^* stands for clustering of documents, whereas G^* represents clustering of index terms. Thus, a feedback diagram implies retrieval through the two alternative classifications.

3 Wishart's mode method and a fuzzy graph

Wishart's clustering using kth nearest neighbor is given by the following algorithm of the steps (W1-W4). The key idea of this method is to judge if each object is a "dense point" in the sense that an object has other k points in its neighbor. In the following, the nearest neighbor of an object x_i means the object x' which has the highest value of similarity $S(x', x_i) = \max S(x_k, x_i)$ for all x_k, $k \neq i$; kth nearest neighbor means the object which has kth highest value of similarity to x_i.

$G_1, G_2, \ldots$ are clusters (clusters in the Wishart method mean mutually disjoint subsets; the union of all the subsets does not always coincide with the object set X). $S(G_i, G_j)$ means similarity between clusters defined as follows. If $G_i = \{x_i\}$ and $G_j = \{x_j\}$, then

$$S(G_i, G_j) = S(x_i, x_j)$$

If G_k and G_ℓ are merged into G_i $(G_i = G_k \cup G_\ell)$, then

$$S(G_i, G_j) = \max[S(G_k, G_j), S(G_\ell, G_j)]$$

Algorithm W (Wishart, 1968)

(W1) Given an integer $k > 1$, calculate the value of similarity c_i for every $x_i \in X$ to its kth nearest neighbor.

(W2) Sort $\{c_i\}$ into the decreasing order. The resulting sequence is denoted by $\{c_{j_1}, c_{j_2}, \ldots, c_{j_n}\}$, i.e., $c_{j_1} \geq c_{j_2} \geq \ldots \geq c_{j_n}$. c_{j_k} corresponds to x_{j_k}.

(W3) Select thresholds $pmin$ from successive c_{j_k} values. At each cycle, introduce a new "dense point" x_{j_k} and test the following (W3.1–W3.3). Repeat (W3) until all points become dense.

(W3.1) If there is a pair of clusters G_p and G_q such that

$$S(G_p, G_q) = \max_{v \in G_p, w \in G_q} S(v, w) \geq pmin \qquad (3)$$

then merge G_p and G_q such as $G_p \cup G_q$ by the level $S(G_p, G_q)$. Repeat the merging until there is no such a pair of clusters that satisfies the above condition (3).

(W3.2) If there is no cluster G_i such that

$$\max_{w \in G_i} S(v_{j_k}, w) \geq pmin,$$

then v_{j_k} generates a new cluster $\{v_{j_k}\}$ that consists of v_{j_k} alone.

(W3.3) If there are clusters, say, $G_1, \ldots, G_\ell$, such that there exists $w_i \in G_i$, $i = 1, \ldots, \ell$, that satisfies

$$S(v_{j_k}, w_i) \geq pmin, \quad i = 1, \ldots, \ell,$$

then the clusters concerned are merged into a new cluster

$$\{v_{j_k}\} \cup G_1 \cup \ldots \cup G_\ell$$

by the level $pmin$.

(W4) When all points become dense and if there are more than one cluster, then repeat the merging process according to the nearest neighbor method.

Next step for the following theorem is to define a new fuzzy graph. In an ordinary fuzzy graph defined by a fuzzy relation, the grade of the relation is represented by a weight on the edge. For example, the value $S(x_i, x_i)$ is represented by the weight on the loop for the vertex x_i. On the other hand, the fuzzy graph considered here does not have any loop for an vertex. Instead, given any α, a vertex x_i with $S(x_i, x_i) < \alpha$ disappears from the α-cut of the fuzzy graph. Formally, we define a fuzzy graph $GS = (X, S)$ derived from a symmetric fuzzy relation S as follows.

A fuzzy graph $GS = (X, S)$ is a collection of $GS_\alpha = (X_\alpha, S_\alpha)$ for $\alpha \in [0, 1]$, where

(a) the vertex set X_α consists of those elements $y \in X$ such that $S(y, y) \geq \alpha$.

(b) the edge set defined by the binary relation S_α consists of unordered pairs $\{y, z\}$ such that $y, z \in X_\alpha$ and $S(y, z) \geq \alpha$.

Let us consider a simple example. Figure 4 shows three graphs. The graph in the left side is a fuzzy graph with membership values on the vertices shown by parentheses. The other two graphs are α-cuts of this fuzzy graph. The graph in the middle is with $\alpha = 0.5$, and the graph in the right side is with $\alpha = 0.3$. The matrix corresponding to the fuzzy graph in this figure is given by

$$S = \begin{pmatrix} 0.4 & 0.7 & 0 \\ 0.7 & 0.8 & 0.6 \\ 0 & 0.6 & 0.9 \end{pmatrix}$$

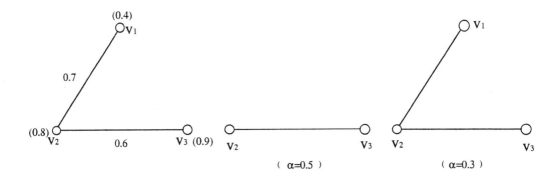

Figure 4: A fuzzy graph with grades on vertices and its α- cuts. The figure in the middle is with $\alpha = 0.5$, and that in the right side is with $\alpha = 0.3$.

Notice that the diagonal elements are not unity. Although the grade of the edge between v_1 and v_2 exceeds 0.5, that edge disappears from the graph in the middle, since the grade on v_1 is less than 0.5 and the vertex v_1 disappears from the same graph. Thus, the main feature of this type of fuzzy graphs is that the vertex set changes with the variation of the value of α.

Now, we obtain a theorem of equivalence.

Theorem. *Given an arbitrary $\alpha \in [0, 1]$, the following three classifications are equivalent, i.e., identical except the orders of the member subsets. (i) A classification generated by $[\{S \wedge (cc^T)\}^*]_\alpha$, i.e., the α-cut of the transitive closure of $S \wedge (cc^T)$. (The elements x_i and x_j are in the same class if and only if ij-entry of the above binary matrix is unity.) (ii) A classification generated by the connected components of the α-cut of the fuzzy graph $GS = (X, S \wedge (cc^T))$. (The α-cut of GS is denoted by $GS_\alpha = (X_\alpha, [S \wedge (cc^T)]_\alpha)$. (iii) Clusters generated by the Wishart's method at the level α. (Namely, merging is stopped at the level α in the algorithm W.)*

(Proof) Since it is easily seen that two classifications (i) and (ii) are equivalent, we concentrate on the proof of equivalence between (ii) and (iii).

For simplicity, the algorithm W and the fuzzy graph GS are called W and GS, respectively. We show, by induction on k, that at the levels $\alpha = c_{j_k}$, $k = 1, ..., n$, clusters by W and connected components of $GS_{c_{j_k}}$ are identical except the orders of the member subsets.

First, let $k = 1$ and consider W. First point x_{j_1} becomes dense and no clusters that contain more than one element are generated. Consider $GS_{c_{j_1}}$. We see only one vertex x_{j_1} in X_α, ($\alpha = c_{j_1}$). Thus, the both define the same cluster $\{x_{j_1}\}$.

Assume that clusters by W and connected components of $GS_{c_{j_k}}$ generate the same classification for $k \leq m - 1$. Consider W when mth dense point is introduced at the level $pmin = c_{j_m}$. Consider the following three cases.

i) If there are clusters that satisfy the condition (3) in (W3.1), then (W3.1) is applied. Note that the clusters are merged according to the way of the nearest neighbor clustering. ($S(G_p, G_q) = \max_{v \in G_p, w \in G_q} S(v, w)$ means the nearest neighbor linkage. See Miyamoto (1990).) Note also that these clusters are equivalent to connected components formed in $GS_{c_{j_{m-1}}}$. Since each point, say x_ℓ, in the clusters concerned satisfies $c_\ell \geq c_{j_{m-1}}$, merging G_p and G_q corresponds to connecting the same components G_p and G_q in $GS_{c_{j_{m-1}}}$ that has an edge $\{v, w\}$, $v \in G_p$, $w \in G_q$, such that $S(v, w) \geq pmin$. Conversely, if there is a pair of connected components G_p and G_q in $GS_{c_{j_{m-1}}}$ that should be connected in $GS_{c_{j_m}}$, then an edge $\{v, w\}$, $v \in G_p$, $w \in G_q$, satisfies $S(v, w) \geq c_{j_m}$. Hence the clusters are merged in W, since the condition (3) holds. Thus, the step (W3.1) generates newly connected components according to the nearest neighbor linkage.

ii) Suppose that, for the point x_{j_m}, there is no point x_ℓ such that $c_\ell \geq c_{j_m}$ and $S(x_{j_m}, c_\ell) \geq pmin = c_{j_m}$, then the step (W3.2) is applied and a cluster with one element $\{x_{j_m}\}$ is generated. Consider GS. Then the last condition means that x_{j_m} is not connected to any other vertices in $GS_{c_{j_m}}$, and vice versa.

iii) Suppose that points $x_{\ell_1}, ..., x_{\ell_t}$ satisfy $c_{\ell_i} \geq c_{j_m}$ and $S(x_{j_m}, x_{\ell_i}) \geq pmin = c_{j_m}$, $i = 1, ..., t$, and there are no other points that satisfy the last condition. Assume that x_{ℓ_i} belongs to a cluster G'_i, $i = 1, ..., t$. Then, the step (W3.3) is applied and we have a new cluster $\{x_{j_m}\} \cup G'_1 \cup ... \cup G'_t$. For GS, this condition means that the vertex x_{j_m} in $GS_{c_{j_m}}$ is connected to $x_{\ell_1}, ..., x_{\ell_t}$. By the inductive hypothesis, x_{ℓ_i} is in the component G'_i, $i = 1, ..., t$. Hence we have the component $\{x_{j_m}\} \cup G'_1 \cup ... \cup G'_t$. It is easy to see the converse is also true.

Note that if there is a tie $c_{j_m} = c_{j_{m+1}} = ... = c_{j_{m+t}}$, we may take $c_{j_{m+t}}$ instead of c_{j_m} and introduce dense points $x_{j_m}, x_{j_{m+1}}, ..., x_{j_{m+t}}$ at a time. Then the above arguments (i–iii) are directly applied with little modification.

Finally, after all points become dense at the level c_{j_n}, the algorithm W proceeds in the same way as the nearest neighbor linkage method. For $\alpha \leq c_{j_n}$, the equivalence property between the classification by the nearest neighbor linkage and that by the connected components of the ordinary fuzzy graph is applied, since $\alpha \leq c_{j_n}$ implies $X_\alpha = X$. (QED)

As a result of the above theorem, we can prove equivalence between the Wishart method and the nearest neighbor method. For this, a new similarity measure

$$S'(x_i, x_j) = \min[c_i, c_j, S(x_i, x_j)], \qquad 1 \leq i, j \leq n$$

is used, where c_i is defined in (W1) of the Wishart algorithm.

Corollary. *For arbitrary $\alpha \in [0, 1]$, the set of clusters that have more than one element which are generated by the Wishart method using the similarity $S(x_i, x_j)$ and the other set* *of clusters of more than one element by the nearest neighbor method using $S'(x_i, x_j)$ are equivalent (i.e., identical except the orders of the member subsets.)*

This result implies that Wishart's mode clustering can be performed by fast algorithms of MST (minimum spanning tree), since the nearest neighbor clustering is known to be equivalent to MST.

Another consequence is that theoretical properties of the nearest neighbor method hold true for the Wishart method. In general, results of agglomerative clustering depend on the ordering of the objects. In other words, when we randomly change the order of the objects in the same set and apply one clustering algorithm to the same collection of objects of different orders, we have different clusters. It is well-known that the result of the nearest neighbor linkage is independent of the ordering. Namely, even if the order is changed, the nearest neighbor clusters remain the same. The above theorem and the corollary show that this property of the independence of the ordering holds true for the Wishart method.

4   Wishart type clustering and information retrieval

As proved in the previous section, the Wishart method is represented by $[S \wedge (cc^T)]^*$ with the definition of c by kth nearest neighbor similarity. This observation implies that the Wishart method can be generalized to a general form of clustering by $[A \wedge (bb^T)]^*$, where A and b are appropriately defined, depending on each application. The vector b need not be the similarity to the kth nearest element. In some cases, the matrix A may not be symmetric, when clustering should be done using nonsymmetric interactions. Expressed as above, it is clear that the Wishart method in its generalized form has two criteria: one is A and the other is b; the two are combined by $A \wedge (bb^T)$, whose geometrical interpretation is the fuzzy graph in Section 3.

Now, we consider how this type of clustering arises in information retrieval with the diagram model using the fuzzy algebra. For this, let us note that the operator $\wedge$ can be removed by transforming the vector b into a diagonal matrix. Let us define a diagonal matrix $D(b)$ by

$$D(b) = diag(b_1, b_2, ..., b_n)$$

Namely, the diagonal elements of $D(b)$ are given by the components of the vector b and the off-diagonal elements are zero. It is clear that

$$A \wedge (bb^T) = D(b)AD(b)$$

is valid, since ij element of the both sides of the above equation are given by the same value of $\min[b_i, b_j, a_{ij}]$.

Two criteria are naturally introduced in information retrieval. One is a collection of relations between information items: the index U and the thesaurus F expressed as matrices are typical. Another criterion less frequently used is the grade of importance or preference. For example, old documents are less important than new ones; articles published in a high quality journal are generally considered to be of some importance. These grades, sometimes used in a real document retrieval system, are determined prior to the issue of a query, according to a user's background or preference. If such a grade, denoted by p_i, is attached to each document d_i, the transformation from a query to the response is

$$r = D(p)Uq$$

where $D(p) = diag(p_1, p_2, ..., p_n)$.

Let us consider a diagram in Fig. 5 which is similar to Fig. 3, except that an additional block $D(p)$ is used. This diagram shows that the grade of preference is introduced in a retrieval system and the feedback is considered. The response is expressed as the infinite sum

$$r = D(p)Uq + D(p)UU^T D(p)Uq + ... + (D(p)UU^T D(p))^k Uq + ...$$

Using the obvious identity $D(p)^2 = D(p)$, the response is expressed in terms of the transitive closure:

$$\begin{aligned} r &= D(p)(I + (D(p)UU^T D(p))^*)Uq \\ &= D(p)(I + (UU^T \wedge (pp^T))^*)Uq \\ &= D(p)U(I + (U^T D(p)U)^*)q \end{aligned}$$

Namely, the Wishart type clustering in the document set is obtained. The former two transitive closures of the above equation use document clustering, whereas the last transitive closure uses the term clustering.

Another criterion that uses preference on terms is also used. In this case, a vector in the term space $t = (t_1, ..., t_m)^T$ is used, as in Fig. 6. In the same way as the previous calculation,

$$r = UD(t)q + UD(t)U^T UD(t)q + ... + UD(t)(D(t)U^T UD(t))^k q + ...$$

We have

$$\begin{aligned} r &= UD(t)(I + (D(t)U^T UD(t))^*)q \\ &= (I + (UD(t)U^T)^*)UD(t)q \end{aligned}$$

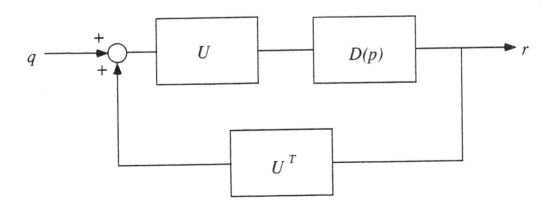

Figure 5: A feedback diagram with a prior grade p

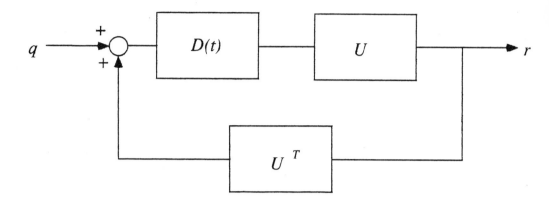

Figure 6: A feedback diagram with a prior grade t on terms

5 Conclusions

Fuzzy relations and fuzzy algebra interrelate two different concepts of agglomerative hierarchical clustering and feedback in information retrieval. The meaning of fuzzy algebra can be understood in terms of reachability among vertices in a fuzzy graph. This observation leads to the proof of the equivalence of the Wishart method and the particular type of fuzzy graphs. The key idea is to introduce grades of membership on vertices. Expression of Wishart's mode clustering by the transitive closure has three implications. First, theoretical properties of the nearest neighbor method are valid for the Wishart method. Second, fast algorithms of MST are applicable with a modification of the similarity. Third, the Wishart method can be generalized to include clustering by two criteria of an association A and a grade b, where the two criteria may be independent. In information retrieval, these two criteria are used. The association means indices; the grade is defined by a user's general preference.

The present result suggests that methods stimulated by A^* and $(A \wedge (bb^T))^*$ should be studied. For example, when A is not symmetric, A^* is not an equivalence relation. Different methods of constructing equivalence classes can be considered from A^* and $(A \wedge (bb^T))^*$. Fast algorithms for calculating these quantities are necessary (See, e.g., Tarjan, 1983).

In the application of clustering to information retrieval, generation of clusters and implementation of them as indices for a considerably large-scale real database should be studied. An interesting study and development in the Science Citation Index is well-known, where citation clusters were generated and were implemented as indices (Garfield, 1979). It should be noted that the clustering applied therein is a simplified version of the nearest neighbor method. By the present argument, application of the Wishart clustering to citation clusters is an interesting problem.

Another problem in the application is efficient implementation of clusters in a retrieval system. For hierarchical classifications, the use of fast algorithms of manipulating trees of the data structures should be studied.

References

ANDERBERG, M. R. (1973), *Cluster Analysis for Applications*, Academic Press, New Nork.

BOOKSTEIN, A. (1985), Probability and fuzzy-set applications to information retrieval, *Annual Review of Information Science and Technology, Vol. 20, 331–342.*

GARFIELD, E. (1979), *Citation Indexing: Its Theory and Application in Science, Technology, and Humanities*, John Wiley and Sons, New York.

HEAPS, H. S. (1978), *Information Retrieval: Computational and Theoretical Aspects*, Academic Press, New York.

MIYAMOTO, S. (1990), *Fuzzy Sets in Information Retrieval and Cluster Analysis*, Kluwer Academic Publishers, Dordrecht.

TARJAN, R. E. (1983), An improved algorithm for hierarchical clustering using strong components. *Information Processing Letters, Vol. 17, 37–41.*

WISHART, D. (1968), Mode analysis: a generalization of nearest neighbour which reduces chaining effects, in A.J.Cole, ed., *Numerical Taxonomy*, Proc. Colloq. in Numerical Taxonomy, Univ., St. Andrews, 283–311.

WISHART, D. (1969), Numerical classification method for deriving natural classes, *Nature, 221,* 97–98.

ZADEH, L. A. (1965), Fuzzy sets, *Information and Control, 8, 338–353.*

The Various Roles of Information Structures

Peter Schäuble, Daniel Knaus

Department of Computer Science, Swiss Federal Institute of Technology (ETH)
CH- 8092 Zürich, Switzerland

Abstract: Information structures represent a domain of discourse in a formal way. The objective of an information structure is to improve the process of retrieving information. The various roles information structures can play in a retrieval system are described. After discussing the role of a controlled indexing vocabulary, we focus on information structures that are aimed at helping users to formulate appropriate queries. Conventional thesauri are often used for this purpose. As an alternative we present a new information structure which we call a similarity thesaurus. In contrast to conventional thesauri, the construction and the maintenance of similarity thesauri is inexpensive and in addition, they can be improved systematically. Finally, we summarize the different aspects that should be considered when augmenting a retrieval system by an information structure.

1 Introduction

represent a domain of discourse in a formal way. The objective of an information structure is to improve a particular retrieval method that is used for retrieving objects stored in a retrieval system. Since there exist many different retrieval methods that can be improved in many different ways, we have to deal with a large variety of different information structures. In this paper, an information structure is a mathematical structure $\langle S_0, \ldots; R_0, \ldots; f_0, \ldots \rangle$ consisting of sets S_i, relations R_j, and functions f_k. In this way, we have a very flexible framework to represent almost every kind of formal knowledge.

Conventional thesauri are considered as information structures because they can be represented as mathematical structures $\langle T; UF, RT, NT \rangle$ representing the structure of a domain of discourse. The set T contains terms which are linked by the relations UF (Used For), RT (Related Term), and NT (Narrower Term). The relations USE (USE instead of) and BT (Broader Term) are considered here as derived relations that are determined by the relations UF and NT respectively. Conventional thesauri are defined in ISO 2788 (1986) and in DIN 1463 (1987). Formal semantics to conventional thesauri is given in Schäuble (1987).

Further examples of information structures are concept spaces (Schäuble, 1989a) that are related to the algebras used in the Generalized Vector Space Model (GVSM) introduced in Wong, Ziarko & Wong (1985). Both the concept space retrieval model and the GVSM are aimed at improving a retrieval function by including term dependencies into the retrieval process. Inference networks (Turtle & Croft, 1990) and knowledge bases (Sembok & van Rijsbergen, 1990) can also be considered as information structures that are aimed at improving a retrieval function.

Furthermore, there are information structures called pseudothesauri. A pseudothesaurus provides a grouping of quasi-synonyms that is used for indexing language normalization purpose during an automatic indexing process (Raghavan & Jung, 1989). Likewise, the fuzzy relation $z(t, s)$ used in the AIR/PHYS system (Fuhr et al., 1991) is also aimed at improving

information structure	aimed at improving
concept spaces algebras used in GVSM inference networks knowledge bases	retrieval functions
pseudothesauri fuzzy relations	automatic indexing methods
conventional thesauri similarity thesauri	manual indexing methods (incl. query formulations)

Table 1: Types of information structures.

an automatic indexing process. The weight $z(t, s)$ is the probability that a document is assigned a descriptor s from a controlled indexing vocabulary provided that the document contains the term t.

Considering the role information structures are playing in retrieval systems, we can distinguish different types of information structures. As a retrieval method consists of a retrieval function and an indexing method, we can distinguish between information structures aimed at improving retrieval functions and information structures aimed at improving indexing methods. Furthermore, we can distinguish between information structures aimed at improving automatic indexing methods and information structures aimed at improving manual indexing methods. We consider queries as (virtual) documents. In this way, the improvement of manual indexing process includes the improvement of query formulations (e.g. by manual query expansion). Table 1 shows the information structures mentioned above and to which type they belong.

Unfortunately, it is not possible to discuss all the different roles information structures can play in a retrieval system. Instead we focus on a few roles we consider as important in future retrieval systems. In section 2, we discuss information structures that are aimed at improving indexing methods by means of a controlled indexing vocabulary. In section 3, we focus on improving query formulation by helping users to find additional search terms. We also discuss, whether conventional thesauri are appropriate for this purpose. In section 4, similarity thesauri are presented as an alternative aid for helping users to formulate their queries. In section 5, some conclusions are drawn.

2 Controlled Indexing Vocabularies

The role of a controlled vocabulary is the most popular role information structures are playing in conventional retrieval systems. Examples of information structures serving as a controlled vocabulary are thesauri satisfying the ISO norm (1986) or the DIN norm (1987). Such thesauri are mostly used in the context of conventional retrieval systems based on manual indexing and Boolean retrieval (Lancaster, 1986).

The objective of a controlled vocabulary is to overcome the problems caused by the many-to-many relationship between terms and their meanings. A term having several meanings, i.e. a polysemic term may cause wrong matches. Conversely, several terms having the same

problem	undesired effect	remedy
polysemy	wrong match	qualifying terms
synonymy	lack of a match	relation "Used For" (UF)

Table 2: Avoiding polysemy and synonymy in thesauri.

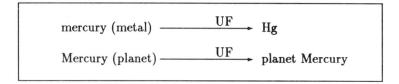

Figure 1: Qualified descriptors and non-descriptors linked by the "Used For" (UF) relation.

meaning, i.e. synonymous terms do not produce a match even though they should do so (Table 2).

When using an ISO or DIN thesaurus, polysemy is avoided by qualifying terms and synonymy is avoided by distinguishing between descriptors and non-descriptors as shown in Figure 1.

The polysemic term "mercury" is qualified which means that an additional term within parentheses has been attached to make its meaning unambiguous. When indexing, the descriptor "mercury (metal)" is Used For (UF) the non-descriptor "Hg." The indexing vocabulary consists of descriptors only and the non-descriptors must not be used for indexing. In this way, the indexing vocabulary does not contain synonymous terms.

Beside the conventional, manual indexing process, there exist also automatic indexing processes that are based on controlled vocabularies. For instance, the AIR/PHYS retrieval system developed for the physics database PHYS of the Fachinformationszentrum Karlsruhe uses a multistage indexing process based on a controlled vocabulary (Fuhr et al. 1991). One stage consists of mapping terms t (single words or phrases occurring in the document collection) to descriptors s of the controlled vocabulary. Instead of the "Used For" relation of conventional thesauri, they are using weighted relationships that can be interpreted in a probabilistic way. The weight $z(t, s)$ is an estimation of the probability that the descriptor s is assigned to a document d_j given that the term t is contained in d_j. These probabilities are estimated by a sufficiently large subcollection $M \subset D$ that has been manually indexed.

$$z(t, s) = \frac{\# \text{ documents containing } t \text{ and } s}{\# \text{ documents containing } t}$$

The indexing process adopted by the AIR/PHYS retrieval system overcomes a major problem of the conventional manual indexing process based on a controlled vocabulary. The domain knowledge consisting of the weights $z(t, s)$ can be used to index any set of documents (Figure 2).

Since the information structure consisting of the weights $z(t, s)$ is an abstraction from specific documents, it is not necessary to add new domain knowledge when new documents are included into the collection. The indexing process, however, is dependent on specific terms and descriptors. Because of this dependence on specific terms and descriptors the subcollection used for estimating the weights $z(s, t)$ must be rather large. The amount of

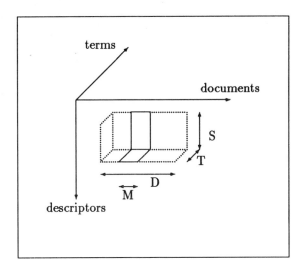

Figure 2: Abstraction of $z(t, s)$ from specific documents.

necessary learning data could be reduced by a new approach that is related to the Darmstadt Indexing Approach (DIA) (Fuhr & Buckley, 1991) and to the approach suggested in Fuhr (1989) where the complex representation of the documents is derived from a thesaurus.

Like the DIA, this new approach would consist of two steps. In the description step, the term t, the descriptor s, and their relationship is described by relevant data (e.g. occurrence frequency, co-occurrence frequency, etc.). In the decision step, $z(t, s)$ is estimated by an estimation function that takes the previously collected data as arguments. In this way, the estimation function is independent of specific terms and descriptors. Hence, the learning set M has not to cover the complete set S of descriptors and the set T of terms. Whether a much smaller learning set would suffice for estimating the weights is an open issue which has to be confirmed experimentally.

3   Improving Query Formulations

In this section, we assume that the more search terms the query contains the better is the usefulness of the answer. Based on this assumption, query formulations can be improved by using an information structure to find additional search terms. In addition to conventional thesauri, there are other information structures for this purpose. Before focusing on such information structures that are aimed at helping users to find additional query terms, we discuss the underlying assumptions.

To make the assumption "the more query terms the better the usefulness of the answers" more evident we performed two experiments with the CACM test collection and the CISI test collection respectively (Fox, 1983). To see whether the retrieval effectiveness depends on the length of the queries, we divided the set of queries into short queries and long queries. By the length of a query we mean the number of different search terms the query contains (without stop words). A query is a short query if the number of different terms is less than the median of the query lengths; otherwise, it is a long query. In the CACM test collection,

recall level	CACM precision			CISI precision		
	short queries	long queries	improvement	short queries	long queries	improvement
0.1	0.5823	0.6551	+24.1%	0.3576	0.5865	+64.0%
0.2	0.3681	0.5438	+47.7%	0.2822	0.5200	+84.3%
0.3	0.2816	0.4744	+68.5%	0.2331	0.3888	+66.8%
0.4	0.2518	0.4066	+61.4%	0.1946	0.3255	+67.2%
0.5	0.1927	0.3386	+75.7%	0.1698	0.2649	+55.9%
0.6	0.1533	0.2281	+48.8%	0.1453	0.2162	+48.8%
0.7	0.1248	0.1839	+47.4%	0.1110	0.1516	+36.7%
0.8	0.0940	0.1528	+62.6%	0.0797	0.1134	+42.4%
0.9	0.0734	0.0881	+20.1%	0.0493	0.0781	+58.3%
1.0	0.0662	0.0809	+22.3%	0.0116	0.0516	+345.2%
average	0.2079	0.3430	+65.0%	0.1735	0.2859	+64.8%

Table 3: Retrieval effectiveness determined by short queries and by long queries.

the median of the query lengths is 9.5 terms and in the CISI test collection, the median of the query lengths is 19.0 terms. In each experiment, the retrieval effectiveness determined by the long queries is compared with the retrieval effectiveness determined by the short queries. The results are shown in Table 3.

The bottom line shows the average precision values at the recall levels 0.25, 0.5, and 0.75 that represent precision oriented, indifferent, and recall oriented searches. The retrieval effectiveness determined by the long queries is consistently higher than the retrieval effectiveness determined by the short queries (65 % improvement for both collections). This seems to indicate that our assumption is true in most cases.

We next discuss conventional thesauri that are frequently used to find additional search terms to improve the query formulations. Such conventional thesauri usually have been constructed according to the ISO norm 2788 (1986) or DIN 1463 (1987). They contain the relations UF, RT, NT, and the derived relations USE and BT as shown in Figure 3. The usage of such thesauri as an aid for improving query formulations is not unproblematic.

There are two major problems when using such thesauri that have been constructed manually according to the ISO or DIN norm. First, the manual construction is rather *expensive*. Even when a thesaurus exists which covers nearly the desired domain, the effort to adapt an existing thesaurus to specific needs is still considerable. The second problem with thesauri is the *lack of a dichotomy* between useful relationships and useless relationships. The norms neither say which kinds of relationships contribute to a higher effectiveness and which do not nor they say how the relationships can be identified systematically. The lack of a systematic construction process can also be characterized by citing Salton & McGill (1983): "Concerning now the actual thesaurus construction method, a manual thesaurus generation process is an art rather than a science." In the next section, we present an alternative information structure which may overcome the problems of conventional thesauri.

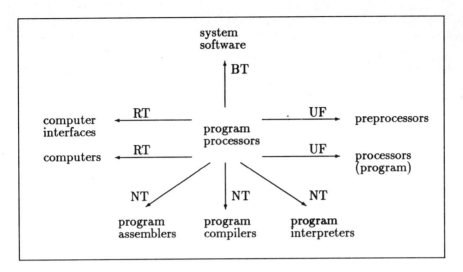

Figure 3: Narrower Terms (NT), Broader Terms (BT), and Related Terms (RT).

search for terms similar to: RETRIEV		
rank	rsv	similar term
1	1.00	RETRIEV
2	0.32	INFORM
3	0.30	SCARC
4	0.30	VALUAT
5	0.28	CONSECUT
6	0.27	QUERY
7	0.27	SALTON
8	0.25	DOCUM
9	0.24	BERGMAN
10	0.23	GOSH

Figure 4: Similarity thesaurus.

4 Similarity Thesauri

In this section, we present a new type of information structure as an attractive alternative to conventional thesauri. This information structure which we call *similarity thesaurus* is based on any kind of retrieval method where the roles of documents and terms have been exchanged: The terms are the retrievable items and the documents are the indexing features of terms. A query consists of a term (e.g. RETRIEV) and the retrieval system determines a ranked list of terms that are in decreasing order of the similarity values (Retrieval Status Values) as shown in Figure 4.

The main advantages of such a similarity thesaurus are the following. First, the construction of a similarity thesaurus is fully automatic, i.e. cheap and systematic. Second, the usefulness of the retrieved terms is optimal according to the probability ranking principle (Robertson, 1977). Finally, domain knowledge can easily be added to a similarity thesaurus when using the Binary Independence Indexing (BII) model together with the Darmstadt Indexing Approach (DIA) (Fuhr & Buckley, 1991).

In what follows, the BII model and the DIA that have been developed originally for document retrieval is adapted to support term retrieval. The readers are assumed to be familiar with the original BII model and DIA as described in Fuhr & Buckley (1991). When using the BII model for retrieving similar terms the BII-formula is as follows.

$$P(R \mid t_h, t_i) = c_h * P(R \mid t_i) \prod_{d_j \in \Phi(t_h) \cap \Phi(t_i)} \frac{P(R \mid d_j, t_i)}{P(R \mid t_i)}$$

Here, the query consists of the term t_h and the answer consists of a ranked list of terms t_i that is in decreasing order of the probabilities $P(R \mid t_h, t_i)$. In the BII model the retrieval status value of t_i with respect to the query t_h is equal to $P(R \mid t_h, t_i)$ which is the probability that when searching for relevant documents, it is useful to augment the query term t_h by the term t_i.

The constant c_h depends on t_h but not on t_i and hence, it does not affect the ranking order of the retrieved terms. The set $\Phi(t_i)$ consists of those documents d_j that contain the term t_i. Finally, the BII formula contains the probabilities $P(R \mid t_i)$ and $P(R \mid d_j, t_i)$. The former is the probability that t_i is an additional useful query term during any search process. The latter is the probability that t_i is an additional useful query term provided t_h is contained in d_j.

The probability $P(R \mid t_i)$ is usually assumed to be constant (Fuhr & Buckley, 1991). Thus, we only have to estimate the probabilities $P(R \mid d_j, t_i)$ to determine the ranking order of the terms. The probabilities $P(R \mid d_j, t_i)$ can be estimated by means of the Darmstadt Indexing Approach (DIA) which consists of two steps. In the *description step*, the vector $\vec{x}(d_j, t_i)$ is determined which contains statistical information on the indexing feature d_j, the item t_i, and their relationship (e.g. feature frequency, inverse item frequency, etc.). In the *decision step*, the probabilities $P(R \mid d_j, t_i)$ are approximated by the probabilities $P(R \mid \vec{x}(d_j, t_i))$ which depend on statistical quantities rather than on specific terms and documents. Finally, a probabilistic learning approach is used to find an estimation function $e : \vec{x} \mapsto e(\vec{x})$ for the probabilities $P(R \mid \vec{x}(d_j, t_i))$. In Fuhr & Buckley (1991), a linear estimation function $e_L : \vec{x} \mapsto \vec{c}^T \vec{x}$ was shown to perform fairly well where the coefficient vector $\vec{c}$ can easily be determined by a least square fit.

We conclude this section by summarizing the main advantages of using the BII model and the DIA for similarity thesauri. First, a BII/DIA based similarity thesaurus is *affordable*. It only needs between 50 and 100 equations for each of five components of the vector $\vec{c}$. An equation is given by two terms t_h and t_i and relevance information whether it is useful to augment the search term t_h by the additional search term t_i or not. Second, the construction of a BII/DIA based similarity thesaurus is *systematic*. Third, the (unknown) probabilistic parameters (i.e. the components of $\vec{c}$) can be estimated *efficiently* by a least square fit. Fourth, the maintenance of BII/DIA based similarity thesauri is *inexpensive*, since new documents and new terms do not require new domain knowledge. The domain knowledge represented by $\vec{c}$ is independent of specific terms and documents. Finally, the search for additional query terms can be augmented *effectively* by using relevance feedback (Salton & Buckley, 1990).

5 Conclusions

It was shown that there exist many different roles information structures can play in a retrieval system. When augmenting a retrieval system by an information structure, the following questions have to be considered. (1) How *useful* is the information structure with respect to the dedicated role(s)? (2) Is there an appropriate *construction process* to build an information structure for a particular domain of discourse? Is there a systematic way to build an information structure. How much manual (expensive) work is required for the construction of an information structure? (3) How big are the expected *maintenance costs?* How much manual (expensive) work is required for the acquisition of new documents?

The first question can be considered from different points of view. From a theoretical point of view, we may ask if the information structure has some desirable algebraic properties as suggested in Schäuble (1989b). From the point of view of evaluation, we may ask for the usefulness of the information structure. In particular, we may ask how the retrieval effectiveness depends on the effort that has been spent on the construction of the information structure. Is this dependency a linear dependency or are there saturation effects? Unfortunately, we lack an appropriate theory of evaluation for information structures. It seems that assessing the usefulness of information structures is much more difficult than assessing the usefulness of conventional retrieval methods (Frei & Schäuble, 1991). Hence, future research should focus on both information structures and their evaluation.

References

DIN (1987). *Erstellung und Weiterentwicklung von Thesauri, DIN 1463.* Deutsches Institut für Normung, Berlin.

FOX, E. A. (1983). Characterization of Two New Experimental Collections in Computer and Information Science Containing Textual and Bibliographic Concepts. Technical Report 83-561, Cornell University, Department of Computer Science.

FREI, H. P., & SCHÄUBLE, P. (1991). Determining the Effectiveness of Retrieval Algorithms. *Information Processing & Management*, 27(2–3), 153–164.

FUHR, N. (1989). Optimum Polynomial Retrieval Functions Based on the Probability Ranking Principle. *ACM Transactions on Information Systems*, 7(3), 183–204.

FUHR, N., & BUCKLEY, C. (1991). A Probabilistic Learning Approach for Document Indexing. *ACM Transactions on Information Systems*, 9(3), 223–248.

FUHR, N., HARTMANN, S., LUSTIG, G., SCHWANTNER, M., TZERAS, K., & KNORZ, G. (1991). AIR/X—a Rule-Based Multistage Indexing System for Large Subject Fields. In *RIAO*, pp. 606–623.

ISO (1986). *Guidelines for the Establishment and Development of Monolingual Thesauri, ISO 2788.* International Organization for Standardization, Geneva.

LANCASTER, F. W. (1986). *Vocabulary Control for Information Retrieval.* Information Resources Press, Arlington, Virginia, second edition.

RAGHAVAN, V. V., & JUNG, G. S. (1989). A Machine Learning Approach to Automatic Pseudo-Thesaurus Construction. Technical Report 89-6-1, The Center for Advanced Computer Studies, University of Southwestern Lousiana, Lafayette, LA 70504.

ROBERTSON, S. E. (1977). The Probability Ranking Principle in IR. *Journal of Documentation*, 33(4), 294–304.

SALTON, G., & BUCKLEY, C. (1990). Improving Retrieval Performance by Relevance Feedback. *Journal of the ASIS*, 41(4), 288–297.

SALTON, G., & MCGILL, M. (1983). *Introduction to Modern Information Retrieval*. McGraw-Hill, New York.

SCHÄUBLE, P. (1987). Thesaurus Based Concept Spaces. In *ACM SIGIR Conference on R&D in Information Retrieval*, pp. 254–262.

SCHÄUBLE, P. (1989a). *Information Retrieval Based on Information Structures*. PhD thesis, Swiss Federal Institute of Technology. VdF-Verlag, Zürich.

SCHÄUBLE, P. (1989b). On the Compatibility of Retrieval Functions, Preference Relations, and Document Descriptions. Technical Report 113, ETH Zürich, Department of Computer Science.

SEMBOK, T. M. T., & VAN RIJSBERGEN, C. J. (1990). SILOL: A Simple Logical-Linguistic Document Retrieval System. *Information Processing & Management*, 26(1), 111–134.

TURTLE, H., & CROFT, W. B. (1990). Inference Networks for Document Retrieval. In *ACM SIGIR Conference on R&D in Information Retrieval*, pp. 1–24.

WONG, S. K. M., ZIARKO, W. & WONG, P. C. N. (1985). Generalized Vector Space Model in Information Retrieval. In *ACM SIGIR Conference on R&D in Information Retrieval*, pp. 18–25.

Classification Properties of Communicating Neural Networks

R. Kree
Institut f. Theoretische Physik
Universität Göttingen, Bunsenstr.11, W-3400 Göttingen, FRG

A. Müller
Institut f. Wirtschafts- und Sozialpsychologie, Universität Göttingen
Humboldtalle 14, W-3400 Göttingen, FRG

Abstract: A rapidly growing body of literature is devoted to applications of neural networks in classification tasks. Most of the authors working in this field consider feed-forward architectures with hidden units, which can be trained via standard algorithms. In the present contribution we discuss possible applications of feedback attractor neural networks to multi-censoring classification problems. We study cooperating networks with self-organizing inter-network connections. In our model, information is transferred from one network to another only if the transmitting node is sufficiently sure (according to some semi-empiric criterion) that it will be able to solve its subtask. In this way it is possible to design networks for multi-censoring tasks which consist of subtasks of widly varying complexity. Furthermore it can be shown that the improved performance of such nets —as compared to fully connected networks— is due to a drastic reduction of connections. This may be a considerable advantage in applications on large data sets.

1   Neural Networks

Networks of formal neurons have been proposed for a variety of applications. Though there has been some continuous work on these models since McCulloch and Pitts (1943), Hebb(1949), von Neumann (1961), Caianiello (1961), Rosenblatt (1962) and others, these topics are studied with renewed and growing interest over the last few years (see e.g. Domany (1992)). The major hope which drives these activities is that artificial neural networks capture features of natural neuronal networks underlying the enormous problem solving abilities of the central nervous systems of higher vertebrates. Whether this is actually true or not is completely unknown up to now; nevertheless there are some very promising features about the existing network models, which make their study worthwhile. First, neural nets are able to learn and generalize from examples without knowledge of rules. Second, they are prototype realizations of parallel distributed processing and third some of the network architectures -in particular attractor neural networks- show an astonishingly high degree of stability against hardware failures, noise and even imprecisely stated problems.

All the neural network models considered so far can be defined as directed graphs with the following properties:

1. A state variable s_i of the neuron at node i

2. A threshold θ_i of the neuron i

3. Real valued weights J_{ij} on each directed link from j to i

4. A transfer function f_i which determines the state of neuron i as a function of all the incoming weights J_{ik}, the states s_k of the other neurons and the threshold θ_i

If the state variables s_i can take on only one of two values they are also referred to as *logical neurons* or *McCulloch-Pitts neurons*. The J_{ij} are usually called *synapses* or *synaptic weights*. Some of the J_{ij} may connect neuron i to receptor or input neurons which are external and remain uninfluenced by other neurons. In most models f_i is assumed to depend only on the so called *postsynaptic potential*

$$h_i = \sum_j J_{ij}s_j - \theta_i \tag{1}$$

The transfer functions f_i may be either deterministic or stochastic. They lead to an updating of the neuronal states s_i and thus to a dynamics of the network defined by:

$$s_i(t+1) = f_i(h_i(t)) \tag{2}$$

There are two distinct modes of updating: either all neurons are updated synchronously and in parallel or they are updated in some sequence, such that after each update of a single neuron the postsynaptic potentials are recalculated. After the network is fed with input from the receptor neurons the result of the network computation is read out from the states s_i the neurons have reached after several cycles of updating.

Besides this neuronal or computational dynamics there is also a synaptic or learning dynamics which affects the synaptic weights J_{ij} after the network is switched to a learning mode and supplied with examples of input-output relations. The rules of this learning dynamics depend strongly on the type of problems to be solved and on the architecture of the network, i.e. the structure of the directed graph. If the neurons can be arranged in layers such that neurons of layer number n get their input only from the neurons of layer $n-1$ the architecture is called *feed-forward network*. To produce an output in the last layer from an input in the first layer it takes exactly as many updating cycles as there are layers in the network. For such networks, which include the *perceptron* of Rosenblatt(1962) and the *multilayer perceptrons with hidden units*, a simple and effective learning dynamics is known under the name of *error backpropagation* (see Rumelhart, Hinton, Williams (1986) as well as Parker (1985) and Le Cun (1986))which allows the network to extract rules from examples.

In the present work we consider neural networks with feedback, i.e. with loops in the defining graph. The prototypes of these networks have a fully connected architecture with symmetric synaptic weights, i.e. $J_{ij} = J_{ji}$ and they go under the name of *Hopfield-Little networks* (see Little (1974) and Hopfield(1982)). Due to the feedback loops, these nets may have an arbitrary number of updating cycles. Therefore the identification of an output becomes an obvious question. The idea of Little (1974) and Hopfield (1982) was that outputs should correspond to attractors of the network dynamics. In this way, a computation becomes a trajectory in the space of neuronal states which starts from an initial condition (*problem*) and eventually ends up in a final state or a final sequence of states (*solution*). Each problem which is in the domain of attraction of the correct solution can be solved by such a network. All networks which operate on this principle are called *attractor neural networks (ANN)*. Before we discuss the advantages of such networks we would like to mention their current weak spot. Up to now we only have learning rules for quite simple problems, all of which are variants of associative recalls or content addressable storage for ANNs. However, the situation is improving quickly (see Schmidhuber (1992)).A number of interesting practical problems exist, however, which can already be attacked with simple learning rules.

On the other hand ANNs have a number of very promising properties some of which have not yet been fully exploited up to now. Among those properties are

1. ANNs are the most robust of the network architectures. E.g. most of the $N(N-1)$ possible synaptic weights can be destroyed without significantly reducing the performance of associative recalls. It has been shown by **Derrida, Gardner, Zippelius** (1987) and by **Kree, Zippelius** (1987) that a strongly diluted network with only finitely many synapses per neuron for an infinite number of neurons is still an efficient content addressable memory.

2. Noise within the updating function of the network makes it possible to avoid misclassifications as was first shown by Amit, Gutfreund and Sompolinsky (1985).

3. ANNs automatically store the results they obtain, because these results are attractors of the dynamics.

In the present work, we want to make special use of the last of these properties and show that it entails an automatic task synchronization in a massively parallel architecture composed of several ANNs which cooperate in solving a complex problem to be described in the next section.

2 An ANN Architecture For Multi-Censoring Problems

We consider the problem of classifying a large set of data according to prescribed class prototypes. In the case of multi-censoring problems we assume that the data set consists of several subsets of data which are of different size and quality. As an example, imagine the diagnosis of a complex machine on the basis of many different testing methods, each with its own limitations of scope and precision. Let us assume that ξ_o is a binary coded data structure which consists of K binary strings $\xi_{o,r}; r = 1 \ldots K$ of lengths $N_1 \ldots N_K$. We want to assign one out of p classes to each input data structure. The classes are chacterized by prototypes $\eta^\mu; \mu = 1 \ldots p$ which have the same structure as the ξ_o. As we have no particular application in mind at this stage of modelling we assume that the classifying distance is simply the Hamming distance. We represent binary elements as ± 1. Instead of the Hamming distance between two binary strings ξ and η it is common in the neural network literature to use the *overlap* which is the difference between the number of coinciding bits and the number of unequal bits divided by the string length, i.e.

$$m(\xi, \eta) = \frac{1}{N} \sum_i \xi_i \eta_i \tag{3}$$

as a measure of similarity. If two strings are equal, their overlap is $+1$ whereas two uncorrelated random strings have an overlap of the order $O(1/\sqrt{N})$ which vanishes for $N \to \infty$.

If all the strings $\xi_{o,r}$ would represent data of the same quality, i.e. with the same level of noise and/or confidence, we could just split the problem into K equivalent problems and use , e.g., an ANN for each subproblem of classifying a string $\xi_{o,r}$. Several learning rules for this problem can be found in the literature. The simplest of these rules is completely sufficient, if the prototypes η^μ are drawn from an uncorrelated random ensemble. It is the so called *Hebb rule* which was suggested as a neurophysiologically plausible form of learning (see Hebb(1949)). The synaptic weights resulting from this learning rule take on the explicit form:

$$J_{ij} = \frac{1}{M} \sum_{\mu=1}^{p} \eta_i^\mu \eta_j^\mu \qquad (4)$$

M denotes the average number of synapses per neuron. It creates p fixed point attractors corresponding to the learnt patterns as long as $p \leq 0.14N$. For the sake of simple and explicit illustrations we will use Hebb's learning rule further on.

The crucial complication which we want to consider in the following arises from the assumption that *the strings $\xi_{o,r}$ represent data of varying noise and/or confidence levels*(multi-censoring). In fact we will assume that some of the strings are of so poor a quality that the corresponding classification subtasks cannot be completed. We want to design a way of communication between the ANNs which allows for a solution of the complete problem even in this situation. The communication between ANN number r and ANN number s is modelled via additional synaptic inter-net couplings $J_{ir,js}$ so that neuron j in ANN s contributes to the postsynaptic potential of neuron i in ANN r. During the learning phase we assume that all inter-net couplings have been modified according to the same learning rule as the intra-net couplings. For Hebb's rule we therefore get

$$J_{ir,js} = \frac{1}{M_s} \sum_{\mu=1}^{p} \eta_{ir}^\mu \eta_{js}^\mu \qquad (5)$$

after the learning phase.

Note that not all the inter-net couplings are really helpful for the performance of the network. If ANN r cannot solve its subtask, all the couplings emerging from this net are in fact disturbing for the others whereas the net r can profit from nets which have already reached the correct prototype as a fixed point of their dynamics.This can be inferred from a simple signal-to-noise analysis of the postsynaptic potential h_{ir}(see Kree and Müller (1992)). As can be shown both by our simulations (see below) and by our analytical methods the disturbing effect of "confused" networks can become large enough to prevent the other ANNs from solving their subtasks. Thus it is not necessarily a good idea to combine all the K ANNs into one large net by introducing additional static synaptic couplings. A better method of communication could be achieved, if each ANN could infer from its own internal state, whether it is "confused" or whether it is on the way of solving its subtask.Using this criterion, the inter-net couplings could then be switched on and off in a self-organizing way which guarantees that only useful information is communicated. Before we propose such a criterion and demonstrate its usefulness let us emphasize that this criterion can only contain information about the neural states $s_r(t), s_r(t-1) \ldots s_r(t-m)$ a finite number m of time steps into the past. The stored prototypes cannot appear in such a criterion unless the neural state trajectory has actually reached them. How can an ANN create a belief that it has something helpful to communicate if it has no other information at hand than its own state a single timestep ago? Our proposed answer to this question can be stated very simply: *the more neurons have changed their state during the last timestep the more confusion there is in the network.* Let us explain the basis of this proposition and then state it in a more precise and quantitative form. In a network with symmetrical synaptic weights and a common deterministic updating function for all neurons there exists a Lyapunov function. For logical neurons,e.g., this function takes on the simple form

$$H = -\frac{1}{2} \sum_{i,j} J_{ij} s_i s_j + \sum_i \theta_i s_i \tag{6}$$

as has been noted by Hopfield (1982). Thus the stored patterns correspond to fixed points of the dynamics and are minima of the function H. Once a fixed point is reached the number of updated neurons in subsequent timesteps is, of course, zero. But while approaching a fixed point the changes in H eventually become smaller and smaller due to a decreasing number of updated neurons per timestep in the vicinity of an extremum of H. To state this conjecture in a quantitative form we express the fraction $\phi(t)$ of updated neurons in an ANN with N nodes in the form

$$\phi(t) = -\frac{1}{2N} \sum_i \{ \langle s_i(t) s_i(t-1) \rangle - 1 \} \tag{7}$$

for McCullogh-Pitts ± 1 neurons. The quantity in angular brackets is just a one-timestep correlation function

$$\begin{aligned} C(t) &= \frac{1}{N} \sum_i \langle s_i(t) s_i(t-1) \rangle \\ &= 1 - 2\phi(t) \end{aligned} \tag{8}$$

The angular brackets in the above expressions denote averages over noise and/or over initial conditions. We have studied the correlation function C(t) for a number of attractor neural network models with and without Lyapunov functions both numerically and analytically (if possible). Our own work together with investigations of Hopfield (1982) and Lewenstein and Nowack (1989) suggest the following proposition:

While approaching a fixed point corresponding to a learnt pattern ξ, C(t) is a monotonically increasing function of the overlap of the network state with ξ

Therefore we let each ANN monitor its own correlation function C(t). This can easily be implemented even on the hardware level because it only requires a memory over one timestep. Then, if $\phi(t)$ decreases beyond a critical value ϕ^* the ANN switches on its couplings transmitting the network state to the other ANNs. In the next section we will demonstrate the effects of this communication strategy.

3 Simulations of Communicating ANNs

In order to analyze the performance of ANNs with self-organizing inter-net couplings of the above described type and to compare it with standard network architectures we undertook numerical simulations as well as analytical statistical mechanics calculations. The latter approach is limited to large networks ($N \to \infty$) and to two special intra-net architectures:

- either a fully connected graph with symmetrical synaptic weights, i.e. a Hopfield network. In this case the number p of stored patterns has to be small so that $p/N \to 0$ for $N \to \infty$.

- or a strongly diluted ANN of the type described in section 1 for which the network is a random, sparse graph with directed links.

In these cases we can derive *exact* (integro-differential) equations of motion for overlaps and correlation functions which can be solved numerically. These results which require some specialized techniques of statistical mechanics will be presented elsewhere.

In our simulations we considered up to $K = 10$ coupled ANNs with total numbers of neurons $N = \sum_r N_r$ ranging from 512 up to 5096. The synaptic connections were randomly diluted (average connectivities have been varied between 1, i.e. fully connected and 10^{-3}) such that the average intra-net connectivities were always at least a factor of 3 higher than inter-net connectivities. The variations of the net sizes N_r were chosen to be within a factor of 3 of the average size. The ± 1 patterns η^μ which represented the classes were drawn from a completely unbiased ensemble, i.e. the probabilities $W(\eta_{ir}^\mu = +1) = 1/2$ for all $\mu = 1 \ldots p$, $r = 1 \ldots K$, $i = 1 \ldots N_r$ independent of each other. This seems an acceptable choice as long as one does not intend to consider special applications. The classes were learned by a so called *clipped* Hebb's rule:

$$J_{ir,js} = \frac{\sqrt{p}}{N_s} sgn \sum_{\mu=1}^{p} \eta_{ir}^\mu \eta_{js}^\mu \tag{9}$$

This learning rule has been studied by van Hemmen and Kühn (1986) and by van Hemmen (1987). The storage capacity is surprisingly high ($\alpha_c \approx 0.1$ for full connectivity) and the computational effort in simulations is significantly reduced as the synaptic weights become ± 1 variables. We averaged our results over up to 124 samples of random classes to obtain estimates of the average performance of the networks. In each run of the simulation, we produced a random initial pattern with fixed overlaps $m_r(t = 0)$. We have not intended to study the dependence on noise systematically. In most runs we fixed the noise strength either to zero or to some value which gave good classification properties.

We like to emphasize that there are two ways to increase the difficulty of a classification task for a single ANN in these simulations. The obvious way is to choose initial conditions with a fairly small overlap. But there is also a more implicit type of difficulty for networks which have a less than average number of neurons or a less than average connectivity. This difficulty is due to the fact that the networks can classify succesfully only if the number of stored patterns p does not exceed a critical fraction α_c of the existing synaptic connections. (For the fully connected Hopfield network equipped with Hebb's learning rule, $\alpha_c = 0.14$ as we already mentioned in section 1.) As p is the same for all K networks, the fraction α will vary among the K networks and may exceed α_c considerably for some "small" ANNs.

The figure 1 shows results from our simulations which illustrate the major finding. The figure shows overlaps $m_r(t)$ and the fractions of updated neurons per timestep $\phi_r(t)$ for three ANNs of different sizes. Net 1 contained $N_1 = 600$ neurons, net 2 $N_2 = 300$ and net 3 $N_3 = 102$. The average connectivity was chosen to be 10^{-1}. Thus the maximum nunmber of synapses is $\approx 10^5$ corresponding to a storage capacity of ≈ 10 classes. Note that the number of classes is low, but the classes are complex objects of 1Kbit. This is a typical situation in diluted networks. There were $\approx 45 * 10^3$ synapses in net 1, $\approx 15 * 10^3$ in net 2 and $\approx 3 * 10^3$ in net 3. Thus if the ANNs were uncoupled net 1 could store ≤ 4500 bit or ≤ 7 classes, net 2 would have a capacity of ≤ 5 classes and net 3 of ≤ 3 classes. The data shown in figure 1 correspond to 5 stored classes. The three pictures -from top to bottom- represent three modes of communication:

- In the upper picture, the ANNs are completely uncoupled

- The central picture represents the communication described in the previous section

- The lower picture shows the performance of a network with *static* inter-net couplings which were randomly diluted to achieve the average connectivity of 10^{-1}

Time is measured in units of updating cycles of the networks. Without any communication, net 1 and net 2 solve classification tasks whereas net 3 fails. Note that $\phi_3(t)$ approaches $\approx 1/2$ which shows that the network has not reached a fixed point. Switching on the communication changes the situation dramatically. Up to time t_1 where ϕ_1 reaches $\phi^* = 0.12$ the three ANNs evolve uncoupled and the initial overlap in net 3 is diminishing rapidly. For $t \geq t_1$ the couplings from ANN 1 to the other two networks are switched on. Obviously this is sufficient to solve the tasks in all three networks. The effects at t_2 where ANN 2 reaches ϕ^* are less dramatic. Each additional synaptic coupling contributes both a signal term *and a noise term due to the other stored classes which are not recognized* to the postsynaptic potential. If the signal to noise ratio in the networks is approximately the same, communication is not helpful. This implies in particular that static couplings are not an optimal communication strategy. In the presented example we *switched off all the inter-net couplings after t_3* when all the nets are on the right track. This improved the final retrieval quality of the stored class prototypes but it did not not work in all cases. In fact it is an interesting and important problem to find a learning strategy for an optimal communication protocol for ANNs. The bottom picture demonstrates that static inter-net couplings do in fact reduce the classification speed and quality. After 15 updating cycles the static network has reached an average overlap of 0.81 whereas the communicating networks reach 0.96. Even more significant is the inhomogeneous distribution of overlaps due to the static inhomogeneous architecture of the network. This demonstrates that communicating networks are better adapted to multi-censoring problems than large static networks.

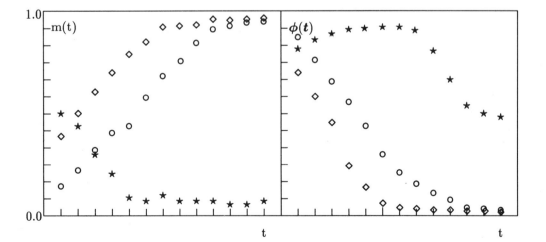

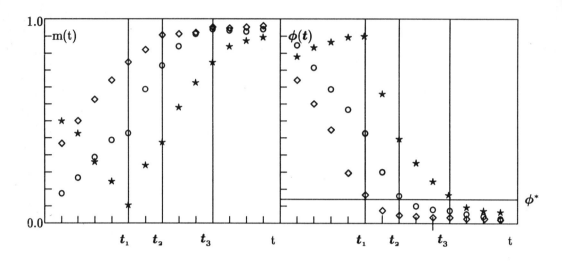

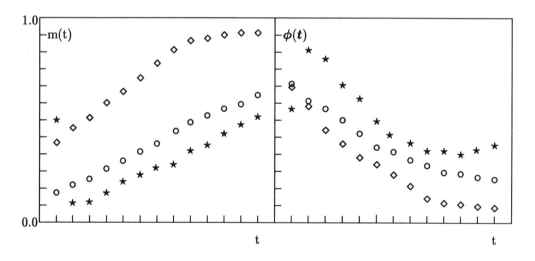

Figure 1: Overlaps m(t) and fraction of updated neurons $\phi(t)$ resulting from simulations. Net $1 = \diamond$, net $2 = \circ$ and net $3 = \star$. The uppermost picture refers to completely uncoupled ANNs, the lower one to static couplings and the central one to self-organized couplings.

4 Perspectives And Open Problems

There is no doubt that this work can only be a first step and that still much has to be done to build large scale practical applications. Nevertheless our preliminary results seem encouraging. In the communicating network architectures most of the synaptic couplings are not needed for most of the time. This can significantly reduce the connectivity problems which plague neural networks hardware implementations if it is possible to realize the communication channels in a soft-wired form. Surprisingly perhaps, there is no price to be paid for the reduction of connections: quite to the contrary, this architecture improves supervised classification for multi-censoring problems.

There may be better communication strategies than the simple one we have proposed and analyzed here. Currently we are trying to construct learning strategies for communication. As a first attempt we tried to optimize the parameter ϕ^* by simple unsupervised trial and error learning. As a guidance for improved strategies, knowledge from social psychology may be useful.

An important factor which can improve the performance of ANNs considerably is the tuning of the noise level incorporated in the updating functions. We have performed some calculations with adaptive noise levels (as proposed by Lewenstein and Nowak (1989)) in communicating neural nets.

More complex architectures can be considered where the ANNs form nodes of a prescribed graph (e.g. a hierarchical structure). Such networks can be useful for complex problems as, e.g. invariant pattern recognition. For work in this direction see C. Fassnacht and A. Zippelius (1991).

References

MCCULLOCH, W.S. and PITTS, W. (1943), A Logical Calculus of the Ideas Immanent in Nervous Activity, *Bull. Math. Biophys. 5, 115*

HEBB, D. O. (1949), *The Organization of Behaviour: A Neurophysiological Theory*, Wiley, New York

VON NEUMANN, J. (1961), *Design of Computers, Theory of Automata and Numerical Analysis, Collected Works Vol V*, A.H. Taub (ed.), Macmillan, New York

CAIANIELLO, E. R. (1961), Outline of a Theory of Thought Processes and Thinking Machines, *J. Theor. Biol., 1, 204.*

ROSENBLATT, F. (1962), *Principles of Neurodynamics*, Spartan, New York.

DOMANY, E., VAN HEMMEN, J.L. and SCHULTEN, K. (Eds.) (1992)*Models of Neural Networks*, Springer, Berlin.

RUMELHART, D. E., HINTON, G. E. and WILLIAMS, R. J. (1986), Learning representations by Back-propagating Errors, *Nature 323, 533.*

PARKER, D. B. (1985), Learning Logic: Casting the Cortex of the Human Brain in Silicon, *MIT Techn. Rep. TR-47.*

LE CUN, Y. (1986), Learning Process in an Asymmetric Threshold Network, in: F. Bienenstock, F. Fogelman Soulié and G. Weisbuch (eds.), *Disordered Systems and Biological Organization*, Springer, Berlin.

LITTLE, W. A. (1974), The Existence of Persistent States in the Brain, *Math. Biosci.*, *19*, *101*.

HOPFIELD, J. J. (1982), Neural Networks and Physical Systems with Emergent Collective Computational Abilities, *Proc. Natl. Acad. Sci. USA*, *79*, *2554*.

SCHMIDHUBER, J. (1992), A Fixed Storage $O(n^3)$ Time Complexity Learning Algorithm for Fully Recurrent Continually Running Networks, *Neural Computation*, *4*, *243*

DERRIDA, B., GARDNER, E. and ZIPPELIUS, A. (1987), An Exactly Solvable Asymmetric Neural Network Model, *Europhys. Lett. 4, 167*.

KREE, R. and ZIPPELIUS, A. (1987), Continuous-Time Dynamics of Asymmetrically Diluted Neural Networks, *Phys. Rev. A 36, 4421*.

AMIT, D. J., GUTFREUND, H. and SOMPOLINSKY, H. (1985), Spin-glass Models of Neural Networks, *Phys. Rev. A 32, 1007*.

KREE, R. and MÜLLER, A. (1992), Communicating Neural Networks, *in preparation*.

LEWENSTEIN, M. and NOWACK, A. (1989), Fully Connected Neural Networks with Self-Control of Noise Level, *Phys. Rev. Lett. 62, 225*.

VAN HEMMEN, J. L. and KÜHN, R. (1986), Nonlinear Neural Networks, *Phys. Rev. Lett. 57, 913*.

VAN HEMMEN, J. L. (1987), Nonlinear Neural Networks Near Saturation, *Phys. Rev. A 36, 1959*.

FASSNACHT, C. and ZIPPELIUS, A. (1991), Recognition and Categorization in a Structured Neural Network With Attractor Dynamics, *Network 2, 63*.

Knowledge Extraction from Self-Organizing Neural Networks

A.Ultsch

Department of Computer Sience, University of Dortmund
P.O. Box 500 500 D-4600 Dortmund 50, Germany

Abstract: In this work we present the integration of neural networks with a rule based expert system. The system realizes the automatic acquisition of knowledge out of a set of examples. It enhances the reasoning capabilities of classical expert systems with the ability of generalise and the handling of incomplete cases. It uses neural nets with unsupervised learning algorithms to extract regularities out of case data. A symbolic rule generator transforms these regularities into PROLOG rules. The generated rules and the trained neural nets are embedded into the expert system as knowledge bases. In the system's diagnosis phase it is possible to use these knowledge bases together with human expert's knowledge bases in order to diagnose a unknown case. Furthermore the system is able to diagnose and to complete inconsistent data using the trained neural nets exploiting their ability to generalise.

1 Introduction

Knowledge acquisition for expert systems poses many problems. Expert systems depend on a human expert formulate knowledge in symbolic rules. It is almost impossible for an expert to describe knowledge entirely in the form of rules. In particular it is very difficult to describe knowledge acquired by experience. An expert system may therefore not be able to diagnose a case which the expert is able to. The question is how to extract experience from a set of examples for the use of expert systems.

Machine Learning algorithms such as "learning from example" claim that they are able to extract knowledge from experience. Symbolic systems as, for example, ID3 [Quinlan (1984)] and versionspace [Mitchell (1982)] are capable to learn from examples. Connectionist systems claim to have advantages over these systems in generalisation and in handling noisy and incomplete data. Queries to expert systems often contain inconsistent data. For every data set the rule based systems have to find a definite diagnosis. Inconsistent data can force symbolic systems into an indefinite state. In connectionist networks a distributed representation of concepts is used. The interference of different concepts allows networks to generalize [Hinton et al. (1986a)]. A network computes for every input the best output. Due to this connectionist networks perform well in handling noisy and incomplete data. They are also able to make a plausible statement about missing components. A system that uses a rule based expert system with an integrated connectionist network could benefit of the described advantages of connectionist systems.

2 System Overview

Figure 1 gives an overview of our system. Case data that are presented to an expert system are (usually) stored in a case database. A data transformation module encodes such cases in a suitable way in order to be learned by neuronal networks. This module performs as follows: first it transforms the data so that the components have equal scopes. One of the possibilities is the use of a z-transformation. The second task of the transformation module is to encode the data into a binary input pattern because some neural networks, as, for example, the competitive learning model [Rumelhart/Zipser (1985)], only processes binary inputs. To do this, the intervals of the components are subdivided into different ranges. These ranges are adapted according to the distribution of the components. So every component of a vector is represented by the range its value belongs to. Depending on the kind of representation, the ranges could be encoded locally, locally-distributed or distributed.

With the so transformed data different neuronal networks with unsupervised learning algorithms, such as Competitive Learning [Rumelhart/Zipser (1985)], ART [Carpenter/Grossberg (1987)] and Kohonen [Kohonen (1984)], are trained. These Networks have the ability to adapt their internal structures (weights) to the structure of the data. In a Rule Generation module the structures learned by the neuronal networks are detected, examined and transormed into expert systems rules. These rules can be inspected by a human expert an added to an expert system. When a case is presented to the expert system, the system first

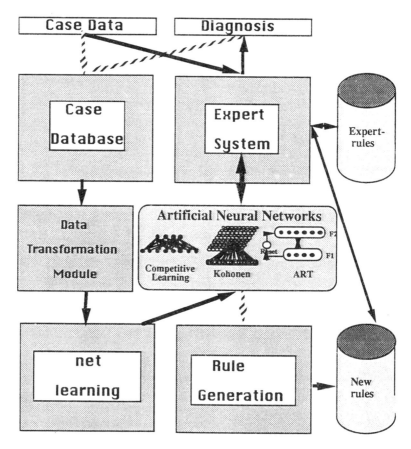

Figure 1: Overview of the System

tries to reason with the rules that have been aquired from an expert in order to produce a suitable diagnosis. If this fails to produce a diagnosis, the new rules produced by the process described above can be used. If the case can be handled in such a way, all steps of the reasoning process may be inspected by and explained to a user of the system.

If the system, however, is not able to produce a suitable diagnosis in this way. Be it, that data is missing, or the input is erroneous. Or no rule fits the data, since such a case has not been considered while building the knowledge base, the expert system can turn the case over to the net-works. The networks, with their aibility to associate and generalize, search for a most suiting case that has been learned before. The diagnosis, that has been associated with that case is then returned as a possible diagnosis.

3 Detecting Structures

One of the network we used was a Kohonen network consisting of two layers. The input layer has n units representing the n components of a data vector. The output layer is a two dimensional array of units arranged on a grid. The number of the output units is determined experimentally. Each unit in the input layer is connected to every unit in the output layer with a weight associated. The weights are initialized randomly taking the smallest and the greatest value of each component (of all vectors) as boundaries. They are adjusted according to Kohonen's learning rule [Kohonen (1984)]. The applied rule uses the Euclidean distance and a simulated mexican-hat function to realize lateral inhibition. In the output layer neighbouring units form regions, which correspond to similar input vectors. These neighbourhoods form disjoint regions, thus classifying the input vectors.

The automatic detection of this classification is difficult because the Kohonen algorithm converges to an equal distribution of the units in the output layer. So a special algorithm, the so called U-matrix method was developed in order to detect classes that are in the data [Ultsch/Siemon (1990)]. For example, using a data set containing blood analysis values from 20 patients (20 vectors with 11 real-valued components) selected from a set of 1500 patients [Deichsel/Trampisch (1985)], the following stucture could be seen: In figure 2 three

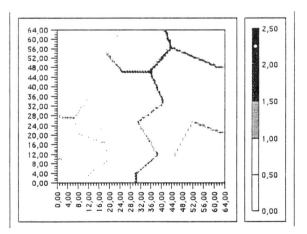

Figure 2: U-Matrix of acidosis data

major classes can be distinguished: the upper right corner the left part and the lower right part. The latter two may be subdivided furthermore into tho subgroups each. It turned out that this clustering corresponded nicely with the different patient's diagnoses, as Figure 3 shows. In summary, by using this method, stucture in the data can be detected as classes.

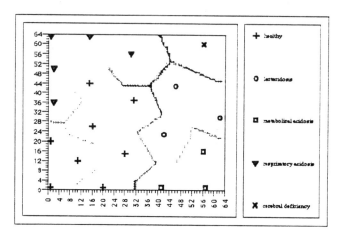

Figure 3: U-matrix with diagnoses

These classes represent sets of data that have something in common. The discovery of the properties of each class and its reformulation in a symbolic form is described in the next chapter.

4   Rule Generation

As a first approach to generate rules from the classified data (see chapter 3) we used a well known machine learning algorithm: ID3 [QUINLAN (1984), Ultsch/Panda (1991)]. While being able to generate rules this algorithm has a serious problem [Ultsch (1991)]: it uses a minimalization criterion that seems to be unnatural for a human expert. Rules are generated, that use only a minimal set of decisions to come to a conclusion. This is not what must be done, for example, in a medical domain. Here the number of decisions is based on the type of the disease. In simple cases, i.e. where the symptoms are unanymous, very few tests are made, while in difficult cases a diagnosis must be based on all available information.

In order to solve this problem we have developed a rule generation algorithm, called sig*, that takes the significane of a symptom (range of a component value) into account [Ultsch (1991)]. One of the data sets we tested the algorithm with, was the diagnosis of iron deficiency. We had a test set of 242 pations with 11 cinical test values each. The rules generated with sig* showed, first,a high degree of coincidence with expert's diagnosis rules and, second, exhibited knowledge not prior known to us while making sense to the experts [Ultsch (1991)].

We have tested this algorithm in another way. The data set was randomly divided into two subsets and only one of them was used to generate rules. The other subset was used to compare the diagnosis generated by the rules with the actual diseases. In 119 out of 121 cases (98%) the system produced the right diagnosis. In eleven cases a wrong diagnosis was

disease	# patients	correct diagnoses	wrong diagnoses
normochr.	23	23	O
hypochrome	15	15	1
Eisenmangel	22	20	2
echter EM	16	16	0
chron.Prozeß	1	1	0
sideroachr.	1	1	1
hyperchr.	5	5	1
Polyglobul.	0	0	0
o.B.	38	38	6
sum	121	119	11

Table 1: sig*-generated diagnoses vs. true diagnoses

given. Most of this cases were additional wrong diagnosis. For example in six cases "o.B."-patients (meaning healthy) were also cosidered to have a slight form of iron deficiency. In other cases, like "sideroacrestic", the data set that has been learnned by the network was to small (one case) in order to produce a meaningful diagnostic rule.

5 Conclusion

The implementation of our system demonstrates the usefulness of the combination of a rule-based expert system with neural networks [Ultsch et al. (1991 b,a), Ultsch/Panda (1991)]. Unsupervised leaning neural networks are capable to extract regularities from data. Due to the distributed subsymbolic representation, neural networks are typically not able to explain inferences. Our system avoids this disadvantage by extracting symbolic rules out of the network. The acquired rules can be used like the expert's rules. In particular it is therefore possible to explain the inferences of the connectionist system.

Such a system is useful in two aspects. First the system is able to learn from examples with a known diagnosis. With this extracted knowledge it is possible to diagnose new unknown examples. Another abilityis to handle a (large) data set for which a classification or diagnosis is unknown. For such a data set classification rules are proposed to an expert. The integration of a connectionist module realizes "learning from examples". Furthermore the system is able to handle noisy and incomplete data. First results show that the combination of a rule based expert system with a connectionist module is not only feasible but also useful. Our system is one of the possible ways to combine the advantages of the symbolic and subsymbolic paradigms. It is an example to equip a rule based expert system with the ability to learn from experience using a neural network.

Acknowledgements

We would like to thank all members of the student research groups PANDA (PROLOG And Neural Distributed Architectures) for their tremendous job in implementing a preliminary version of the system. H.-P. Siemon has implemented the Kohonen algorithm on a transputer. This work has been supported in part by the Forschungspreis Nordrhein-Westfalen.

References

CARPENTER, G.A., GROSSBERG, S. (1987), *Self-Organization of Stable Category Recognition Codes for Analog Input Patterns*, Applied Optics, Vol. 26, S. 4.919–4.930.

DEICHSEL, G., TRAMPISCH, H.,J. (1985), *Clusteranalyse und Diskriminanzanalyse*, Gustav Fisher Verlag, Stuttgart 1985.

KOHONEN, T. (1984), *Self-Organization and Associative Memory*, Springer Verlag, Berlin.

MICHALSKI, R., CARBONELL, J.,G., MITCHELL, T.,M. (1984), *Machine Learning - An artificial intelligence approach*, Springer Verlag, Berlin.

MITCHELL, T.,M. (1982), Generalization as Search, in: *Artificial Intelligence*, 18, p. 203–226.

QUINLAN, J.,R. (1984), Learning Efficient Classification Procedures and their Application to Chess End Games, in: MICHALSKI et al. (eds) (1984), *Machine Learning - An artificial intelligence approach*.

RUMELHART, D.,E., MCCLELLAND, J.,L. (1986), *Parallel Distri- buted Processing: Explorations in the Microstructure of Cognition*, Volume 1: Foundations, MIT Press, Cambridge (Massachusetts).

RUMELHART, D.,E., ZIPSER, D. (1985), *Feature discovery by Com- petitive Learning*, Cognitve Science vol. 9, pp.75–112.

SIEMON, H.P., ULTSCH, A. (1990), *Kohonen Networks on Transputers: Implementation and Animation*, Proc. Intern. Neural Networks, Kluwer Academic Press, Paris, pp 643–646.

ULTSCH, A. (1991), *Konnektionistische Modelle und ihre Integration mit wissensbasierten Systemen*, Habilitationsschrift, Univ. Dortmund.

ULTSCH, A., HALMANS, G., MANTYK, R. (1991 b), *CONCAT: A Connectionist Knowledge Ackquisition Tool*, Proc. IEEE International Conference on System Sciences, January 9-11, Hawaii, pp 507–513.

ULTSCH, A., HALMANS, G., MANTYK, R. (1991 a), *A Connectionist Knowledge Acquisition Tool: CONCAT*, Proc. International Workshop on Artificial Intelligence and Statistics, January 2-5, Ft. Lauderdale FL.

ULTSCH, A., PANDA, PG. (1991), *Die Kopplung konnektionistischer Modelle mit wissensbasierten Systemen*, Tagungsband Expertenystemtage Dortmund, Februar 1991, VDI Verlag, pp 74–94.

ULTSCH, A., SIEMON, H.P. (1990), *Kohonen's Self Organizing Feature Maps for Exploratory Data Analysis*, Proc. Intern. Neural Networks, Kluwer Academic Press, Paris, pp 305–308.

Self-Organizing Neural Networks for Visualisation and Classification

A. Ultsch

Department of Computer Science, University of Dortmund, P.O. Box 500 500
D-4600 Dortmund 50, FRG

Abstract: This paper presents the usage of an artificial neural network, Kohonen's self organizing feature map, for visualisation and classification of high dimensional data. Through a learning process, this neural network creates a mapping from a N-dimensional space to a two-dimensional plane of units (neurons). This mapping is known to preserve topological relations of the N-dimensional space. A specially developed technique, called U-matrix method has been developed in order to detect nonlinearities in the resulting mapping. This method can be used to visualize structures of the N-dimensional space. Boundaries between different subsets of input data can be detectet. This allows to use this method for a clustering of the data. New data can be classified in an associative way. It has been demonstrated, that the method can be used also for knowledge acquisition and exploratory data analysis purposes.

1 Introduction

Artificial neural networks may be classified according to their learning principles into two big classes: networks that learn in a supervised and networks that learn in an unsupervised way. Supervised learning means the adaptation of a network's behaviour to a given input-to-output relationship. The approximation of a (eventually nonlinear) function is an example for a typical task for a supervised learning network. The popular back-propagation algorithm (Rumelhart, McClelland (1986)) is a supervised algorithm.

Unsupervised learning networks adapt their internal structures to structural properties (e.g. regularities, similarities, frequencies etc.) of the input. Networks like Competitive Learning, ART (Grossberg (1987)) and Kohonen's self-organizing feature maps (Kohonen (1982)) belong to this class. The purpose of such networks is to detect structural properties of the input space and adapt their structures to those properties.

In this paper we describe how an unsupervised learning neural network model, Kohonen's self-organizing feature map, may be used for visualisation. A special algorithm, the so called U-matrix method is proposed to detect the structural features inherent in the (often several thousand) weights of the neural network. This algorithm produces a three dimensional landscape which represents structural properties of the high dimensional input space. Using this visualisation technique, clustering of the data is possible in a natural way. The method may also be used for exploratory data analysis and classification.

2 Kohonen's Self-Organizing Feature Maps

In this section a short and informal description of Kohonen's self-organizing feature maps (SOFM) is given. For a more detailed description see for example Ritter, Martinez, Schulten (1990) or Ultsch (1991).

SOFM consist essentially of two layers of neurons: the input-layer I and the unit layer U. Input to the network are n-dimensional vectors of real numbers. Each component of such an input vector is fed into one neuron of the input layer.

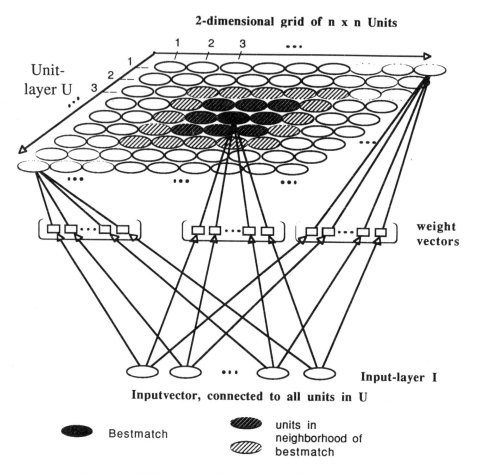

Figure 1: SOFM with 2-dimensional grid as unit-layer

The neurons in the unit-layer U are arranged in a certain topology, i.e. there is a notion of neighbourhood among the neurons. In this paper we assume that the neurons in U are arranged in a 2-dimensional grid (see Figure 1). All input neurons are connected to all neurons in U. The strength of this connections can be represented as a n-dimensional weight vector of real values associated with each neuron in U.

If an input vector is presented to the network, all neurons in U compare their weight vectors (in parallel) to the input vector. The neuron in U with the most similar weight vector is called the bestmatch. This bestmatch and all neurons in a neighbourhood of a certain range are allowed to "learn" the input vector. I.e. their weight vectors become more similar to the given input vector.

Inputs are presented one at a time in a random sequence. During this process the range of the neighbourhood is changed such that a bestmatch initially has many units in its neighbourhood and at the end of the learning process it possesses only a few or no neighbours.

The effect of this process is, that the positions of the bestmatches for input vectors become topologically ordered on the unit layer. This ordering is such that topological relations of the input vectors are represented on the unit layer.

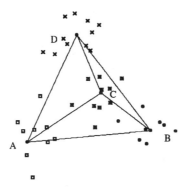

Figure 2: Sample Input Space

In Figure 2 an example of an input space is given. Four clusters of ten three-dimensional data points are generated such that they are in the vicinity of the edge points of a regular tetrahaedron. After learning, the bestmatches on a 64x64 Unit-layer are arranged as the following picture shows.

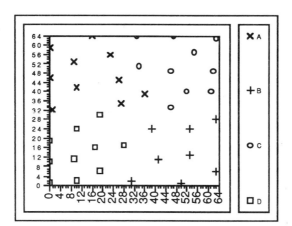

Figure 3: Bestmatches of tedrahedron data

Each data cluster is concentrated in a different corner of the unit-layer and the topological relationships of the clusters are represented as best as possible. If the clusters would not be known beforehand, the position of the bestmatch would however not suffice to detect the clusters. The U-matrix method described in the next section is however able to display the clustering information.

3   The U-matrix Method

Each neuron u possesses the eight immediate neighbours depicted in Figure 4. Associated with each unit is a n-dimensional weight value. The same metric by which the bestmatch to a input vector is found can now be used to determine a distance between u and its immediate neighbours (n, no, o, so, ... in Figure 4).

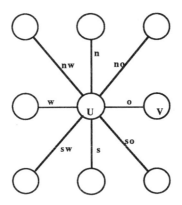

Figure 4: Neuron u and its immediate neighbours

A method to unify these distances into a measure that can be represented as a height value in a matrix that corresponds to the unit-layer is called "Unified distance matrix (U-matrix-) method". It can be used to display the similarity/dissimilarity structure of all units and their neighbours.

One U-matrix method is displayed in Figure 5. For each neuron in U exist three distances dx, dy and dxy in the U-matrix with dx = o, dy = n and dxy = 0.5 * (no + nw(V)). Where nw(V) denotes the nw-distance of neuron v in Figure 4.

U-matrices give a picture of the topology of the unit-layer and therefore also of the topology of the input space. Altitude in the U-matrix encodes dissimilarity in the input space. Valleys in the U-matrix (i.e. low altitudes) correspond to input-vectors that are quite similar. Figure 6 shows a U-matrix for the tetrahaedron data described in the last section.

Viewed from above, the four clusters contained in the input are easily detected. One data point that is rather far from its own cluster is also identifiable (see figure 7).

4   Implementation and Application

We have implemented our method on a combination of transputers and graphical machines (Ultsch, Siemon (1990)). The SOFM learning process and the calculation of the U-matrix takes place on the transputers in parallel. The U-matrix is then visualized on a SUN Machine with special graphic hardware or on a MacIntosh.

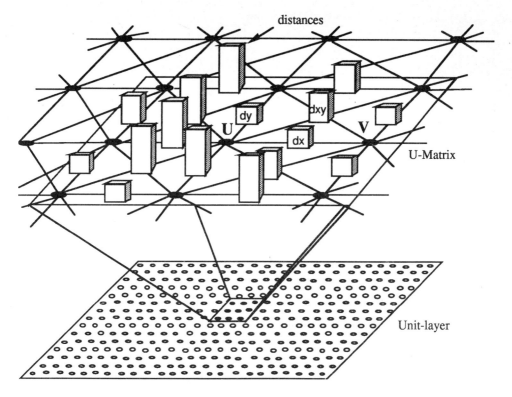

Figure 5: U-matrix

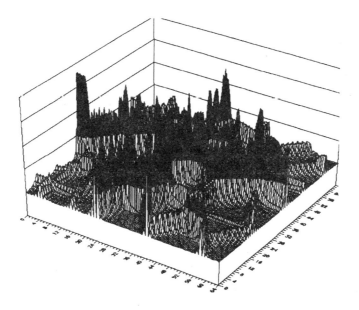

Figure 6: U-matrix of tetrahaedron data

312

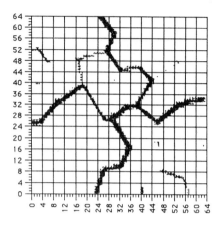

Figure 7: U-matrix of tetrahaedron data seen from above

We have used our method successfully for several different applications. One is the classification of patients with disturbances in their blood pH level (Ultsch, Panda (1991)). In this case the dimensionality of the input space is around 10. Another example is the diagnosis of iron deficiency in Ultsch (1991). The classification found for this application has been used for the diagnosis of cases that have not been used to train the network. 89% of all new cases were correctly identified. It is also possible to integrate our method with expert system technology. We have shown that it is possible to extract rules from U-matrices that may be used in an expert system (Ultsch et al. 1991c))

5   Conclusion

Our U-matrix method recodes topological relationships in a n-dimensional space to a three dimensional landscape. The basic structure is determined by a self-organizing feature map that is known to preserve as best as possible topological features of the input space. The resulting picture is such that dissimilarities among the input date result in mountains or "walls" between different sets of input data points. This feature makes it easy for humans to detect clusters and to classify the data. The method has been successfully used to detect structures in input spaces with a dimensionality ranging from three to several thousand.

Acknowledgements

H.-P. Siemon has implemented the Kohonen algorithm on a transputer. G. Pietrek has implemented some of the visualisation techniques on a SUN machine. This work has been supported in part by the Forschungspreis Nordrhein-Westfalen.

References

GROSSBERG, S. (1987), Competitive Learning: From Adaptive Activation to Adaptive Resonance, *Cognitive Science, 17,* 23–63.

KOHONEN, T. (1982), Clustering, Taxonomy, and Topological Maps of Patterns. in: Lang,

M. (Ed.), *Proceedings of the Sixth International Conference on Pattern Recognition*, Silver Spring, MD, IEEE Computer Society Press, 114–128.

RITTER, H., MARTINEZ, T., SCHULTEN, K. (1990), *Neuronale Netze*, Addison Wesley.

RUMELHART, D.E., MCCLELLAND, J.L. (1989), *Parallel Distributed Processing: Explorations in the Microstructures of Cognition, Volume 1: Foundations*, MIT Press, Cambridge.

ULTSCH, A. (1991), *Konnektionistische Modelle und ihre Integration mit wissensbasierten Systemen*, Habilitationsschrift, Univ. Dortmund.

ULTSCH, A. (1991a), *The Integration of Neuronal Networks with Expert Systems*, Proceedings Workshop on Industrial Applications of Neural Networks, Ascona, Vol III, 3–7.

ULTSCH, A., PALM, G., RÜCKERT, U. (1991a), Wissensverarbeitung in neuronaler Architektur, in: Brauer, Hernandez (Eds.): *Verteilte künstliche Intelligenz und kooperatives Arbeiten*, GI-Kongress, München, 508–518.

ULTSCH, A., HANNUSCHKA, R., HARTMANN, U., MANDISCHER, M., WEBER, V. (1991b), *Optimizing Logical Proofs with Connectionist Networks*, Proc. Intl. Conf. Artificial Neural Networks, Vol I, Helsinki, 585–590.

ULTSCH, A., HALMANS,G., MANTYK, R. (1991c), *A Connectionist Knowledge Acquisition Tool: CONCAT*, Proc. International Workshop on Artificial Intelligence and Statistics, January 2-5, Ft. Lauderdale.

ULTSCH, A., HALMANS, G. (1991), *Data Normalization with Self-Organizing Feature Maps*, Proc. Intl. Joint Conf. Neural Networks, Seattle, Vol I, 403–407.

ULTSCH, A., HALMANS, G. (1991a), *Neuronale Netze zur Unterstützung der Umweltforschung*, Symp. Computer Science for Environmental Protection, Munich.

ULTSCH, A., PANDA, PG. (1991), *Die Kopplung konnektionistischer Modelle mit wissensbasierten Systemen*, Tagungsband Expertensystemtage, Dortmund, VDI Verlag, 74–94.

ULTSCH, A., SIEMON, H.P. (1990), *Kohonen's Self Organizing Feature Maps for Exploratory Data Analysis*, Proc. Intern. Neural Networks, Kluwer Academic Press, Paris, 305–308.

HyDi: Integration of Hypermedia and Expert System Technology for Technical Diagnosis[1]

Ralph Traphöner

tecInno - Gesellschaft für innovative Software-Systeme und Anwendungen mbH
Sauerwiesen 2, 6750 Kaiserslautern 25

Frank Maurer
Uni Kaiserslautern, AG Prof. Richter
Postfach 3049, 6750 Kaiserslautern, e-Mail: maurer@informatik.uni-kl.de

Abstract: We are developing an expert-system shell for diagnostic tasks basing on the MOLTKE-System and the Hypertext Abstract Machine HAM. The HAM is implemented on the object-oriented database management system GemStone. It stores the knowledge-base and supplies contexts for static structuring. A filter mechanism which is combined with a net oriented query language provides dynamic structuring. The tool HyperCAKE for the definition of application specific shells is implemented basing on the HAM. It supplies a rule-interpreter and mechanisms for the creation of nodes and arcs with specific semantics. HyDi uses this framework to represent diagnostic primitives like symptoms, strategies and diagnoses.

1 System Architecture

Here we give an overview on the overall architecture of the knowledge acquisition tool HyperCAKE, which is the basis of the HyDi-System. HyperCAKE uses the object-oriented database management system GemStone for the management of the objects of the knowledge base. GemStone guarantees the integrity of the knowledge base in a multi-user-environment by its transaction concept. Based on the object management system we implemented a hypertext abstract machine (HAM) following the ideas of (Campbell and Goodmann 1988). The implementation language is Smalltalk-80.

1.1 The Extended HAM

The question, how semantic is linked to information, leads us to an extension of the HAM with an (object-oriented) type concept (types in the meaning of abstract data types, where data is linked with operations). Based on the polymorphism of the used implementation language we can replace generic methods (e.g. show, edit, followLink, getSuccessors, getPredecessors, set:, etc.) by specializations. This allows to define new semantics for special information classes.

Nodes. Nodes store atomic information units as texts, graphics, tables, concept descriptions, knowledge sources etc. Nodes are the general means for the representation and management of knowledge. Our System includes a hierarchy of predefined classes for nodes. To define node classes with a special meaning in a new application context (in our case: technical diagnosis) this hierarchy is extended and the default methods are redefined.

---

[1]This work is partialy funded by the Ministerium für Wirtschaft und Verkehr/Rheinland-Pfalz

Links. To represent relations and associations between information nodes we use links. The present version of the HAM supports directed, binary links. The starting point of a link is (a part of) a node. Target of a link is always an entire node. To implement semantics for different kinds of links we support a set of link classes. Links describe static relations between information nodes and are explicitly stored. Additionally, the HAM allows to define dynamic links which are computed if required based on the actual structure of the network. Dynamic links represent associations between nodes which are not direct neighbors, i. e. there exists no direct link between them.

A basic assumption for the development of HyperCAKE is, that there is a huge amount of information to be managed within a expert system project. Therefore, we use a the database management system GemStone for storing all objects. Additionally, the knowledge base is structured by so called contexts. So far, the implemented HAM allows the representation of knowledge by a semantic network. The nodes of the semantic network may contain multimedia information. The "semantic" is defined by the implementation of the functionality for the node and link classes.

1.2 The Knowledge Acquisition Tool HyperCAKE

The HyperCAKE system is built to support the knowledge acquisition for hypermedia-based expert systems. The difference to conventional expert systems is the integration of multimedia data, the difference to conventional hypermedia systems is the integration of an inference engine. HyperCAKE is not built to extend the capabilities of knowledge based systems by expert system techniques but by the integration of hypermedia. In addition to the conventional textual descriptions, symbolic representations (e.g. concept descriptions or attribute values) can be coupled with graphics or audio signals. These can be referenced in the communication with the user of the expert system. The integration of additional media is absolutely needed if the normal representation of knowledge in a domain uses other media than text (e.g. architecture, town or area planning, mechanical engineering). Here we give a short overview on HyperCAKE. For further information see (Maurer 1992a) or (Maurer 1992b). With HyperCAKE we developed a tool CoMo-Kit (Conceptual Model Construction Kit) which supports the knowledge engineering process from raw data (e.g. protocols of interviews with the expert) via an intermediate (semi-formal) representation up to an operational specification of the expert system. CoMo-Kit is based on our extension of the KADS methodology (Wielinga et al. 1992)

Concept Hierarchies. HyperCAKE contains an object system for the development of application-specific shells which

- allows the definition of classes and slots
- supports slot typing and defaults,
- supports inheritance,
- allows to create instances,
- permits to define methods.

The objects are represented as HAM nodes.

Rule Interpreter. We developed a rule interpreter for the processing of the knowledge. The present version includes forward propagation and read/write access to slot values. The rule interpreter uses attribute nodes, which have symbols or numbers as contents. The actual values of attribute nodes are assigned by user input or by rule firing. New values are propagated through a rete-like network which results in additional rule firings. For example the rule "IF (valve5y2 = 0) THEN relais5r2 := SWITCHED" means that if the contents of attribute node valve5y2 is set to zero then the attribute node relais5r2 is set to SWITCHED.

Rules are interpreted as logical implications: If the logical value of the precondition becomes false, the (formerly set) slot is reset to unknown. This simple rule interpreter is sufficient for the requirements of our present application domains (see below).

1.3 The Interface to the Knowledge Base

On top of the knowledge management system we developed a configurable user interface for HyperCAKE according to the needs of each application domain.

Editors for Multimedia Objects. HyperCAKE includes an editor for texts with normal functionality for the creation and formatting. Graphics are drawn with standard software and imported into a hypertext node. We developed facilities for recording, playing and cutting of audio-data. In addition to the described editors, HyperCAKE includes mask-oriented interfaces for the input of knowledge structures (e.g. objects, slots, attributes).

Definition of User-Dependent Views. Normally, the information stored in a network is too voluminous to show it entirely to the user. Therefore, HyperCAKE allows to define views based on a simple user model. The user model consists of the name of the user, his abilities (a knowledge engineer needs a different kind of access to a knowledge base from that of the domain expert or the unexperienced user), a filter, and a layout. The filter describes declarativly which information is shown (e.g. HAMFilter on: HyDiNet condition: '[:e | e isKindOf: Diagnosis]' lists all instances of node class Diagnosis which are stored in the net HyDiNet) while the layout describes how it is shown.

Generic Interface for the Domain Layer for Knowledge Elicitation. A conceptual model describes concepts and relations of the application. Knowledge elicitation fills the defined structures with the domain knowledge. This task shall be carried out by the domain experts because they are able to check the correctness of the decisions of the expert system. Normally, domain experts are not used to computers. Therefore, a knowledge acquisition interface must support the easy development of domain specific interfaces. The effort for interface construction must be small to support rapid prototyping. HyperCAKE contains a generic interface for the acquisition of domain instances. If this is not sufficient, the interface can be enhanced by aid of

- Network browsers

- Texteditors

- Views for graphics and (in a later step) video

- Possibilities to define domain dependent menus.

The interface tool described above allows also the realization of consultation interfaces. Figure 1 shows an example user interface which was build with the facilities of HyperCAKE. The possibilities described so far are used to develop the HyDi system.

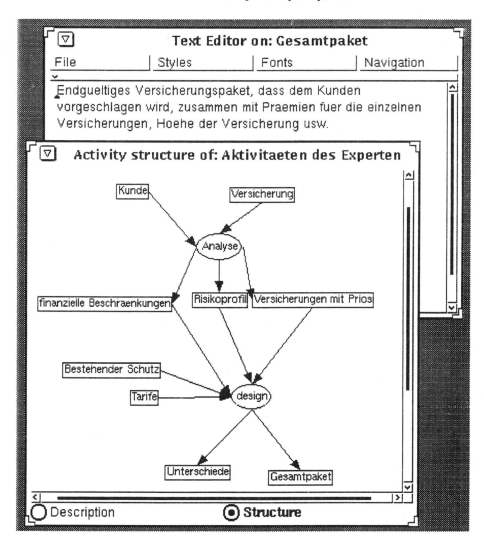

Figure 1: A user interface built with HyperCAKE

2 The HyDi System

HyDi is developed following the MOLTKE methodology of technical diagnosis (Althoff et al. 1990), (Richter 1992). It uses a simple model of diagnostics:

$$diagnostics = classification + selection\ of\ tests$$

2.1 Knowledge Representation

From this model we can conclude some basic primitives for the representation of the required knowledge structures.

Symptom. A symptom assigns a name to a range of values (i. e. a data type), comments and tests. Values may have many different depictions. We can think of them as abstract symbols or as concrete sounds.

Diagnosis. A diagnosis represents a failure and instructions for repair or maintenance. Diagnoses have different degrees of exactness. In the simplest case it is the statement '*The machine is defect*'. They are arranged as a directed non-cyclic graph. The links of this graph are labeled with formulas that state the condition under which a diagnosis is a refinement of another one. Thereby a link represents a conjunctive precondition. If a Diagnosis has different links pointing on it, these links form a disjunctive precondition. The graph structure represents the classification aspect of our model of diagnostics. Figure 2 shows an example of such a graph.

Strategy. Each diagnosis is associated with a strategy. Strategies determine the procedure of selection of tests, i. e. of the inquiry of symptoms. They are described as a set of rules or as a decision tree. The diagnostic process consists in finding a path through the graph of diagnoses. Figure 2 shows a small example. It was taken from the domain of CNC-machining-centers. Starting at the diagnosis `machine failure` the strategy is to test the symptom `error code` (ec), i.e. the diagnosis node (see below) `machine failure` contains the ruleset {IF (true) THEN TEST (ec)}. The instruction TEST executes the test procedure which is associated with the symptom `ec`. If the result of this test is the value i59 we can follow the link to the diagnosis `tool changer defective` because the condition ec = i59 holds. This diagnosis contains the ruleset {IF (true) THEN TEST (valve5y2), IF (valve5y2 = 0) THEN TEST (out35), ...} as its strategy. The result of the evaluation process of this set could be `valve5y2 = 0` and `out35 = 1`. Then, the final diagnosis `i/o-card defective` is reached. Searching through the graph this way is a continuing refinement of diagnoses by alternating test selection and classification. A final diagnosis is reached, if the actual node is a leaf of the graph structure. The algorithm can be described informally as follows:

```
choose the root node as the actual node.
REPEAT
    call the rule interpreter of HyperCAKE which
    fires an applicable rule of the actual node
    and execute its test.
    IF there is a link which has the actual node
        as its source and has a formula which
        evaluates to true
```

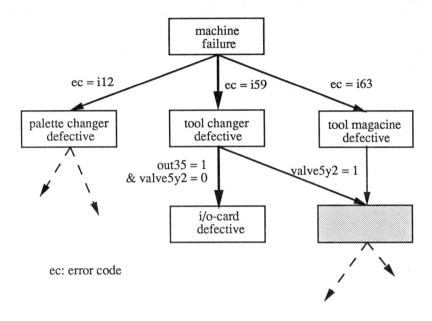

Figure 2: A graph of diagnoses

```
THEN choose the destination of this link
     as the new actual node
   ENDIF.
UNTIL the actual node is a leaf.
```

2.2   Integration into the Hypermedia Framework

Using HyperCAKE the previously described knowledge structure is easily implemented as a hypermedia network. All primitives are represented as nodes of the network. The diagnostic process then corresponds to a knowledge based navigation through this network. We introduced some new node and link classes which are listed partly in table 1. ConditionLinks connect DiagnosisNodes. If the condition of a link holds, it can be traversed to the next DiagnosisNode. We found that using hypermedia networks is very suitable for implementing the MOLTKE methodology of technical diagnosis.

Because of the object oriented implementation the whole hypertext functionality is inherited from the object classes of the HyperCAKE system. This way the integration is in fact a refinement and extension of the facilities provided by HyperCAKE. The resulting network is very flexible, i. e. it is easily extended by additional nodes to represent help information or for documentation tasks. Each type of node in the hierarchy of HAMNodes, e. g. TextNode or SoundNode, is accessible.

HAMNode	class	content
	DiagnosisNode	a ruleset
	RepairNode	instructions for repair
	SymptomNode	a value
	TestNode	a test procedure
	PresentationNode	a depiction of a value
	....	
HAMLink		
	ConditionLink	a formula

Table 1: Node and link classes of HyDi

2.3 The Utilization of the Hypermedia Capabilities

A technical diagnosis system has to provide support for different kinds of users. Support means a user-specific access to technical information on machinery. Three possible categories of users are machine operators, service technicians and experts like machine engineers. In the case of a machine failure the machine operator is first confronted with the problem. A diagnosis system has to enable him to fix simple failures by himself or at least to localize possible reasons for it.

A hypermedia based system like HyDi is able to do so by supplying additional technical instructions like videos, sounds and graphics. Thereby many situations in which usually the service technician is brought into action are handled by the machine operator. This reduces cost for technical service and decreases the loss of productivity caused by the machine failure. A service technician needs intelligent access on technical documentations of the machine. Knowledge-based hypermedia systems like HyDi can provide such access. Additionally the diagnostic capabilities support the navigation. The use of different user models and data views take into account the mentioned different user categories. E.g., the category of machine operators will access the diagnosis and instruction components of the graph through its special filter. The filter mechanism provides the possibility to define different data views on the network (see also section 1.3: the definition of user dependent views). The layout of the presentation of this view is described by browsers. Figure 3 and Figure 4 show some of the possibilities. The use of browsers and the so called Model-View-Controller concept comes from our implementation language Smalltalk-80.

3 Conclusions and Further Work

HyperCAKE is fully implemented. The work on HyDi will be finished in December of this year. The approach of using a hypermedia network to implement a knowledge based system has shown advantages. The multimedia capability of the system allows the construction of user friendly interfaces and increases the acceptance of such systems. The resulting structures are very flexible and extendable. A difficulty is the fast growing complexity of the networks. In addition to the filter mechanism there has to be a facility which garuantees the correctness and the completeness of the structure according to some topological constraints. Further work will be on the field of this knowledge base maintenance problem. In addition we are working on the extension of the presented mechanisms by case based reasoning and induction techniques within the ESPRIT project 6322 (INRECA).

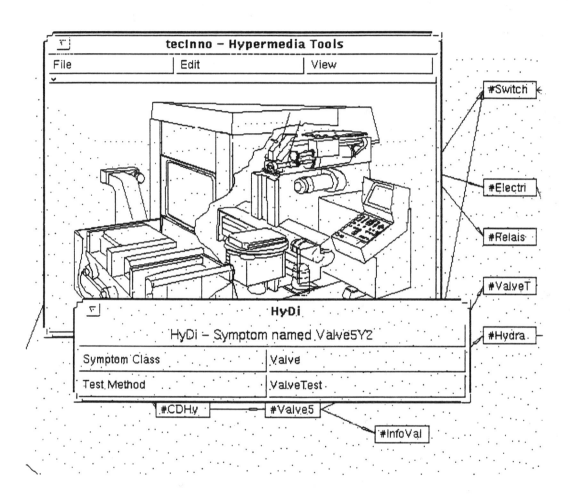

Figure 3: Some node views

322

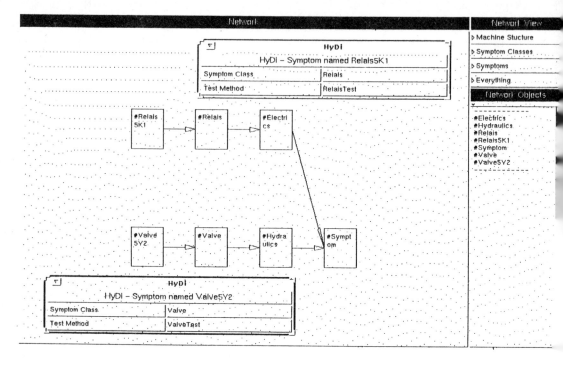

Figure 4: A network view

References

ALTHOFF, K.-D., MAURER, F., REHBOLD, R. (1990): Multiple Knowledge Acquisition Strategies in MOLTKE, *in: Proc. EKAW 90*, 1990

CAMPBELL, B., and GOODMAN, J. (1988): HAM: A General Purpose Hypertext Abstract Machine, *Communications of the ACM*, July 1988, Vol. 31, No. 7

HALASZ, F. G. (1988): Reflections on Notecards: Seven Issues for the Next Generation of Hypermedia Systems, *Communications of the ACM*, July 1988, Vol. 31, No. 7

MAURER, F. (1992): HyperCAKE - Ein Wissensakquisitionssystem fuer hypermediabasierte Expertensysteme, *in: J. Biethahn, R. Bogaschewsky, U. Hoppe, M. Schumann: Wissensbasierte Systeme in der Betriebswirtschaft - Anwendung und Integration mit Hypermedia*

MAURER, F. (1992): HyperCAKE - A Knowledge Acquisition Tool for Hypermedia-Based Expert Systems, , *SEKI-Report SR-92-04 (SFB)*, Uni Kaiserlautern

RICHTER, M. M. (1992): MOLTKE - Methoden zur Fehlerdiagnose in technischen Systemen, *to appear*

WIELINGA, B. J., SCHREIBER, A. TH., BREUKER, J. A. (1992): KADS: A Modelling Approach to Knowledge Engineering, *Knowledge Acquisition Journal, Special issue on KADS*

Classification and Learning of Similarity Measures

Michael M. Richter

Fachbereich Informatik, Universität Kaiserslautern and
Deutsches Forschungszentrum für Künstliche Intelligenz, Kaiserslautern, FRG

1   Introduction

The background of this paper is the area of case-based reasoning. This is a reasoning technique where one tries to use the solution of some problem which has been solved earlier in order to obtain a solution of a given problem. As example of types of problems where this kind of reasoning occurs very often is the diagnosis of diseases or faults in technical systems. In abstract terms this reduces to a classification task. A difficulty arises when one has not just one solved problem but when there are very many. These are called "cases" and they are stored in the case-base. Then one has to select an appropriate case which means to find one which is "similar" to the actual problem. The notion of similarity has raised much interest in this context. We will first introduce a mathematical framework and define some basic concepts. Then we will study some abstract phenomena in this area and finally present some methods developed and realized in a system at the University of Kaiserslautern.

We consider a universe U which is partitioned into a disjoint union of subsets called classes and we refer to the elements of U as objects. Each object has a structure; for simplicity we take this as a fixed number of n attribute-value pairs. This allows a twofold description of objects:

a) The objects are coded as vectors of length n of real numbers, each coordinate represents an attribute;

b) the objects are described as conjunctions of unary predicates $P(a)$ where P stands for an attribute and a for a value.

We call a) an *analytic* and b) a *logical* representation of the objects.

The task is to determine for a given object its class. The available information may be, however, incomplete in two respects:

1. The object itself is only partially known;

2. only for a restricted number of objects the class where its belongs to is known.

In order to predict the class of an object one assumes an underlying regularity in the formation of the classes which has to be determined or at least approximated on the basis of the available information. In machine learning one considers mainly two basic ways to achieve this:

a) The logical approach: Classes are described by formulas in predicate logic using the attributes. These may e.g. be rules which have conjunctions of attribute formulas or their negations as premises and class names as conclusions.

b) The analytic approach: There is a distance function d in $\mathbb{R}^n$ and the class of some presented vector a is the class of that particular vector b from the already classified vectors for which the distance $d(a, b)$ is minimal.

With both approaches a number of concepts are connected. In order to discuss the interrelations between them from a mathematical point of view we make use of a number of results in economics, in particular utility theory. This stems from the fact that the notion of similarity shares some mathematical properties with the notion of a preference order. In utility theory one studies objects which may be more or less preferrable; we will employ the mathematical analogy between the partial orderings coming from similarity and preference.

Both, the classifying rule system and the distance function have to be built up in the training phase. The algorithms for this task have some (sometimes hidden) common properties. A fundamental problem is to exhibit the the connections between the distance function and the classification problem. In a nutshell this reads as follows:

How to construct a distance function d such that for sufficiently small $d(a, b)$ the objects a and b are in the same class?

This is essentially an a posteriori problem which can principally only be answered after the class of the objects is known. From this principal point of view this asks for an adaptive approach. Nevertheless one has first to explore the basic aspects and concepts of distance functions and the related similarity measures. This attempt focusses the attention on problems which should also be approached (at least presently) in an empirical way. The PATDEX- system discussed in section 6 realizes a number of essential tasks from a practical point of view.

2   Basic concepts

Each object is given by the values of a fixed number of attributes. If A is such an attribute with value a then this is denoted by $A(a)$. We describe objects alternatively as vectors where each coordinate corresponds to an attribute and the entry to its value. An object description is like an object except that instead of the value of an attribute a variable may occur (indicating that the value is unknown). The universe of our object descriptions is U. In general we do not distinguish between objects and object descriptions.

There are different ways to represent similarity which we will introduce now.

1. A binary predicate $SIM(x, y) \subseteq U^2$ meaning "x and y are similar";

2. a binary predicate $DISSIM(x, y) \subseteq U^2$ meaning "x and y are not similar";

3. a ternary relation $S(x, y, z) \subseteq U^3$ meaning "y is at least as similar to x than z is to x";

4. a quaternary relation $R(x, y, u, v) \subseteq U^4$ meaning "y is at least as similar to x than v is to u";

5. a function $sim(x, y) : U^2 \rightarrow [0, 1]$ measuring the degree of similarity between x and y;

6. a function $d(x, y) : U^2 \rightarrow \mathbb{R}$ measuring the distance between x and y.

The obvious questions which arise here are:

(i) How to axiomatize these concepts, i.e. which laws govern them?

(ii) How are the concepts correlated, which are the basic ones and which can be defined in terms of others?

(iii) How useful are the concepts for the classification task?

There are split opinions about the properties of the various concepts. It is certainly agreed that SIM is reflexive. There are arguments that SIM should neither be symmetric nor transitive. A typical example to support the first claim is that one could say 'my neighbor looks similar to the president' but one would not use the reverse phrase. This argument, however, says nothing about the truth or falsity of the similarity relation; it is only concerned with its pragmatics. For this reason we will accept that SIM is symmetric. In order to reject the transitivity of SIM one gives examples like 'a small circle is similar to a large circle and a large circle is similar to large square but a small circle is not similar to a large square'. The reason for this effect is that one deals here with two different similarity relations, one concerning size and another concerning form. A basic problem is how one can amalgamate two different similarity relations into one. A second type of counter argument arises when the objects are partially unknown. Suppose we have three such objects a, x and b where $SIM(a,b)$ does not hold and x is partially unknown. An opportunistic view could then assume both $SIM(a,x)$ and $SIM(x,b)$, violating transitivity. As a consequence, we will not accept transitivity for SIM.

A next observation tells us that we should distinguish $DISSIM(x,y)$ from $\neg SIM(x,y)$. The latter means simply that there is not enough evidence for establishing $SIM(x,y)$ but that may not be sufficient to claim $DISSIM(x,y)$; we have here the same distinction as one has between the negation in classical and intuitionistic logic. The deeper reason for this is that similarity between objects is not given as a relation with truth values 0 and 1 but as something to which the terms 'more or less' apply. We will therefore not consider SIM and $DISSIM$ anymore but the arguments given above do also apply to the remaining concepts.

In the sequel we will encounter several preorderings. A preordering $\geq$ on a set U is a reflexive and transitive binary relation. $\geq$ is called complete if $y \geq z \lor z \geq y$ holds. Such a relation can always be decomposed into two parts:

(i) $y > z \leftrightarrow y \geq z \land \neg(z \geq y)$, this called the strict part of the relation;

(ii) $y \sim z \leftrightarrow y \geq z \land (z \geq y)$ (indifference).

'$>$' is always asymmetric and transitive and '$\sim$' is an equivalence relation.

The relation $S(x,y,z)$ induces for each x a binary relation $y \geq_x z$. We assume:

(i) $\geq_x$ is a complete preorder (with $>_x$ as its strict part and $\sim_x$ as the indifference relation);

(ii) $y >_x z$ implies $y >_x u$ or $u >_x z$;

(iii) $x \geq_x z$.

(iii) refers to the reflexivity of SIM; the symmetry of SIM has no counterpart here. A further axiom is often required where the structure of the objects is involved:

Monotonicity Law: If y' agrees at least on one more attribute with x than y does, then $y' \geq_x y$ holds.

We will not require this law in general because it includes a kind of independence between the values of the attributes. If the attributes depend on each other then the same value can have a different meaning in different contexts so that more agreement on the attribute values can mean less similarity.

The relation S allows to define the concept 'y is most similar to x': For some set $M \subseteq U$ some $y \in M$ is called *most similar* to x with respect to M iff

$$(\forall z \in M)S(x,y,z)$$

This notion is essential in case-based reasoning.
For the relation R we assume the axioms

(i) $R(x,x,u,v)$;

(ii) $R(x,y,u,v) \leftrightarrow R(y,x,u,v) \leftrightarrow R(x,y,v,u)$.

(i) and (ii) are the counterparts of the reflexivity and of symmetry of SIM, resp.

The relation $R(x,y,u,v)$ induces a partial ordering $\geq$ on pairs of objects by $(x,y) \geq (u,v) \leftrightarrow R(x,y,u,v)$. $\geq$ can be decomposed as above and we assume the same axioms as for $>_x$. R also induces a relation S_R by $S_R(x,y,z) \leftrightarrow R(x,y,x,z)$.

The basic axioms for a similarity measure *sim* are:

(i) $sim(x,x) = 1$ (reflexivity);

(ii) $sim(x,y) = sim(y,x)$ (symmetry).

The dual notion is that of a *distance measure* $d(x,y)$ which may attain arbitrary nonnegative values. In the corresponding axioms reflexivity reads as $d(x,x) = 0$. One does not require, however, the triangle inequality and allows $d(x,y) = 0$ for $x \neq y$ which means that d is neither a metric nor even a pseudo- metric. The argument for skipping the triangle inequality is the same as the one for not requiring transitivity for SIM.

One says that d and *sim* correspond to each other iff there is an order reversing one-one mapping

$$f : range(d) \rightarrow range(sim)$$

such that $f(0) = 1$ and $sim(x,y) = f(d(x,y))$; we denote this by $d \equiv_f sim$.
Popular candidates for f are:
$f(z) = 1 - \frac{z}{1+z}$ for unbounded d or $f(z) = 1 - \frac{z}{max}$ if d attains a greatest element max.

Some interrelations between the introduced concepts are immediate. If d is a distance measure and *sim* a similarity measure then we define

$$R_d(x,y,u,v): \iff d(x,y) \leq d(u,v)$$
$$R_{sim}(x,y,u,v): \iff sim(x,y) \geq sim(u,v)$$

and

$$S_d(x,y,z): \iff R_d(x,y,x,z)$$
$$S_{sim}(x,y,z): \iff R_{sim}(x,y,x,z)$$

We say that d and sim are compatible , iff

$$R_d(x, y, u, v) \iff R_{sim}(x, y, u, v) ;$$

compatibility is ensured by $d \equiv_f sim$ for some f.

As usual in topology the measures also define a neighborhood concept. For $\in > 0$ we put

$$V_\in(x) := V_{d,\in}(x) := \{y | d(x, y) \le \in\},$$

and analogously $V_{sim,\in}(x)$ is defined; if d is a metric then these sets are ordinary closed neighborhoods. $S_d(x, y, z)$ expresses the fact that each neighborhood of x which contains z also contains y. In order to be useful for the classification task the neighborhood system has to be compatible with the partition into classes in the sense that the neighborhood should group the elements of the classes 'closely together'

3 Ordinals and cardinals

The concepts presented in (1) to (6) of Section 2 contain in an increasing order more and more information about the similarity of object descriptions. Least informative are SIM and $DISSIM$ and most informative are the measures and distance functions. The latter ones define the relations R_d and R_{sim} as indicated above in such a way that their axioms are satisfied. From R we obtain the relation S; again the axioms for S follow from those for R. S finally can, using some threshold, define relations SIM and $DISSIM$.

Comparing first $y \ge_x z$ and $sim(x, y)$, $sim(x, z)$ the additional information provided by sim is that it tells us *how* more similar y is to x than z is to x. S contains only an ordinal information while sim has also a cardinal aspect.

In the application to classification the main use of this cardinal aspect is that one forms differences like $|sim(x, y) - sim(x, z)|$. Such a difference is of interest when one searches the object y most similar to x. If $|sim(x, y) - sim(x, z)|$ is small, then one could choose z instead of y with a small error only; for the classification task this may be sufficient. From this point of view $R(x, y, u, v)$ contains some cardinality information. Another type of implicit cardinality information is contained in the *sensibility potential*, cf. Wagener (1983).

The reverse way from the ordinal to the cardinal view is more involved. First, the relations SIM and $DISSIM$ carry very little information about the relation S. Given S, one has for every object description x the preorder $\ge_x$. In order to obtain R from S we proceed in several steps:

1) Define: $R_1(x, y, x, z) \leftrightarrow S(x, y, z)$;
2) obtain R_2 from R_1 by adding the tuples (x, x, y, z);
3) define $\ge_3$ as the transitive closure of $\ge_2$;
4) obtain $\ge$ from $\ge_3$ by extending it to a complete preorder in such a way that $y >_3 z$ implies $y > z$ (this is always possible).
5) Define $R(x, y, u, v) \leftrightarrow (x, y) > (u, v)$.

If we define from this R as above the relation S_R the strict parts of the preorders may, however, be different. This is due to the fact that in step 1) where essentially the join $\cup \ge_x$ of the preorders $\ge_x$ was formed some cycles in the strict parts of the join may occur which means that elements are now indifferent which were strictly ordered before. Therefore we require that this cannot happen and call it the *compatibility condition* on S.

The step from R (or $\geq$) to a measure or distance function is done by embedding $\geq$ into $\mathbb{R}^2$. This is possible because our universe is finite.

We emphasize again that for our classification task the relation S is the one which is used. To be of interest the compatibility condition has to be satisfied. This is essentially the step to the relation R which, as remarked above, has additional benefits. In our learning process below we will learn the measure directly but will essentially use information about relation S.

4   The amalgamation of similarity measures

Suppose we are given different experts E_i who are confronted with a fixed object x and a number of objects which may be more or less similar to x. The task for these experts is to arrange the objects according to their similarity to a, i.e. to establish an ordering $\geq_a^i$. Each expert is supposed to represent a certain aspect and will come up with his individual arrangement. Furthermore, there is a general manager who takes these individual ratings and whose task is to amalgamate the different ratings into a general ordering of the objects under consideration.

A very simple method for integrating such orderings is to use a number assignment according to the orderings and sum up these numbers. This is Borda's method which he invented in 1781. We give an example with 5 participating objects t, y, z, u and v and 5 experts (representing 5 aspects) :

	t	y	z	u	v
1	4	3	2	1	0
2	2	4	3	0	1
3	3	2	1	0	4
4	4	3	0	2	1
5	1	4	2	3	0
Sum	14	16	8	6	6

The winner, i.e. the object most similar to x is y, followed by t, z etc. Suppose now that we want to remove the objects z and u from the database because they are perhaps not of great interest anyway. Then we are left with three objects and we apply the same method to rank them. We get the following table:

	t	y	v
1	2	1	0
2	1	2	0
3	1	0	2
4	2	1	0
5	1	2	0
Sum	7	6	2

The result is that the final ordering of the remaining objects is changed and that now t is the winner. This effect is very undesirable because the elimination of uninteresting objects leads to a change of the ordering of the remaining objects; the whole data base is subject to a global analysis in order to recompute the similarity relation. We will explain now that this is not an accident which is due to the special method but that there is an underlying deeper phenomenon.

We start with a set U of object descriptions s.t. $|U| \geq 3$ and an index set $M \neq \emptyset$. We consider partial orderings as introduced in section 2. Let S be the set of such orders on U and $F = \{f|f : M \to S\}$. M represents the different aspects and F the orderings (i.e. the strict part) with respect to similarity to the reference object according to these aspects. What one looks for is a mapping $\sigma : F \to S$ which amalgamates the individual orderings into a universal one. The function σ has to satisfy certain very plausible conditions:

(a) If $y\ f(m)z$ for all $m \in M$, then $y\ s(f)z$;

(b) if f and g coincide on y and z, then $\sigma(f)$ and $\sigma(g)$ coincide on y and z too.

(c) There is no $m \in M$ such that for all y and z in U we have:
If $y\ f(m)z$, then $y\sigma(f)z$.

These conditions have a clear motivation. (a) says that the universal ordering should not contradict all aspects. (b) was discussed above and (c) says that one cannot reduce the problem to one aspect.

Theorem: There is no function f satisfying (a), (b) and (c).

This theorem is due to Arrow (cf. Arrow (1963)) and well known in the area of social choice functions. There the partial orderings are preference orderings, M is the set of voters, (a) is the principle of democracy and (c) excludes dictatorship. The function σ combines the individual votes. Arrow's impossibility theorem is also called the theorem of the dictator and was considered as somewhat paradoxical. Slight variations of the condition do not change the validity of the theorem. The crucial and most discussed condition is (b). It is also important for our situation; according to the theorem changes in the data base have other consequences. The most we can hope for is that these consequences have a local character.

5 General forms of distance functions and similarity measures

We consider objects which are defined in terms of boolean valued attributes and study their relations using distance functions only. There is a great variety of distance functions and an enormous amount of literature. When distance functions are used for classification purposes they cluster the objects in such a way that the cluster coincide with the given classes as much as possible. If this is the case then one can say that the function contains some knowledge about the classes. Different applications lead to different types of classes and therefore to different kinds of distance functions; this explains mainly the richness of this area. In our approach we are not so much interested in our introducing a particular clever distance function but rather in showing how some general knowledge can be improved by an adaptive process. The type of functions we introduce is general enough to study these techniques but many other distance functions would have worked as well. We will restrict ourselves here to Boolean attributes, i.e. we have values 0 and 1 only. The most simple distance measure is the Hamming distance. A generalization of the Hamming distance is given by the *Tversky-Contrast model* (cf. Tversky (1977)). For two objects x and y we put

$A :=$ The set of all attributes which have equal values for x and y ;
$B :=$ the set of all attributes which have value 1 for x and 0 for y ;
$C :=$ the set of all attributes which have value 1 for y and 0 for x ;
The general form of a Tversky distance is

$$T(x,y) = \alpha \cdot f(A) - \beta \cdot f(B) - \gamma \cdot f(C)$$

where α, β and γ are positive real numbers. Most of the other possible distance functions are located between the Hamming and the Tversky measure with respect to the information which they can contain. In PATDEX (see below) we start out with a measure for which we need some notation. An object description from the case base is denoted by x and an arbitrary one by x_{act} (indicating that this is the actual description for which we want a similar one from the base). We put

$x_{act} = (w_{i_1}, ..., w_{i_k})$, $x = (v_{r_1}, ..., v_{r_j})$; here we list only the coordinates with a known value.

$H = \{i_1, ..., i_k\}$,
$K = \{r_1, ..., r_j\}$;
$E = \{i | i \in H \cap K, w_i = v_i\}$, the set of attributes where the values agree;
$C = \{i | i \in H \cap K, w_i \neq v_i\}$, the set of attributes with conflicting values;
$U = H \setminus K$, the set of attributes with a known value for x but unknown value for the actual object;
$R = K \setminus H$, the set of attributes with a redundant value for x_{act}.
The measure used is of the form

$$sim_{PAT}(x_{act}, x) = \frac{\alpha \cdot |E|}{\alpha \cdot |E| + \beta \cdot |C| + \gamma \cdot |U| + \eta \cdot |R|} .$$

The parameters α, β, γ and δ can be chosen ; presently we use:

$$\alpha = 1, \qquad \beta = 2, \qquad \eta = 1/2, \qquad \gamma = 1/2 ;$$

which gives

$$sim_{PAT}(x_{act}, x) = \frac{|E|}{|E| + 2 \cdot |C| + 1/2 \cdot |U| + 1/2 \cdot |R|} .$$

This measure pays special attention to attributes with missing values. On the other hand, it abstracts from the Tversky measure in so far that it sees only the cardinality of sets instead of the sets themselves.

6   Learning similarities and the PATDEX system

The difficulty with the similarity measure is that its quality is related to the final success of the whole reasoning procedure; this is an *a posteriori* criterion. A priori it is not clear what the criteria for similarity of objects should be; they do not only depend on the objects themselves but also on the pragmatics of reasoning. In case-based reasoning it is usually clear whether a solution for a given problem (in our situation a classification problem) is correct but is far from clear what it means that two problems are similar enough so that the solution for one problem also works for the other one. Looking at the object descriptions only one neither knows a suitable general form of the measure nor has one an indication how the parameters should be determined. An even more serious difficulty arises when the world of problems is continuously changing. This suggests that the similarity should not be defined in some fixed way but instead be the result of an adaptive learning process. This will be carried out in the PATDEX- System.

PATDEX is a part of the MOLTKE-System (cf. Althoff (1992)) which was developed in the past years at the University of Kaiserslautern. Its domain is the fault diagnosis of technical systems. Here we are only concerned with the aspect that diagnosis can be regarded as a classification task and we will suppress the other aspects. For this reason we modify the present terminology of PATDEX. The system accepts a description of an object as an input; this description may be partial, some attribute values may be unknown. The basic instrument for the classification is the *case base*; a *case* is a pair (Object x, class(x)) where class(x) is the class to which x belongs.

The first version of PATDEX is PATDEX/1. It contains the basic structures which have been extended later on. It is convenient to describe it first. As basic techniques, PATDEX/1 applies learning by memory adaptation and analogical reasoning. The toplevel algorithm of PATDEX reads as follows:

Input: The actual object description x **Output:** a class C or failure

1) Find a case in the case base with an object x' most similar to x. If there is no case with an object at least '*minimally similar*' to x then stop with failure.

2) If x and x' are '*sufficiently similar*' then accept the class C of x' also for x and goto 4).

3) Otherwise select an attribute with unknown value and determine its value in order to obtain an improved situation and goto 1).

4) If the class is correct then add the case (x, C) to the case base and stop with success.

5) If the class is not correct then cancel temporarily (i.e. for the actual problem) all cases with class C and goto 3).

Here we need an external teacher who says whether a class is correctly chosen or not. We also have to explain '*minimally similar*' and '*sufficiently similar*'. For this we need a partition of the case base which is given after the introduction of the similarity measure.

For object descriptions PATDEX we introduced as a first proposal the similarity measure sim_{PAT} in section 5 with parameters $\alpha = 1$, $\beta = -2$, $\gamma = \eta = -1/2$. This special choice of the parameters is at the moment mainly motivated by experimental results. It has a defensive, pessimistic character. A high negative contribution to the measure is given for conflicting attribute values, i.e. we strongly wish to avoid false classification.

For the partition of the case base we choose real numbers $\in$ and δ such that $0 < \in < \delta < 1$ and define:

Def.: The object descriptions x_1 and x_2 are called
(i) indistinguishable $\Leftrightarrow sim(x_1, x_2) = 1$;
(ii) sufficiently similar $\Leftrightarrow \delta \leq sim(x_1, x_2) < 1$;
(iii) at least minimally similar $\Leftrightarrow \in \leq sim(x_1, x_2) < \delta$;

(iv) not minimally similar $\Leftrightarrow 0 \leq sim(x_1, x_2) < \in$;

The lower bound $\in$ is called the hypothesis threshold, a case succeeding here is said to be qualified for further processing. If the value exceeds an upper bound δ it is even qualified as providing the classification (classification threshold). If, for a given case, the similarity value equals 1 this case is said to be proven. The thresholds are locally defined for each case of the case base, i.e. we have the possibility to make the numbers $\in$ and δ dependent on the respective cases.

It is an important feature of PATDEX that it supports for an object description the selection of an attribute with an unknown value. An optimal or at least good choice of such an attribute is crucial for an efficient classification procedure. We will, however, not deal with this question.

The use and analysis of PATDEX has lead to the conclusion that its performance concerning the classification problem showed some weaknesses. Ultimately this was a problem of the similarity measure in two respects as already indicated. First, the type of the measure (as an abstraction of the Tversky measure) was too simple in order to reflect information of the objects which are necessary for the classification. Secondly, even if the type of the measure would have been optimal one would still face the problem of chosing the parameters of the measure. To overcome this problem a learning process will be introduced.

We will first describe the structural improvements of the measure. They get their motivation from the actual use of the system for diagnostic purposes rather than from purely mathematical considerations. The information reflected by the improvements is usually available in the intended applications. The improvements are contained in the system PATDEX/2 (cf. Weß90, Weß91).

The underlying pattern of the new features in PATDEX/2 is that not all attributes are equally important for determining the class of an object description. This leads to the notion of relevance. The relevances are numbers $w_{ij} \in [0, 1]$; where the index i points to an attribute A_i resp. its value and the index j refers to a class C_j. The w_{ij} should indicate the degree with which a_i points at C_j. The relevances give rise to the *relevance* matrix $R[w_{ij}]$. The main problem is now to determine the entries (also called weights) of the relevance matrix. These weights are exactly the elements which will be learned later on.

It is convenient to normalize the matrix such that

(i) For all i and j $0 \leq w_{ij} \leq 1$ holds;

(ii) For all j we have $\sum_{i=1}^n w_{ij} = 1$.

We will now discuss the possibilities for the weights. This leads to some changes in the computation of sim_{PAT}.

(1) <u>Local and global weights:</u> Global weights satisfy $w_{ij} = w_{ik}$ for all j and k; otherwise weights are called local. Global weights are less precise but easier to determine.

(2) <u>Conflicting attribute values:</u> If two objects have different values for an attribute A with domain D then the form of the difference should play a role. This can be achieved by introducing a function $\omega : D^2 \rightarrow [0, 1]$ which has to represent the similarities of the attribute values. If one of the values is unknown then the similarity ω_i evaluates to zero.

(3) <u>Redundant attribute values:</u> Redundant attribute values for the actual object description count negative in the measure. This has the undesired effect of decreasing the similarity by the acquisition of more and e.g. completely uninteresting attribute values. This leads to the notion of *classifying* and *not classifying* attributes for redundant attributes, depending on their values. This division of the attributes has to be made by the user; in applications

to diagnosis this is usually not so difficult because classifying attributes there correspond to attributes with an abnormal value. The impact on the measure is that only classifying attributes enter the computation of R in sim_{PAT}.

(4) <u>Unknown attribute values:</u> Unknown values for the actual object description also count negatively in the computation of the measure. This may not be justified because the known values may determine, at least with some probability, the missing ones. Hence for such unknown values a value should be substituted which has a probability above some (user defined) threshold θ. The probability can be estimated by the frequencies in the base of object descriptions.

These remarks lead to a redefinition of the similarity measure. For the similarity between values the user chooses a threshold λ. We put

$x_{act} = (w_{i_1}, ..., w_{i_k}), x = (v_{r_1}, ..., v_{r_j})$; here we list only the coordinates with a known value or where the value in xact can be predicted with probability $\geq \theta$.

$H = \{i_1, ..., i_k\}$,

$K = \{r_1, ..., r_j\}$;

$E' = \{i | \omega(w_i, v_i) \geq \lambda\}$, the set of attributes with sufficiently similar values;

$C' = \{i | \omega(w_i, v_i) < \lambda\}$, the set of attributes with not sufficiently similar values;

$U' = H \setminus K$, the set of attributes with a known or estimated value for x but unknown value for x_{act}.

$R' = K \setminus H$, the set of attributes with a redundant and classifying value for x_{act}.

Using this we define

$$E_0 = \sum_{i \in E'} w_{ij} \cdot \omega(w_i, v_i);$$

$$C_0 = \sum_{i \in C'} w_{ij} \cdot (1 - \omega(w_i, v_i));$$

$$R_0 = |R'|;$$

$$U_0 = \sum_{i \in U'} w_{ij} \ .$$

This leads finally to the measure of PATDEX/2:

$$sim_{PAT/2}(x_{act}, x) = \frac{\alpha \cdot |E_0|}{\alpha \cdot |E_0| + \beta \cdot |C_0| + \gamma \cdot |U_0| + \eta \cdot |R_0|} \ .$$

α, β, γ and η can be chosen as before. The partially user defined parameters are a step towards the idea of the Tversky measure. The approach takes into account that the precise form of the measure is a priori (i.e. when the problem is given) not available; the user can fill in as much knowledge as he has about the problem. Given a base of correctly classified object descriptions experiments with PATDEX/2 showed that the similarity measure did not even classify the cases from the base correctly. This was expected and here a learning process starts. What is learned are the weights, i.e. the entries of the relevance matrix. This process has an initial phase and a learning phase; the training set is the case base.

<u>Initial phase:</u> The initial weights w_{ij} are determined according to the observed frequencies in the base.

<u>Learning phase:</u> The cases (x_{act}, C) are taken from the case base. The system selects the most similar case (x, D) from the case base (similarity of cases means similarity of their object descriptions). If $C = D$, then nothing will be changed. For $C \neq D$ we distinguish two possibilities.

(1) x contains less known attribute values than x_{act}. Here the class D was obviously only correct by accident and the case (x,D) is eliminated from the case base.

(2) In all other situations (x,D) remains in the case base but the weights are updated. The numerical form of the learning rule is not of interest; the leading principles are:

- $sim_{PAT/2}(x_{act}, x) < \delta$ should be achieved, they are not anymore sufficiently similar;

- weights for attributes in C' and U' are increased;

- weights for attributes in E' are decreased;

- weights for attributes in R' remain invariant;

- the weights w_{ij} are still normalized according to $\sum_{i=1}^{n} w_{ij} = 1$ for each j.

Rules of this type are known in unsupervised neural networks; an example is the Grossberg rule resp. the rule in competive learning, cf. Rumelhart, Zipser (1985).

After each erroneous diagnosis the weights of the relevance matrix are changed. In summary, the measure sim (and therefore the relation $S(x, y, z)$ has been built up in two steps:
a) The first approximation is done by modifying the measure sim_{PAT} using knowledge about the classification task.
b) The result of a) is the starting point for an adaptive learning process where only the success in the classification task plays a role.

Acknowledgement: The author thanks Stefan Weß for helpful discussions and the referee for useful remarks.

References

ALTHOFF, K.-D. (1992), Lernen aus Fallbeispielen zur Diagnose technischer Systeme, in: Doctoral Dissertation, University of Kaiserslautern.

ARROW, K.J. (1963), Social Choice and Individual Values, New York.

RICHTER, M.M., WEß, S. (1991), Similarity, Uncertainty and Case-Based Reasoning in PATDEX, in: *Automated Reasoning, Essays in Honor of Woody Bledsoe* (ed. R.S. Boyer), Kluwer Acad. Publ..

RUMELHART, D.E., ZIPSER, D. (1985), Feature Discovery by Competitive Learning, *Cognitive Science* 9, 75-112.

TVERSKY, A. (1977), Features of Similarity, *Psychological Review* 84, p.327-352.

WAGENER, M. (1983), Kardinalität in der Nutzentheorie, *Mathematical Systems in Economy* 81.

WEß, S. (1990), PATDEX/2: Ein System zum adaptiven, fallfokussierenden Lernen in technischen Diagnosesituationen. Masters Thesis, Kaiserslautern.

WEß, S. (1991), PATDEX/2 - Ein System zum fallfokusierenden Lernen in Technischen Diagnose Situationen, Seki-Report SWP-01-91

WEß, S., JANETZKO, D., MELIS, E. (1992), Goal-Driven-Similarity Assessment, Seki-Report, Universität Kaiserslautern.

Context Sensitive Knowledge Processing

Klaus Witulski

Chair for Systems Theory and Systems Engineering

Department of Urban Planning, University of Dortmund

Box 500 500, D-4600 Dortmund 50

Abstract: In the technical, legal, administrative and especially in the planning domain we often face the problem to distinguish between members of a class. Concepts as ideas underlying a class of things at the same time are our substantial elements for building knowledge bases. Especially rules are made up from concepts and expressions, that are connected by the logical functors. As any concept of a rule may stand for a whole class of elements, but in a given situational context only a selected set of combinations of the components of the concept forms valid examples of its intension, extension of knowledge based systems by facilities that allow processing of contextual information is desirable.

We have developed a model that enables us to represent contextuation in knowledge bases of hypertext systems and expert systems. The system is part of a shell that is implemented in PROLOG. The features supplied are context sensitive inference and context sensitive browsing. Although our proposal of context processing is not application specific - it relies mainly on classical t valued logic - especially planning tasks where legal prescriptions rule classes of problems seem to be able to benefit from it. It turns out that context sensitive knowledge processing can be regarded as a serious and viable alternative to the attempt to model 'vague' knowledge by means of fuzzy logic.

1 Introduction

The influence of an existing situation on the interpretation of a logical formula today seems to be a concern of many domains. In the field of legal knowledge processing, for instance, isolation of concepts from their context necessitates reasonable efforts to describe even simple cases [Fied84]. Many linguistic studies are devoted to a precise definition of operation and meaning of context in language processing [Hall89][Givo89]. Social scientists emphasize that the mediation of hermeneutic knowledge always occurs through pre- understanding stemming from the interpreters initial situation [Albr85]. In urban planning [Fore89] outlines the context dependency of practical strategies and plans. In computer science especially artificial intelligence research is concerned with the influence of contexts on knowledge processing. However, while professionals concerned with theorem proving tend to present their work on context logic in a form that makes transition to application problems quite difficult [Ohlb89], hypertext research still seems to be lacking a consential view to the representation and interpretation of contextual information [HYPE87][Barr88][Pars89][Gloo90][Jone91]. Our aim here is to clarify the notion of context and its interpretation in a way that is independen of a specific domain and implementation and thus can be applied to any of the fields mentioned, can be easily understood and handeled and at the same time is suitable to support knowledge processing in hypertext systems and rule based expert systems.

In the sequel we describe our approach to context oriented knowledge processing. Context oriented knowledge processing comprises the correspondent representation, a context sensitive interpreter working on it, a context sensitive hypertext browser which will turn out to be a specialisation of the context interpreter as well as the appropriate context sensitive update module in order to accomplish changes of the knowledge base in a way that allows our specialized interpreters to make use of contextuation. In this paper we focus on the aspects of the hypertext browser and interpreter, but all of these components have been implemented in Clocksin-Mellish compatible PROLOG as an extension to the existing knowledge based environment DOFLEX. The latter has been developed until 1989 [Witu89] as an expert system shell to support planning tasks by processing frequently changing prescriptions, but has by now been extended to make possible explicit modelling of the contextual aspect of knowledge.

2 Representation

The design of our context sensitive representation is guided by some assumptions as follow:

- Contextuation in principle is auxiliary or subsidiary. Contexts of rules are only created and maintained if necessary. A contextuated rulebase is an imperfect stadium of a knowledge base, a matter of temporary necessity and thus of only limited importance. The goal state is a uncontextutated set of PROLOG rules that can be run without a meta interpreter.

- Contextuation is only done on structures of atomic plans that contain disjunctions. If no OR-cases of atomic plans are in the base, then there cannot be any restrictive contextuation. This restrictive type of contextuation has to be distinguished from the why-context of a concept, which is not discussed in this paper.

- For contextuation only direct, one-level ancestors of any concept are taken into account. Longer sequences of contextual elements are not supported by us because of their bad ratio of cost as compared to benefit.

- Each atomic object of the base can become a potential context to any other, more specialized concept.

- Consideration of context entries during interpretation means, that the expansion of an OR-case of a rulehead may depend on the choice of the concept of the direct ancestor. In other words, to where we are specializing on expansion may depend on from where we come in terms of parent concepts.

For any contextuated OR-rule, one context may lead to a number of possibilities of expansion. Vice versa, an OR-case or possiblity of expansion may be reached from a number of parent-concepts or contexts. So, existence of a contextuation entry for a rulehead always limits or narrows down the possibilities of expansion amongst the OR-cases of the rulehead. In this sense, contextuated knowledge bases or plans are narrowed views to a problem domain.

Most of these features are reflected by Fig.1. Here C1 to C4 as well as A, A1-A4 and a-d all represent concepts, that can be thought of to be atomic plans, entries in a table of contents of a book as well as ruleheads of a knowledge-base represented in PROLOG and restricted to two valued classical logic. The different letter notation at the various levels of the example base has only been chosen to faciliate an easier discussion. Solid lines represent the AND/OR-net of the knowledge base in its uncontextuated appearance. So far this is also exactly the same representation as used in [Witu89].

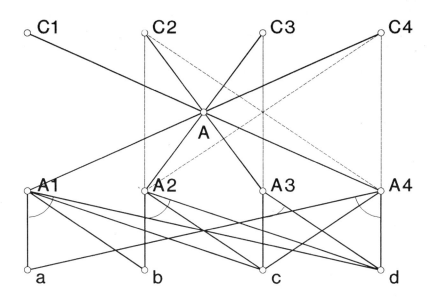

Figure 1: Contextuation in a Knowledge Base with two Levels of Specialisation

Dotted lines stand for contextuation. The knowledge structure or plan of the example comprises two levels of abstraction, which is the minimum number of levels in order to be able to do contextuation. Note, that only AND-expansion in our representation forms a level on its own. The OR-differentiation of A to OR-possibilities A1-A4 forms no change of level on its own, because actually each A, A1-A4 stands for one and the same object, atomic plan or concept or entry in a table of contents. The indices of A1-A4 are only inserted to name the different combinations of (a,b,c,d). Therefore, in the PROLOG representation of Fig.2 A1-A4 will all be represented by A. As usual, arcs mark the logical conjunction, for example (A :- a,b,c,d.), while their absence between expansion lines marks the logical disjunction (A:- a,b,c,d. A:- b,c,d.).

Because in the both figures A is the only concept that possesses at least one alternative or OR-case, it is the only concept that can be contextuated in this knowledge base. The possible contexts only one level upward are C1-C4. They can be thought of as atomic plans, points in time, persons, locations etc.. In dependency of if we consider A under the aspect or context of one of C1-C4, the extension or meaning or expansion of A may be different in each case. Seen from C1, all four interpretations of A are possible, since C1 does not impose contextual restrictions on the meaning of A. Thus this is the uncontextuated case of a concept, in which why-contexts may exist as superordinate, more general concepts, but these do not influence interpretation in a special way. In the example this means that C1 is a context under which all possibilities of interpretation of A are considered to be valid. Vice versa, interpretation of A1 is not appropriate in any of the restrictive explicit contexts of C2-C4.

The other extreme to this maximum of choices if no restrictive contextuation is imposed can be seen if A is regarded in the light or context of C3. Under the aspect of C3, only A3

```
rules('C1') :- rules('A'), !.
rules('C2') :- rules('A'), !.
rules('C3') :- rules('A'), !.
rules('C4') :- rules('A'), !.
rules('A') :- user(a), user(b), user(c), user(d), !.
rules('A') :- user(b), user(c), user(d), !.
rules('A') :- user(c), user(d), !.
rules('A') :- user(a), user(c), user(d), !.


r('C1',[]). r('C2',[]). r('C3',[]). r('C4',[]). r('A',['C1','C2','C3','C4']).


u(a,['A']). u(b,['A']). u(c,['A']). u(d,['A']).


c('C2','A',[],[b,c,d]).
c('C2','A',[],[a,c,d]).
c('C3','A',[],[c,d]).
c('C4','A',[],[b,c,d]).
c('C4','A',[],[a,c,d]).
```

Figure 2: PROLOG-Representation of the Contextuated Knowledge Base

is a correct, true expansion or extension of A, that is, A here only contains the conceptual components (c,d).

C2 limits interpretation of A to either A2 or A4. That is, within the context of C2 there is now an alternative interpretation of A possible by either (b,c,d) or (a,c,d). Further, A2 is a meaningful interpretation not only within the scope of the restrictive environment C2 but also in context C4. Moreover, in context C2 as well as in context C4, A has the same possible extensions.

Figure 2 gives the correspondent representation of this knowledge base or plan in PRO-LOG. Note, that predicates rules/1, r/2 and u/2 are identical in meaning and syntax to those described in [Witu89], where the rules entries index the rules by their ruleheads, r/2 indexes the goals of a rule to be able to efficently access the set of parent rules of a given rulehead and u/2 provides the same service in order to find the parents to a given leaf of a knowledge base. For contextuation, however, the c/4 entry is added to the PROLOG database. We use for representation of each restrictive context that explicitly guides interpretation one fact of c/4. The f argument contains the parent- or context concept, under which another successor concept that is said to be contextuated is considered. The second argument represents the concept or atomic plan to be contextuated. A c/4 entry exists only for those concepts that need contextuation. If no c/4 entry exists for a concept, its interpretation will not be dependent on or be restricted by any context. The third argument collects the goals of the contextuated concept, that have rules-type, that is, collects the conceptual components that are further specified within the scope of the knowledge base. The list of the fourth argument shows the user-goals of the contextuated concept, that is, of those concepts whose contents is no further differentiated in the knowledge base.

Read from left to right, the meaning of a c/4 entry is: A is contextuated. In the context C2 as one of its possible contexts, A is first expanded to no concepts that can be further expanded in this knowledge base and to concepts [b,c,d] that cannot be further expanded in this knowledge base or plan.

Note, that the one or several root nodes of the net structure of a knowledge base never possess a context within that same base. However, they themselves can be a context to the other more specialized objects. On the other hand, leaf nodes can be contextuated, but are no context to any node of the base, if the base is not, as a whole, circular. In such circles, root nodes and leaf nodes would have to be determined by decision. Of course, any node with the exception of the root nodes can be contextuated in the way we have shown.

We further view our representation as suitable to analyze the relationships between statistics and knowledge processing. Please imagine statistical treatment of a number of observations or cases as can be collected by a questionaire. We may be interested in how many percent of the cases have criteria a and b, how many b and c and so on. For example, assume that we are interested in stating that officials (=A) have two children (a) and mid-income (b) or three children (d) and high income (e). What we usually do in statistics is: we take from all our observations those whose profession is official and check, if the cases with two children as well as those with mid-income are the same. If we cannot exclude independency, dependency of the variables is often assumed and rules are formulated. Examples would be: 'official, if two children (60%)' or 'official, if mid income (80%)'. These would be fuzzy rules, that have to be always based on grounds on a big enough number of observations. If we cannot supply such a number of cases, our percentages are quite worthless, because they then may be purely subjective guesses, which as candidates for heuristics may even lack the foundation of sufficient experience.

Moreover, in statistical representation, the single case usually is pressed into a flat structure in form of a linear record. Opposed to this, in a knowledge based representation the single case can be modelled in a net structured, non flat manner. The identity or individual structure of the modelled object does not get lost, as contexts can be used to differentiate the members of the sample, which are corresponding to the OR-cases of A in the previous example. For each A, its properties can be known exactly at the individual level and structurally be represented in our contextual model, while the statistical treatment with statements about aggregates of the representative level tends to make the knowledge of the single individual case inaccessible.

In short, our representation allows the structured modelling of individual cases, retaining the contextual identity of the objects if desired by reflecting the knowledge of the single case and keeping it accessible. It allows, in difference to the statistical approach to knowledge handling, to select and view the structured single case in the framework of its restrictive context or contexts. Probability theory thus from our knowledge based point of view seems to operate on the OR-cases of a class of unstructured objects anonymously, while the knowledge based representation allows to see the knowledge, not the probabilities, of properties of the single structured object in its context.

Thus, by means of our contextual representation, we might also have pointed to a conceptual, less statistical view to a theory of fuzzy conceptual objects v contextuation of rules. For us, a fuzzy concept can be free of statistical confidences, but a contextuated one, containing different extensions in dependency of the contextual view imposed on it. Thus, fuzzy logic can possibly be redefined at the level of the non aggregated individual in terms of context logic - that is, at the level of the individual case - in a more prec manner than the statistical view allows for.

3 Context Sensitive Hypertext Browser

Considering the logical formalism or representation of knowledge, the knowledge base of an expert system as well as a hypertext as well as a plan may share one identical representation. We assume that our framework of Figure 2 can integrate each of the three system aspects [Witu90][Witu91]; additionally it incorporates the contextuation quality. A hypertext browser for us is a means of associative access to such a knowledge base in a step-by-step manner [Barr88][Witu92a][Witu92b]. Contextuation serves the purpose to limit the possibilities a concept can be extensionally expanded to by making use of the situational information of the environment from where we have come.

A hypertext browser basically is a software tool that faciliates the transition from a start concept to a goal concept. Only, if the goal concept is disjunctive a context sensitive browser can come into operation and effect. Its task is to show such an element of the disjunctive concept first whose extension is appropriate in the existing immediate context formed by the parent concept from where we came.

We define a context sensitive hypertext browser as a hypertext browser that, on the walk from a start concept, first expands to a context correct alternative of the goal concept. Such a browser helps the user to distinguish context correct extensions from those that are irrelevant for the context chosen on the hypertext walk. The hypertext browser of DOFLEX has been designed and extended in order to allow for context sensitive positioning on a transition from one hypertext state to the next as well as to show and distinguish context correct possibilities of expansion of a disjunctive rule from the other disjunctions of the goal concept.

```
walkFromEntry(StartHead) :-
      collectOrRules(StartHead,BodyList),
      length(BodyList,Len),
      Len > 1, !,
      ((tryToFindAContextCorrectOrCase(StartHead,BodyList,
        CaseNrForGivenContext), !,
        walkOrRules(StartHead,BodyList,CaseNrForGivenContext));
       (!, walkOrRules(StartHead,BodyList,1))).
```

Figure 3: Integration of Context Sensitive Case Positioning in walkFromEntry

Figure 3 shows the context sensitive part of the DOFLEX-predicate walkFromEntry, whose task it is to perform one transition or browser step from a given concept or state to the goal concept of StartHead. As a restrictive context differentiation is only meaningful if OR-cases exist for the concept or goal of the one step walk, first the possible expansions are collected and checked for a length of the resulting list greater than one. If the goal concept is no disjunction, backtracking occurs to another walkFromEntry alternative performing the usual, context independent transition. Else the next step is to find amongst the OR-cases the position of the first context satisficing or context correct OR-case. If the goal concept is not contextuated, the predicate tryToFindAContextCorrectOrCase whose code is shown in Fig.4 fails at clause/2, and the subsequent alternative within walkFromEntry initiates the further walk, positioning on the first OR-rule of the goal concept as a default. In the

```
tryToFindAContextCorrectOrCase(StartHead,BodyList,StationHeads,
                               CaseNrForGivenContext) :-
    rev(StationHeads,[LastStationHead|_]),
    clause((c(LastStationHead,StartHead,RGoals,UGoals)),_), !,
    append(RGoals,UGoals,RU1Goals),
    quiSort(RU1Goals,SortedRU1Goals),
    countOrNumberForThisContext(BodyList,SortedRU1Goals,
                                CaseNrForGivenContext).
```

Figure 4: PROLOG Implementation of tryToFindAContextCorrectOrCase

uncontextuated alternative, the first OR-case of a rule amongst the others in DOFLEX is displayed first, as it is considered to be the most important one.

On the other hand, if tryToFindAContextCorrectOrCase succeeds in finding a context entry for the goal concept - that is, if an entry c(AContext,StartHead,RulesGoals,UserGoals) exists - we know by this that we are able to position on a context correct OR-case. Now we only have to find the position of this first context correct OR-case in the sequence of OR-cases of the goal-concept. We use the user- and rules goals which we received as a result of our call to c(AContext,StartHead,RulesGoals, UserGoals) to determine the position. The predicate countOrNumberForThisContext accomplishes this task by comparing from the first to the last of the OR-cases of the goal-concept the rules- and user goals with the rules- and user goals of the contextuated entry, incrementing at each attempt of comparison by one until the position number is found where both sorted lists show the same elements. This number is used by walkOrRules to show the user the first context correct rule, if any exists.

Example 1: Consider Fig.1. We walk from C1 to A. A is contextuated, but no context entry C1 exists for A. So we are positioned at the first OR-case of A, which is A1.

Example 2: We are at C4 and want to walk to A. Amongst the OR-cases of A, the first context entry of C4 points to the rules goals [] and the user goals [b,c,d]. This combination is compared to the sequence of rules/1 entries of A in the knowledge base, and the position of the case A2 is determined to be 2. As Fig.5 shows, DOFLEX-walkFromEntry thus positions directly to A2, expanding A to ([],[b,c,d]). Now further walks are possible amongst the OR-cases of A as well as a backwalk to C4 or upwalks to C1, C2, C3. As (a,b,c,d) are all leaves of the knowledge base, no walk is possible to them, unless they would be expanded by updating the knowledge base of the plan.

4 Context Sensitive Interpreter

While in hypertext systems browsing is the predominant way to retrieve information, expert systems use systematic search strategies. For example, PROLOG interpreters as inference engines of expert systems run left-right depth-first search with backtracking. As we do not feel entitled to write PROLOG interpreters ourselves, our approach to a context sensivite interpreter is that to specify it as a front-end to PROLOG, that is, as a meta interpreter. As we shall see, a context sensitive interpreter working with the PROLOG search strategy as shown in Fig.6 necessitates only slight modifications as compared to the context sensitive hypertext browser we have already discussed.

```
----------------------------------------------------------------
C4
----------------------------------------------------------------
C4 :-
   1 A
----------------------------------------------------------------
q,1-n,u,b,m,e,i,c,s,d,o,r,p,pd,t,so,l,g,mv,ld,lr,lc,dt,lt,h       **1

------------------------------------------------------ 2/4 OR - Rules
C4
A
----------------------------------------------------------------
   1  C1
   2* C2
   3  C3
   4* C4
A :-
     b
     c
     d
LEAF OF TREE --------------------------------------------- FLEX 260991
q,<,>,u,b,m,e,i,c,s,d,o,r,p,pd,t,so,l,g,mv,sq,ld,lr,lc,dt,lt,h   **q
```

Figure 5: One Step of Context Sensitive Browsing Using DOFLEX

In principle, the part of software that makes the meta interpreter a context sensitive one is implemented as a filter, which is pointed out in Fig.6 as well. If the concept to be expanded turns out to be uncontextuated, the filter behaves as if it were not present at all, allowing expansion of all OR-cases of the concept if forced by backtracking. If the concept to be interpreted is actually restrictively contextuated, the filter onl allows expansion of those OR-cases of the concept that fit the given immediate context. Given the immediate context as well as the rules- and user-goals to be expanded, if the concept in question possesses any context, the rules- and user-goals to be expanded have to be amongst the entries of contexts. If not so, then ifRuleContextuatedFilterContextFittingGoals causes backtracking to clause/2 in order to generate and test another possibility for agreement with the immediate context in question.

In order to determine the concept that forms the immediate context, the recursive predicate execARule/6 maintains a LineOfReasoning as path from the goal of the consultation to the concept whose context correct expansion is attempted. The immediate context is found as the last element of the LineOfReasoning. Note that the whole line of reasoning at any time of interpretation contains the correct why-context from the concept to be expanded up to the level of the consultation goal in full length. This line of reasoning is prolongated by the concept to be expanded as soon as a context correct OR-case has been encountered via the backtracking attempts at clause/2.

Backtracking may be forced if the user fails a user-goal of the tree that is to be inter-

```
execARule(RuleHead,ActualDepth,MaxDepth,ForwardFrom,ForwardTo,
        LineOfReasoning) :-
    clause((rules(RuleHead)),Body),
    scan(Body,Rules,Users,_),
    ifRuleContextuatedFilterContextFittingGoals(LineOfReasoning,
                                        RuleHead,Rules,Users),
    append(LineOfReasoning,[RuleHead],LongerLineOfReasoning),
    execGoals(Rules,LongerLineOfReasoning),
    execUsers(Users,LongerLineOfReasoning), !.

ifRuleContextuatedFilterContextFittingGoals(_,RuleHead,_,_) :-
    not(clause((c(_,RuleHead,_,_)),_)), !.
ifRuleContextuatedFilterContextFittingGoals(LineOfReasoning,RuleHead,
                                        Rules,Users) :-
    getLastElemOfList(LineOfReasoning,ParentOfRule), !,
    clause((c(ParentOfRule,RuleHead,RGoalsOfThisContext,
            UGoalsOfThisContext)),_),
    checkTwoRuleBodiesEquality(Rules,Users,RGoalsOfThisContext,
                            UGoalsOfThisContext).
```

Figure 6: Implementation of Context Sensitive Interpretation in DOFLEX

preted. Rules-goals themselves can only fail indirectly by means of failed user goals, as the representation of the knowledge base and its maintanence in any case guarantee at least one possibility of expansion for any concept of rules-type. The clause/2 predicate is the only one that provides points of choice in this context sensitive interpreter. It generates the possibilities that are to pass the check for context correctness as well as further possibilities if the user failed a leaf node of the overall AND-OR consultation tree.

Unlike with the context-browser routine, where we first checked that at least one OR-case exists for the concept to be expanded and then we tried to determine the number of the first context correct OR-case, in this context sensitive interpreter we implicitly make use of the assumption that only OR-rules can be contextuated by c/4 entries and that the contextuated knowledge-base at any time consistently reflects this design issue.

Example 1: If in Fig.1 we would have chosen C1 as a consultation goal, the following consultation would be successful: a:no causes backtracking to A2. b:no causes backtracking to A3. c:yes, d:yes satisfy A by A3 and thus C1 is true. As C1 imposes no restrictions on - or, in other words - is no restrictive context to any object of the knowledge base, in this example only the bypassing of the context filter can be observed.

Example 2: In the next example, however, the consultation goal C2 is used as a context that limits possibilities of expansion: b:no disables A2, but then a,c,d:yes satisfy A4 and thus A and the consultation goal C2. Note, that expansions of A1 and A3 are blocked by the predicate ifRuleContextuatedFilterContextFittingGoals, because - seen from the point of view of C2 - A is restrictively contextuated and only those expansions of A2 and A4 can be considered for backtracking with the possibility to satisfy the context of C2 at the same time.

5 Conclusions

Context sensitive knowledge processing is considered to be an extension of knowledge processing that itself relies largely on classical two valued logic. I value is seen in the possibility to model knowledge that still is in its pre- consential stage or - as it is the case with homonyms - cannot be modelled adequately by classical two valued logic.

We have proposed a way to express a contextual reserve of the subjective perspective via knowledge structures of plans, political programs, software programs and presumably in general of any logical structure of concepts, expressions, propositions or objects. The model seems to be compatible with the basic principles of representation and operation of current knowledge based systems, especially of expert systems and hypertext systems.

Contextuated knowledge structures can model intersubjective disparities of understanding, vage knowledge as well as a certain kind of fuzzyness of rules: all of these items are estimated to import the aspect of inconsistency or open- endedness into pure classical logic, aiming to widen its range of pragmatic application in doing so. To be able to experiment with such a kind of weakened knowledge structures, a hypertext browser and an interpreter that work on this type of pragmatic knowledge representation have each been specified.

However, contextuated knowledge lacks generality in principle because of the extreme subjectivity that accompanies the reservation of the singular contextual environment. Thus, for intersubjective planning as well as for other tasks of consential classification a common context for the coordinating plan, as it is given in legal regulations in an idealized form, has to be achieved. Thus, contextuated knowledge is considered to be knowledge in its early, hypothetical state.

References

[Albr85]   J. Albrecht, Planning as Social Process, Verlag Peter Lang, Frankfurt, 1985

[Barr88]   E. Barrett: Text, Context and Hypertext: writing with and for the computer, MIT Press, Cambridge, 1988

[Givo89]   T. Givon: Mind, Code and Context, Erlbaum, Hillsdale, 1989

[Fied84]   H. Fiedler, T. Barthel, G. Voogd: Untersuchungen zur Formalisierung im Recht als Beitrag zur Grundlagenforschung juristischer Datenverarbeitung, Westdeutscher Verlag, Opladen, 1984

[Fore89]   J. Forester: Planning in the Face of Power, University of California Press, Berkeley, 1989

[Gloo90]   P.A. Gloor, N.A. Streitz (ed.): Hypertext und Hypermedia, Informatik Fachbericht Nr.249, Springer Verlag, Berlin-Heidelberg-New York, 1990

[Hall89]   M.A. Halliday, R. Hasan: Language, context, and text: aspects of language in a social-semiotic perspective, Oxford University Press, Oxford, 1989

[HYPE87]   Hypertext '87: Proceedings of the Hypertext Workshop 1987 November 13-15 in Chapel Hill, North Carolina, Association for Computing Machinery, New York, 1989

[Jone91]   S. Jones: Text and Context, Springer Verlag, Berlin, 1991

[Ohlb89]   H.J. Ohlbach: Context Logic, SEKI Report, Universitaet Kaiserslautern, Kaiserslautern, 1989

[Pars89]   K. Parsaye, M. Chignell, S. Khoshafian, H. Wong: Intelligent Databases, Wiley, New York, 1989

[Witu89]   K. Witulski: Entwicklung einer Expertensystemumgebung zur Verarbeitung sich häufig ändernder Vorschriften, DUV, Wiesbaden, 1989

[Witu90]   K. Witulski: Hypertextsysteme und Expertensysteme - eine Methode systematischer Vorschriftenverarbeitung, in: W. Abramowicz, A. Ultsch (ed.): Proceedings des Internationalen Workshops "Integrated Intelligent Information Systems" der Akademie fuer Oekonomie Poznan / Polen, der Universitaet Dortmund Fachbereich Informatik, Lehrstuhl VI, und der Universitaet Oldenburg, Poznan, 1990

[Witu91]   K. Witulski: Integrating Hypertext System, Expert System and Legal Document Database, in: Proceedings der Dortmunder Expertensystemtage '91 in Verbindung mit dem 3. Anwenderforum Expertensysteme des KI-Verbundes NRW, Verlag TUV Rheinland, Koeln, 1991

[Witu92a]   K. Witulski: Improving Access to Legal Information in Document Databases Exploiting Techniques of Knowledge Based Systems, in: M. Schader (ed.), Analyzing and Modeling Data and Knowledge, Proceedings of the 15. Annual Conference GfKl, Springer Verlag, Berlin, 1992

[Witu92b]   K. Witulski: Knowledge Based Route Selection in Public Transport, Second OECD Workshop on Knowledge-Based Expert Systems in Transportation June 15-17 1992, Montreal (submitted)

An Efficient Application of a Rule-Based System

R. Kiel

Institut für Informatik, Universität der Bundeswehr Hamburg
Holstenhofweg 85, 2000 Hamburg 70, GERMANY

Abstract: In WIMDAS, a system for knowledge-based analysis of marketing data, two formalisms are used to express domain knowledge. Both user and knowledge engineer communicate with the system by way of an external knowledge representation, for instance the knowledge engineer uses it to formulate the rules of the WIMDAS knowledge base. WIMDAS' inference engine uses an internal knowledge representation. The connection between internal and external knowledge representation is established by means of the maintenance module which applies several transformations to the set of rules to increase the efficiency of consultations of WIMDAS's knowledge base. This paper discusses some of these transformations.

1   Introduction

WIMDAS (**W**issensbasiertes **M**arketing-**D**atenanalysesystem) is a knowledge-based system to support data analysis in marketing, which is jointly developed by research groups in Hamburg, Karlsruhe and Mannheim (see Hantelmann (1991), Kiel, Schader (1992b) Baier, Marx (1992) and Marx, Rundshagen (1992) for recent reports.) The current version of WIMDAS is designed as a distributed system consisting of several components, which communicate among each other using the UNIX socket mechanism and/or RPC-calls.

A *graphical user interface* manages the entire interaction of the user with the system, providing an easy to use tool to select input data and goals for the analysis to be performed. Based on the input of the user the *knowledge-based component* proposes the application of several algorithms, which are generally organized like a network where the result of one algorithms may serve as the input of another. When the user decides to perform such a proposed analysis the *analysis component** is activated which gets the input data to be used from the *data management component*. This component provides a secure and efficient way of storage and manipulation of raw and calculated data matrices. (A detailed discussion of this component can be found in Marx, Schader (1991) and Baier, Marx (1991).) Finally, the results of an analysis can be visualized by the *graphical subsystem*. (For details see Hantelmann (1990) and Hantelmann (1991b).)

Since the user conducts the session with WIMDAS by the graphical user interface, this component can be seen as a central control of the entire system, but note that the data transmission between two components is performed directly and does not involve the user interface.

In WIMDAS we intent to support the ordinary user (normally an expert for the application of data analysis methods) of the system in the task of enlarging and maintaining the

---

Research for this paper was supported by the Deutsche Forschungsgemeinschaft

*In earlier papers about WIMDAS this component was called the *method component*. Because the demands made on this component changed rapidly, we have decided also to change the name of this component. For example, the old method component used only one computer to run a data analysis procedure, whereas now the analysis component distributes this task over a computer network. (An object-oriented analysis of this component is represented in Marx, Rundshagen (1992).)

knowledge base. Because this user is not an expert in encoding the knowledge base of an AI-system one needs a tool to assist in this work. From the same reason it follows that this tool has to provide an easily understandable and human readable formalism to represent the knowledge. Furthermore it must check new rules against existing ones, e.g., it has to search for inconsistencies in the rule base (see Kiel, Schader (1990, 1991)).

Unfortunately, an easily handable knowledge representation formalism will in general not be an equally good way to represent the knowledge in a rule base for automatic reasoning by a computer, e.g. with respect to runtime behavior. To match the requirements of both man and machine one needs a human readable form but has to encode knowledge in an efficient and compact way, optimized for the machine. In WIMDAS we have therefore decided to use two distinct knowledge representations. The *external* knowledge representation is used in the dialog between the user and the system for instance by the explanation module or the maintenance module of the knowledge-based component. Otherwise an *internal* knowledge representation is used for instance in the inference engine, the consistency checker and also by the explanation module, if it analyzes the work of the inference engine. (A detailed description of the external knowledge representation can be found in Kiel, Schader (1992b).) On entry of new knowledge specified in the external representation, WIMDAS's maintenance module of the knowledge-based component translates them into the internal representation. (First concepts, how this module will handle inconsistencies and how to enter a description of the set of correct matrix characterisations into the knowledge base are discussed in Kiel, Schader (1990, 1991) and Kiel, Schader (1992a, 1992c), respectively.)

In the reminder of this paper we discuss some of the transformations which are performed during the mapping from the external to the internal representation, in order to make the knowledge better usable by WIMDAS' inference engine. To show concepts and problems more clearly we use the formalisms of a prototype of the knowledge-based component, the so called *proposal generator* (its rule formalism was introduced in Baier, Gaul (1990) and improved in Kiel (1992)). Note, that the implementation of the knowledge-based component has changed in the meantime.

2 Rules in the Proposal Generator

For every data analysis method supported by WIMDAS the *knowledge-based component* contains at least one rule describing to which type of data this method can be applied and which kind of data the method produces. To this purpose data matrices in WIMDAS are characterized by attributes like 'Data type', 'Relation type', 'Scale type', Way1, Way2,.... This characterization of data matrices is described, e.g., in Arabie, Hubert (1992) and Baier, Gaul (1990) and is based on a suggestion given in Carroll, Arabie (1980). Every raw data matrix has to be characterized in such a way and is stored along with this information by the data management component (see Baier, Marx (1992) for a discussion).

How to apply the (exemplary) method method-1 to construct a new matrix from two given matrices, can informally be told by the following rule:

```
If you have a first matrix characterized by
                Data type:      Representation,
                Scale type:     Absolute,
                Relation type:  Pair,
                Way 1:          {<Mode1>,<Mode4>}
```

```
             Way 2:          <Mode3>,
     and a second matrix characterized by
             Data type:      Preference,
             Scale type:     Absolute,
             Scale min:      0,
             Scale max:      unlimited,
             Relation type: Pair,
             Way 1:          <Mode1>,
             Way 2:          <Mode2>,
             Way 3:          <Mode2>,
     then you can apply method-1
     generating a matrix characterized by
             Data type:      Representation,
             Scale type:     Absolute,
             Relation type: Pair,
             Way 1:          {<Mode1>,<Mode2>,<Mode4>},
             Way 2:          <Mode3>.
```

In the above rule `<Mode1>`, `<Mode2>`, `<Mode3>` and `<Mode4>` stand for variables. In the proposal generator such a rule is encoded by a Prolog clause of the form

```
data([method-1, Description1, Description2],...)
  :- data(Description1,...),
     data(Description2,...).
```

The parameters of the data predicates that are omitted in this example are used to characterize the type of data matrices in question, they correspond to the entries of `Data type`, `Scale type`,...of the above stated informal rule. The first argument of the `data`-predicates (the *description argument*) tells us, -if it is instantiated-, how to construct the matrix which is described by this `data`-literal.

In the example above we assume that the variables `Description1` and `Description2` are instantiated to descriptions telling us how to generate matrices which match the characterizations given by the first and second `data`-literal. Such a description could for example be simply a reference to a data matrix already stored by the data management component or an instruction how to compute a data matrix of this type by applying various methods to given matrices.

In the given situation the first argument of the rule's conclusion states that one has to apply `method-1` or `method-1-2` to the matrices whose constructions are given by `Description1` and `Description2` to get a matrix of the type of the conclusion.

The knowledge base of the proposal generator consists of a set of rules of the following scheme:

```
data(...) :- data(...),...,data(...).
```

For each method we have at least one such rule. The number of conjuncts of a rule's premise corresponds to the number of matrices the described method needs as input data.

To reach a proposal for analysis of given data matrices the proposal generator uses the ordinary Prolog inference strategy which has several drawbacks. A lack of this strategy is the absence of any mechanism to avoid unnecessary inferences, such as computing identical parts of a solution more than once. For instance, if the proposal generator fails to find a solution to a subproblem during a consultation, and if it has to solve an identical subproblem later on, it tries again to find a solution. It would be better to have the possibility to reuse an already found solution instead of recomputing it every time it is used.

3 Transformations Increasing Efficiency

It may be impossible to avoid all redundant computations during a consultation of a proposal generator but we will try to avoid as much as possible of unnecessary effort. To this aim the system's knowledge base is reorganized automatically by WIMDAS' maintenance module by applying several transformations. These transformations can be divided in two groups. The first group contains all specific transformations which can only be applied for the specific WIMDAS rule base and the other group contains transformations which may be applicable to other rule bases as well.

3.1 A General Transformation

For functional logic programs (i.e. programs which do not produce more than one distinct output for a single input) Debray, Warren (1986) and Nakamura (1986) developed optimizing strategies. Since we generally can infer different proposals for analyzing a given set of data matrices from the proposal generator's knowledge base, these strategies cannot be applied to optimize our rule base. Therefore, we have to develop our own optimizing strategies.

The Basic Idea

To motivate the first transformation consider a rule base wich contains several rules with the same conclusion A and whose premises have an identical part B. In short we have n rules

$A :\text{-} B, C_1.$

$\vdots$

$A :\text{-} B, C_n.$

Now, assume that we want to infer A and the subgoal B can be established by a complex deduction. Furthermore, assume that the subgoals $C_1, \ldots, C_{n-1}$ cannot be inferred but the subgoal C_n can be deduced. To establish A, the Prolog interpreter tries the rules in turn, first using the rule $A :\text{-} B, C_1$. After a complex computation the result B is established but Prolog fails to deduce the result C_1. Therefore the next rules are tried and because they all fail except the last, the result B is n times deduced by the same complex computation.

To prevent this unnecessary computation of the subgoal B in the given situation the first idea is to swap the occurrences of the literals B and $C_1, \ldots, C_n$ in the rule set yielding:

$A :\text{-} C_1, B.$

$\vdots$

$A :\text{-} C_n, B.$

This remedies the above mentioned situation, but now assume that we are interested in all possibilities to infer A, for instance if A has to be deduced in the context of another inference process. In this *backtracking* case the Prolog interpreter will try all rules for the conclusion A in turn. Assume furthermore that all the literals $C_1,\ldots,C_{n-1}$ *can be inferred*, then the literal B will be deduced whenever a rule is used, since the first premises never fail. Consequently, we again end up by computing the subgoal B n times.

To avoid this unnecessary effort even in the backtracking case we split the deduction of A into a chain of two deduction steps by transforming the original rule set into

A :- B,C.
C :- C_1.
$\qquad\vdots$
C :- C_n.

To construct this new rule set, we have introduced the new literal C, which *must not* unify with any literal already occurring in the rule base we want to optimize.

In summary, we have reduced the n rules with conclusion A to only one by adding n less complex auxiliary rules for C. This idea is also applicable in much more complicated situations as we will see in the remainder of this section.

The General Case

In the section above we have not yet considered variables which may occur in the literals A, B and C_i of our original rule set. First, we will discuss a small example to show the different situations we have to take into account. Then we will present a general method for performing the transformation.

Consider the two rules

```
a(Y₁,Y₂,[Y₃,Y₄])
    :- b(Y₁,Y₅,[Y₃,Y₆]),
       c₁(Y₂,Y₅,[Y₃,Y₄,Y₆]).
```

and

```
a(Y₁,Y₂,[Y₃,Y₄])
    :- b(Y₁,Y₅,[Y₃,Y₆]),
       c₂(Y₁,[Y₄,Y₇,Y₇]).
```

To apply the above introduced transformation to these two rules we have to construct a suitable c-literal which respects the dependencies between the conclusions and their premises and also among the premises alone. Such a dependency exists whenever the same variable is used because they must be instantiated with the same value when the rule is used by the Prolog interpreter. The new rules we want to construct have to be of the form

```
a(Y₁,Y₂,[Y₃, Y₄])
    :- b(Y₁,Y₅,[Y₃,Y₆]),
       c(...).
```

```
c(...) :- c₁(Y₂, Y₅, [Y₃, Y₄, Y₆]).
```

and

$c(\ldots):- c_2(Y_1, [Y_4, Y_7, Y_7])$.

For the second rule of our transformed rule set the c-literal has to establish a correspondence of the variables Y_2, Y_3, Y_4, Y_5 and Y_6 occurring in this rule to the variables with identical names in the first rule of this rule set. The third rule of this rule set requires us to establish a correspondence from the variable Y_1 and Y_4 to the identically named variables in the first rule. This requirements can be satisfied if we transform these variables to parameters of the c-literal we want to construct. Finally we have the rule set:

$a(Y_1,Y_2,[Y_3,Y_4])$
    $:- b(Y_1,Y_5,[Y_3,Y_6])$,
        $c(Y_1,Y_2,Y_3,Y_4,Y_5,Y_6)$.

$c(Y_1,Y_2,Y_3,Y_4,Y_5,Y_6):- c_1(Y_2,Y_5,[Y_3,Y_4,Y_6])$.

and

$c(Y_1,Y_2,Y_3,Y_4,Y_5,Y_6):- c_2(Y_1,[Y_4,Y_7,Y_7])$.

To transform any given rule set the following algorithm can be applied. Since the identifier of variables in literals which are essentialy the same may not be identical (different programmers prefer different names) steps 1. and 2. are introduced in addition.

1. Take the rules

   $A_1 : -B_1, C_1.$
   $\vdots$
   $A_n : -B_n, C_n.$

   whose premises A_i and first literals B_i are variants of each other from the data base to be transformed.

2. Rename the variables occurring in these rules such that they have identical conclusions A and first literals B in their premisses. This modified rule set is denoted by R. It consists of the n rules

   $A: -B, C_1'.$
   $\vdots$
   $A: -B, C_n'.$

3. Compute the parameters of the newly to be constructed c-literal. This variable set V is given by

   $$V = \{X_1,\ldots,X_k\} = (var(A) \cup var(B)) \cap \bigcup_{i=1}^{n} var(C_i'),$$

   where $var(\ldots)$ denotes the set of free variables occurring in a literal.

4. Add the new rules

$$A: -B, c(X_1, \ldots, X_k).$$
$$c(X_1, \ldots, X_k): -C'_1.$$
$$\vdots$$
$$c(X_1, \ldots, X_k): -C'_n.$$

to the rule set R.

5. Repeat this construction until the data base do not contain any two rules for which the condition of the first step holds.

The discussed transformation method can be easily extended to rules with more than two literals in their premises. We collect the literals of each premise in two conjunctions in which one consists of the literals which are equal in all rules' premises and the other consists of the literals which are different. The first corresponds to the literal B and the other to the literals $C_1, \ldots, C_n$ and we can, again, apply the above transformation.

Application to the Proposal Generator

At the first glance, the rule set of the proposal generator does not seem to be a candidate for this transformation because the description arguments of any two rules are always different. But the idea is applicable in a modified form. The situation we have to handle can be illustrated by the following rule set.

```
data([m₁, D₁, D₂],...)
    :-data(D₁,...),
      data(D₂,...).
    ⋮
data([mₙ, D₁, D₂],...)
    :-data(D₁,...),
      data(D₂,...).
```

This set describes n different data analysis methods $m_1, \ldots, m_n$ producing matrices with identical characterization. The first literals of the premises are the same and -with respect to the data characterization- the conclusions are the same for each rule. The second literals in the premises of these rules may be different.

First, we ignore the different methods in the description argument of the conclusions of our rules by substituting the methods by a new variable M. Then we can apply the above introduced transformation to produce the following rule set:

```
data([M, D₁, D₂],...,)
    :-data(D₁,...),
      data2(D₂,X₁,...,Xₖ).
data2(D₁,X₁,...,Xₖ) :- data(D₁,...).
    ⋮
data2(D₂,X₁,...,Xₖ) :- data(D₂,...).
```

In the original rule set, the methods m_i in the conclusions depend only on the second literals of these rules. In the new rule set we can therefore instantiate the variable M by matching the data2-literal of the first rule's premise with one of the data2-conclusions of the other rules. Consequently the variable M becomes an additional parameter of the data2-predicate in the first rule's premisse and the methods m_i become additional parameters in the data2-conclusion of the other rules. Finally, we have the following rules

```
data([M, D₁, D₂],...,)
    :-data(D₁,...),
       data2(M,D₂,X₁,...,Xₖ).
data2(m₁,D₁,X₁,...,Xₖ)  :- data(D₁,...).
    ⋮
data2(mₙ, D₂,X₁,...,Xₖ)  :- data(D₂,...).
```

3.2   A Special Transformation

This section covers a problem which depends strongly on the special task for which the WIMDAS knowledge based component is designed for but similar situations may occurr during the development of rule bases for other systems. It also shows that under certain circumstances it may be necessary to modify the scope of duties for a rule based system because the originally given task is too complex to handle.

As mentioned in the introduction, the knowledge based component of WIMDAS has to produce proposals on the application of analysis methods supported by WIMDAS' analysis component to a given problem. As the methods avaible are arranged in form of a net a proposal can be seen as an bipartite directed graph $P = (K, E)$. Where the set of nodes K consists of the set of data matrices D and the set of applied methods M, i.e. $K = D \cup M$. For the set of the edges E we have $E \subset D \times M \cup M \times D$. (There is no edge between two methods or two data matrices.) Herein methods are only inner nodes, the source nodes correspond to the input data and the sinks are the results.

In the proposal generator every $m \in M$ describes, in fact, exactly one algorithm. During developing the knowledge base we encountered that many rules differ only in the first parameter of their conclusions or, in other words, they describe methods which have exactly the same input/output behavior with respect to WIMDAS' data characterization. This property results a combinatorial explosion when the rule base is consulted. As an illustration consider a proposal which uses seven methods ($|M| = 7$). If for each of these methods the rule base contains only one additional method with the same input/output behavior, the proposal generator would produce $2^7 = 128$ different proposals.

Under these circumstances we have decided to change the knowledge representation and the form of results which are produced by the knowledge-based component. Now, different methods which can be described by rules which differ only in the first argument of their conclusions are described by a single rule. For a specific inquiry the old version of the proposal generator crashes after producing about 500 proposals, because memory is exhausted. For an analogous inquiry the new version of this component produces 14 proposals.

Consequently, the nodes $m \in M$ in a graph representing a proposal of the knowledge–based component now describe classes of methods instead of one single method. The proposals which are described by such a graph are called *conceptual proposals*. Before the analysis component can start a data analysis process based on such a proposal, a *concrete proposal* whose nodes describe exactly one data analysis method has to be constructed from the original conceptual proposal. One possibility to perform this task is to ask the user to choose one method from every method class in the conceptual proposal containing more than one methods. Since the method classes in a conceptual proposal are generally not independent from each other, the system must take care that the user is only permitted to choose admissible combinations of methods during transforming a conceptual into a concrete proposal. (Detail of this construction are discussed in Kiel (1992).)

Because we cannot expect that a person who maintains the WIMDAS knowledge base, knows which methods can be described by a single rule, the maintenance module has to support him in doing this work. The maintenance module can automatically find rules which are identical with exception of the first arguments of these rules conclusions. But, to construct a single rule describing how to apply these methods the maintenance module has to know a general name for those methods. If this name is entered by the user then the resulting rule can be automatically constructed.

4   Outlook

In the future, we will elaborate additional strategies for increasing the efficiency of consulting the WIMDAS knowledge base. For example, we intend to explore whether it is advantageous to use classification methods to examine the structure of WIMDAS' knowledge base. Such information could possibly be used to formulate the rules with different predicates instead of using only the data-predicate. Then it would not be neccessary to test all rules in the knowledge base to find a rule firing in a given situation.

During a consultation WIMDAS' knowledge–based component produces more than only one proposal for the analysis of a set of given data matrices. Therefore, it may be useful to store partial proposals on one hand and to store and mark not solvable subgoals on the other hand in order to avoid recomputing identical parts in different proposals. This information may not only be used for making the inference process more efficient but also to explain to the user why a solution could not be found.

The implementation of this idea is not as straightforward as it may seem. Since it is impossible to store all of those partial solutions and unsolvable subgoals we have to develop criteria helping us to select those which should be stored. After solving this more theoretical problem, we have to integrate this concepts into the inference engine of WIMDAS knowledge based component or into the maintenance module for WIMDAS' knowledge base, respectively.

References

ARABIE, P., and HUBERT, J.L. (1992), Combinatorial data analysis, to appear in: *1992 Annual Review of Psychology.*

BAIER, D., and GAUL, W. (1990), Computer-assisted market research and marketing enriched by capabilities of knowledge-based systems, in: *New Ways in Marketing and Market Research, EMAC/ESOMAR Symposium, Athens,*202–226.

BAIER, D., and MARX, S. (1992), Data management in a knowledge–based system for marketing research, in: M. Schader, (ed.), *Analizing and Modeling Data and Knowledge,* Springer-Verlag, Berlin, pp. 189–197.

CARROLL, J. D., and ARABIE, P. (1980), Multidimensional scaling, in: *Annual Review of Psychology 31*, 607–649.

DEBRAY, S. K., and WARREN, D. S. (1986), Detection and optimization of functional computations in prolog, in: *Third International Conference on Logic Programming*, 490–504.

HANTELMANN, F. (1990), Grafische Präsentation statistischer Daten unter UNIX: Ein Konzept zur Gestaltung von Grafiksubsystemen für KI-Umgebunge, in: *Operations Research Proceedings 1989*, Springer-Verlag, Heidelberg.

HANTELMANN, F. (1991a), Applikations-Schnittstellen für KI-Umgebungen am Beispiel des WIMDAS-Projekts, in: *Proceedings of the DGOR Annual Conference*, Springer-Verlag Berlin, Heidelberg, New York.

HANTELMANN, F. (1991b), Graphical data analysis in a distributed system. in: E. Diday and Y. Lechevallier (eds.), *Proceedings of the International Conference of Symbolic-Numeric Data Analysis and Learning*, Nova Science Publishers New York, 565–575.

KIEL, R., and SCHADER, M. (1990), Dialogkontrollierte Regelsysteme Definition und Konsistenzbetrachtungen. Tech. rep., Institut für Informatik der Universität der Bundeswehr Hamburg.

KIEL, R., and SCHADER, M. (1991), Detecting Inconsistencies in Dialog-Controlled Rule Systems, in: G. H. Schildt and J. Retti (eds.), *Dependabilty of Artificial Intelligence Systems*, North Holland, Amsterdam, New York, Oxford, Tokyo.

KIEL, R. (1992), Ein prototypischer Handlungsplangenerator für ein System zur wissensbasierten Entscheidungsunterstützung bei der Auswertung von Marketing-Daten. Tech. rep., Institut für Informatik der Universität der Bundeswehr Hamburg.

KIEL, R., and SCHADER, M. (1992a) Dialog-gesteuerte Regelsysteme als Hilfsmittel zur Definition konsistenter Datenobjekte. Tech. rep., Dept. of Information Science, University of Mannheim.

KIEL, R., and SCHADER, M. (1992b) Knowledge representation in a system for marketing research, in: M. Schader (ed.), *Analizing and Modeling Data and Knowledge*,Springer-Verlag, Berlin, 179–187.

KIEL, R., and SCHADER, M. (1992c), Using dialog-controlled rule systems in a mainenance module for knowledge bases, submitted.

MARX, A., and RUNDSHAGEN, M. (1992) Systemanalyse nach Coard und Yourdon am Beispiel der Methodenkomponente von WIMDAS. Tech. rep., Dept. of Information Science, University of Mannheim.

MARX, S., and SCHADER, M. (1992), On the database component in the knowledge-based system WIMDAS. in: P. Ihm and H. Bock (eds.), *Classification, Data Analysis, and Knowledge Organization* Springer-Verlag, Heidelberg, 189–195.

NAKAMURA, K. (1989), Control of logic program execution based on the funktional relation. in: E. Shapiro (ed.), *Third International Conference on Logic Programming*, 504–512.

Acquisition of Syntactical Knowledge from Text

Jürgen Schrepp

Linguistische Datenverarbeitung, Universität Trier, Postfach 3825
W-5500 Trier, FRG

Abstract: The outline of a system is described which is designed to infer a grammar from a finite sample of linguistic data (corpus). It is inspired by the research on inductive inference in the sense of Gold(1967). After tagging the corpus, an incremental learning algorithm is used to produce a sequence of grammars which approximates the target grammar of the data provided. In each step, a small set of sentences is selected in a way which reduces the danger of overgeneralization. The sentences selected are analysed by a modified Earley parser which allows to measure the "distance" between the language generated by the actual grammar G and sentences not covered by G. The sentence which minimizes the "inductive leap" for the learner is selected to infer a new grammar. For this sentence several hypotheses for completing its partial structural description are formulated and evaluated. The "best" hypothesis is then used to infer a new grammar. This process is continued until the corpus is completely covered by the grammar.

1 Introduction

In most Natural Language Processing applications a parser plays a central role. Parsing a sentence includes the tagging of each word with its part-of-speech and the generation of an appropriate phrase structure for the whole sentence. Despite the success of this approach for restricted domains the performance for unrestricted text is still unsatisfactory.

Normally, the grammars used for parsing mainly base on a *context-free* grammar component which encodes linguistic intuitions about a particular language. To increase computational power a unification component is often added. But hand-written grammars suitable for practical purposes tend to be very large, are often incomplete and highly ambiguous. This tedious task should therefore be delegated to a computer equipped with the "right" discovery procedures.

The proposed approach to the acquisition of syntactical rules from positive examples bases on the paradigm of *inductive inference* as established by Gold(1967). It is aimed at the development of a system S that when confronted with a collection of linguistic data, called Corpus from now on, generates a grammar G for Corpus, i.e. Corpus is taken as a finite sample of the language to be learned and a minimal requirement for G output by S is that Corpus $\subseteq L(G)$. Actually, in most cases Corpus $\subset L(G)$. S is basically composed of:

- A *tagging* component TC which maps the sentences in Corpus onto sequences of tags (i.e. lexical categories).

- A *learning* component LC which, while processing Corpus, produces a sequence of grammars $G_1, G_2, \ldots, G_n$. Each grammar in this sequence determines a language which incorporates a greater portion of Corpus than its predecessors. G_n is called the *target grammar*, i.e. G_n is the grammar finally output by S.

In the past few years quite an amount of work has been done on the development of reliable tagging systems (see e.g. Garside et.al., 1987). The systems proposed so far typically employ a huge number of unstructured tags. The tagging component used here employs feature-structures for the representation of lexical and grammatical categories which offer advantages both for the tagging of sentences and the grammar generated.

The learning component LC, which is described in this paper, operates in the following way: A small set **B** of sentences is selected from **Corpus**. The computation of **B** is guided by a modified Earley parser, which allows to measure the "distance" between a sentence and the language generated by the actual grammar G_t. For sentences in **B**, which are *not* captured by G_t, hypotheses for completing the partial structural descriptions delivered by the parser are formed. The most plausible hypothesis is selected and used to infer G_{t+1}. This process is continued until a grammar covers **Corpus**.

The grammars inferred by the learning algorithm are context-free or a subclass of context-free grammars depending on the compression algorithm used. For example, the algorithm of Yokomori (1989) allows to infer *reversible context-free* grammars in polynomial time from a finite structural sample which are particular well suited for bottom-up parsing.

2 Theoretical Foundations

The extraction algorithm is motivated by the paradigm of *inductive inference* established by Gold (1967) where inductive inference is viewed as the hypothesizing of general rules from examples[1]. In his study on language learning Gold introduced the concept of *identification in the limit*. Let L be a (formal) language and M an inductive inference algorithm (the *learner*). At time t=1, 2, ... the learner will be presented a unit of information i_t concerning the unknown language L, which means M will be confronted with an infinite sequence of examples. After receiving i_t, M has to make a guess G_t about L based on all the information units recognized so far, i.e. M performs the function

$$G_t = M(i_1, i_2, \ldots, i_t)$$

The guesses made by M are grammars, and after some finite time t the guess G_t should be the target grammar with $L(G_t) = L$ and never changed for time $t + i$ ($i \geq 1$).
Within this learnability framework Gold distinguished between two different kinds of information presentation:

- TEXT: at each time t, M is presented a word $w \in L$ *(positive examples)*

- INFORMANT: the learner is provided with positive *and* negative examples

One of his results was that even the class of regular languages is *not* identifiable in the limit from TEXT. The reason for the weakness of TEXT is the fact that, if the unknown language L is a proper subset of another language L', then every finite sequence of words $w \in L$ is also a valuable sequence for L'. So if M guesses L' to be the target language, M will never recognize without negative examples that his guess was too general.

In 1980 Angluin proposed some conditions for the inference from TEXT which avoid the above mentioned "overgeneralizations".

---

[1]For a good survey of inductive inference see: D. ANGLUIN & C. H. SMITH (1983), Inductive inference: Theory and methods, *Computing Surveys, 15(3)*, 237-69.

Condition 1 (Angluin, 1980) *A family* $\mathcal{L} = \{L_1, L_2, \ldots\}$ *of nonempty languages* satisfies Condition 1 (the **Subset Principle**[2]) *iff there exists an effective procedure which on any input* $i \geq 1$ *enumerates a set of strings* T_i *such that*

1. $|T_i| < \infty$

2. $T_i \subseteq L_i$, *and*

3. $\forall j \geq 1 : T_i \subseteq L_j \Rightarrow \neg(L_j \subset L_i)$

Theorem 1 (Angluin, 1980) *A family of nonempty recursive languages is inferrable[3] from positive data iff it satisfies the Subset Principle.*

Normally the Subset Principle is used to establish a linear order on the space of hypotheses so that they are *maximally disconfirmable* by positive examples (i.e. hypothesis i+1 should not be a proper subset of hypothesis i (Berwick, 1986)). I want to go in the opposite direction and sort the positive data (i.e. the sentences of a corpus) given to a learner according to Condition 1.

Let V be an arbitrary alphabet (e.g. a set of tags), $L \subseteq V^*$ the unknown language and Corpus a finite sequence of words ("sentences") $\alpha \in L$, i.e.

$$\mathsf{Corpus} := (\alpha_\nu)_{\nu=1}^n = \alpha_1, \ldots, \alpha_n \; ; \quad \alpha_\nu \in L$$

Now imagine a learner M who should infer a grammar on the input of Corpus. It will be easier for M to avoid overgeneralizations if the complexity of the sentences increases gradually from α_1 to α_n. The *inductive leap* M has to perform in every step $i \to i+1$ should be minimized, i.e. Corpus has to be sorted in a way that this goal can be achieved.

Definition 1 *Define* $\mathbf{TS} := \{p \mid p$ *is a permutation of* $\mathsf{Corpus}\}$. *The elements* $ts \in \mathbf{TS}$ *are called* **training sequences** *for* L. *For* $ts = (\beta_\nu)_{\nu=1}^n$ *the subsequence from the jth to the kth element is denoted by* $ts_{jk} = (\beta_\nu)_{\nu=j}^k$.

Assume $\mathbf{G}$ to be an arbitrary set of grammars (i.e. the *hypotheses space*) containing the grammar $G_\emptyset = (\{S\}, V, S, \emptyset)$, and define an inference algorithm

$$\mathbf{F} : \mathbf{G} \times V^* \to \mathbf{G}, \quad \mathrm{F}(G, \alpha) \mapsto G' \tag{1}$$

so that $L(G) \cup \{\alpha\} \subseteq L(G')$ holds. F makes an inductive leap iff the strict subset relation holds. The effort F has to make in hypothesizing a new grammar from an input (G, α) depends on the "distance" between G and α.

A preliminary version of the extraction algorithm can be described by the function f : $\mathbf{G} \times \mathbf{TS} \to \mathbf{G}$ with

$$f(G, ts) := \begin{cases} \mathrm{F}(G, ts) & , \text{ if } |ts| \leq 1 \\ f(\mathrm{F}(G, ts_{11}), ts_{2|ts|}) & , \text{ if } |ts| > 1 \end{cases} \tag{2}$$

The sentences of the input sequence ts will be processed by f in incoming order. Because f is *order-dependent* the resulting grammar depends heavily on the order of ts. As a consequence,

---

[2] The term is adopted from Berwick(1986).

[3] The distinction between Angluin's concept of inferrability and Gold's identification in the limit is of minor importance in this context and is therefore not taken into consideration.

the training sequence given as an input to f should be sorted in a way that the above mentioned "distance" is minimized in each step. Assume G' to be the grammar generated by f after processing a subsequence of ts, and ts' to be the rest of the original input. Then the next sentence $\beta \in \{ts'\}^4$ has to be chosen in a way that $L(\mathrm{F}(G', \beta))$ is equal to the intersection of all possible languages wrt G' and ts' generated by F, i.e.

$$L(\mathrm{F}(G', \beta)) = \bigcap_{\alpha \in \{ts'\}} L(\mathrm{F}(G', \alpha))$$

Now the Subset Principle can be established for training sequences as follows.

Condition 2 $ts = \beta_1, \ldots, \beta_n \in \mathbf{TS}$ satisfies Condition 2 *wrt an inference algorithm* F *and a grammar* G *iff*

$$\forall 1 \leq j < n, \nexists 1 \leq k \leq n - j : L(\,f(G, \underbrace{\beta_1 \ldots \beta_{j-1}}_{ts_{1\,j-1}} \beta_{j+k})\,) \subset L(\,f(G, ts_{1j})\,)$$

Unfortunately Condition 2 cannot be verified in general because the inclusion problem is only decidable for the class of regular languages. With the help of a suitable complexity measure Φ_{F}, Condition 2 can be approximated by a computationally tractable criterion. For this reason let

$$\Phi_{\mathrm{F}} : \mathbf{G} \times V^* \to \mathbf{N} \tag{3}$$

be an arbitrary *complexity measure* wrt F (cf Eq.1) which yields a natural number on the input of a grammar G and a sentence α.

Condition 3 $ts = \beta_1, \ldots, \beta_n \in \mathbf{TS}$ satisfies Condition 3 *wrt an inference algorithm* F *and a grammar* G *iff*

$$\forall 1 \leq j < n, \nexists 1 \leq k \leq n - j : \Phi_{\mathrm{F}}(\,f(G, ts_{1\,j-1}), \beta_{j+k})\,) < \Phi_{\mathrm{F}}(\,f(G, ts_{1\,j-1}), \beta_j\,)$$

Definition 2 $ts \in \mathbf{TS}$ *is called* **optimal training sequence** *wrt a complexity measure* Φ_{F} *and a grammar* G *iff* ts *satisfies Condition 2.*

It is clear that such an optimal ordering cannot be computed in advance because in each step the next sentence to be chosen depends on the actual grammar.
Therefore the choice of the next sentence has to be computed according to Φ_{F} every time after processing a sentence. The final version of f is formulated in pseudocode using a PASCAL-like notation (cf Fig.1)

3   Extraction of Rules

3.1   Structuring the Corpus

As mentioned in the introduction, the TEXT used is a tagged corpus. Let us consider the while-loop of the algorithm shown in Fig.1. In the first step, the whole corpus has to be scanned in order to find a "best" sentence according to Φ_{F}(cf Eq.3). Although having polynomial time complexity (provided that Φ_{F} is polynomial bounded) this scheme is *not*

---

[4]$\{a_\nu\}$ denotes the associated set with the sequence (a_ν).

```
program DACS (G : grammar, corpus : corpus)
  var ts : training-sequence;
      G' : grammar;
      i : index
  begin
    ts := corpus;
    G' := G;
    while  |ts| ≠ 0
      do
        i := [choose i in a way that Φ_F(G', α_i) is minimal for ts];
        G' := F(G', α_i);
        ts := α_1, ..., α_{i-1}, α_{i+1}, ..., α_{|ts|}
      od
  end.
```

Figure 1: The extraction algorithm

very practicable for large corpora. Based on the assumption that the length of a sentence is proportional to the complexity of its structural description, it is supposed that

$$\forall G \in \mathbf{G}, \forall \alpha, \beta \in V^* : |\alpha| \ll |\beta| \Rightarrow \Phi_F(G, \alpha) < \Phi_F(G, \beta)$$

holds. As a consequence

1. the items of the corpus are sorted in a length increasing order(i.e. for all i holds $|\alpha_i| \le |\alpha_{i+1}|$), and

2. only a window of k sentences will be searched for in each pass where k is fixed.

The value of k is determined by the probability distribution of the sentence length in regard to the actual corpus.

3.2   The complexity measure

The method proposed for measuring the distance between a grammar G and a sentence α is called "correction". It transforms α into a new sentence β using single-character insertion, deletion, or change operations which are weighted with different costs, i.e $\Phi_F(G, \alpha)$ are the minimal costs for changing α into $\beta \in L(G)$. The correction of a sentence α into a language L at minimal costs is known as the *least-error correction* problem. By defining Φ_F this way, the value of Φ_F depends only on the language L and the sentence α, but *not* on the grammar which generates L (i.e. $\exists n \in \mathbf{N}, \forall G \in \mathbf{G} : L(G) = L \Rightarrow \Phi_F(G, \alpha) = n$).

Definition 3 (Wagner & Seiferas, 1978) *Let V be an alphabet.*
(a) *The elements of*

$$\Delta_V := \{[a \to b] \mid a, b \in V \cup \{\epsilon\} \wedge ab \ne \epsilon\}$$

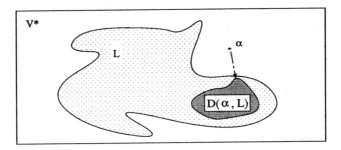

Figure 2: The least-error correction problem

are called **edit operations** over V. $[a \rightarrow b]$ is called **insertion, deletion,** or **change** if $a = \epsilon \neq b$, $a \neq \epsilon = b$, or $a \neq \epsilon \neq b$, respectively. An edit sequence $E \in \Delta_V^*$ **edits** $\alpha \in V^*$ to $\beta \in V^*$, denoted $\alpha \xrightarrow{E} \beta$, iff

either $\quad E = \epsilon$ and $\alpha = \beta = \epsilon$

or $\qquad$ the following hold for some $[a \rightarrow b] \in \Delta_V$, $E' \in \Delta_V^*$, $\alpha', \beta' \in V^*$:

$\qquad E = [a \rightarrow b]E' \wedge$

$\qquad \alpha = a\alpha' \wedge \beta = b\beta' \wedge$

$\qquad \alpha' \xrightarrow{E'} \beta'$

(b) For $L \subseteq V^*$, $E \in \Delta_V^*$ **corrects** α **into** L (denoted by $\alpha \xrightarrow{E} L$) iff $\alpha \xrightarrow{E} \beta$ holds for some $\beta \in L$.

According to the above definition, every character in α is edited exactly once and no inserted character is ever edited again.

Definition 4 Let $w : \Delta \rightarrow \mathbb{N}$ be a function[5]. w is called an **edit cost function** if w satisfies the two following conditions:

1. $w([a \rightarrow a]) = 0$

2. $w([a \rightarrow c]) \leq w([a \rightarrow b]) + w([b \rightarrow c])$

Using the rule $w(EE') = w(E) + w(E')$, w can be extended to a cost function $w : \Delta^* \rightarrow \mathbb{N}$.

The second condition ensures that no editing cost could be saved by editing some character several times.

Definition 5 For any edit cost function w as above and any nonempty language $L \subseteq V^*$, define

$$c_w(\alpha, L) := \min\{w(E) \mid E \in \Delta_V^* \wedge \alpha \xrightarrow{E} L\} \tag{4}$$

as the minimal costs needed for correcting α into L, and

$$D_w(\alpha, L) := \{\beta \mid \beta \in L \wedge \exists E \in \Delta_V^*(\alpha \xrightarrow{E} \beta \wedge w(E) = c_w(\alpha, L))\} \tag{5}$$

as the set of sentences to which α can be corrected using a minimal edit sequence.

---

[5] w can be a *partial* or *total* function. In the following, w is assumed to be total.

Obviously, the set $\{E \in \Delta_V^* \mid \alpha \overset{E}{\to} L\}$ is nonempty for every sentence $\alpha \in V^*$, since α can be corrected into any $\beta \in L \neq \emptyset$ using $|\alpha|$ deletions and $|\beta|$ insertions. If w is total then c_w exists and the set D_w is nonempty.

The problem of correcting an arbitrary sentence α into any fixed context-free language L can be solved in $O(|\alpha|^3)$. The proposed methods base on Earley's parser without look-head where the Earley items are extended by an error count (for details see Aho & Peterson, 1972; Lyon, 1974). The advantage of using a parser for this task is that a parser delivers structural descriptions for all sentences in D_w which can be used to build hypotheses concerning the unknown structural description[6] of the sentence α.

Example 1 *Let $V := \{n, v, adj, det, prep, adv, conj\}$ and w an edit cost function with*

$$w([a \to b]) := \begin{cases} 0, & \text{if } a = b \\ 1, & \text{otherwise} \end{cases}$$

The set of productions R of the grammar G acquired so far consists of the following rules:

$$S \overset{(1)}{\to} NP\ VP \qquad NP \overset{(2)}{\to} adj\ n \qquad VP \overset{(3)}{\to} v$$
$$NP \overset{(4)}{\to} det\ n \qquad VP \overset{(5)}{\to} v\ NP$$

Consider now the sentences $\alpha_1 = det\ adj\ n\ v$ and $\alpha_2 = det\ n\ v\ prep\ n$.

α_1: *$c_w(\alpha_1, L(G)) = 1$ and $D_w(\alpha_1, L(G)) = \{[[det\ n]\ [v]],\ [[adj\ n]\ [v]]\}$*

α_2: *$c_w(\alpha_2, L(G)) = 1$ and $D_w(\alpha_2, L(G)) = \{[[det\ n]\ [v\ [det\ n]]],\ [[det\ n]\ [v\ [adj\ n]]]\}$*

Both sentences can be mapped onto two sentences in L(G) using single-character deletion for α_1 and single-character change for α_2. The structural descriptions of the corrected sentences can now be used to generate hypotheses concerning the structural descriptions of the α_i in order to extend G in a way that the new grammar G' covers both sentences.

3.3 Generation and evaluation of hypotheses

Let G be the grammar inferred at time t. In each step, a sentence α from the rest of Corpus has to be found in order to minimize $c_w(\alpha, L(G))$. Two cases have to be considered:

1. $\alpha \in L(G) \implies c_w(\alpha, L(G)) = 0$, i.e G does *not* have to be changed

2. $\alpha \notin L(G) \implies c_w(\alpha, L(G)) > 0$, i.e. G has to be extended to cover α

In the first case, only some statistical information (first-order Markov model) from the structural description of α is gathered. In the second case, hypotheses for a complete structural description of α have to be formulated and a "best" hypothesis h_α has to be chosen to infer a new grammar. Like in the first step, the statistical information of h_α is gathered.

Generating hypotheses. The main idea is to map the complete structural description of a "similar" sentence $\beta \in D_w(\alpha, L(G))$ onto α. Based on the assumption that the syntactic knowledge is correct up to this point this mapping should preserve as much as possible of

---

[6]For building hypotheses these structural description are non-annotated, i.e. strings which are generated by a parenthesis grammar.

the structure of β. The problem of finding a suitable mapping is governed by two conflicting needs: on the one hand, the size of the set of hypotheses **H** should be small for computational reasons; on the other hand, **H** should include the "right" hypotheses.

Let E be the edit sequence which corrects α into β ($\alpha \overset{E}{\to} \beta$). In the first step, the insertion and change errors are disregarded and the structural description of β is mapped onto α, e.g.

$$\left. \begin{array}{l} \alpha = adj \; n \; v \; adv \\ \beta = [[det \; n] \, [v]] \\ E = [adj \to det][n \to n][v \to v][adv \to \epsilon] \end{array} \right\} \Rightarrow \bar{\alpha} = [[adj \; n] \, [v] \; adv \;]$$

The result is a partial structural description for α where the deleted constituents are not yet integrated.

Let $E' = [a_1 \to \epsilon][a_2 \to \epsilon] \cdots [a_n \to \epsilon]$ be a subsequence of the edit sequence E and K be the smallest closed constituent of β where the deletion errors occur. Then the hypotheses which can be build must obey the following conditions:

C1: The constituent boundaries of K are *not* changed, i.e. restructuring of constituents which are error-free is excluded.

C2: The constituent boundaries of the subconstituents of K can only be altered by inserting some neighboring categories of the deletion errors, i.e. restructuring is restricted to the subconstituents of K which delimit the erronous sequence captured by E'.

C3: If $n > 1$ only such hypotheses are built which do not conflict with the structural information for $a_1 \cdots a_n$ already present in the grammar.

Evaluating hypotheses. After the generation of **H** a "best" hypothesis has to be chosen. Depending on the size of $D_w(\alpha, L(G))$, **H** can get very large. As a consequence the evaluation procedure consists of two steps:

1. The probability $P(h_i)$ that h_i is the "best" hypothesis in **H** is computed for each $h_i \in$ **H**. If $P(h_i)$ is lesser than a threshold δ, then h_i is disregarded. After this filtering process we obtain $H_{best} := \{h_i | h_i \in \text{II} \wedge P(h_i) \geq \delta\}$. δ is dynamically updated to insure that H_{best} is not empty.

2. For all hypotheses $h_j \in H_{best}$ a new grammar G_j is inferred from the actual grammar G using a grammar compression algorithm so that $L(G) \subseteq L(G_j)$ holds. Then each new grammar is used to compute c_w for *all* sentences in the actual "window" (**Window** $:= (\beta_\nu)_{\nu=1}^k$), and the sum of the errors is computed, i.e.

$$sum(G_j) := \sum_{\nu=1}^{k} c_w(\beta_\nu, L(G_j))$$

The grammar G_j with minimal sum is chosen, i.e. h_j is the "best" hypothesis. h_j is then used to update the statistical information.

The obtained grammar is therefore "best" wrt the statistical information gathered from the analysed sentences *and* the parsebility of the sentences in **Window**.

3.4 Extending the grammar

During the evaluation process it is necessary to infer a new grammar from a hypothesis h and the actual grammar G. For this purpose several algorithms have been designed (e.g. Yokomori, 1989; Crespi-Reghizzi, 1972) which are very similar. In the first step, the primitive context-free grammar $G_p(h)$ is computed, and in the second step $G_p(h) \cup G$ is compressed by merging nonterminals according to certain criteria. The advantage of such a compression algorithm is that recursive rules can be inferred. This can *not* be achieved when using the rules from the hypothesis directly. A shortcoming of these algorithms is that they cannot be used directly for linguistic purposes because the compression rate is too large so that in most cases the grammar is not very adequate.

The development of plausible lingistic criteria can be guided by several principles depending on the results to be obtained (see e.g. Morgan, 1986). The problem one is confronted with is how to find a plausible compromise between the automaton which accepts V^* and the grammar which generates only the given positive examples.

4 Conclusions

An algorithm for the extraction of a context-free grammar from a finite linguistic corpus has been proposed. After tagging each word in the sample with its part-of-speech, an incremental learning algorithm guided by heuristic principles acquires a grammar for the language presented. The result is a context-free grammar where the nonterminals are complex feature-structures.

First tests on a small sample have shown that the evaluation of the structural hypotheses has to be improved. The statistical information and parsibility within the Window don't provide enough cues which hypothesis is really the best one. The set of "best" hypotheses remains too large. So a small amount of explicit linguistic knowledge seems to be necessary. In a second step, statistical information has to be integrated in the compression algorithm to justify the merging of nonterminals especially when recursive rules are generated.

References

AHO, A. V. & T. G. Peterson (1972), A minimum distance error-correcting parser for context-free languages, *SIAM Journal on Computing, 1(4), 305-12.*

ANGLUIN, D. (1980), Inductive inference of formal languages from positive data, *Information and Control, 45, 117-35.*

BERWICK, R. C. (1986), Learning from positive-only examples, in: R. S. Michalski, J. G. Carbonell & T. M. Mitchell (eds), *Machine Learning-Vol.II*, Morgan Kaufmann, Los Altos, 625-45.

CRESPI-REGHIZZI, S. (1972), An effective model for grammar inference, in: B. Gilchrist (ed), *Information Processing 71*, Elsevier North-Holland, 524-29.

GARSIDE, R., G. LEECH & G. SAMPSON (1987), *The computational analysis of English*, Longman, New York.

GOLD, E. M. (1967), Language identification in the limit, *Information and Control, 10, 447-74.*

LYON, G. (1974), Syntax-directed least-errors analysis for context-free languages: A practical approach, *Communications of the ACM, 17(1), 3-14.*

MORGAN, J. L. (1986), *From simple input to complex grammar*, The MIT Press, Cambridge, MA.

WAGNER, R. A. & J. L. SEIFERAS (1978), Correcting counter-automaton-recognizable languages, *SIAM Journal on Computing, 7(3), 357-75.*

YOKOMORI, T. (1989), Learning context-free languages efficiently, in: K. P. Jantke (ed), *Analogical and inductive inference*, Springer, Berlin-Heidelberg, 104-23.

Generating Topic–Based Links in a Hypertext–System for News

Heinz J. Weber

Department of Computational Linguistics, University of Trier,
P.O. Box 3825, D–5500 TRIER (Germany),
phone: 0651/201–2253; e–mail: weber@utrurt.uucp.de

Abstract: This paper is concerned with a model for generating links between texts. These links are used for transferring unstructured samples of newspaper articles into a hypertext–like organization. The generation of links is carried out on the basis of a structural description of each individual text comprising a hierarchy of topics. Topic–structures are the result of partial text–parsing. Intertextual links are generated by relating different texts according to their topic–structure. Linking of texts in such a way leads to a more adequate arrangement of news than an accumulation in a stack of unconnected files.

1 News as Hypertext

News are not always presented as isolated pieces of information but show a rather large variety of intertextual relations enabling the recipients to collect and integrate information by crossing the boundaries of different texts. Basically, it depends on the reader's interests and efforts which points are connected. But there are also landmarks and signposts which can be used for orientation. Thus, relations between different texts in the same issue or in an earlier one may be indicated explicitly by giving the co–ordinates of the respective texts (e.g. page number, date of issue, title). But very often, text–text–relations in news are given only implicitly, e.g. by indicating that a topic has been treated previously - without providing co–ordinates of the respective text - or by connecting several topics in one text which already have been treated separately. Texts which are explicitly or implicitly related to each other can be represented by a structure called 'hypertext'. A hypertext is an arrangement of information units (e.g. texts or parts of texts) in non–linear order. Connections between information units can be established intellectually or can be generated automatically. Automatic generation of links between texts or parts of texts requires a system for language–processing, i.e. a text parser and a link–generator. This paper presents an outline of a system (t–X–t) which is to generate links between news topics. These links are used for converting unstructured samples of newspaper articles into a hypertext–like organization. This is to be carried out in three steps.

The first step is a conventional keyword–based text retrieval which leads to a heap of unconnected texts. Texts may be sorted in an arbitrary manner (e.g. according to the frequency of the keywords, according to their dates of publication or to the respective publishers etc.).

The second step is a topic analysis of individual texts. Very often, news deal with more than one topic. Moreover, topics can occur on different levels of narration, as reported speech, as subordinate topic, as annotation. In this case, texts have to be cut down to chunks which are parsed and labelled according to their narrative level. These labelled chunks are regarded as the basis for topic analysis and are used as the nodes of the hypertext–structure.

The third step is the conversion of retrieved and parsed texts into a hypertext–structure by the generation of links between these nodes. Links are defined on the basis of an inter–textual comparison of topic representations and chunk labels. Similarity or dissimilarity of topics and labels leads to a typology of links, such as continuation, retracing, upgrading or downgrading connection.

2 Description of Individual Texts: Chunks, Chunk Labels and Topic Profiles

Provided that there is an accessible database for newspaper articles and given a combination of keywords to start with (e.g. 'FCS' and 'East–Sylvanian People's Chamber'), a system for text retrieval is to deliver a sample of articles ordered by one or more criteria, e.g. chrono–logically or according to the respective publishers. For experimental reasons, the subsequent sample is restricted to three texts. The texts are semi–authentic; they have been anonymized to avoid distraction.

T1, *S1: Chaired by Mr. Schulz a delegation of the FCS parliamentary group arrived in Gladz.*

S2: There, talks began with delegates of the East–Sylvanian People's Chamber. ...

T2, *S1: A delegation of the FCS parliamentary group visited the East–Sylvanian People's Chamber in Gladz.*

S2: In a speech Mr. Schulz, chairman of the parliamentary group, gave expression to the hope that the border between the two Sylvanian states will be one without walls and barbed wire.

T3, *S1: Members of the FCS parliamentary group visited the East–Sylvanian People's Chamber in Gladz.*

S2: It is the first official meeting of parliamentarians of both Sylvanian states.

S3: Meanwhile, Mr. Meyer, the speaker of the CSF–Party, expressed doubts concerning official contacts of such a kind.

S4: The visit of a West–Sylvanian delegation might touch questions of status that should not be taken too easily.

As for newspaper articles, a conventional presentation (as indicated afore) leads to a rather artificial assortment. Besides that, it eventually conceals relations between articles which are more substantial than the mere existence of common keywords. A hypertext–like organization, however, might render an arrangement of texts which bears more analogy to the connection and integration of news by readers. In t–X–t, the decisive point for integrating retrieved texts into a hypertext–structure will be the similarity of topics. A topic is conceived as a construct which is the result of a process of textual analysis. This analysis will be outlined in the following.

2.1 Cutting Texts to Chunks

The sentences of each text are combined to more extended information units on the basis of cohesive means like connectives or anaphora in initial position (e.g. 'there' in T1), recurrence of expressions (lexemes, proper names) and embedding constructions (e.g. in cases of reported speech, cf. S3 & S4 in T3). Recurrence is defined in a narrow sense as occurrence of the same lexeme or name in adjacent sentences (e.g. 'parliamentary group' in T2: S1 & S2) or as occurrence of members of the same lexical family (e.g. 'delegation' - 'delegate' in T1: S1 & S2).

On the other hand, absence of cohesion indicators (i.e. connectives, recurrent expressions or speech markers) leads to cuts within a text (e.g. between sentences S2 and S3 of text T3):

T3: *chunk 1:*
S1: Members of the FCS parliamentary group visited the East–Sylvanian People's Chamber in Gladz.
S2: It is the first official meeting of parliamentarians of both Sylvanian states.

chunk 2:
S3: Meanwhile, Mr. Meyer, the speaker of the CSF–Party, expressed doubts concerning official contacts of such a kind.
S4: The visit of a West–Sylvanian delegation might touch questions of status that should not be taken too easily.

It is possible that a text consists of only one chunk or that it has as many chunks as sentences. In our text corpus, there are - on an average - four chunks per text and three sentences per chunk. Cutting a text into chunks in such a way leads to a provisional basis for topic analysis, i.e. a textual punctuation which makes possible the next steps, the parsing and labelling of chunks and the construction of a topic hierarchy. With respect to the parsing of text chunks, we know that there are more indicators in the text as the afore–mentioned, both for connectivity (e.g. semantic related lexemes like 'fight' – 'hostility') or for cuts (e.g. topic discontinuities suggested by indefinite articles or by complex expressions indicating topical shifts). But at the moment, we try to cope with the problem without regarding these more elaborate indicators and to get along with reliable and robust information.

2.2 Parsing and Labelling of Text Chunks

After cutting down a text to pieces, the chunks are analyzed according to structural criteria. Analysis results in a typology of chunk labels which can be used both for the structuring of individual texts and the generation of intertextual links. The following chunk labels are provided: Main topic, (connected) minor topic, annotation, reported speech. Though being identified by rather simple structural criteria, each of the chunk labels can be interpreted on a more abstract level of description, e.g. as indicating a special mode of textualization or the origination from a particular text–source (author or character).

Main topic and (connected) minor topic(s) can be regarded as textualized in the mode 'narration', both originating from the primary text–source, the so–called 'author'. Narration indicators for newspaper articles are present tense or past tense and third person of verbs.

The main topic is placed as a rule in text–initial position. The differentiation of main topic and minor topic as well as the cuts between minor topics are established on the basis of a cumulation of factors:

Minor topics often occur in non–initial position, are modified by initial adverbials or subjunctions (e.g.'meanwhile', T3), show tense–contrasts (past tense vs. pluperfect), introduce new referents (characters, places; e.g. 'Meyer', 'CFS–Party', T3). At the moment, two of four factors are sufficient for making a decision.

T3: *chunk 1:*
| main topic: | *Members of the FCS parliamentary group visited the East–Sylvanian People's Chamber in Gladz. ...*

chunk 2: | Minor topic: | *Meanwhile, Mr. Meyer, the speaker of the CSF–Party, expressed doubts concerning official contacts of such a kind.*

Annotations can be regarded as textualized in the mode 'commentating', also originating from the 'author'. Annotation indicators are: Tense shift from past tense to present tense (e.g. S2 in T3: 'It is the first ...'). Annotations can as well be connected with the main topic and minor topic(s) or embedded into main or minor topics.

Reported speech can be regarded as originating from a secondary, non–autonomous text source (e.g. a text character) and as being textualized in the mode 'quotation'. Quotation indicators are expressions for speech activities (e.g. 'expressed doubts'), subjunctive mode, orthographic markers (e.g. quotation marks). Chunks labelled as reported speech can be subdivided again into main topic, minor topics etc. with the possibility of recursion via reported speech of a third text source (with main topic, reported speech etc.). The identification and isolation of reported speech within texts is one of the most important points for text parsing of narrative text–types, like news or stories. Besides the possibility of embedded events which is an obstacle for straight schema–directed text parsing, the difference between author–initiated, i.e. original text, and character–initiated text, i.e. quoted text, corresponds with different shadings of factuality in a text. This difference might be crucial in many respects, especially in the context of information retrieval and knowledge extraction.

T3: *chunk 1:* | Main topic: | *Members of the FCS parliamentary group visited the East–Sylvanian People's Chamber in Gladz.*
| Annotation: | *It is the first official meeting of parliamentarians of both Sylvanian states.*

chunk 2: | Minor topic: | *Meanwhile, Mr. Meyer, the speaker of the CSF–Party, expressed doubts concerning official contacts of such a kind.*
| Reported speech: | *The visit of a West–Sylvanian delegation might touch questions of status that should not be taken too easily.*

2.3 Representation of Topics

The next step after the cutting of texts and the parsing of chunks is to find out the respective topics and to bring them into a format of representation. Therefore, chunks are melt down by deletion of all so–called structure words. The remaining content words are arranged in an alphabetical list:

T1: $\boxed{\text{main topic:}}$ *arrive - begin - chair - Chamber - delegate - East-Sylvanian - FCS - Gladz - group - parliamentary - People - Schulz - talk*

Of course, this list of words cannot be regarded as a full representation of the content of a text topic, like a macro–proposition, for example. It is regarded only as the surface of a hypothetical topic, a topic profile, and its range of application is restricted only for news. But in the limited context of a labelled chunk, it seems to be sufficient for a decision concerning the similarity or dissimilarity of (hypothetical) topics in different news texts. Concerning the similarity of topics, decisions are based again on a cumulation of factors. Two or more profiles are regarded as representations of the same topic:

- If the set of keywords is included in all profiles (e.g. FCS, East–Sylvanian, People's Chamber);

- or if the number of common content words in a topic profile amounts to 33 % of all its words (e.g. T2: 7 identical words out of 15)

- and if the set of common words includes at least one proper name (e.g. Gladz).

At the moment, this measure for topic similarity in news is only a rough formula which needs more experimental and conceptional corroboration. On the one hand, the intertextual comparison of (hypothetical) topics on the basis of a topic profile looks very simple compared to the identification or comparison of topics via schemata (e.g. schemata for political meetings, traffic accidents etc.), on the other hand, it is a robust and reliable method of topic analysis for purposes like ours. Notwithstanding the fact that the knowledge–based identification of topics via schemata is costly in many respects, it is very often inadequate for the required textual descriptions. Many texts are not enough stereotyped and homogeneous to be parsed in accordance with schemata. Even texts supposed to be rather plain, like short newspaper articles, can be complex montages of sequences of events intertwined with passages of reported speech which itself may be very complex. With respect to the non–predictability of news topics, it seems therefore to be more promising to part with knowledge–based topic analysis by means of prefabricated topic schemata and to rely on genuine linguistic and textual indicators.

2.4 Topic Structure

The description of individual texts is completed by constructing a text–tree representing a topic hierarchy. Text–trees are built up according to the following structure rules:

1. $TEXT \rightarrow$ *main topic*(&*minor topic*∗)(&*annotation*∗)

2. *main/minor topic* $\rightarrow$ *topic*(&*annotation*)(&$TEXT$)

In the following, Figure 1 presents the topic hierarchy in text T3. The topic–structure is converted into a hypertext–like design (cf. Figures 2 and 3).

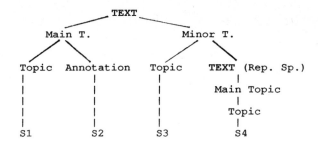

Figure 1: Topic–structure of text T3 (text–tree)

3   Linking of Topics: Some Types of Text–Text–Relations

Intertextual links are generated by comparison of topic profiles in relation to their chunk-labels. Starting–point is the similarity of topic profiles in different texts. Depending on the respective chunk–labels of identical topics, various types of intertextual links can be established. Configurations of linked texts with identical topics can provide interesting information on text–text–relations which would not be obvious in a conventional assortment of texts. E.g. the sequencing of texts in chronological order in connection with a topic-based linking can be interpreted as presentation of the life–cycles of news topics, i.e. their continuous expansion and gradual fading, their sudden appearance and disappearance, their coexistence with better–off or less prominent topic neighbors, their transition into new topics. These are the most frequent types of links in our corpus: Links between identical topics on the same level of narration, i.e. continuation or retracing of topics (cf. Figure 2); author-initial connection of topics in different texts), e.g. upgrading connection (cf. Figure 3).

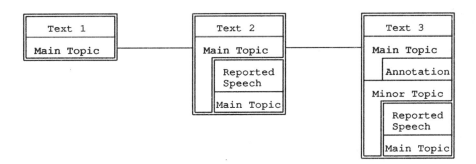

Figure 2: Text 1 (main topic) preceeds texts 2 and 3 (main topic): continuation or retracing of topics

Topic continuation (Fig.2):

Text 1 | *main topic:* | *Chaired by Mr. Schulz, a delegation of the FCS parliamentary group arrived in Gladz. There, talks began with delegates of the East–Sylvanian People's Chamber.*

Text 2 | *main topic:* | *A delegation of the FCS parliamentary group visited the East–Sylvanian People's Chamber in Gladz. In a speech Mr. Schulz, chairman of the parliamentary group gave expression to the hope*
| *reported speech:* | *that the border between the two Sylvanian states will be one without walls and barbed wire.*

Text 3 | *main topic:* | *Members of the FCS parliamentary group visited the East–Sylvanian People's Chamber in Gladz.*
| *annotation:* | *It is the first official meeting of parliamentarians of both Sylvanian states.*
| *minor topic:* | *Meanwhile, Mr. Meyer, the speaker of the CSF-Party, expressed doubts concerning official contacts of such a kind.*
| *reported speech:* | *The visit of a West–Sylvanian delegation might touch questions of status that should not be taken too easily.*

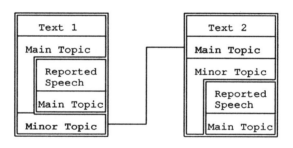

Figure 3: Text 1 (minor topic) linked with text 2 (main topic): upgrading connection

Upgrading connection (Fig.3):

Text 1: | *Main topic:* | *All the opposing war parties of Belanon agreed*
| *reported speech:* | *to participate in a reconciliation conference planned to begin in Saulanne on Monday.*
| *minor topic:* | *In the meantime, fights in Reibout and the nearby Shlouf mountains are continuing ...*

Text 2: | *main topic:* | *Fights in the Belanese capital and in the Shlouf mountains are still enduring.*
| *minor topic:* | *Nonetheless, President Megayel has made an appeal*
| *reported speech:* | *to cease hostilities ...*

4 Conclusion

Experiments concerning text parsing and link generation are based on a corpus of 300 texts (news) in German. The text parser (in C) exists in the major part: the modules for Reported Speech Analysis, Connection and Recurrence Analysis are working. As for the link generator, programs are still in a provisional stage. Text–Linking is executed on XCard, a UNIX–based hypertext–system. We are confident that the way of identifying and structuring topics as well as the linking of texts on the basis of such a textual description is not restricted to news, but is also applicable to other text types with a narrative textualization mode, e.g. reports, testimonies, short stories.

References

CONKLIN, E.J. (1987), Hypertext: An Introduction and Survey. *IEEE Computer 2, 9 (Sept.1987), 17–41.*

KUHNS, R.J. (1988), A News Analysis System. *Proceedings of COLING 1988 Budapest, 351–355.*

POLANYI, L./SCHA, R.J.H. (1983), On the Recursive Structure of Discourse. In: Ehlich, K. /van Riemsdijk, H. (eds.), *Connectedness in Sentence, Discourse and Text. 141–178.* Tilburg: University

WEBER H.J./THIOPOULOS, C. (1991), Ein Retrievalsystem in Hypertext. *LDV–Forum 8, 1/2, 21–27.*

Interactively Displaying Ultrametric and Additive Trees

G. De Soete[1]
Department of Psychology, University of Ghent,
Henri Dunantlaan 2, B-9000 Ghent, Belgium

J. Vermeulen
Department of Psychology, University of Ghent,
Henri Dunantlaan 2, B-9000 Ghent, Belgium

Abstract: Recent clustering procedures provide an ultrametric or additive tree representation of proximity data. Various operations can be performed on such trees without affecting the goodness of fit of the representation. In an ultrametric tree, subtrees can be rotated without changing the ultrametric tree distances. Additive trees are usually displayed as rooted trees, although the placement of the root is arbitrary. Commonly, a user wants to find the rotations of the subtrees and—with additive trees—the placement of the root that lead to the best interpretation of the tree representation. While some analytic procedures have been presented in the literature, none of them is entirely satisfactory. In this paper, a software system is described that allows the user to interactively manipulate an ultrametric or additive tree until—in his or her view—the best representation is found. This software system is implemented in a portable way using the X Window System.

1 Introduction

A hierarchical cluster analysis of a set of proximity data yields an ultrametric tree representation (see, for instance, Johnson, 1967). As is well known, an ultrametric tree is a rooted tree in which a nonnegative weight (or height) is attached to each node such that (a) the terminal nodes have zero weights, (b) the root has the largest weight, and (c) the weights attached to the nodes on the path from any terminal node to the root constitute an increasing sequence. Each of the objects on which the proximities are defined is represented by a terminal node of the tree. The ultrametric tree distance between any two objects i and j is then defined as the largest weight assigned to the nodes on the path connecting the two terminal nodes that represent objects i and j. The goodness of fit of a hierarchical clustering is usually formulated in terms of the correspondence between the original proximities and the derived ultrametric tree distances.

Ultrametric trees are commonly displayed as dendrograms as in Fig. 1. A dendrogram is an inverted tree with the root placed at the top and the terminal nodes representing the objects placed at the bottom. The height of each node corresponds to the weight that is attached to the node. The ultrametric tree distances do not uniquely determine the horizontal ordering of the objects. The left–right order of the two subtrees emanating from each nonterminal node can be changed without affecting the ultrametric tree distances and, thus, the goodness of fit of the ultrametric tree representation. The two ultrametric trees in

---

[1]Supported as "Bevoegdverklaard Navorser" of the Belgian "Nationaal Fonds voor Wetenschappelijk Onderzoek".

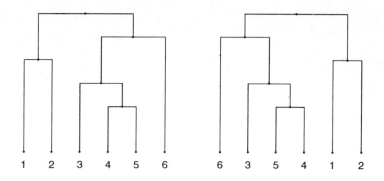

Fig. 1: Example of Two Equivalent Ultrametric Trees.

Fig. 1, for instance, define identical ultrametric distances, although the horizontal orderings of the objects are quite different. The tree on the right hand side can be obtained from the tree on the left hand side by interchanging the two subtrees emanating from some of the internal nodes. In the sequel, this interchange operation will be referred to as a *subtree rotation*. Two ultrametric trees that define identical ultrametric tree distances will be called *equivalent ultrametric trees*.

While two equivalent ultrametric trees always constitute representations that are equally good in terms of goodness of fit, one tree might give a more informative representation than the other, as argued and demonstrated by Gruvaeus and Wainer (1972). That is, a particular ordering of the objects in an ultrametric tree might more easily suggest a certain interpretation. Hence, it might be useful to attempt to find through subtree rotations the ordering of the objects that yields the most interpretable solution. An analytic method for finding such an ordering was proposed by Gruvaeus and Wainer (1972). However, this method works only when the data cannot be perfectly represented by an ultrametric tree. Critchley and Heiser (1988) present some results that—in the case of imperfect data—can also be used to define a particular ordering. Degerman (1982) devised a procedure for ordering the terminal nodes of an ultrametric tree to maximize correspondence with a given external criterion. In this paper, a different avenue is explored. We present a procedure that allows the user to interactively find—through subtree rotations—the ordering that in his or her view yields the best interpretable representation.

In biology and psychology, clustering techniques are used that yield a representation of the proximity data by means of a type of tree that is somewhat more general than an ultrametric tree, viz., an additive tree (see, for instance, Sattath and Tversky, 1977). An additive tree is an unrooted tree in which a nonnegative weight is assigned, not to the nodes, but to the links of the tree. The additive tree distance between any two nodes is defined as the sum of the weights assigned to the links on the path joining the two nodes. As in an ultrametric tree, the terminal nodes correspond to the objects on which the proximities are defined. Additive trees are usually displayed as rooted trees in parallel form (Sattath and Tversky, 1977), as exemplified in Fig. 2. In this form, only the horizontal links are relevant; the vertical lines are dummy links. The length of a horizontal link is proportional to the weight that is attached to it. Thus, the additive tree distance between two nodes is equal to the sum of the *horizontal* links on the path connecting the nodes. As in ultrametric trees, subtrees can be rotated without affecting the additive tree distances. When representing an additive tree in parallel form as in Fig. 2, a root is arbitrarily introduced, although additive trees are in principle unrooted. While the introduction of a root does not affect the distances defined by an additive tree, the placement of the root clearly induces a particular hierarchy of partitions or clusters and, hence, influences the interpretation of the tree. The two additive trees in Fig. 2 define identical additive tree distances, but obviously suggest a different clustering of the objects. Two additive trees that define the same distances, will be referred to as *equivalent additive trees*.

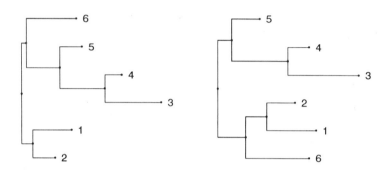

Fig. 2: Example of Two Equivalent Additive Trees.

Clearly, the placement of the root (and to a lesser extent, the rotations of the subtrees) greatly affects the interpretation of an additive tree. Sattath and Tversky (1977) suggest to place the root at the point that minimizes the variance of the distances from the root to the terminal nodes. While this choice can lead to an interpretable tree, one often likes to experiment with different placements of the root to see which representation is most interpretable. In this paper, an interactive program is presented that allows the user to perform subtree rotations and to change the placement of the root in a dynamic way to arrive at the most interpretable additive tree representation.

The procedures presented in this paper allow the user to select in a interactive fashion that tree in a family of equivalent trees that yields the most interpretable representation. This usage can be compared to the situation of two-way multidimensional scaling where each orthogonal rotation of the configuration gives an equally good account of the data and where the user wants to find an orientation of the axes that provides the best interpretable solution. From a measurement–theoretic point of view (e.g., Roberts, 1979), such an approach can be criticized. Only features that are invariant under the permissible transformations (in the case of ultrametric trees, subtree rotations, and in the case of additive trees, subtree rotations and root placement) can be meaningfully interpreted. Even if one adheres to such a philosophy, the current procedure is still useful because it allows the user through interactive manipulation to detect the properties of a tree representation that are invariant under the permissible transformations. Moreover, such an interactive procedure can be gainfully used when comparing tree representations of the same data set produced by different ultrametric or additive tree clustering algorithms, or when comparing tree representations (produced by the same method) of two different proximity data sets about the same stimuli.

In Section 2, a library for manipulating ultrametric and additive trees is presented. This library is implemented in the programming language C (Kernighan and Ritchie, 1988). In Section 3, an interactive graphical program for dynamically transforming trees is described.

This program uses the library presented in Section 2. Finally, in Section 4, some concluding comments are made and a possible extension of the current approach is mentioned.

2   A C Library for Manipulating Trees

Without loss of generality it may be assumed that ultrametric and additive trees are binary trees. (Dummy nodes and links can be introduced in case more than three links are incident in a single internal node.) The following C data structures will be used to store an ultrametric or additive tree. All information about a node in a tree is stored in the following structure:

```
struct _NODE
    {
    char *label;
    double order;
    double weight;
    LINK *upper;
    LINK *left;
    LINK *right;
    } NODE;
```

Except for the root, each nonterminal node points to the three links that are incident in the node through the **upper**, **left** and **right** link pointers. The **upper** link refers to the link that connects the node with its ancestor node (i.e., the node that is closer to the root); while the **left** and **right** links connect the node to its two child nodes. The **upper** link pointer of the root node is NULL. In the case of terminal nodes, the **left** and **right** link pointers are NULL. The **label** pointer is used only by the terminal nodes to store the object label. Internal nodes are unlabeled. The position of a node is represented by the **order** and **weight** variables. In an ultrametric tree, **order** refers to the horizontal position of the node (scaled between 0 and 1), and **weight** refers to its vertical position. In an additive tree, **order** stores the vertical position of the node, while **weight** contains the horizontal position.

Each link is represented by a LINK structure:

```
struct _LINK
    {
    double length;
    NODE *upper;
    NODE *lower;
    } LINK;
```

The **length** variable is only used in additive trees to store the link weight.

Finally, a tree is stored in the following structure as a collection of linked nodes:

```
struct _TREE
    {
    TreeType type;
    NODE *root;
    int n_objects;
    } TREE;
```

The **type** field specifies whether the tree is an ultrametric or an additive tree. The **root** field points to the root of the tree. Through this root, all nodes and links of the tree can be reached. The variable **n_objects** stores the number of objects represented in the tree.

The library includes a number of routines that can be used to manipulate both ultrametric and additive trees. The routine

```
TREE *ReadTree(char *filename)
```

reads an ultrametric or additive tree as produced by tree fitting programs such as LSULT and LSADT (De Soete, 1984) from the indicated file or, if **filename** is a NULL pointer, from the standard input. **ReadTree** dynamically allocates the appropriate structures for storing the nodes and the links and returns a pointer to a **TREE** structure. All memory allocated by **ReadTree** can be released by invoking

```
FreeTree(TREE *tree)
```

The following function rotates the subtree of a tree at a specified (internal) node:

```
NODE *RotateTree(TREE *tree, NODE *node)
```

RotateTree interchanges at the indicated node the pointers to its left and right children and recalculates the **order** values of all nodes whose horizontal position changed. **RotateTree** returns a pointer to the highest node in the tree of which the horizontal position was altered.

To change the placement of the root in an additive tree, two functions are provided:

```
void SetRoot(TREE *tree, LINK *link, double interpolation)
void SetBestRoot(TREE *tree)
```

SetRoot puts the root of the tree on edge **link**. The variable **interpolation** holds a value between 0 and 1, indicating the relative location of the root on the indicated link. After removing the old root and inserting the new root on the selected link, **SetRoot** reorders all the nodes on the path from the old root to the new root and recalculates the horizontal and vertical positions of all the nodes in the tree. The function **SetBestRoot** places the root at the point in the tree that minimizes the variance of the distances from the root to the terminal nodes, as suggested by Sattath and Tversky (1977).

In addition to the functions listed here, the library contains some other utility routines. The data structures and routines in the library allow one to easily devise a procedure for drawing an ultrametric or additive tree (or a subtree thereof). Such a routine can be easily implemented by recursion and would typically have the following form:

```
DrawTree(NODE *node)
{
    if (node->left)
        DrawTree(node->left->lower);
    if (node->right)
        DrawTree(node->right->lower);
    DrawNode(node);
}
```

where **DrawNode** is the (device-dependent) function that would carry out the actual drawing.

3   A GUI for Interactively Displaying Trees

Using the library described in Section 2, a GUI (Graphical User Interface) has been implemented for interactively displaying ultrametric and additive trees, based on the X Window System (Scheifler and Gettys, 1986; Scheifler, Gettys, and Newman, 1988). The decision to base the graphical interface on the X Window System was motivated by a number of considerations. First of all, a program was aimed at that was portable and would work without changes on a large variety of machines. Not only does the X Window System provide a device-independent and portable system for implementing graphics, but it has also become the *de facto* standard for windowing systems on workstations. Secondly, the client-server model on which the X Window System is based, works well in a distributed, possibly heterogeneous computing environment.

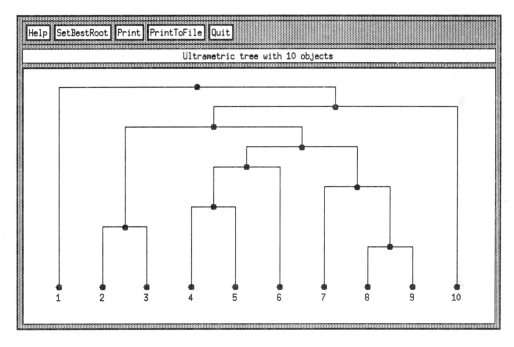

Fig. 3: Screen Dump of **Xtree** Showing an Ultrametric Tree.

The GUI, called **Xtree** has been implemented using a toolkit based on the *Xt intrinsics*. More specifically, the so-called HP Widget Set, fully described in Young (1989), was used because it has been placed in the public domain and is widely available. Changing to another *Xt* based toolkit is fairly straightforward and does not require any major changes.

Fig. 3 shows a screen dump of **Xtree** displaying an ultrametric tree. The main window is a bitmap showing the ultrametric tree. A user can dynamically perform subtree rotations by clicking with the mouse on the appropriate node (using the left mouse button). Once a good ordering of the objects is found, the tree can be printed on a PostScript (Adobe Systems, 1985) printer by clicking on the **Print** button. An encapsulated PostScript file can be generated by clicking on **PrintToFile**. The PostScript output is not merely an enlarged image of the bitmap displayed on the screen, but entails a resolution-independent redrawing

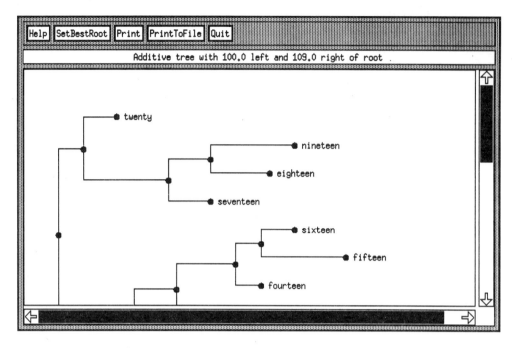

Fig. 4: Screen Dump of **Xtree** Showing an Additive Tree.

of the tree using true PostScript primitives. The bitmap on which the tree is drawn can be larger than the window in which it is displayed. In such a case, the user can select which region of the bitmap that he or she wants to see by manipulating the scroll bars (see Fig. 4). When the size of the bitmap equals the size of the window the scroll bars disappear. The size of the bitmap, as well as a variety of other parameters (such as color, font size, etc.) can be specified using *X resources*. The use of a scrollable region enables **Xtree** to be used with fairly large trees without affecting the readability of the displayed tree. A screen dump of **Xtree** showing a part of a large additive tree is presented in Fig. 4. When an additive tree is displayed, the user can carry out subtree rotations by clicking on internal nodes (using the left mouse button). In addition, the placement of the root can be changed by clicking on a link of the tree using the right mouse button. By clicking on the **SetBestRoot** button, the root is put at the point that minimizes the variance of the distances from the root to the terminal nodes.

When the user presses the left mouse button, the action routine **RotateTree_Action** is invoked:

```
void RotateTree_Action(TREE *tree)
{
    NODE *selected_node;

    if ((selected_node = SearchNode(tree->root)) == NULL)
        return;
    DrawTree(RotateTree(tree, selected_node));
}
```

SearchNode determines in a recursive fashion whether the mouse was on (or near) an internal node when the user pressed the left button:

```
NODE *SearchNode(NODE *node)
{
    NODE *selected_node = NULL;

    if (MouseOnNode(node))
        selected_node = node;
    if (node->left && !selected_node)
        selected_node = SearchNode(node->left->lower);
    if (node->right && !selected_node)
        selected_node = SearchNode(node->right->lower);
    return(selected_node);
}
```

MouseOnNode returns TRUE, if the mouse location was near vertex node when the mouse button was pressed.

When Xtree displays an additive tree, the root can be changed by clicking on a link of the tree using the right mouse button. To determine whether the mouse was on (or near) a link when the right mouse button was pressed, the recursive function SearchLink is called:

```
LINK *SearchLink(NODE node)
{
    LINK *selected_link = NULL;

    if (node->upper && MouseOnLink(node->upper))
        selected_link = node->upper;
    if (node->left && !selected_link)
        selected_link = SearchLink(node->left->lower);
    if (node->right && !selected_link)
        selected_link = SearchLink(node->right->lower);
    return(selected_link);
}
```

MouseOnLink determines whether the mouse pointer was near the indicated link when the mouse button was hit. SearchLink is used by SetRoot_Action which is invoked whenever the user presses the right mouse button:

```
void SetRoot_Action(TREE *tree)
{
    LINK *selected_link;

    if (tree->type != Additive)
        return;
    if ((selected_link = SearchLink(tree->root)) == NULL)
        return;
    SetRoot(tree, selected_link, GetInterpolation());
    DrawTree(tree->root);
}
```

The function `GetInterpolation` computes the relative position of the root on the selected link, depending on the location of the mouse on the screen.

4   Conclusion

In this paper, a graphical procedure was presented for dynamically displaying ultrametric and additive trees. The procedure allows the user to perform all permissible operations on the tree that do not affect the tree distances. The tree obtained in such an interactive session can be stored in an encapsulated PostScript file for inclusion in documents or can be directly printed on a PostScript printer. Unlike the printer plots proposed by Kruskal and Landwehr (1983) and Rousseeuw (1986), the plots generated by **Xtree** are fully graphical and resolution-independent.

The current procedure could be extended to deal with so-called two-class ultrametric and additive trees (cf. De Soete, DeSarbo, Furnas, and Carroll, 1984). As discussed by De Soete et al. (1984), the indeterminacies in these types of trees are much more intricate than in the usual one-class ultrametric and additive trees. Hence, an interactive procedure along the lines of **Xtree** might be very useful for interpreting and comparing such trees.

Upon request, the source of the **Xtree** procedure and associated library can be obtained free of charge via electronic mail. Enquiries and requests should be mailed to `geert@crunch.rug.ac.be`.

References

ADOBE SYSTEMS (1985), *PostScript Language Reference Manual*, Addison-Wesley, Reading, Massachusetts.

CRITCHLEY, F., and HEISER, W. (1988), Hierarchical Trees Can be Perfectly Scaled in One Dimension, *Journal of Classification, 5, 5–20*.

DEGERMAN, R. (1982), Ordered Binary Trees Constructed Through an Application of Kendall's Tau, *Psychometrika, 47, 523–527*.

DE SOETE, G. (1984), Computer Programs for Fitting Ultrametric and Additive Trees to Proximity Data by Least Squares Methods, *Behavior Research Methods, Instruments, & Computers, 16, 551–552*.

DE SOETE, G., DESARBO, W.S., FURNAS, G.W., and CARROLL, J.D. (1984), The Estimation of Ultrametric and Path Length Trees from Rectangular Proximity Data, *Psychometrika, 49, 289–310*.

GRUVAEUS, G., and WAINER, H. (1972), Two Additions to Hierarchical Cluster Analysis, *British Journal of Mathematical and Statistical Psychology, 25, 200–206*.

JOHNSON, S.C. (1967), Hierarchical Clustering Schemes, *Psychometrika, 32, 241–254*.

KERNIGHAN, B.W., and RITCHIE, D.M. (1988), *The C Programming Language*, Prentice-Hall, Englewood Cliffs, New Jersey.

KRUSKAL, J.B., and LANDWEHR, J.M. (1983), Icicle Plots: Better Displays for Hierarchical Clustering, *American Statistician, 37, 162–168*.

ROBERTS, F. (1979), *Measurement Theory*, Addison-Wesley, Reading, Massachusetts.

ROUSSEEUW, P. J. (1986), A Visual Display for Hierarchical Classification, in: E. Diday, Y. Escoufier, L. Lebart, J. Pages, Y. Schektman and R. Tomassone (eds.), *Data Analysis and Informatics, IV,* North-Holland, Amsterdam, 743–748.

SATTATH, S., and TVERSKY, A. (1977), Additive Similarity Trees, *Psychometrika, 42, 319–345.*

SCHEIFLER, R.W., and GETTYS, J. (1986), The X Window System, *ACM Transactions on Graphics, 5, 79–109.*

SCHEIFLER, R.W., GETTYS, J., and NEWMAN, R. (1988), *The X Window System,* DEC Press, Massachusetts.

YOUNG, D.A. (1989), *X Window Systems Programming and Applications with Xt,* Prentice Hall, Englewood Cliffs, New Jersey.

Anaglyphen 3D — A Program for the Interactive Representation of Three–Dimensional Perspective Plots of Statistical Data

Jürgen Symanzik
Fachbereich Statistik, Universität Dortmund, Postfach 50 05 00
W–4600 Dortmund 50, FRG

Abstract: Two projections of a three–dimensional figure, usually drawn in the complementary colours red and green, are looked at through filter glasses of the corresponding colours. Then the onlooker obtains the impression of a three–dimensional object. This kind of representation is called anaglyphs. Although anaglyphs have been used in geometry as in Pal (1974) and Fejes– Toth (1965), in architecture, and in chemistry for a long time, only fast computers with a high–resolution colour–screen allow an easy calculation and fast representation of the two required projections especially if in motion. The use of anaglyphs for analyzing statistical data has been proposed by Hering in 1987 and has been realized by von der Weydt (1988/89) and Symanzik (1990/91). In contrast to a single scatter plot, where only a pair of the factors of multi–dimensional data is to be seen in a single graph, and in contrast to the well–known static scatter plot matrices, anaglyphs allow the presentation of three factors at the same time in motion pictures. Further, the use of anaglyphs allows a good plastic impression of bivariate probability density functions, e. g. to emphasize the differences between bivariate normal–, Cauchy–, and t–distributions. Based on a few operations on matrices to calculate the red and green biplots, presented in Graf (1987), the program *Anaglyphen 3D* has been developed for Sun and Sparc workstations, allowing interactive work on anaglyphs.

1 Introduction

Due to new improvements in the area of computer hardware, especially concerning the speed of evaluation and the amount of main memory, new perspectives open up for computer software. There exists a lot of programs for graphical representation of statistical data, and further improvements on this software are made to achieve even better releases.

However, there is one limitation in common to all these programs: each representation of multi–dimensional data is confined to the size of the medium they are presented on, e. g. paper or screen, and thus, all graphics must be drawn on the two–dimensional plane.

On the other hand, there exists a technique first described by Rollmann (1853) and meanwhile well–known in geometry (comp. Pal (1961, 1974) and Fejes–Toth (1965)), but also used in architecture (comp. Schwenkel (1972)) and in chemistry (comp. Klages (1965)), to yield a real three–dimensional picture. For this method, called anaglyphs, two pictures have to be drawn, usually in the complementary colours red and green, and then they are looked at through filter glasses of the corresponding colours. Obviously, the adaptation of this technique to statistical data might be very powerful. At least, a better "real world" impression of 3D data will simplify graphical data analysis and probably more adequate hypotheses for later hypotheses testing will be found in this first step of the analysis. Concerning the use of anaglyphs in statistics Huber (1987), p. 450, notes: "Statisticians still lag behind other scientists in their use of stereo pairs. [..] The statisticians' efforts (e. g., the plates in Wegman and DePriest 1986) come late and pale in comparison."

Aside the use of anaglyphs as a means of exploratory data analysis, they may be used to visualize bivariate functions $z = f(x, y)$. Although there exist several statistical packages such as Huber's ISP or the S–Language (comp. Becker, Chambers and Wilks (1988)) producing grids of connected points, none of these programs succeeds in obtaining such a good three–dimensional impression of the interesting function as anaglyphs do.

In a first phase of work a program for anaglyphs in statistics has been developed for PC's. Results are reported in Hering and von der Weydt (1989). Although the goal of a three–dimensional data presentation has been achieved, there were limitations caused by the computer hardware. Therefore, a new version of the program, now called *Anaglyphen 3D*, has been implemented on Sun and Sparc workstations with a high–resolution colour–screen using the graphics environment Suntools. A report has been given in Hering and Symanzik (1992).

2 Main Properties of the Program *Anaglyphen 3D*

During planning and implementation of the program, two aspects have been considered as most important:

- The regular user of *Anaglyphen 3D* will be a statistician or even a student not very familiar with computers. Easy handling is necessary. Therefore, the graphics environment Suntool was choosen to give full support to the mouse as main input device. Meanwhile, a version is available based on the XView toolkit and thus running on all machines supporting the XWindow System.

- New pictures should be available in a part of a second to yield a cinematic impression during rotations of the data. The actual version of the program allows [1] the recalculation and presentation of anaglyph pictures composed of 400 single data points for 60 times a second. This fulfills the requirements for many not too large sample analyses, but the program is still an excellent tool for didactics and demonstrations even for larger data sets. E. g. the well known Anderson (1935) iris data set which often is used to demonstrate features of statistical graphic software, only contains 150 data points.

Figure 1 shows the main menu [2] of *Anaglyphen 3D*.

A summarizing description of the main features follows:

Via the "Winkel" [Angle] submenu and the "Rotiere x" [Rotate x], "Rotiere y" [Rotate y], "Rotiere z" [Rotate z] buttons the viewpoint, number of rotation steps, and angle of each step are fixed, "Stop" terminates a rotation. Hardware parameters are set using "Default".

The "Funktion" [Function] submenu, on the one hand, enables access to external programs creating data, and on the other hand, it integrates external data available in ASCII files into the system. As an example a link to the S–PLUS Language incorporates data valid for S–PLUS's three–dimensional perspective plots command "persp" into *Anaglyphen 3D*.

In "Dichten" [Density Functions] an internal library of ten bivariate pdf's taken from Mardia (1970) is available. This part gives a plastic impression of surfaces which are hard to imagine by formulae, and also emphasizes differences and identities of several distributions,

---

[1] This was measured on a Sparc 2 workstation otherwise unloaded.

[2] A plot of the bivariate normal pdf with parameters $\mu_1 = \mu_2 = 0, \sigma_1^2 = \sigma_2^2 = 1, \rho = 0$ is shown for $x, y \in [-5, 5]$.

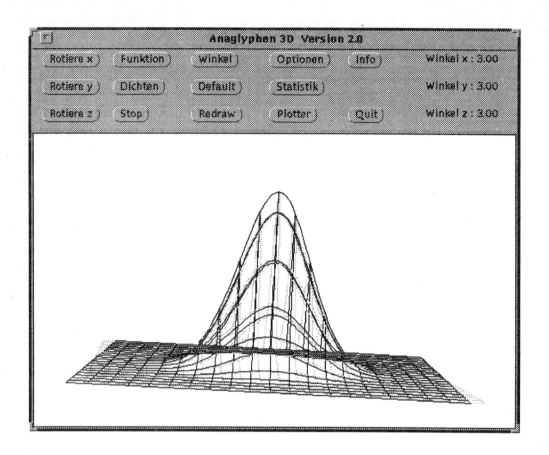

Figure 1: Main Menu of *Anaglyphen 3D*

e. g. of bivariate normal–, Cauchy–, and t–distributions. Furthermore, the visualization of pdf's in combination with a minimum of statistical information noted in the "Statistik" [Statistics] window allowed the removal of printing bugs from literature at first glance. As an example, the bivariate Type II distribution noted in Mardia (1970), p. 93–94, as

$$h(x,y) = \frac{1}{2\pi\sigma_1\sigma_2\sqrt{1-\rho^2}} \frac{n+1}{n+2} \left(1 - \frac{1}{2(n+2)(1-\rho^2)}\left(\frac{x^2}{\sigma_1^2} - \frac{2\rho xy}{\sigma_1\sigma_2} + \frac{y^2}{\sigma_2^2}\right)\right)^n$$

where $-c\sigma_1 < x < c\sigma_1, -c\sigma_2 < y < c\sigma_2, c^2 = 2(n+2)$ and $n > 0$, was easy to correct. Obviously, the support of this pdf must be restricted to the ellipsoid

$$\frac{x^2}{\sigma_1^2} - \frac{2\rho xy}{\sigma_1\sigma_2} + \frac{y^2}{\sigma_2^2} < 2(n+2)(1-\rho^2),$$

since otherwise, $h(x,y)$ yields negative values in the rectangle outside the ellipsoid if n is odd. Moreover, for $n = 0$, a setting neglected by Mardia, this pdf reduces to a uniform distribution over the given ellipsoid, and it converges towards the bivariate normal distribution for $n \to \infty$. Good graphical results of convergence are obtained even for small n.

Using "Plot", an external file is created. It includes HPGL commands to create anaglyph pictures on a pen plotter. "Info" shows information on the actual version of the program.

Finally in "Optionen" [Options] the plot style can be changed, "Redraw" recalculates changes on the last graphic, and "Quit" leaves the program.

3 Basic Mathematics for Anaglyphs

Only a few operations on matrices, presented in Graf (1987), are necessary to calculate anaglyph pictures. Therefore, consider the model that an object consisting only of edges, but without surfaces, is suspended between the onlooker's eyes and a computer screen. Now draw the image of the projection from each eye onto the computer screen. Two different pictures will appear as in Figure 2.

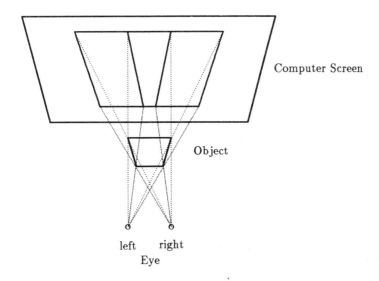

Figure 2: Visual Models for Anaglyphs

When removing the object, but retaining the projection on the screen, the same object would reappear, if each eye could see only its part of the projection. Since filter glasses of complemantary colours solve this part of the problem, only a few formulae must be known.

To simplify trigonometrical calculations, assume the case of Figure 3.

The onlooker's left eye is at a distance d from the center of the screen, representing the origin of the coordinate system. Let a be the distance between the observer's two eyes, then the angle α between the two projections approximately equals $\alpha \approx \arctan \frac{a}{d}$, since $d \gg a$.

As is well known, the matrices

$$d_1(\beta) := \begin{pmatrix} 1 & 0 & 0 \\ 0 & \cos\beta & -\sin\beta \\ 0 & \sin\beta & \cos\beta \end{pmatrix}, d_2(\beta) := \begin{pmatrix} \cos\beta & 0 & \sin\beta \\ 0 & 1 & 0 \\ -\sin\beta & 0 & \cos\beta \end{pmatrix}, d_3(\beta) := \begin{pmatrix} \cos\beta & -\sin\beta & 0 \\ \sin\beta & \cos\beta & 0 \\ 0 & 0 & 1 \end{pmatrix}$$

yield rotations about an angle β on the x_1-, x_2-, and x_3-axis when multiplied from the left hand side to a vector p. Arbitrary rotations are obtained by multiplications of these

388

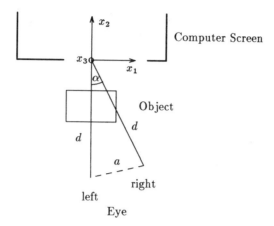

Figure 3: Important Parameters for Trigonometrical Calculations

matrices, and especially the two different projections result, when the points of the first picture are transformed via the matrix $d_3(\alpha)$.

Finally, in the case of a central projection each three-dimensional point $p := \begin{pmatrix} p_1 \\ p_2 \\ p_3 \end{pmatrix}$ must be transformed into a two–dimensional point $\begin{pmatrix} x \\ y \end{pmatrix} := f \cdot \begin{pmatrix} p_1 \\ p_3 \end{pmatrix}$, where $f := \frac{d}{d+p_2}$ indicates a transformation of the depth value from the x_2–axis. These new coordinates are to be mapped onto the computer screen using their corresponding red and green colours.

Acknowledgements

The author would like to thank Prof. Dr. Franz Hering, Dortmund, for his helpful comments and hints.

References

ANDERSON, E. (1935) : The Irises of the Gaspe Peninsula, *Bulletin of the American Iris Society*, **59**, 2–5.

BECKER, R.A., CHAMBERS, J.M. and WILKS, A.R. (1988) : *The New S Language*, Wadsworth & Brooks/Cole Advanced Books & Software, Pacific Grove, California.

FEJES–TOTH, L. (1965) : *Reguläre Figuren*, Akademiai Kiado, Budapest.

GRAF, M. (1987) : Drehen und Wenden, Ein Verfahren zur Manipulation räumlicher Objekte, *c't*, Heft **7**, 126– 130.

HERING, F. and VON DER WEYDT, S. (1989) : *Interaktive Anaglyphendarstellungen als Hilfsmittel zur Analyse mehrdimensionaler Daten*, Forschungsbericht 89/7, Fachbereich Statistik, Universität Dortmund.

HERING, F. and SYMANZIK, J. (1992) : *Anaglyphen 3D — Ein Programm zur inter-*

aktiven Anaglyphendarstellung, Forschungsbericht 92/1, Fachbereich Statistik, Universität Dortmund.

HUBER, P.J. (1987) : Experiences With Three–Dimensional Scatterplots, *Journal of the American Statistical Association*, **82**, 448–453.

KLAGES, F. (1965) : *Einführung in die organische Chemie*, 2. Auflage, de Gruyter, Berlin.

MARDIA, K.V. (1970) : *Families of Bivariate Distributions*, Griffin, London.

PAL, I. (1961) : *Darstellende Geometrie in Raumbildern*, 1. Auflage der deutschen Übersetzung, Lindauer, München.

PAL, I. (1974) : *Raumgeometrie in der technischen Praxis*, Akademiai Kiado, Budapest.

ROLLMANN, W. (1853) : Zwei neue stereoskopische Methoden, *Annalen der Physik und Chemie*, **Band 90**, 186–187.

SCHWENKEL, D. (1972) : Räumlich perspektive Darstellung von Bauwerken durch automatisch gezeichnete Anaglyphenbilder, *Bildmessung und Luftbildwesen*, **40**, 144–147.

WEGMAN, E.J. and DE PRIEST, D.J. (Editors) (1986) *Statistical Image Processing and Graphics*, Marcel Dekker, New York.

Part III

Applications and Special Topics

Discovering Consensus Molecular Sequences

William H. E. Day
Department of Computer Science
Memorial University of Newfoundland
St. John's, NF A1C 5S7, CANADA

F. R. McMorris
Department of Mathematics
University of Louisville
Louisville, KY 40292, USA

Abstract: Consensus methods are valuable tools for data analysis, especially when some sort of data aggregation is desired; but although methods for discovering consensus sequences play a vital role in molecular biology, researchers often seem inattentive to the features and limitations of such methods, and so there are risks that criteria for discovering consensus sequences will be misused or misunderstood. To appreciate better the issues involved, we survey methods for discovering consensus sequences such as those based on frequency thresholds, voting strategies, heuristics, neighbourhoods, and measures of inhomogeneity or information content.

1 Introduction

For centuries past, consensus methods have been effective tools to bring about political or social change by rational means — the pioneering contributions of Borda (1781) and Condorcet (1785) spring immediately to mind — and so it seems natural to develop basic concepts of consensus in a simple voting context (Day (1988)). Let there be an electorate of k voters in an election involving a slate S of alternatives. The votes cast in the election are specified by a profile $P = (p_1, \ldots, p_k)$ in which p_i denotes the alternative selected by voter i. For each alternative $a \in S$, let $n_a(P)$ denote the number of occurrences of a in P, while $N(P) = \max_{a \in S} n_a(P)$. The set of all profiles is the k-fold Cartesian product S^k; the set of all electoral outcomes is the power set $\Pi'(S)$ of all subsets of S. The criterion of obtaining consensus by *majority rule* (Campbell (1980, 1982); May (1952); Straffin (1977)) is a function $Maj : S^k \to \Pi'(S)$ such that for each profile $P \in S^k$, $Maj(P) = \{a \in S : n_a(P) > k/2\}$. Majority rule's appealing features are that, when it exists, the winner is unique and is supported by more than half the voters; but with slates of more than one alternative, majority rule need not select a winner. One generalization, the criterion of obtaining consensus by *plurality rule* (Richelson (1975, 1978); Roberts (1991a, 1991b)), is a function $\wp_1 : S^k \to \Pi'(S)$ such that for each profile $P \in S^k$, $\wp_1(P) = \{a \in S : n_a(P) = N(P)\}$. Although plurality rule always returns at least one winner, such winners may be supported by only a minority of the voters. A third consensus criterion identifies the set of median (or middlemost) alternatives among those in S. To compare alternatives of S, let d be a metric defined on S. The criterion of obtaining consensus by *median rule* (Kemeny (1959), Kemeny and Snell (1962)) is a function $Med : S^k \to \Pi'(S)$ such that for each $P = (p_1, \ldots, p_k) \in S^k$, $Med(P) = \{a \in S : \sum_{1 \le j \le k} d(a, p_j)$ is the minimum$\}$. A particularly natural specification of a median rule $M : S^k \to \Pi'(S)$ replaces d by the identity metric $\iota : S^2 \to \{0, 1\}$ for which

$\iota(x, y) = 0$ if and only if $x = y$, so that $M(P) = \{a \in S : \sum_{1 \le j \le k} \iota(a, p_j)$ is the minimum$\}$. In this context, $\wp_1 = M$ so that plurality rule is a special case of a median rule. These rules — majority, plurality, and median — play fundamental roles in consensus theory.

In contexts relevant to biologists, consensus concepts are being developed and applied with considerable success. Barthélemy and Monjardet (1981, 1988) explore the median rule's pervasive influences in cluster and data analyses. Margush and McMorris (1981), McMorris and Neumann (1983), and Barthélemy and McMorris (1986) apply consensus concepts, such as the majority rule, in analyses of rooted trees with labeled leaves. An issue of the *Journal of Classification* has featured papers about consensus classifications (Day (1986)), notable among them the review by Barthélemy, Leclerc, and Monjardet (1986) of how ordered set theory provides both framework and tools for investigating problems on the consensus and comparison of classifications. Papers presented at the First Conference of the International Federation of Classification Societies consider consensus methods as tools for data analysis (Day (1988)) in the biological sciences (Faith (1988)) and the social sciences (Leclerc (1988)).

Researchers have also investigated problems involving consensus molecular sequences (strings, words, motifs, patterns). These problems are of two major types. In the *consensus matching problem*, we are given a set of variants of a known consensus pattern, and we must analyse a new sequence to locate subsequences that are similar to (*i.e.*, match) the variants (Staden (1988), Stormo (1987, 1988, 1990)). In the *consensus discovery problem*, we are given a set of related molecular sequences, and we must analyse them to discover frequently occurring subsequences (if they exist) that may not be conserved precisely in either location or pattern. This is the problem on which we focus. Its character, and even the complexity of its solution, varies according to the underlying assumptions being made.

(*i*) If the unknown pattern for which we search appears in every sequence and in contiguous positions (*e.g.*, XBOCKYZ, XYBOCKZ, BOCKYZX), the consensus discovery problem simplifies to searching for a longest common substring (*e.g.*, BOCK). Although this variant can be solved efficiently, we do not consider it further because its hypotheses are too strong to be useful in molecular contexts.

(*ii*) If the unknown pattern need not appear perfectly in every sequence, but appears (perhaps imperfectly) in the same contiguous positions of every sequence (*e.g.*, XHBOCKZ, YHBOZKX, ZXBOCKY), the consensus discovery problem becomes one of searching intervals of aligned sequences for consensus substrings (*e.g.*, HBOCK in positions 2-6). The complexity of such a problem depends on the consensus criteria being applied within the sequence intervals. One questionable (but often-made) assumption is that aligned positions within molecular sequences can be treated independently of each other, so that consensus patterns can be discovered by analysing aligned positions, one by one. Section 2.1 surveys such methods.

(*iii*) If the unknown pattern appears in every sequence but need not occupy contiguous positions (*e.g.*, XBOCKYZ, XBOYCKZ, BYOZCXK), the consensus discovery problem becomes one of searching for a longest common subsequence (*e.g.*, BOCK). For two sequences this problem is solvable by efficient dynamic-programming algorithms (Hirschberg (1975, 1983); Needleman and Wunsch (1970)). If the number of sequences is not constrained and is part of the problem specification, the decision problem of finding a longest common subsequence is NP-complete (Maier (1977)) and thus is not likely to be solved efficiently. The fourth case, although not more tractable, has received considerably more attention.

(*iv*) If the consensus pattern (*e.g.*, HBOCK) need not appear identically in every sequence (*e.g.*, XHBOCKZ, XYHBOCZ, HBOCKXZ), the consensus discovery problem is a generalization of the problem to find a longest common subsequence (Day and McMorris (1992d)) and, as before, it is not likely to be solved efficiently. Researchers typically further restrict such problems by fixing the length of the pattern being sought, or by fixing the width of a larger sequence window in which the search will occur. Section 2.2 surveys such methods.

The consensus discovery problem bears a dual relationship to that of aligning molecular sequences so that aligned positions represent homologous biological characters. Consider, for example, two fragments of 5S ribosomal RNA

$$E.\ coli: \qquad \text{UGCCUGGCGGCCGUA...}$$
$$\text{Human:} \qquad \text{GUCUACGGCCAUA...}$$

We might align them to obtain (where φ represents a gap)

$$E.\ coli: \qquad \text{UGCCUGGCGGCCGUA...}$$
$$\text{Human:} \qquad \varphi\text{GUCUA}\varphi\text{CGGCCAUA...}$$

and thus a consensus sequence: $\qquad$ - G - CU - - CGGCC - UA...

Alternatively, we might find a longest subsequence common to the two sequences (e.g., GCUCGGCCUAI) and align the remaining positions relative to that consensus sequence. Since alignment problems are ubiquitous in the applied sciences, their theory and methodology have been studied intensively. Kruskal (1983) gives an excellent tutorial on sequence alignment, while the book edited by Sankoff and Kruskal (1983) summarizes the state of alignment research as of 1983. More recently, Waterman (1989b) and Waterman, Joyce, and Eggert (1991) review mathematical and algorithmic aspects of sequence alignment problems.

2 Methods for Discovering Consensus Sequences

Our survey of methods for the consensus discovery problem is based on cases *(ii)* and *(iv)* in Section 1.

2.1 When Sequences are Aligned and Positions are Analysed Independently

The consensus methods in this class summarize the distributions of alternatives (e.g., nucleic acid bases, amino acids) at each single position of an aligned set of molecular sequences. Typically they assume: the analysis is a multi-stage process in which the alignment of molecular sequences precedes the discovery of consensus sequences; an alignment has already been obtained; each aligned position can be treated independently of the others (Day and McMorris (1992a)). Thus the problem to find a consensus of k aligned molecular sequences, in which n aligned positions have been identified, can be viewed as a set of n simpler problems, each to find a consensus of k symbols at an aligned position. To model the simpler problem, let a *consensus method* be a function $cm : U \to V$ which maps each element of its domain U to a suitable element of its codomain V. To specify U and V, let S be a finite set of symbols of interest. Although S might be the set of amino acids, or any other finite set, we usually take S as the set {A,C,G,T} of nucleic acid bases so as to place the consensus problem in a DNA context. Each element of U represents a possible k-tuple of bases appearing at a given aligned position of k molecules. As before, for any positive integer k, let S^k denote the set of profiles of S of length k. We will use S^k as the domain of the consensus methods we investigate. The codomain V typically is a set of ambiguity codes such as those proposed by the Nomenclature Committee of the International Union of Biochemistry (1985). However, in order to emphasize the constituent bases, and their number, we will represent each such code by the subset of its bases. Thus (with a minor abuse of set-theoretic notation) we define the set of nonempty subsets of S to be

$$\Pi'(S) = \{A, C, G, T, AC, AG, AT, CG, CT, GT, ACG, ACT, AGT, CGT, ACGT\},$$

where AG denotes a purine base, CT denotes a pyrimidine base, and so on. Letting $\Pi(Z)$ denote the set of all subsets of Z, we will use $\Pi(\Pi'(S))$ as the codomain of the consensus methods we investigate. Consequently each consensus method is a function $cm : S^k \to \Pi(\Pi'(S))$ which maps a profile P of length k to a (possibly empty) subset of ambiguity codes from $\Pi'(S)$.

Threshold Methods. When defining a threshold method's result for a profile P, it is convenient to specify the frequency of occurrence of each base in P, and furthermore to relabel the bases of P in such a way that frequencies have a standard order. These assumptions cause no loss of generality since threshold methods depend on the frequencies, but not on physical or chemical properties of particular bases. Throughout, when referring to $S = \{A, C, G, T\}$, we adopt the alphabetical order $A < C < G < T$ as a standard order of the bases, and we specify the frequencies with which bases occur in P by a vector $f(P) = (f_1, f_2, f_3, f_4)$, with $1 = f_1 + f_2 + f_3 + f_4$, where bases in the profile are relabeled so that $f_1 \geq f_2 \geq f_3 \geq f_4 \geq 0$, f_1 is the frequency of A, f_2 of C, f_3 of G, and f_4 of T. A method's decision to return an ambiguity code for P may employ threshold criteria of the forms $f_i > c$ or $f_i \geq c$, where c is a constant (Yamauchi (1991)), gap criteria of the form $f_i/f_{i+1} > c$ (Choo et al. (1991)), or combinations of the two (Cavener (1987), Shapiro and Senapathy (1987)). But when these methods return ambiguity codes of different lengths (i.e., codes representing different numbers of bases), they may apply different criteria depending on the lengths of the codes being returned. Day and McMorris (1992b) conduct a critical comparison of eight threshold consensus methods (including the ones just mentioned). They also introduce a parameterized threshold consensus method (1992e) that is based on a majority-rule voting principle and that uses a single criterion to return ambiguity codes of different lengths.

Plurality-rule Methods. Consider a consensus method $\wp_1 : S^k \to \Pi(\Pi'(S))$ such that for each profile $P \in S^k$, $\wp_1(P) = \{a \in S : n_a(P) = N(P)\}$. Informally, $\wp_1$ is a consensus method in which each of the k molecules votes in an election with the slate $S = \{A, C, G, T\}$ of candidates. If the base receiving the largest number of votes is unique, $\wp_1$ returns it as the consensus; if several bases receive the largest number of votes, $\wp_1$ returns them all as consensus results. Since $\wp_1$ is a plurality-rule voting strategy, much is known about its formal properties (Day and McMorris (1992a); Roberts (1991a, 1991b)). Since the definition of $\wp_1$ requires it to return ambiguity codes of length at most one (as indicated by the subscript), Day and McMorris (1992a) propose a generalization $\wp_j$, where $1 \leq j \leq |S|$, which is based on the plurality-rule concept but which returns ambiguity codes of length at most j. When $S = \{A, C, G, T\}$, Day and McMorris compare $\wp = \wp_4$ with threshold methods (1992b) and interpret its results (1992c), while Day and Mirkin (1992) establish that $\wp$ returns exactly 26 nonequivalent consensus results.

Weighted methods. Methods to discover consensus sequences can be based on scoring strategies that were developed for the consensus matching problem. Such methods use the observed frequency $n_a(P)$ with which a base $a \in S$ occurs in a profile P in order to give a score to a; then all bases having the maximum score in P are returned as the consensus result for P. Informally, these consensus methods are variants of $\wp_1$ in which scores are substituted for observed frequencies $n_a(P)$ in the specification of $\wp_1$. Mulligan *et al.* (1984) use a score which is the distance between a base's observed and expected frequencies, with distance being measured in standard deviations (assuming Poisson statistics). Staden (1984) uses a score for $a \in S$ which is the logarithm of $n_a(P)/k$ and so estimates the probability of finding a in P. Stormo (1990) and others use a specificity score which is the logarithm of $n_a(P)/(kp_a)$, where p_a estimates the probability of finding a in the molecules being analysed. This score can be rationalized using information theory, thermodynamics, or likelihood statistics (Schneider *et al.* (1986); Stormo (1988, 1990)).

A Heuristic Method. Bains describes a heuristic algorithm to align multiple DNA sequences (Bains (1986)) or protein sequences (Bains (1989)). A novel feature of the algorithm is that it iterates between an alignment step, which generates an alignment of the molecular sequences using the current consensus sequence, and a consensus step, which generates a consensus sequence using the current sequence alignment. When $S = \{A, C, G, T\}$ Bains (1986) applies seven heuristically-developed rules to generate a consensus at each aligned position; this consensus method could be applied independently of the alignment program for which it was developed.

2.2 When Sequences are Unaligned and Positions are Grouped into Words

Neighbourhood Methods. Consider a set of R DNA sequences that are approximately aligned with respect to some biological feature. Place a window of width W over the sequences. Suppose we wish to discover a consensus word w^*, of fixed length k, which might occur in many of the sequences. Since w^* need not appear perfectly in a sequence, we use the notion of a neighbourhood of w^* to determine when w^* is realized by a word in the sequence. To specify this, let d be the maximum number of mismatches permitted between w^* and any realization w_i in the ith sequence; informally, d defines a neighbourhood of w^* in which w_i must lie in order to be considered a realization of w^*. Since the sequence alignment need not be exact, we assume only that any realization of w^* must occur within the window. Once the user has specified the parameters W, k, and d, the problem is to find a consensus word w^* of length k, and its realizations w_i, $i = 1, \ldots, R$, such that the total number of mismatches between w^* and its realizations is minimized. Waterman, Arratia, and Galas (1984) describe a general method for discovering such consensus words, and they develop estimates of the statistical significance of such words. Galas, Eggert, and Waterman (1985) use the method to analyse promoter sequences from *E. coli*. Waterman extends the method to align multiple sequences by consensus (1986, Waterman and Jones (1990)) and to discover long consensus words and consensus palindromes (1989a). Waterman and Jones (1990) generalize the method so as to discover consensus words in proteins. Taylor, Rosenberg, and Samsonova (1991) use the neighbourhood concept in a method which uses bitwise operations and careful encodings of data to discover long consensus words. Staden (1989b) describes a method to maintain an indexed list of frequencies of similar words so as to permit searches for the commonest or best-defined subsequences, or for those that occur in the greatest number of different sequences.

Statistical Methods. Consider a set of R functionally related DNA sequences that are approximately aligned with respect to some biological feature. Place a window of width W over the sequences. To discover highly descriptive consensus words of fixed length k, Mengeritsky and Smith (1987) propose a method based on using a χ^2 statistic to measure the relative inhomogeneity of a word's empirical distribution as the window moves over the sequences. At each position of the window, the algorithm calculates the observed frequencies of all k-long words in the window; from these frequencies it calculates the observed frequencies of all k-long words that contain ambiguity codes. With these frequencies, the algorithm then calculates a χ^2 value for every observed k-long word. Any k-long word with a large χ^2 value is a possible consensus word, and it is associated with those window positions at which its observed frequency is a local maximum. Although it may be infeasible to use the method when $k > 5$, consensus words from neighbouring window positions can be combined to obtain consensus sequences longer than k.

Smith, Annau, and Chandrasegaran (1990) describe a method to find highly conserved 3-amino acid patterns in large sets of functionally related proteins. The approach is unusual because the patterns being sought have the form $aa1 - d1 - aa2 - d2 - aa3$, where the aa_i are amino acids and the user can fix the maximum values of the distances separating them. The observed frequencies of all such patterns are accumulated in a (very large) array. Those patterns that occur frequently are scored in order to distinguish significant patterns from random background patterns. Segments of the proteins containing significant patterns can then be used to align the proteins. The method works well for large numbers of sequences with few gaps located in significant patterns.

An Information-theoretic Method. Stormo and Hartzell (1989) and Hertz, Hartzell, and Stormo (1990) describe a method to discover a consensus word of length k in a set of R functionally related DNA sequences that are approximately aligned with respect to some biological feature. The method's idea is that, if the sequences were aligned on a functionally important feature of length k, such as a recognition pattern for a DNA-binding protein, the feature would maximize information content (*sensu* Schneider *et al.* (1986)) within a local window of width k. The information content of such a feature could be described by a matrix $F = (f_{ij})$ in which f_{ij} is the frequency with which base i appears at position j of the feature. Since the actual alignment is unknown, the problem is to search for a matrix F that maximizes the sum of the information content at each of the k positions. The algorithm adds sequences to the analysis, one by one. At each stage it maintains a pool of many possible candidate matrices, each representing a possible alignment of the current sequences at the feature being sought. As more sequences enter the analysis, a matrix with a high information content emerges from the pool of candidates. When the final matrix F has been identified, any of the consensus methods in Section 2.1 could be applied at each position to describe the consensus sequence: Stormo and Hartzell (1989) favour the specificity score (Stormo (1990)) since it is based, as well, on analyses by information content.

3 Discussion

The classification scheme of Section 2 does not easily accommodate some methods for discovering consensus sequences. Smith and Smith (1990), for example, construct a binary dendrogram for a set of related protein sequences, and then they analyse its nodes in order from leaves to root. At each interior node they align the sequences of the node's children and thus generate a consensus sequence, or 'covering pattern', for the corresponding cluster of sequences. Vingron and Argos (1991), on the other hand, use dot plots to depict pairwise comparisons of similarities among functionally related protein sequences. They describe a novel procedure, based on matrix multiplication, to discover consistently alignable sequence fragments (*i.e.*, consensus words).

Complementary to the problem of discovering consensus sequences is that of assessing their significance. Waterman (1989a, 1989b) discusses how results from extreme value theory and the theory of large deviations can be used to estimate the significance of consensus words discovered by the neighbourhood method of Waterman, Arratia, and Galas (1984). Staden (1989a) uses probability-generating functions to calculate the probabilities of finding a consensus word in nucleic acid or protein sequences when the word is defined in any of nine ways. Stückle *et al.* (1990) describe a Markov chain model which permits the simultaneous analysis of several linear nucleic acid sequences with unequal base frequencies and Markov order other than zero. Day and McMorris (1992c) use partitions of integers to calculate the probabilities of obtaining plurality-rule consensus results (Day and McMorris 1992a) in analyses of up to 100 sequences.

4 Acknowledgements

This work was supported in part by the Natural Sciences and Engineering Research Council of Canada under grant A-4142, and by the US Office of Naval Research under grant N00014-89-J-1643. The first author is an Associate in the Program in Evolutionary Biology of the Canadian Institute for Advanced Research.

References

BAINS, W. (1986), MULTAN: A Program to Align Multiple DNA Sequences, *Nucleic Acids Research*, 14(1), 159–177.

BAINS, W. (1989), MULTAN (2), A Multiple String Alignment Program for Nucleic Acids and Proteins, *Computer Applications in the Biosciences*, 5(1), 51–52.

BARTHÉLEMY, J.-P., LECLERC, B., and MONJARDET, B. (1986), On the Use of Ordered Sets in Problems of Comparison and Consensus of Classifications, *Journal of Classification*, 3(2), 187–224.

BARTHÉLEMY, J.-P., and McMORRIS, F. R. (1986), The Median Procedure for *n*-Trees, *Journal of Classification*, 3(2), 329–334.

BARTHÉLEMY, J.-P., and MONJARDET, B. (1981), The Median Procedure in Cluster Analysis and Social Choice Theory, *Mathematical Social Sciences*, 1, 235–267.

BARTHÉLEMY, J.-P., and MONJARDET, B. (1988), The Median Procedure in Data Analysis: New Results and Open Problems, in: H. H. Bock (ed.), *Classification and Related Methods of Data Analysis*, Elsevier Science, Amsterdam, 309–316.

BORDA, Jean-Charles de (1781), Memoire sur les Élections au Scrutin, Histoire de l'Academie Royale des Sciences, Paris.

CAMPBELL, D. E. (1980), Algorithms for Social Choice Functions, *Review of Economic Studies*, 47, 617–627.

CAMPBELL, D. E. (1982), On the Derivation of Majority Rule, *Theory and Decision*, 14, 133–140.

CAVENER, D. R. (1987), Comparison of the Consensus Sequence Flanking Translational Start Sites in Drosophila and Vertebrates, *Nucleic Acids Research*, 15(4), 1353–1361.

CHOO, K. H., VISSEL, B., NAGY, A., EARLE, E., and KALITSIS, P. (1991), A Survey of the Genomic Distribution of Alpha Satellite DNA on all the Human Chromosomes, and Derivation of a New Consensus Sequence, *Nucleic Acids Research*, **19**(6), 1179–1182.

CONDORCET, M. J. A. N., CARITAT, Marquis de (1785), Essai sur l'Application de l'Analyse a la Probabilite des Decisions rendues a la pluralite des voix, L'Imprimerie Royale, Paris.

DAY, W. H. E. (1986), Foreword: Comparison and Consensus of Classifications, *Journal of Classification*, **3**(2), 183–185.

DAY, W. H. E. (1988), Consensus Methods as Tools for Data Analysis, in: H. H. Bock (ed.), *Classification and Related Methods of Data Analysis*, Elsevier Science, Amsterdam, 317–324.

DAY, W. H. E., and McMORRIS, F. R. (1992a), Consensus Sequences Based on Plurality Rule, *Bulletin of Mathematical Biology*, to appear.

DAY, W. H. E., and McMORRIS, F. R. (1992b), Critical Comparison of Consensus Methods for Molecular Sequences, *Nucleic Acids Research*, **20**(5), 1093–1099.

DAY, W. H. E., and McMORRIS, F. R. (1992c), Interpreting Consensus Sequences Based on Plurality Rule, *Mathematical Biosciences*, to appear.

DAY, W. H. E., and McMORRIS, F. R. (1992d), On the Computation of Consensus Patterns in DNA Sequences, *Mathematics and Computer Modeling*, accepted.

DAY, W. H. E., and McMORRIS, F. R. (1992e), Threshold Consensus Methods for Molecular Sequences, *Journal of Theoretical Biology*, accepted.

DAY, W. H. E., and MIRKIN, B. G. (1992), On the Existence of Constrained Partitions of Integers, *Journal of Computing and Information*, accepted.

FAITH, D. P. (1988), Consensus Applications in the Biological Sciences, in: H. H. Bock (ed.), *Classification and Related Methods of Data Analysis*, Elsevier Science, Amsterdam, 325–332.

GALAS, D. J., EGGERT, M., and WATERMAN, M. S. (1985), Rigorous Pattern-recognition Methods for DNA Sequences: Analysis of Promoter Sequences from Escherichia coli, *Journal of Molecular Biology*, **186**, 117–128.

HERTZ, G. Z., HARTZELL, G. W., and STORMO, G. D. (1990), Identification of Consensus Patterns in Unaligned DNA Sequences Known to be Functionally Related, *Computer Applications in the Biosciences*, **6**(2), 81–92.

HIRSCHBERG, D. S. (1975), A Linear Space Algorithm for Computing Maximal Common Subsequences, *Communications of the Association for Computing Machinery*, **18**(6), 341–343.

HIRSCHBERG, D. S. (1983), Recent Results on the Complexity of Common-subsequence Problems, in: D. Sankoff and J. B. Kruskal (eds.), *Time Warps, String Edits, and Macromolecules: The Theory and Practice of Sequence Comparison*, Addison-Wesley, Reading, Massachusetts, 325–330.

KEMENY, J. G. (1959), Mathematics Without Numbers, *Daedalus (Boston)*, **88**, 577–591.

KEMENY, J. G., and SNELL, J. L. (1962), Preference Ranking: An Axiomatic Approach, in: *Mathematical Models in the Social Sciences*, Ginn, New York, 9–23.

KRUSKAL, J. B. (1983), An Overview of Sequence Comparison: Time Warps, String Edits, and Macromolecules, *SIAM Review*, **25**, 201–237.

LECLERC, B. (1988), Consensus Applications in the Social Sciences, in: H. H. Bock (ed.), *Classification and Related Methods of Data Analysis*, Elsevier Science, Amsterdam, 333–340.

MAIER, D. (1977), The Complexity of Some Problems on Subsequences and Supersequences, *Journal of the Association for Computing Machinery*, **25**(2), 322–336.

MARGUSH, T., and McMORRIS, F. R. (1981), Note: Consensus n-Trees, *Bulletin of Mathematical Biology*, **43**(2), 239–244.

McMORRIS, F. R., and NEUMANN, D. (1983), Consensus Functions Defined on Trees, *Mathematical Social Sciences*, **4**, 131–136.

MAY, K. O. (1952), A Set of Independent, Necessary and Sufficient Conditions for Simple Majority Decision, *Econometrica*, **20**, 680–684.

MENGERITSKY, G., and SMITH, T. F. (1987), Recognition of Characteristic Patterns in Sets of Functionally Equivalent DNA Sequences, *Computer Applications in the Biosciences*, **3**(3), 223–227.

MULLIGAN, M. E., HAWLEY, D. K., ENTRIKEN, R., and McCLURE, W. R. (1984), Escherichia coli Promoter Sequences Predict in vitro RNA Polymerase Selectivity, *Nucleic Acids Research*, **12**(1), 789–800.

NEEDLEMAN, S. B., and WUNSCH, C. D. (1970), A General Method Applicable to the Search for Similarities in the Amino Acid Sequences of Two Proteins, *Journal of Molecular Biology*, **48**, 444–453.

NOMENCLATURE COMMITTEE OF THE INTERNATIONAL UNION OF BIOCHEMISTRY (NC-IUB) (1985), Nomenclature for Incompletely Specified Bases in Nucleic Acid Sequences — Recommendations 1984, *European Journal of Biochemistry*, **150**, 1–5.

RICHELSON, J. (1975), A Comparative Analysis of Social Choice Functions, *Behavioral Science*, **20**, 331–337.

RICHELSON, J. (1978), A Characterization Result for the Plurality Rule, *Journal of Economic Theory*, **19**, 548–550.

ROBERTS, F. S. (1991a), Characterizations of the Plurality Function, *Mathematical Social Sciences*, **21**(2), 101–127.

ROBERTS, F. S. (1991b), On the Indicator Function of the Plurality Function, *Mathematical Social Sciences*, **22**(2), 163–174.

SANKOFF, D., and KRUSKAL, J. B. (eds.) (1983), *Time Warps, String Edits, and Macromolecules: The Theory and Practice of Sequence Comparison*, Addison-Wesley, Reading, Massachusetts, xii + 382 pp.

SCHNEIDER, T. D., STORMO, G. D., GOLD, L., and EHRENFEUCHT, A. (1986), Information Content of Binding Sites on Nucleotide Sequences, *Journal of Molecular Biology*, **188**, 415–431.

SHAPIRO, M. B., and SENAPATHY, P. (1987), RNA Splice Junctions of Different Classes of Eukaryotes: Sequence Statistics and Functional Implications in Gene Expresseion, *Nucleic Acids Research*, **15**(17), 7155–7174.

SMITH, H. O., ANNAU, T. M., and CHANDRASEGARAN, S. (1990), Finding Sequence Motifs in Groups of Functionally Related Proteins, *Proceedings of the National Academy of Sciences of the USA*, **87**, 826–830.

SMITH, R. F., and SMITH, T. F. (1990), Automatic Generation of Primary Sequence Patterns from Sets of Related Protein Sequences, *Proceedings of the National Academy of Sciences of the USA*, **87**, 118–122.

STADEN, R. (1984), Computer Methods to Locate Signals in Nucleic Acid Sequences, *Nucleic Acids Research*, **12**(1), 505–519.

STADEN, R. (1988), Methods to Define and Locate Patterns of Motifs in Sequences, *Computer Applications in the Biosciences*, **4**(1), 53–60.

STADEN, R. (1989a), Methods for Calculating the Probabilities of Finding Patterns in Sequences, *Computer Applications in the Biosciences*, **5**(2), 89–96.

STADEN, R. (1989b), Methods for Discovering Novel Motifs in Nucleic Acid Sequences, *Computer Applications in the Biosciences*, **5**(4), 293–298.

STORMO, G. D. (1987), Identifying Coding Sequences, in: M. J. Bishop and C. J. Rawlings (eds.), *Nucleic Acid and Protein Sequence Analysis: A Practical Approach*, IRL Press, Oxford, 231–258.

STORMO, G. D. (1988), Computer Methods for Analyzing Sequence Recognition of Nucleic Acids, *Annual Review of Biophysics and Biophysical Chemistry*, **17**, 241–263.

STORMO, G. D. (1990), Consensus Patterns in DNA, in: R. F. Doolittle (ed.), *Molecular Evolution: Computer Analysis of Protein and Nucleic Acid Sequences. Methods in Enzymology*, **183**, 211–221.

STORMO, G. D., and HARTZELL, G. W., III (1989), Identifying Protein-binding Sites from Unaligned DNA Fragments, *Proceedings of the National Academy of Sciences of the USA*, **86**, 1183–1187.

STRAFFIN, P. D., Jr. (1977), Majority Rule and General Decision Rules, *Theory and Decision*, **8**, 351–360.

STÜCKLE, E. E., EMMRICH, C., GROB, U., and NIELSEN, P. J. (1990), Statistical Analysis of Nucleotide Sequences, *Nucleic Acids Research*, **18**(22), 6641–6647.

TAYLOR, P., ROSENBERG, P., and SAMSONOVA, M. G. (1991), A New Method for Finding Long Consensus Patterns in Nucleic Acid Sequences, *Computer Applications in the Biosciences*, **7**(4), 495–500.

VINGRON, M., and ARGOS, P. (1991), Motif Recognition and Alignment for Many Sequences by Comparison of Dot-matrices, *Journal of Molecular Biology*, **218**, 33–43.

WATERMAN, M. S. (1986), Multiple Sequence Alignment by Consensus, *Nucleic Acids Research*, **14**(22), 9095–9102.

WATERMAN, M. S. (1989a), Consensus Patterns in Sequences, in: M. S. Waterman (ed.), *Mathematical Methods for DNA Sequences*, CRC Press, Boca Raton, 93–115.

WATERMAN, M. S. (1989b), Sequence Alignments, in: M. S. Waterman (ed), *Mathematical Methods for DNA Sequences*, CRC Press, Boca Raton, 53–92.

WATERMAN, M. S., ARRATIA, R., and GALAS, D. J. (1984), Pattern Recognition in Several Sequences: Consensus and Alignment, *Bulletin of Mathematical Biology*, **46**(4), 515–527.

WATERMAN, M. S., and JONES, R. (1990), Consensus Methods for DNA and Protein Sequence Alignment, in: R. F. Doolittle (ed.), *Molecular Evolution: Computer Analysis of Protein and Nucleic Acid Sequences. Methods in Enzymology*, **183**, 221–237.

WATERMAN, M. S., JOYCE, J., and EGGERT, M. (1991), Computer Alignment of Sequences, in: M. M. Miyamoto and J. Cracraft (eds.), *Phylogenetic Analysis of DNA Sequences*, Oxford University Press, New York, 59–72.

YAMAUCHI, K. (1991), The Sequence Flanking Translational Initiation Site in Protozoa, *Nucleic Acids Research*, **19**(10), 2715–2720.

Alignment and Hierarchical Clustering Method for Strings

A. Guénoche

G.R.T.C - C.N.R.S., 31 Ch. J. Aiguier, 13402 Marseille Cedex 9

Abstract: We develop a conceptual clustering method for strings to realize an multiple alignment of biological sequences. We associate to each cluster a common subsequence of its strings. Unfortunately, the longest common subsequence problem is NP-hard as soon as there are more than two strings. To avoid this difficulty, we present :

(i) a greedy alignment method,

(ii) some improvements of the Hirschberg algorithm to build a maximum subsequence common to two strings,

(iii) an ascending clustering method, which provides a common subsequence that is longer than the one given by the greedy algorithm.

1 Introduction

In molecular biology, proteins and nucleic acids can be represented as strings over an alphabet. Macromolecules of DNA are sequences of four bases and so the alphabet has four characters {A, T, G, C}. An important question that arises in Biology is to determine which are, among a set of strings, those that are very close, because one can suppose they have a common ancestor. Clustering methods provide a large set of technics to analyse and organize a given set of sequences according to a proximity measure. A subsequence μ' of a given string μ is also a string of characters that are present in μ in the same order; one can see μ' as the result of deletions of characters in μ. In biology, subsequences of common characters are used to realize alignments, that is, to write strings in such a way that identical characters are on a vertical line, even if we must insert "." to shift characters. We will use the term *optimal alignment* when the number of aligned characters is equal to the length of a maximum common subsequence. In fact, we use clustering methods to realize alignments.

Throughout this text we use a proximity measure between strings based on a longest common subsequence of these strings. This measure is very close to the editing distances (Levenshtein (1966), Needleman and Wunsch (1970), Sellers (1974)), that transform a string into another using operations such as substitution, insertion or deletion (the two last being called indel) of characters. It is well known that if we ignore substitutions or if their costs are greater than or equal to the cost of two indels, the distance value is proportional to the number of indels needed to align a maximum number of identical characters in both strings. In fact, the longer is the common subsequence, the shorter is the distance, since it corresponds to the number of characters that are not aligned.

In an ascending clustering method, at each step we join together the two most similar clusters. To use this proximity measure, we have to build for any pair of clusters a subsequence common to all the strings of their union. In addition, if we want a maximum length subsequence, it has to be one of the two following cases. In the first case, there are two strings and the problem has a polynomial complexity. In the other case, there are more than two strings and the problem is NP-hard (Maier (1975)). Therefore, in the last case, we have two possibilities. In the first we use a fast heuristic method to build a common subsequence which may be not optimal. In the second, we associate to each cluster a string and we build for any pair of clusters that will be joined a maximal common subsequence of the two associated strings. In both cases the final subsequence obtained, after a series of $n-1$ fusions, is shared by all the strings. The aim of this text is to compare the two methods.

In the first part, we present such a fast heuristic method to align any number of strings. In the second part, we detail improvements of the famous Hirschberg algorithm (1975) to produce an optimal alignment of two strings. In the third part, we propose a hierarchical ascending clustering method that can be qualified as *conceptual*, since each cluster is characterized by a common subsequence. It leads not only to a hierarchy of clusters, but also to a subsequence common to all the strings and so to an alignment.

In the following, we consider p strings μ_i for $i = 1, \ldots, p$, over an alphabet $\mathcal{A}$. The length of a string is its number of characters, $m = |\mu|$, and the maximum length is denoted L. Let $\mu[i, j]$ be the substring in μ included between indices i and j. If $i = 1$, $\mu[1, j]$ is a prefix, if $j = m$, $\mu[i, m]$ is a suffix and if $i = j$ we more simply note $\mu[i]$ the character with index i.

2   Heuristic method to align strings

It consists in a greedy algorithm that realizes the first possible alignment, then the next one, and so on without analyzing consequences of these choices. So, we read simultaneously all the strings, character after character, from left to right. For each string μ_i, we save the index of the first instance of each character of the alphabet, and we sum the number of strings for which this character has been encountered. We need as many tables as strings plus one for the whole set; all these tables are indexed over the alphabet. The first character that is found in all the strings belongs to the subsequence that we are building. Then, for each string we start again from the saved index after we have initialized all the tables to 0. We describe more precisely one step:

Greedy heuristic to align strings
For each character char Do
        For each string μ_i Do $T(i, \text{char}) = 0$
        Sum(char) = 0
Endo
While for any character char, Sum(char) < NbStrings Do
        For each string μ_i Do
            Ind(μ_i) = Ind(μ_i) +1 : char = $\mu_i[\text{Ind}(\mu_i)]$
            If T(i,char) = 0 Do
                T(i,char) = Ind(μ_i): Sum(char) = Sum(char) +1
        Endo
End of While
Let char be such that Sum(char) = NbStrings
For each string μ_i Do Ind(μ_i) = T(i,char)

Obviously, this algorithm is order dependent; the number of resulting aligned characters will not be the same if we read strings from left to right or from right to left. In the latter case, we just start from the end of the strings and make the indices decrease. We decide to read strings in both way and to keep the best alignment.

Example 1: Consider these two random nucleotide sequences

$$\mu = \text{AGTGTGAAATCTTCAGAGATGAATT} \text{ and } \nu = \text{TTAAGTGTCAAATCTACAGAGA}.$$

Reading strings from left to right, the method yields 13 alignments

```
A G T G T G A A A T C T T C A G A G A T . G A . A T T
|   |   | |   |   |   | |   |   | |       |   |
T . T . A A G T G T . C A . A . A T C T A C A G A G A
```

The reverse method yields 17 alignements, which is optimal.

```
  A G T G T G A A A T C T T C A G A G A T G A A T T
  | | . | |   | | | | |   |   |   | | |   |   |
T T A A G T G T C A A A T C . T . A C A G A . G . A
```

We now study the complexity of this algorithm. At each iteration, we find a common character and we start from the next one, but this character may have been read previously. So we must modify the data structure; for each string we link together all the instances of a character that is $O(p \cdot L)$. Then choosing a character consists in comparing $|\mathcal{A}|$ possibilities in each string and taking the first that occurs. Therefore, if we have p strings, with maximum length L, this algorithm has a complexity $O(|\mathcal{A}| \cdot p \cdot L)$ and is very fast.

To evaluate its efficency applied to pairs of strings with the same length, over a four letters alphabet, we perform random simulations. We measure the ratio of the number of aligned characters, using this heuristic, divided by the length of a maximum common subsequence. These estimated proportions of an optimal alignment are computed over 50 trials. We also indicate the lowest results obtained.

$m \times n$	100×100	200×200	500×500
% length of a maximum alignment	90%	88%	85%
lowest percentage	77%	77%	81%

3 Maximum subsequence common to two strings

Since we only have two sequences in this paragraph, we note them μ and ν. Let $A(i,j)$ be the maximum number of characters that can be matched on $\mu[1,i]$ and $\nu[1,j]$ (that is the length of maximal subsequences common to prefixes), $Lp(j)$ the length of a maximal subsequence common to μ and prefixes $\nu[1,j]$ and $Ls(j)$ the length of a maximal subsequence common to μ and suffixes $\nu[j+1,n]$.

The values of matrix A, initialized to 0, are computed using a dynamic programming algorithm which is similar to the one used to evaluate distances between strings:

- $A(i,j) = A(i-1,j-1) + 1$ if characters $\mu[i]$ and $\nu[j]$ are identical,

- $A(i,j) = \max\{A(i-1,j), A(i,j-1)\}$ else

The values of array Lp are the maximum of the columns of A:

$$Lp(j) = \max_{i=1,\ldots,m} A(i,j)$$

The values of array Ls are obtained in the same way, but the values of A must be computed in the decreasing index order, that is, reading strings from right to left.

To compute $A(m,n)$ we must compare all the characters of μ to all the characters of ν. Thus this algorithm has a complexity $O(n \cdot m)$. If we compute the values of A row by row to obtain those of row i we just need those of row $i-1$. Therefore, it is enough to keep in memory two rows of n values. In the same way, the values of Lp or Ls can be updated at each row. But to hexibit a subsequence, that is the list of the indices of the identical characters, if matrix A has been memorized, it is very easy to build back from cell (m,n) a single path that leads to cell $(0,0)$. If characters $\mu[i]$ and $\nu[j]$ are identical the path goes through $(i-1,j-1)$, otherwise it can come from $(i-1,j)$ or $(i,j-1)$. This very simple algorithm uses matrix A, and can be applied only to short strings. Hirschberg (1975) defines a method which is always $O(n \cdot m)$ in time complexity and linear in memory places, so it is efficient, and a subsequence common to two strings of a thousand characters can be computed in a reasonable time, even on a microcomputer. We now describe two improvements of the Hirschberg algorithm to increase its efficiency.

3.1   About the length of a maximal common subsequence

The Hirschberg algorithm uses intensively the values of arrays Lp and Ls to select one character of a longest subsequence common to μ and ν. Our first improvement deals with the computing of these arrays. We will describe it only for Lp. Rather than computing the values of A row by row in two arrays, and storing the maximum value of column j in $Lp(j)$, we will calculate directely array Lp.

First, we notice that two consecutive values in Lp can only differ in one unit. So Lp is defined by the positions where Lp increases. We shall use an array Pos to store these indices, that is $Pos(j+1) = 1$ if $Lp(j+1) = Lp(j)+1$, and $Pos(j+1) = 0$ if $Lp(j+1) = Lp(j)$. In a first step, for each character in μ, considered from left to right, we update the array Pos. Then, in a second step, we compute Lp, adding the values of Pos: $Lp(j+1) = Lp(j) + Pos(j+1)$, starting with $Lp(1) = Pos(1)$.

To realize the first step we link together all instances of the same character in string ν. This operation is $O(n)$. Suppose we know the values of $Pos(j)$, for $j = 1,\ldots,n$, after having examined the $i-1$ first characters in μ. We now consider $c = \mu[i]$. A value in Lp can only change from an occurence of c in ν, and we point to it directly with the linked data structure. Let j be the index of the first occurence of c ($\nu[j] = c$). Two cases occur:

If $Pos(j) = 0$ any subsequence common to $\mu[1, i-1]$ and $\nu[1, j-1]$ can be extended by c; Lp increases in position j and now $Pos(j) = 1$. Let k be the nearest position greater than j such that $Pos(k) = 1$. It was in this position that we get a subsequence with the same length. Now there is no increase for Lp, so we pose $Pos(k) := 0$. After we seek for another occurence of c at a position greater than k.

If $Pos(j) = 1$ this occurence of c is already used in a longest subsequence common to $\mu[1, i]$ and $\nu[1, j]$; the value of Pos remains unchanged and we seek for the next occurence of c. So we apply the following procedure:

$k := 0$
While there exists j smallest index such that $j > k$ and $\nu[j] = c$ Do
    If $Pos(j) = 0$ Do
        $Pos(j) := 1; \quad k := j + 1$
        While $k < n$ and $Pos(k) = 0$ Do $k := k + 1$
        if $k \leq n$ Then $Pos(k) := 0$
    Else $k := j$
End of while

Numerous simulations with random strings over a four letters alphabet have shown that the computing time is two and a half lower than the time required when we calculate Lp with two rows of the A matrix (Gempp (1992)), even if only the useful band of A is computed (Spouge (1991)).

3.2   About the end condition for the algorithm

The Hirschberg method is based on this obvious proposition: If $c_1 c_2 \ldots c_p$ is a maximal subsequence common to μ and ν, and c_k corresponds to indices i in μ and j in ν, (that is, $\mu[i] = \nu[j] = c_k$), then $c_1, \ldots, c_k$ is a maximal subsequence common to $\mu[1, i]$ and $\nu[1, j]$ and $c_{k+1}, \ldots, c_p$ is a maximal subsequence common to $\mu[i + 1, n]$ and $\nu[j + 1, n]$.

In order to use this splitting proposition, we compute the Lp values for strings $\mu[1, (m/2)]$ and $\nu[1, n]$ and the Ls values for $\mu[(m/2) + 1, m]$ and $\nu[1, n]$. Therefore, we arbitrarily cut μ in two equal parts and the question is to find an index j after which we can cut ν as well, in such a way that a common maximal subsequence is the concatenation of two maximal subsequences; the first one is common to $\mu[1, (m/2)]$ and $\nu[1, j]$ and the second one is common to $\mu[(m/2) + 1, m]$ and $\nu[j + 1, n]$. According to the definition of Lp and Ls, the answer is very simple. It suffices to cut after any index j such that $Lp(j) + Ls(j + 1)$ is maximum.

The Hirschberg method is strongly recursive. After having cut each string in two parts, two identical sub-problems remain, one with the strings $\mu[1, (m/2)]$ and $\nu[1, j]$ and the other with the strings $\mu[(m/2) + 1, m]$ and $\nu[j + 1, n]$. This decomposition procedure continues until all substrings of μ are reduced to 0 or 1 character.

But we observe that, generally, it is not necessary to wait for this condition to stop. For each pair of substrings $\mu[m_1, m_2]$ and $\nu[n_1, n_2]$ for which we seek a common subsequence, we know the length of this subsequence which is equal to $Lp(n_2)$ or $Ls(n_1)$ depending whether it is a begining or an ending substring. If the number of characters in one of these substrings is equal to the length of the common subsequence, it is useless to cut it again; the whole substring belongs to the subsequence common to μ and ν.

In the same way, if the common subsequence has length 1, it is sufficient to find a single character common to $\mu[m_1, m_2]$ and $\nu[n_1, n_2]$. That can be done by just counting the number of characters instances in both strings, and so the number of operations needed is proportional to their lengths.

Example 2: We consider again strings μ and ν as in example 1. The first subdivision gives:

```
j     =  1  2  3  4  5  6  7  8  9 10 11 12 13 14 15 16 17 18 19 20 21 22
Lp(j) =  1  2  3  4  4  5  5  6  6  6  7  8  9 10 11 11 11 11 11 11 11 11
                                                    \  \  \
Ls(j) = 10 10  9  9  9  9  9  9  8  8  8  8  8  7  7  6  6  5  4  3  2  1
```

The maximum value 17 for $Lp(j) + Ls(j + 1)$ occurs three times. If we choose the last one $(j = 16)$, we obtain for the left side substrings $\mu[1, 12]$ and $\nu[1, 16]$ with an unknown common subsequence of length 11, and on the right side substrings $\mu[13, 25]$ and $\nu[17, 22]$ with a common subsequence of length 6. But suffix $\nu[17, 22]$ has length 6 so they realize the common subsequence of the 13 last characters of μ; it is useless to split the right side problem. In the second step, we just look for a subsequence with length 11 common to $\mu[1, 12]$ and $\nu[1, 16]$. The second subdivision yields:

```
Lp(j) =   1 2 2 2 3 3 4 5 5 5 5 5 5 5 5 5
                         \ \
Ls(j) =   6 6 6 6 6 6 6 6 6 6 5 4 3 2 1 1
```

Again we have some choice, but whatever the choice is, the 6 characters of $\mu[7, 12]$ are a subsequence of $\nu[9, 16]$. At the third step, we seek for a subsequence of length 5 common to $\mu[1, 6]$ and $\nu[1, 8]$. The third iteration yields:

```
Lp  =   1 1 1 1 2 3 3 3
                 \
Ls  =   3 3 3 3 3 2 2 1
```

This leads to an exact decomposition with two substrings $\mu[1, 3]$ and $\nu[7, 8]$ having the appropriate lengths. Finally, the maximal subsequence common to μ and ν is $\mu[1, 3] + \nu[7, 8] + \mu[7, 12] + \nu[17, 22]$, that is, **A G T G T A A A T C T C A G A G A**. It enables realizing another optimal alignment:

```
    A G T G T G A A A T C T T C A G A G A T G A A T T
    | | | | |   | | | | | |   | | | | | |
T T A A G T G T C A A A T C T A C A G A G A
```

4   Conceptual clustering method

As far as I know, the term *conceptual clustering* has been introduced by Michalski, Diday and Stepp (1981). This paper initiated the methodological connection between Similarity Based Learning and Data Analysis. In France, we also use the term Symbolic-Numeric methods to denote problems they share. This terminology consists in emphasizing that Learning expects treatments of quantitative knowledge in its reasonings, without loosing its explicative vocation, and that Data Analysis deals with symbolic data, generally using combinatorial methods. We do not develop further this matter. For more details, one can read Gascuel and Guénoche (1990).

The adjective conceptual added to a large family of methods in Data Analysis is there to recall some constraints. The problem in clustering is not only to build a partition or a hierarchy of clusters of a set of objects, but also to associate each cluster with a *characterization* in the *representation space* where the objects are described. In learning the result of a concept characterization, (that is, some condition for an instance to belong to this concept) is called *generalization*. One solution inherited from Data Analysis consists in using a clustering method, through some distance provided on the representation space. Then after clusters have been defined, some characterization using methods as Discriminant Analysis or Decision trees or Conceptual Discriminant Analysis (Guénoche (1991)) is built.

Another possibility, illustrated here, is to realize simultaneously these two operations, that is, to characterize clusters together with their appearance.

Essentially, strings are symbolic data; they are a coding of events, in finite number and linearly ordered, using one symbol, a character of an alphabet, for each event. The set of finite words over this alphabet is the representation space for the given strings. The notion of generalization of a set of strings can receive several meanings. We chose to characterize a cluster by a common subsequence of maximal length. Other options are possible as longest common word or a common order of several common words. We retain the first one because it seems relatively close to practices observed in Molecular Biology. The notion of common subsequence is included in the alignments realized to put in evidence some mutation from a species to another. Sometimes the quality of an alignment is not only its length, but which characters are common. We use maximal common subsequence here only to limit the number of possible characterizations of a cluster.

4.1 An ascending clustering method

The ascending or descending methods in hierarchical clustering are also greedy algorithms to optimize some criteria over partitions with a fixed number of clusters. In an ascending method, at each step we have a partition with p clusters and among the partitions with $p-1$ clusters obtained by joining two of them, we take the one that optimizes a given criteria.

Let us define the homogeneity of a cluster as the length of a maximal common subsequence divided by the average of the lengths of the strings. We note $\alpha(\mu, \nu)$ a maximal subsequence common to μ and ν and we define a similarity measure S between the strings μ and ν as the homogeneity of this pair:

$$S(\mu, \nu) = \frac{2 \cdot |\alpha(\mu, \nu)|}{(|\mu| + |\nu|)}$$

In fact two strings are very close if they share a long subsequence, and this measure is bounded by 0 (different characters) and 1 (identical strings). If several pairs have the same subsequence length, the most similar is the shortest couple; so this similarity function gives some advantage to strings that have the same length.

By joining in each step two clusters with a maximal similarity, we obtain an ascending hierarchical clustering method. We associate with each class a subsequence common to its strings. As we do not want, for complexity reasons, to compute a longest one, we determine, using the Hirschberg method, a longest subsequence common to the two strings associated with the clusters to join. Thereafter, we calculate the similarity of this new cluster with any of the unchanged clusters.

Ascending conceptual clustering algorithm

Step 1:   Compute the similarity between any pair of strings.

Step 2:   While there are more than one cluster

Choose a pair of clusters with maximum similarity, each one being associated with a string.

Join these clusters and compute a maximal common subsequence $\alpha(\mu, \nu)$ using the improvements described in section 3.

For any other cluster associated with string π, compute the similarity $S(\alpha(\mu, \nu), \pi)$.

End of While

Step 3: Reorder initial strings for drawing the dendrogram.
For any string decide the position of each character such that characters of the common subsequence are vertically aligned.

4.2 Complexity

Let p be the number of strings and L the length of the longest one.

The first step is $O(p^2 \cdot L^2)$, since there are $p(p-1)/2$ pairs of strings and the calculation of the length of a maximal common subsequence remains $O(L^2)$.

In the second step, there are $p-1$ iterations. For the k-th, we look for a maximum value among $(p-k)(p-k+1)/2[O(n^2)]$, then we determine a maximal common subsequence $[O(L^2)]$, and after that we compute the similarity between this subsequence (considered as a string) and the $p-k$ remaining strings $[O(p \cdot L^2)]$. Generally L is much greater than p, so each iteration is in $O(p \cdot L^2)$ and the second step is in $O(p^2 \cdot L^2)$.

In the third step, we first build a linear order for the $2p-1$ leaves of the classification tree (Barthélemy and Guénoche (1991)). It is performed using a depth first search of the tree, which is $O(p)$. Then we realize the alignment using a procedure that will not be completely described here (Ebabil (1992)). Given the subsequence, we must locate, in all the strings, one instance of each common character. To determine the necessary gap between two successive aligned characters, we count the maximum number of characters that separate them in one string. Then, to write each string, it is sufficient to respect these gaps. This editing method is $O(p \cdot L)$.

Example 3: Rather than selecting several random strings, we start back from the two strings of example 1. We duplicate and slightly modify them. Therefore, odd numbered strings can be seen as mutations of the first one and even numbered strings as mutations of the second one. We obtain the following file:

```
1 :    AGTGTGAAATCTTCAGAGATGAATT
2 :    TTAAGTGTCAAATCTACAGAGA
3 :    AGCGTGAAGTCTTCAGAGATGAAT
4 :    TTAAGCGTCAAATCTACTGAGAC
5 :    AGTGTGATATCTTCAAAGATGAATTC
6 :    TTAGCTGTCAAATCGACAGAGA
7 :    AGTGTGAACTCTTCAGAGATGAA
```

Using function S, the first step gives the following similarity array. It will be updated at each step:

```
      1   2   3   4   5   6
2 :   72
3 :   90  65
4:    62  89  64
5 :   90  67  84  65
6 :   68  91  65  84  62
7 :   92  71  89  61  86  67
```

The successive iterations yields the following clusters with their associated strings:

```
Cluster  1  and  7  joined in cluster  8
Maximal common subsequence : A G T G T G A A T C T T C A G A G A T G A A
Cluster  2  and  6  joined in cluster  9
Maximal common subsequence : T T A G T G T C A A A T C A C A G A G A
Cluster  8  and  3  joined in cluster 10
Maximal common subsequence : A G G T G A A T C T T C A G A G A T G A A
Cluster  9  and  4  joined in cluster 11
Maximal common subsequence : T T A G G T C A A A T C A C G A G A
Cluster 10  and  5  joined in cluster 12
Maximal common subsequence : A G G T G A A T C T T C A A G A T G A A
Cluster 12  and 11  joined in cluster 13
Maximal common subsequence : A G G T A A T C C G A G A
```

It remains only one cluster that contains all the strings; the algorithm stops. Strings are reorderd and aligned. A sign * is placed over columns that contain exclusively identical characters for the seven strings. A sign + is placed over characters that are identical in cluster 11 (even indices) and 12 (odd indices) but that are different from one to the other.

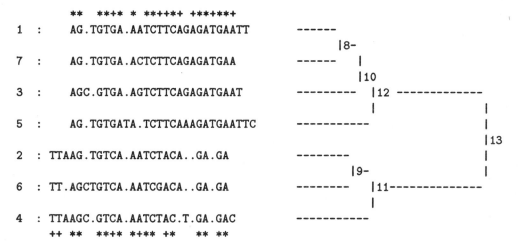

This alignment is to be compared to the common subsequence A G T A A T C C A A obtained applying the greedy algorithm of section 2:

```
        * *   *    * * **   *  * *
1  :  ..A.G..TGTGA.AATCTTC..AGAGATGAATT
2  :  TTAAG..TGTCA.AATCTAC..AGAGA
3  :  ..A.GCGTG..A.AGTCTTC..AGAGATGAAT
4  :  TTAAGCGTC..A.AATCTACTGAGAC
5  :  ..A.G..TGTGATA.TCTTC..A.AAGATGAATTC
6  :  TTA.GC.TGTCA.AATCGAC..AGAGA
7  :  ..A.G..TGTGA.ACTCTTC..AGAGATGAA
        * *   *    * * **   *  * *
```

The greedy subsequence is shorter, a fact which is not surprising if we compare the complexity orders. We have seen that conceptual clustering is $O(p^2 \cdot L^2)$, instead of the heuristic which is $O(|\mathcal{A}| \cdot p \cdot L)$. It is easy to conjecture that being more parsimonious in computational effort implies being less accurate.

An illustration based on an example is not a demonstration of this conjecture. We now have to experiment with real data to confirm our results and to compare them with other alignment methods. I will finish this paper claiming that clustering methods are certainly good methods to realize subsequences common to a set of strings.

Acknowledgments

This work was supported in part by the IMABIO program (CNRS).

References

BARTHELEMY, J.P. and GUENOCHE, A. (1991), *Trees and Proximity representations*, John Wiley, Chichester.

EBABIL, T. (1992), *Méthode de subdivision pour l'alignement multiple*, Mémoire de DEA, Université d'Aix-Marseille II.

GASCUEL, O. and GUENOCHE, A. (1990), Approche symbolique-numérique en Apprentissage, in: B. Bouchon Meunier (ed.), *Journées nationales du PRC-IA*, Hermes, Paris, 91–112.

GEMPP, T. (1992), *Sous-séquences communes maximales: méthodes exactes et approchées*, Mémoire de DEA, Université d'Aix-Marseille II.

GUENOCHE, A. (1991), Optimization in Conceptual Clustering, in: R. Gutiérrez, M. Valderrama (eds.), *Proceedings of the 5th. International Symposium on Applied Stochastic Models and Data Analysis*, Grenade, World Scientific, 302–314.

HIRSCHBERG, D.S. (1975), A linear Space Algorithm for computing Maximal Common Subsequences, *Communications of the ACM*, 18, 6, 341-343.

KRUSKAL, J.B. (1983) An overview of sequence comparison, in: D. Sankoff, J.B. Kruskal (eds.), *Time wraps, string edits and macromolecules: the theory and practice of sequence comparison*, Addison-Wesley, Reading, 1–42.

LEVENSHTEIN, V.I. (1966), Binary code capable of correcting deletions, insertions and reversals, *Cybernet. Control Theory*, 10, 707–710.

MAIER, D. (1975), The complexty of some problems on subsequences and supersequences, *J.A.C.M.*, 25, 2, 322–336.

MICHALSKI, R.S., DIDAY, E. and STEPP, R.E. (1981), A recent advance in data analysis: Clustering Objects into Classes Characterized by Conjunctive Concepts, in: L.N. Kanal and A. Rosenfeld (eds.), *Progress in Patern Recognition*, North-Holland, 33–56.

NEEDLEMAN, S.B. and WUNSCH C.W. (1970), A general method applicable to the search of similarities in the amino acid sequence of two proteins, *J. of Molecular Biology*, 48, 443–453.

SELLERS, P.H. (1974), An algorithm for the distance between two finite sequences, *J. of Comb. Theory*, 16, 253–258.

SPOUGE, J.L. (1991), Fast optimal alignment, *CABIOS*, 7, 1, 1–7.

More Reliable Phylogenies by Properly Weighted Nucleotide Substitutions

Michael Schöniger

Theoretical Chemistry, Technical University Munich,
Lichtenbergstr. 4, W-8046 Garching, Germany

Arndt von Haeseler

Institute for Zoology, University of Munich,
Luisenstr. 14, W-8000 Munich 2, Germany

Abstract: The efficiency of the neighbor-joining method under a variety of substitution rates, transition-transversion biases and model trees is studied. If substitution rates vary considerably and the ratio of transitions and transversions is large, even a Kimura (1980) two-parameter correction cannot guarantee reconstruction of the model tree. We show that application of the combinatorial weighting method by Williams and Fitch (1990) together with the Jukes-Cantor (1969) correction significantly improves the efficiency of tree reconstructions for a wide range of evolutionary parameters. Advantages, as well as limitations, of this approach are discussed.

1   Introduction and the Problem

Whenever a phylogenetic tree is reconstructed based upon sequence data it is desirable to know which part of the sequence data is most informative. In the case of low similarities there might be "noisy" information caused by a great deal of parallel and back substitutions. The underlying evolutionary process may have been governed by a substitution bias, *e.g.* the well known preference of transitions *vs.* transversions. For distance matrix methods these problems have been tackled with distance corrections proposed by Jukes and Cantor (1969), and by Kimura (1980). Although, in some examples both methods turned out to be successful in finding the true tree (Jin and Nei 1990) there is a trade-off: they increase the variability of the corrected distances thus leading to wrong phylogenies in other cases (Vach 1991).

For any correction method more complicated than the one parameter model by Jukes and Cantor it is necessary to guess which bias was present during the evolution of the sequences under study. We overcome this lack of objectivity and suggest a distance estimation that will detect automatically any bias in the data. This is accomplished by assigning lower weights to more frequent, less informative substitutions and higher weights to rare, more informative ones. Thereby the sequence distances decrease. Hence, they may be corrected with a lower increase of variability.

We studied the performance of the neighbor-joining method by Saitou and Nei (1987) under a variety of evolutionary parameters. A number of extensive computer simulations was conducted to elucidate the effects of different weighting schemes on the success of the tree reconstruction under various transition-transversion biases, model trees, and rates of substitution.

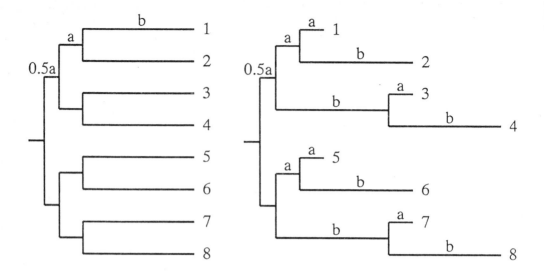

Figure 1: Model trees with substitution rates *a* and *b*. Left: constant rates of substitutions (molecular clock), right: varying rates of substitutions (no molecular clock).

2   Generation of Sequence Data

Evolution of sequences was performed using either constant (*i.e.* existence of a molecular clock) or varying (no molecular clock) rates of nucleotide substitutions. Our simulations are identical to those described by Saitou and Nei (1987). The model trees are shown in Figure 1. We used the following sets of parameters: I: $a=0.01$, $b=0.07$; II: $a=0.02$, $b=0.19$; III: $a=0.03$, $b=0.42$. The sequence lengths were 1000.

In order to obtain a transition-transversion bias we used Kimura's two-parameter model (Kimura 1980) instead of the simple Jukes-Cantor (1969) model. In the simulations transitions happened eight times (arbitrarily chosen) as frequently as transversions, *i.e.* $B=0.8$, where B defines the proportion of transitional changes among total changes. $B=0.33$ corresponds to the Jukes-Cantor model.

3   Tree Reconstruction

We used only neighbor-joining as a fast and efficient example for distance matrix methods.
The following measures for pairwise sequence distances were studied:

a) Simple Hamming distances, using observed fractions of sites showing different nucleotides (HAM).

b) Jukes-Cantor corrected Hamming distances (HAM+JC).

c) Kimura's two-parameter corrected distances, using observed fractions of nucleotide sites showing transitions and transversions (TPC).

d) Existential weighting (Williams and Fitch 1990): using weights for substitutions between nucleotides i and j inversely proportional to the number of overall alignment positions containing both i and j (EX).

e) Existential weighting and subsequent Jukes-Cantor correction (EX+JC).

f) Combinatorial weighting (Williams and Fitch 1990): using weights for substitutions between nucleotides i and j inversely proportional to the number of occurrences of pairs (i, j) in the overall alignment (COMB).

g) Combinatorial weighting and subsequent Jukes-Cantor correction (COMB+JC).

4 Results

For each set of parameters we carried out 1000 simulations. We evaluated only the empirical probability P_c of constructing the correct tree as Sourdis and Nei (1988) observed a strong correlation between P_c and the average topological distance from the model tree.

We conducted a great variety of simulations exploring many combinations of evolutionary parameters as model trees, transition-transversion biases, rates of substitution, and sequence lengths. In this paper we only report a small, but representative selection of results. In any case the sequence length is equal to 1000. This number is somewhere in between typical lengths of DNAs and the lengths required to reconstruct a statistically sound tree (Churchill et al. 1992). It is well known (Saitou and Nei 1987, Sourdis and Nei 1988) that P_c increases with sequence length.

Figure 2 shows the results under a strict molecular clock. At a first glance one can see that P_c drops with increasing rates of substitution a and b. Especially for set III with a ratio b/a of 14 which corresponds to quite a bush-like evolutionary pattern the efficiencies in finding the true tree are small. This result depends neither on the weighting scheme nor the bias B.

Correction of observed distances for multiple substitutions using Kimura's two-parameter model (TPC) does not improve the efficiency. On the contrary, for $B=0.8$ a significant deterioration is found, though TPC was designed for that mode of evolution.

The most striking feature is that simple weighting schemes like EX, EX+JC, COMB, or COMB+JC significantly improve the performance of neighbor-joining as long as the substitution rates are not too small (II, III). The decrease of P_c for COMB and COMB+JC in case I is due to sampling problems when the weight matrix is calculated. On the contrary, for high rates of substitution (III) COMB is superior to EX weighting since the EX weight matrix already approaches saturation, i.e. the HAM measure.

None of the applied weighting schemes affects significantly the probability of obtaining the modeled tree in case $B=0.33$, i.e. when the sequences evolved under a Jukes-Cantor pattern. This is a reasonable result.

Summarizing, one realizes that under a molecular clock it is better to avoid any distance correction like JC, although the impairment is not serious.

Figure 3 shows the results if there is no molecular clock. Most evident here is that distance corrections like JC and TPC are inevitable as soon as substitution rates are not too small (II, III). Again TPC, invented to compensate for skewed substitution rates, i.e. large biases B, cannot cope with $B=0.8$. EX+JC and COMB+JC perform best when the sequences evolved under absence of a molecular clock and with a strong transition-transversion bias.

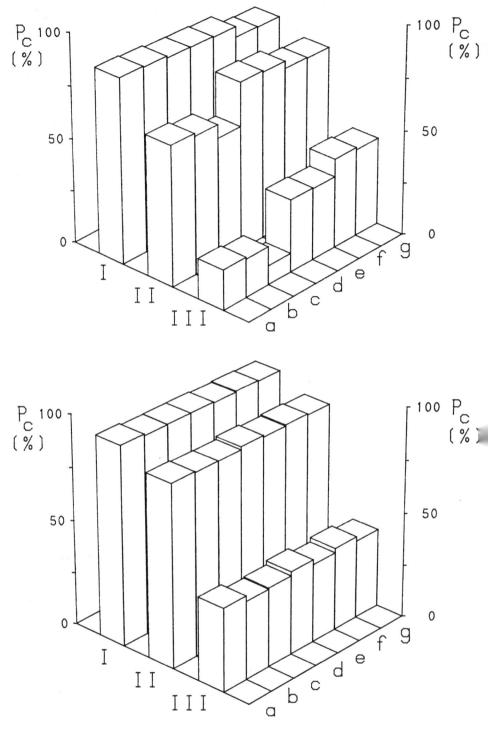

Figure 2: Plot of percentages P_c of correctly reconstructed trees under a molecular clock. Top: transition-transversion bias $B=0.8$, bottom: $B=0.33$. Three sets of parameters I-III, and seven different weighting schemes a-g were used (details see text).

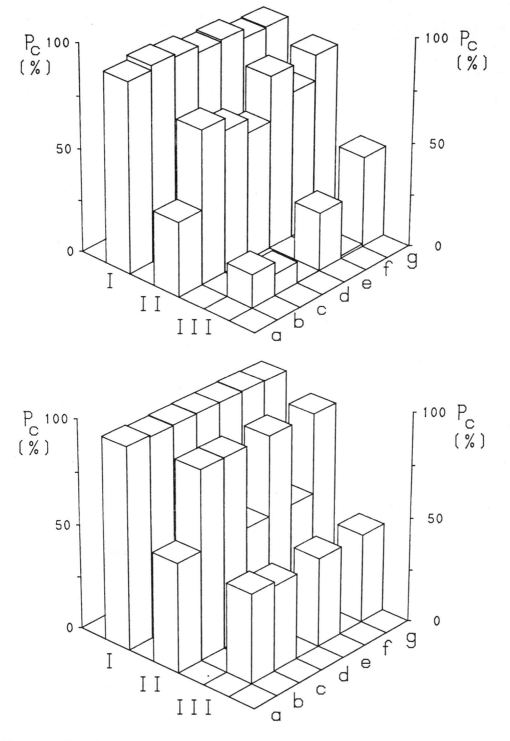

Figure 3: Plot of percentages P_c of correctly reconstructed trees without molecular clock. Details see Figure 2.

5 Discussion and Extensions

Our simulations indicate that there is no general solution to improve the reliability of tree reconstruction methods. Some of the applied weighting schemes perform well under clock-like evolution and small substitution rates (*e.g.* HAM, EX). But those methods loose their efficiency when substitution rates are very different or too large. On the other hand HAM plus JC correction performs well under conditions as non clock-like evolution, no strong transition-transversion bias, and small substitution rates. Due to an increased variability of estimated corrected distances (Kimura 1980) the method looses efficiency if substitution rates are large (Vach 1991).

Combinatorial weighting and subsequent application of the Jukes-Cantor correction has been proven to be surprisingly powerful for a large group of tree models. The use of the COMB matrix has two effects. First, it accounts for the transition-transversion bias in the data. Secondly, it generally reduces the distance between two sequences because the frequently occurring changes (*i.e.* transitions) receive a lower weight. Therefore, subsequent application of the Jukes-Cantor correction, which accounts for multiple substitutions, is less affected by an increase in the variability of the corrected distances.

The transition-transversion bias and substitution rates may vary considerably without affecting the efficiency of this method. While currently used weighting methods are only efficient for a narrow range of parameters, the method suggested here works in a wide area.

For small substitution rates existential weighting performs better than combinatorial weighting due to sampling problems. However, for high substitution rates existential weighting is close to saturation, thus providing a matrix similar to the HAM matrix. Hence, given a transition-transversion bias existential weighting is no longer appropriate for detecting the true tree. Combinatorial weighting is therefore best for high rates of change.

Our results suggest that one should use COMB weighting together with JC correction. However, one should keep in mind that only very simple models of sequence evolution were investigated. If the true evolution of sequence families is considerably different from our simulation studies we may end up with erroneous trees. Most tree reconstruction methods provide a tree no matter whether the data are really tree-like or not. Other methods at least indicate how far away from an additive tree the result is, *e.g.* the minimum number of misplaced quartets method by Dress *et al.* (1986). In perfectly additive data the percentage of misplaced quartets equals zero (Dress *et al.* 1986). If this number is too big one would simply not trust the reconstructed tree. Neighbor-joining does not provide us with an easy decision criterion whether to accept the tree. If one gets the impression that the data do not fit a tree it is certainly advisable to use more difficult methods to analyze the sequence relationships (*cf.* Bandelt and Dress 1990, Dopazo *et al.* 1990). Unfortunately, it is generally not true that a small number of misplaced quartets in a reconstructed tree implies that this tree is close to the true tree. Therefore, procedures which try to optimize the tree-likeness measured by some parameters (Eigen *et al.* 1988) generally do not improve the efficiency of tree reconstruction methods (Bandelt *et al.* 1991, Vach 1991).

As mentioned above, our suggested method works best for varying rates of nucleotide substitution and for large substitution rates if one studies model trees with a clock. For real data those parameters are generally not known. In order to avoid applying an inappropriate weighting scheme, the subsequent table is helpful:

substitution rates	molecular clock	
	yes	no
slow	EX	EX + JC
fast	COMB	COMB + JC

The test for the existence of an evolutionary clock is not crucial because the efficiency of tree reconstruction methods under COMB+JC weighting does not heavily depend on the existence of the clock. More important is the correct estimation of the substitution rates.

We have tested our method also for sets of 16, or even 64 sequences. Although it is very hard to reconstruct the true tree for 64 sequences, we obtained considerable improvements towards the model tree if the appropriate method from the table above is used. Another study investigated positional variability: some sequence positions changed much faster than others. All our simulations led to P_c values somewhere in between the P_c values for the extreme substitution rates depending on the ratio of fast to slow changing positions. This showed how robust our weighting method is. Another advantage of COMB weighting is that it is not restricted to transition-transversion biases because it automatically detects any substitution bias in the data. We therefore believe that our method of improving the reliability of tree reconstructions is widely applicable and useful.

Some studies were dedicated to sequences with not uniformly distributed nucleotides (e.g. high GC content). As long as such deviations remain in the range of naturally occurring ones (about 70 % GC) the efficiency of EX or COMB weighting is not affected. This observation changed drastically when we studied larger alphabets such as amino acids. We used observed amino acid frequencies and replacement probabilities as derived by Dayhoff (1978). The frequencies of the stationary distribution show remarkable deviations from uniformity (e.g. Alanine 8.6 %, Tryptophan 1.3 %). In those cases any weighting scheme to compute pairwise dissimilarities should take into account the differences in the frequencies of the corresponding pair of amino acids. However, when dealing with large alphabets, weighting and distance correction approaches loose their importance since back and parallel replacements become rare events compared to nucleotide sequences.

References

BANDELT, H.-J., and DRESS, A. W. M. (1990), A canonical decomposition theory for metrics on a finite set, *Preprint 90032, Sonderforschungsbereich 343: Diskrete Strukturen in der Mathematik, Universität Bielefeld*.

BANDELT, H.-J., VON HAESELER, A., BOLICK, J., and SCHÜTTE, H. (1991), A comparative study of sequence dissimilarities and evolutionary distances derived from sets of aligned RNA sequences, *Mol. Biol. Evol. submitted*.

CHURCHILL, G. A., VON HAESELER, A., and NAVIDI, W. C. (1992), Sample size for a phylogenetic inference. *Mol. Biol. Evol. accepted*.

DAYHOFF, M. O., *Atlas of protein sequence and structure, Vol. 5, Supplement 3*, National Biomedical Research Foundation, Washington, D. C., 1978.

DOPAZO, J., DRESS, A., and VON HAESELER, A. (1990), Split decomposition: a new technique to analyze viral evolution, *Preprint 90037, Sonderforschungsbereich 343: Diskrete Strukturen in der Mathematik, Universität Bielefeld*.

DRESS, A., VON HAESELER, A., and KRÜGER, M. (1986), Reconstructing phylogenetic trees using variants of the four point condition, *Studien zur Klassifikation, 17, 299-305*.

EIGEN, M., WINKLER-OSWATITSCH, R., and DRESS, A. (1988), Statistical geometry in sequence space. A method of quantitative comparative sequence analysis, *Proc. Natl. Acad. Sci. USA, 85, 5913-5917*.

JIN, L., and NEI, M. (1990), Limitations of the evolutionary parsimony method of phylogenetic analysis, *Mol. Biol. Evol., 7, 82-102*.

JUKES, T. H., and CANTOR, C. R. (1969), Evolution of protein molecules, in H. N. Munro (ed.), *Mammalian protein metabolism*, Academic Press, New York, 21-132.

KIMURA, M. (1980), A simple method for estimating evolutionary rates of base substitutions through comparative studies of nucleotide sequences, *J. Mol. Evol. 16, 111-120*.

SAITOU, N., and NEI, M. (1987), The neighbor-joining method: a new method for reconstructing phylogenetic trees, *Mol. Biol. Evol., 4, 406-425*.

SOURDIS, J., and NEI, M. (1988), Relative efficiencies of the maximum parsimony and distance-matrix methods in obtaining the correct phylogenetic tree, *Mol. Biol. Evol., 5, 298-311*.

VACH, W. (1991), The Jukes-Cantor transformation and additivity of estimated genetic distances, in M. Schader (ed.), *Analyzing and modeling data and knowledge*, Springer, Berlin, 141-150.

WILLIAMS, P. L., and FITCH, W. M. (1990), Phylogeny determination using dynamically weighted parsimony method, in R. F. Doolittle (ed.), *Methods in enzymology Vol. 183*, Academic Press, San Diego, 615-626.

Caminalcules and Didaktozoa: Imaginary Organisms as Test-Examples for Systematics

Ulrich Wirth

Biologisches Institut I (Zoologie), Universität Freiburg, Albertstr. 21a
D-7800 Freiburg i.Br., Germany

Abstract: There are several artificial data sets created for theoretical considerations or didactic purposes in systematics (sect.1). Most famous are the Caminalcules (Figs.1-4). They have been used for three decades for taxonomic exercises (mainly in numerical taxonomy) and subjected to intensive analysis (sect.2). Their "phylogeny" ("true tree") is claimed to be representative for real organisms. Hence it was argued that the taxonomic procedures proved to be successful for Caminalcules could be used for living organisms with success as well.

The present paper analyses the "evolution" of the Caminalcules from the viewpoint of a morphologist and evolutionary biologist and shows that the Caminalcules are   n o t   representative of real organisms in many respects (sect.3): the tree topology (monophyla vs. paraphyla etc.), the character states in the stem-species, the absence of any serious problem of homologizing, the low number of convergences, and esp. the high number of living fossils. In Fig.2, the phenetic approach (overall similarity) and the cladistic approach are contrasted in easily understandable form.

The "Didaktozoa" (Figs.5-7) are introduced and discussed (sect.4). They are more "handier" than Caminalcules, having only 12 terminal taxa. Also, they were created by rules most biologists will agree upon, and some problems with convergences are included to simulate the situation systematists are frequently confronted with. By giving alternative trees (Fig.7 vs. Fig.6) it is demonstrated that ingroup-analysis alone is usually not sufficient, rather that additional information from ontogeny and outgroups should be incorporated. Although not directly involved with molecular systematics, this article may be of relevance to molecular systematists as well (sect.5).

1   Introduction

Progress and Controversies in Systematics: In recent decades, the theory and practice of systematics has made much progress; for historical review, see Mayr (1982: 209-250), for textbook treatment of phylogenetics (cladistics) see Wiley (1981), Ax (1984). Nevertheless, there is still much debate with respect to phylogenetic evaluation of molecular data, highlighted for example by the number of letters to the editor after the publication by Gorr, Kleinschmidt, and Fricke (1991), claiming that the "living fossil" Latimeria (and not the lungfishes) is the closest relative of the tetrapods.

Artificial Data Sets: There are several artificial data sets developed for theoretical discussions or didactic purposes: a. Simple data matrices (e.g. Lorenzen and Sieg 1991) or abstract figures. b. Complex data matrices generated by computer and not directly visualized (e.g. "Genesis" by Heijerman 1992). c. Simple imaginary organisms, e.g. the genus "Planta", the "Cookophytes", the insect genus Hypotheticus, some imaginary amphibians (Klob 1981) and gastropods (Wirth 1984a). d. Complicated imaginary organisms like Rhinogradentia (Stümpke 1961), Experimentalia (Riedl 1975) and Caminalcules. The Rhinogradentia were created to deal with functional morphology in an amusing manner, and Dixon (1981) described animals of the future. All the other "organisms" were created for didactic purposes to illustrate principles of systematics (classification and phylogenetics).

Caminalcules as the Most Discussed of all Imaginary Organisms: The most famous imaginary organisms are the Caminalcules, created by the late Prof. Joseph H. Camin, Department of Entomology, University of Kansas. They were created according to rules "which were believed to be consistent with what is generally known of transspecific evolution", ("genetic continuity was accomplished by tracing the drawings of the animals from sheet to sheet", Camin and Sokal 1965). They were "generated artificially according to principles believed to resemble those operating in real organisms", but (simulation of the evolutionary process) "by rules that have not been made explicit" (Sokal 1983: 159). They have been used for taxonomic exercises (e.g. Moss 1971, Sokal and Rohlf 1980), treated thoroughly in "Numerical Taxonomy" (Sneath and Sokal 1973), considered in popular articles (e.g. Sokal 1966) and in textbooks for undergraduates (e.g. Sengbusch 1974). – In a series of four papers, Sokal (1983) analysed the Caminalcules intensively in many respects, also applying several computer programs. In a plenary lecture at our Gesellschaft für Klassifikation, Sokal (1984) summarized that the Caminalcules are good test-examples for systematics and called them "taxonomische Lehrmeister" (taxonomic mentors). He asked opponents to consider them accordingly: "Die Beweislast ist nun jenen Kritikern aufgebürdet, die über die Caminalcules als bedeutungslos für die biologische Systematik hinweggehen möchten" (the burden of proof is now on those critics who prefer to disregard Caminalcules as having no relevance for biological systematics).

Aim of this Article: In the course of several of my seminars on systematics, students were asked to present cladograms (=trees with arguments) for the Caminalcules. Very diverse cladograms were presented, and no student came up with the true tree. The question arose whether cladistics is incapable of handling such a problem, or whether additional information is necessary to establish the character polarity, e.g. through knowledge from outgroups, ontogeny, function, or any other pre-knowledge like Sokal (1983:187,254), who used his knowledge of the ancestor to root a tree he obtained with the "Phylip Wagner" procedure. – My contribution will not treat aspects directly connected with current issues of molecular systematics nor will I discuss the results of the sophisticated computations of Sokal (1983: 170: summary of formulas: lengths in character state changes, path lengths, dendritic index, consistency index, deviation ratio, etc.). Instead, as a morphologist and evolutionary biologist, I will examine the basic question *whether the evolution of the Caminalcules is representative for real living organisms,* i.e. whether they have a phylogeny similar to real organisms at all.

2 Caminalcules: Classification and Cladistic Attempts

Morphology: The Caminalcules constitute an assemblage of 29 recent species (Fig.1) numbered randomly. The 48 fossil species remained unpublished until 1983; some of them can be seen in Fig.3. They remind us most of tetrapod vertebrates, so terms like eye, trunk, leg, arm, digit, etc. can be used. The slender types are somewhat reminiscent of squids (Loligo, Cephalopoda).

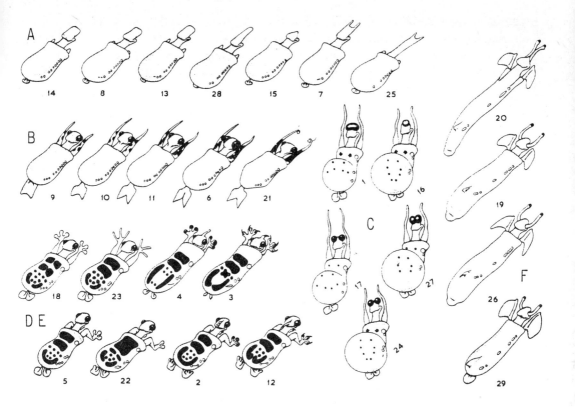

Fig. 1: Caminalcules: Images of the 29 recent species, grouped as 5 "genera". (Sokal 1983: 161, newly arranged).

Classification: When subjects of different age, different education, and different profession were asked to classify the Caminalcules "by whatever principle seemed natural to them" (Sokal and Rohlf 1980), almost all persons suggested the same 5 groups. For simplicity, these groups can be called A (Acephala), B (blotched arms), C (circular trunks), DE (digital extremities), F (flippers). It is now customary to treat these 5 groups as "genera". Sometimes, an additional 6th group has been suggested for those with "elbows" (# 2,5,12,22) or those with a pointed head (# 3,4). – Why is the classification so unambiguous? The answer is clear: There are obvious gaps *between* the groups, and *within* the groups there is a high degree of homogeneity (invariance); e.g. all members of F possess flippers, a slender trunk, stalked eyes, and a ship's keel; and none of these features occur outside F. In this respect, the Caminalcules are not representative of real organisms (see discussion on convergences and reductions below in sect.3).

Cladistics vs. Phenetics: Sokal (1983) presented a standard phenogram and admitted (1983, 1984) that cladistic procedures (e.g. "Phylip Wagner" program) matched the true tree more accurately than the phenogram. Only one example will be discussed (Fig.2): the position of #29 within the genus F. The UPGMA-clustering put this species more centrally within the group, whereas the cladistic attempt gives the correct position by using the *outgroup-criterion*: If we look at all the other Caminalcules as "outgroups", possessing legs seems to be the ancestral (plesiomorphic) character state. So it is more parsimonious to split off #29

424

very early and assume that the evolutionary loss (leg reduction) by the other species of F occurred only once, i.e. assume this reduction is a synapomorphy for these 3 species. – It is suggested to use more than one outgroup, i.e. the supposed sister-group (adelphotaxon = the closest relatives), the other relatives, and also other taxa farther away within the provisional phylogenetic hypothesis (Wirth 1984a: 12).

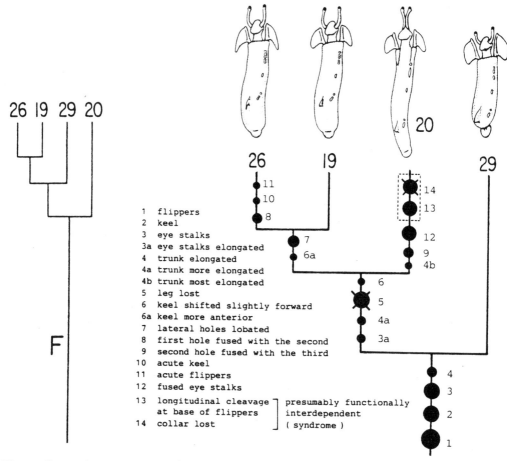

Fig. 2: Caminalcules: Genus F (flippers). Left: From the standard phenogram (Sokal 1983: 168). – Right: Argumentation scheme (cladogram with list of apomorphies). Black circles are apomorphies, crossed circles are evolutionary losses (reversals), larger size of symbol indicates greater weighting. (Orig.)

3   Critical Analysis of the Phylogeny of the Caminalcules

Some evolutionary trends and rules (sometimes called "laws") exist, also some consensus on stem-species and tree topology, and there are always problems with homology vs. convergence (analogy, homoplasy). Every systematist is confronted with these problems. Let us now evaluate the true tree (Figs.3,4) of the Caminalcules with respect to these aspects.

Anagenesis (Specialization and Differentiation): It is generally assumed that in many phylogenies evolution proceeds with specialization and differentiation. Clear examples are serial structures, where anagenesis is also often combined with fixation and reduction of numbers, e.g. appendages of Arthropoda (many similar legs in Myriapoda → fewer and differentiated extremities in Insecta: antennae, mouth feeding parts, walking legs, abdominal leg rudiments, gonopods) and the teeth of Vertebrata (homodont → heterodont). (Of course this is no strong law, only a rule, and the systematist must consider the possibility of regressive evolution, e.g. in parasites and cave animals!) The Caminalcules follow this principle, e.g. lateral holes of the trunk, and armstumps of the stem-species #73 → hands in group DE / → flippers in group F.

Key Characters: In real organisms, the evolution towards a new grade is often accomplished by "invention" of "Schlüsselmerkmale" (key characters), e.g. feathers in birds, dentition with new jaw articulation in mammals. The Caminalcules evolved partly by this principle, e.g. digits in #43, the stem-species for DE + C (2 digits reduced).

Extension of Function and Rearrangement of Body Plan: In real organisms radical changes of life habits and morphology can occur, and the changed features may pose great problems to the systematist. Obvious examples occur among burrowing animals and parasites, where sometimes only ontogeny can tell us the systematic position, e.g. for the rhizocephalan Sacculina, or the Pentastomida (see Wirth 1984b: 337 vs.332). Such radical changes do not occur within the Caminalcules.

Homology: A real problem of any systematic work is that of homology. With the Caminalcules there are only a few examples for this, e.g. the "arms" in genus C: the tip is actually an elongated digit, a fact which can be seen only from the fossil line # 43-65-35-63 and not by analysing the recent species. (To understand it from the recent species alone, some additional information would be necessary, esp. from ontogeny.) To simulate realistic problems with homologizing, we could draw fossil #61 (an extinct offshoot with a pair of arms and a pair of lateral fins) with only one pair of appendages pointing obliquely. Then we would be forced to look more closely whether any additional evidence can tell us whether these lateral appendages are derived from arms or whether new lateral protrusions have evolved.

Convergence (Analogy, Parallelism, Homoplasy): Convergences are perhaps the most serious of all problems for the systematist. Among Caminalcules there are only minor examples of convergence, and in these few cases it is always within a genus: The eyes are reduced three times convergently within genus A; within genus DE there is a convergent evolution of finger-nails and claws (#2/4, #12/3); fusion of the eyes occurred twice within genus C. Having 5 convergences (concerning 3 characters) and a few other ones altogether among 77 species with over 100 characters is unrealistically low!

Stem-Species: In the evolution of real organisms we often encounter step-wise evolution from one level (grade) to another, e.g. in Chordata: continuous fins of Branchiostoma → several fins in fishes → pentadactyl extremities in most Tetrapoda → hoofed legs in horses and other ungulates. (That is, by the way, also an additional example for anagenesis.) Corresponding to this step-wise evolution is the fact that most stem-species are neither "Nullwert-Ahnen" (zero-ancestors, Remane 1952: 261) nor a "Zentral-Typus" (Remane 1952: 152), and the phylogenetic trees are often very asymmetrical. The stem-species of the Caminalcules (#73) is something like a zero-ancestor: almost no arms (what about the function of the little stumps?), very small eyes, no real head. Also, the stem species is fairly symmetrically located in the tree, and therefore it is not surprising that midpoint-rooting and similar phenetic procedures will yield results close to the true phylogeny.

Paraphyla and Tree Topology: The relatively "primitive" ancestral members of a group are often *paraphyla* (explanation: Fig.7), e.g. "Gymnospermae" among the Spermatophyta, "Apterygota" among the Insecta, "Agnatha" (Petromyzon and Lampetra), "Pisces" and "Anamnia" among the Vertebrata, and "Prosimiae" among the Primates. These groups are mainly characterized by the absence of a prominent feature, and

careful phylogenetic analysis must establish whether this absence is primary (only then would we have paraphyla characterized by symplesiomorphy = common ancestral character states) or secondary (in this case we must distinguish between synapomorphy and convergence, which would correspond to monophylum and polyphylum). (Sequence of argumentation steps see Wirth 1984b: 319). Actually, all "genera" of Caminalcules are monophyla and are not test-examples for this problem!

Living Fossils: Real "living fossils" are rare and therefore famous and of peculiar interest to the systematist. Within the tree of the Caminalcules (Fig.4: double lines) there are a very high number of living fossils and surviving stem-species (e.g. #18 is identical with the ancestor of all 8 species of DE). Such a situation is extremely unlikely for real organisms!

In summary, we may state that the phylogenetic tree of the Caminalcules is in the main not representative for organismic trees. It should be emphasized that this is not a critique on the late Prof. Camin, who created them. Afterwards, I looked at the first version of my Didaktozoa and also found very few convergences. This applies to most other didactic animals mentioned in sect.1, as well. It is understandable that the creator avoids problems, more or less unconsciously, when he draws organisms for didactic purposes, hoping the students will be able to come up with the true tree. My only critique is of the claim that tests applied to the Caminalcules are relevant for phylogenetics in general. The results (esp. the tests with different phylogenetic computer programs) are transferable only to those (presumably few) real organisms which have a tree topology and few convergences and pronounced gaps between groups like the Caminalcules.

4 Didaktozoa

Images and Phylogenetic Tree: Based on my critique, I have drawn a new set of imaginary animals (Fig.5), whose "true phylogeny" is shown as a cladogram in Fig.6. The respective evolutionary novelties (apomorphies) are indicated as black symbols, evolutionary losses (secondary absence, reversal) are crossed additionally (discussion of symbols see Wirth 1984a: 24). There are 12 recent species with 25 characters (some are multi-state characters, see list on Fig.7). The total sum of character changes (evolutionary steps) is 38: 14 synapomorphies and 24 autapomorphies (=apomorphies of the terminal taxa, 15 among them convergently).

Functional Considerations: Didaktozoon L has developed a giant sucking mouth which can be used for locomotion as well (inspired by rotifers who use their wheel organ for feeding and for swimming). Consequently the arms (used by the ancestors for climbing between algae or corals) could be reduced. (A more sophisticated didactic addition would be to say that during ontogenesis rudiments of arms develop and then give rise to the lateral parts of the sucking mouth, which would then be homologous to mouth + head + arms.) – B and K are characterized by 4 apomorphic features not shared by any other group (absence in arms also in L). Thereby, on first inspection, the "Vermiformia" seem to be a well-substantiated monophylum. But we must admit that these 4 apomorphies belong to a "syndrome" (func-

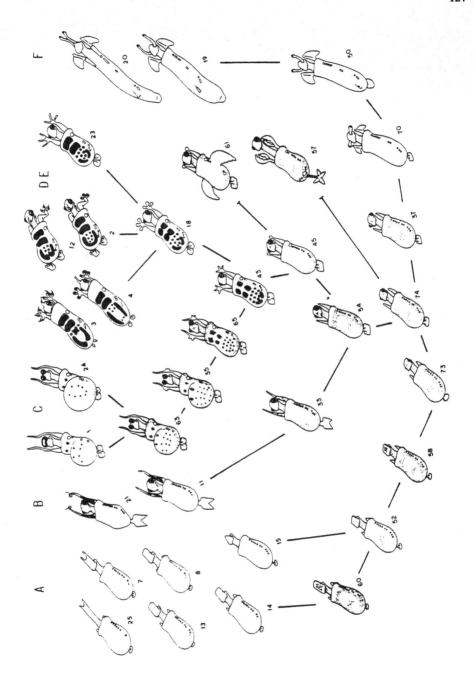

Fig. 3: Caminalcules. Phylogenetic tree ("true tree"). Only 35 representatives of the 29 recent and 48 fossil species are shown. (Newly combined from images in Sokal 1983). The crossed lines to #57 and #61 indicate extinct lineages. For the complete tree see Fig.4!

428

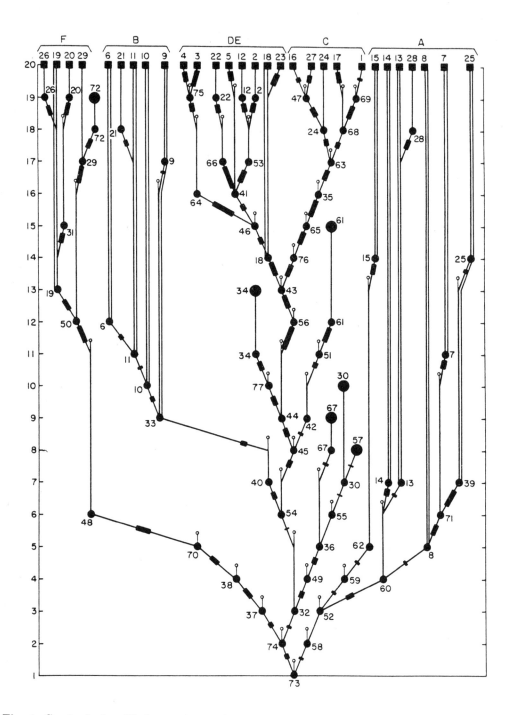

Fig. 4: Caminalcules. Phylogenetic tree ("true tree") with a relative time-scale. Black circles
are fossil, squares are recent species. Length of the thickened bar indicates the amount of
evolutionary change. Double lines indicate living fossils. (Modified from Sokal 1983: 169).

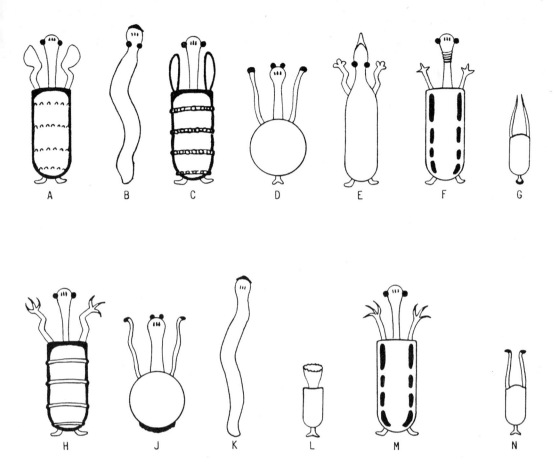

Fig. 5: Didaktozoa. Images of the 12 recent species (A-M). N is a fossil species. For additional information from ontogeny see Fig.6!

tional complex) associated with the specific mode of living, burrowing within the substrate. Once an organism changes his life habits in evolution, such a series of features may evolve relatively easy. Counting these 4 apomorphies as only one evolutionary step, convergent evolution is equally probable; actually, this is the situation in the true tree, and it can be guessed from study of the ontogeny! With this example of the diphyletic "Vermiformia" it is shown graphically that the decision on the character polarity (plesiomorphy vs. apomorphy) is logically independent of the next step of analysis, the decision synapomorphy vs. convergence. (For some real examples from electron microscopy see Wirth 1984b: 311-320; 1991).

Ontogeny: In consideration of actual problems in phylogenetics some relations are shown which can be uncovered mainly or solely by evaluating ontogeny, e.g. the "Vermiformia" (preceding paragraph). Another example is Didaktozoon C, where we have to decide whether the opercula are modified arms (Fig.7) or evolved as modifications of the armour by substitution of larval arms (Fig.6). (For estimating the relevance of recapitulation for phylogenetics see

430

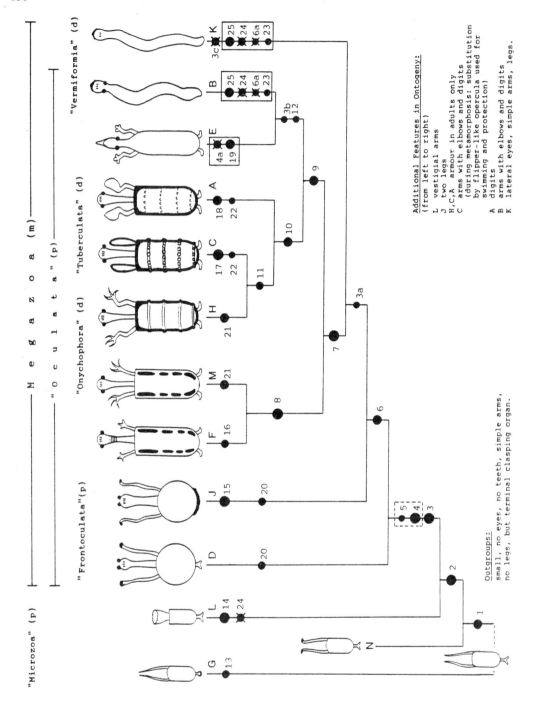

Fig. 6: Didaktozoa. Cladogram representing the "true tree". Apomorphies are marked with black circles, evolutionary loss is shown by additional crossing. – Classification at top shows some monophyletic (m), paraphyletic (p), and diphyletic (d) groups.

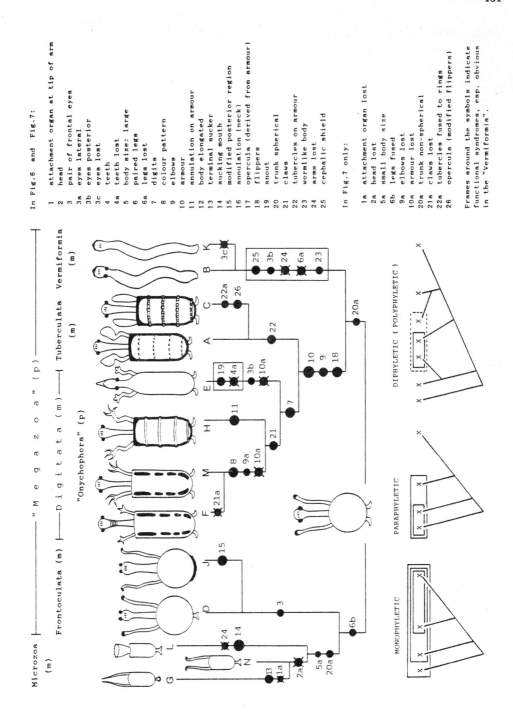

431

Fig. 7: Didaktozoa. Alternative cladogram using the same characters and symbols as in Fig.6. List of characters for both figures. – Inset: Explanation of monophyletic (m), paraphyletic (p), and diphyletic (d) groups.

Riedl 1975, Osche 1982, and more sceptically Gould 1977).

Alternative Cladogram: Whereas Fig.6 represents the "true tree", Fig.7 is another cladogram. This cladogram is even more parsimonious (34 evolutionary steps: 20 synapomorphies, 14 autapomorphies), esp. by taking the Vermiformia as monophyletic. The ancestor in this version is similar to D / J, and is more like a "Zentral-Typus" (see under stem- species in sect.3), and the tree topology is more symmetrical. Without knowledge from ontogeny and outgroups most students and presumably all computer programs would come up with trees like Fig.7 and not the true tree of Fig.6.

5   Conclusion

I endorse partly the statements of Sokal (1983: 159) that *Caminalcules* and similar artificial organisms "have the advantage over other simulations in presenting a visual record to the investigator, illustrate a variety of evolutionary phenomena and are therefore of considerable pedagogical and heuristic value". Then, however, we should not concentrate on evaluating only the 106 x 77 data matrix (Sokal 1983: 166), but take full advantage of the fact that we have "organisms" which can be interpreted with biological knowledge.

After criticizing the Caminalcules, I introduced the *Didaktozoa*, created according to rules most biologists will agree upon and hence more realistic than the Caminalcules. They have only 12 terminal taxa, they possess 25 characters, and their "true tree" has 38 evolutionary events, thus they are easier to handle than the Caminalcules. They are so constructed that they provide opportunities to discuss functional aspects. They illustrate graphically the relevance of ontogeny and outgroups for phylogenetics. Finally, they contain many convergences thus simulating the situation every systematist is confronted with, again making the Didaktozoa better suited for taxonomic exercises than the Caminalcules.

Prof. H.-J. Bandelt (Hamburg) gave a tutorial *"Phylogenetic Analysis of Molecular Data"* during this conference. He advocated that anyone using computer programs should try to use the principles of his programs also by hand on small data sets; this would help him to evaluate properly the results his programs are offering him. McKenna (1987: 57) showed that even extensive sequence data can be evaluated cladistically by hand. Conversely, it may be useful if some readers apply their routine procedures on the data matrix of the Didaktozoa. Then it may be easier to see which of the competing programs are truly phylogenetic in the sense of Hennig (see e.g. Wiley 1981, Ax 1984, Wirth 1984a), namely that they attempt to determine the character polarity (ancestral / plesiomorph → derived / apomorph) and to distinguish between synapomorphy and convergence – and which programs rely more on overall-similarity and midpoint-rooting.

References

AX, P. (1984), *Das Phylogenetische System*, Gustav Fischer, Stuttgart.

CAMIN, J.H., and SOKAL, R.R. (1965), A method for deducing branching sequences in phylogeny, *Evolution 19, 311-326.*

DIXON, D. (1981), *After Man - A Zoology of the Future*, Harrow House, London.

GORR, Th., KLEINSCHMIDT, T., and FRICKE, H. (1991), Close tetrapod relationships of the coelacanth Latimeria indicated by haemoglobin sequences, *Nature 351, 394-397.*

GOULD, St.J. (1977), *Ontogeny and Phylogeny*, Harvard University Press, Cambridge, Mass.

HEIJERMAN, Th. (1992), Adequacy of numerical taxonomic methods, *Z. zool. Syst. Evolut.-forsch. 30, 1-20.*

KLOB, W. (1981), Modellaufgaben zur Stammbaumentwicklung für Grund- und Leistungskurs Biologie, *Praxis Naturwiss, Biol. 1981(11), 346-349.*

LORENZEN, S., and SIEG, J.(1991), PHYLIP, PAUP, and HENNIG 86 - how reliable are computer parsimony programs used in systematics? *Z. zool. Syst. Evolut.-forsch. 29, 466-472.*

McKENNA, M.C. (1987), Molecular and morphological analysis of high-level mammalian interrelationships, in: C. Patterson (ed.), *Molecules and Morphology in Evolution: Conflict or Compromise?* Cambridge University Press, Cambridge, 141-176.

MAYR, E. (1982), *The Growth of Biological Thought*, Harvard University Press, Cambridge, Mass.

MOSS, W. W. (1971), Taxonomic repeatability: an experimental approach, *Syst. Zool. 20, 309-330.*

OSCHE, G. (1982), Rekapitulationsentwicklung und ihre Bedeutung für die Phylogenetik - Wann gilt die "Biogenetische Grundregel"?, *Verh. naturwiss. Ver. Hamburg (NF) 25, 5-31.*

REMANE, A. (1952), *Die Grundlagen des natürlichen Systems, der vergleichenden Anatomie und der Phylogenetik*, Akad. Verlagsges. Geest & Portig, Leipzig.

RIEDL, R. (1975), *Die Ordnung des Lebendigen*, Paul Parey, Hamburg und Berlin.

SENGBUSCH, P. von (1974), *Einführung in die Allgemeine Biologie*, Springer, Heidelberg. (3. Aufl. 1985)

SNEATH, P.H.A., and SOKAL, R.R. (1973), *Numerical Taxonomy*, Freeman, San Francisco.

SOKAL, R.R. (1966), Numerical taxonomy, *Sci. Amer. 215(6), 106-116.*

SOKAL, R.R. (1983), A phylogenetic analysis of the Caminalcules I.-IV., *Syst. Zool. 32, 159-184, 185-201, 248-258, 259-275.*

SOKAL, R. R. (1984), Die Caminalcules als taxonomische Lehrmeister (Lessons from the Caminalcules), in: H.-H. Bock (Hrg.), *Anwendungen der Klassifikation: Datenanalyse und numerische Klassifikation (= Studien zur Klassifikation 15)*, Indeks Verlag, Frankfurt, 15-31.

SOKAL, R.R., and ROHLF, F.J. (1980), An Experiment in Taxonomic Judgment, *Syst. Bot. 5, 341-365.*

STÜMPKE, H. (1961), *Bau und Leben der Rhinogradentia*, Gustav Fischer, Stuttgart.

WILEY, E.O. (1981), *Phylogenetics. The Theory and Practice of Phylogenetic Systematics*, John Wiley, New York.

WIRTH, U. (1984a), Die Phylogenetische Systematik (Das Prinzip von Hennig), *Mitt. dtsch. malakozool. Ges. 37, 6-35.*

WIRTH, U. (1984b), Die Struktur der Metazoen-Spermien und ihre Bedeutung für die Phylogenetik, *Verh. naturwiss. Ver. Hamburg (NF) 27, 295-362.*

WIRTH, U. (1991), Cladistic Analysis of Sperm Characters, in: B. Baccetti (ed.), *Comparative Spermatology 20 Years After (Serono Symposia 75)*, Raven Press, New York, 1025-1029.

Multivariate Analysis of the Process of Acclimation of Physiologic Variables

Paul Schwarzenau, Peter Bröde, Peter Mehnert, Barbara Griefahn
Institute for Occupational Health
Dept. of Environmental Physiology and Occupational Medicine
(Director: Prof. Dr. Barbara Griefahn)
Ardeystr. 67, D-4600 Dortmund 1, Fed. Rep. Germany

Abstract:

Fundamentals: The influence of heat causes considerable effects on several physiologic functions such as circulation, core temperature etc.. This strain decreases gradually during repeated exposure. The process of acclimation is characterized by a decrease of heart rates, core and skin temperatures and by an increase of sweat rates. It is terminated if the physiologic variables reach a steady state (leveling off); the organism is then regarded as acclimated.

Objectives: The aim of the paper was to classify (the data of) the process of acclimation using as few data as possible.

Methods - experimental procedures: 11 male subjects, 19 to 23 years of age were exposed during 15 to 23 days to a hot-humid climate while exerting a defined physical work (treadmill, 4 km/h). During the up to six 30-minutes lasting working periods heart rates, core- and skin-temperatures were continuously recorded and sweat rate was assesed by weighting the subjects every 30 minutes.

Model: A model for the process (non linear regression) was developped on the basis of the descriptive plots. These processes are described by an ascending e-function with the parameters: initial size, steepness of the ascent and final size (value of the steady state).

1 Introduction

The maintenance of body temperatures becomes difficult if man must work in the heat. Sweat production increases but only to an insufficient degree thus leading to elevated temperatures of the core and of the skin. As heart rates increase concomitantly physical performance is remarkably reduced.

However, if man is repeatedly exposed to heat sweat rate increases considerably, where core and skin temperatures as well as heart rates decrease. The rate of change is largest within the first few days and becomes then gradually less. If the physiologic functions do not alter anymore, if they reach a steady state, the organism is regarded as acclimated and full physical performance is regained.

To avoid detrimental effects on health man should not work at his limit before the point of acclimation is reached. It is therefore necessary to assess the duration of this period which depends on the climatic factors and on the individual.

2   Objectives

The aim of the present paper was to classify the process of acclimation and to predict the point of the steady state with as few data as possible.

3   Experimental procedure

11 healthy male subjects 19 to 23 years of age took part in the experiments. They were exposed to a hot humid climate at 15 to 23 days, while excerting a defined physical work on a treadmill (4 km/h). Weekends and holidays were free thus simulating a real occupational situation. Each session consisted of up to 6 working periods of 30 minutes each. However, exercise was terminated as soon as the core temperature exceeded 38.3°C. During the working periods, heart rates, core and skin temperatures were continuously recorded and sweat rate was assessed by weighing the subjects every 30 minutes.

The subjects were divided into 2 groups according to the experimental conditions (see table 1).

Table 1: Subjects and experimental conditions

group	n	trials	inclination[degree]	temperature[°C]	humidity[%]
1	4	13-16	2-3	35	50-80
2	7	14-21	0	38-40	65-70

4   Application of classification procedures

The processes of acclimation were then described by a non linear growth model. The parameters of the model - particularly the time when the steady state (acclimation) was reached - were estimated from the data. To characterize both the experimental groups the parameters were averaged by weighting them with the reciprocals of their variances. The parameters of both these groups were then compared using a weighted analysis of variance.

4.1   Descriptive plots

The averages of the sweat rate and of the core temperature of the 3rd working period were then used for the following presentations and calculations. Sweat rates (SR) and core temperatures (T_c) are plotted in figures 1 and 2.

4.2   Modelling

An exponential growth model with constraint increase was chosen for the modelling of the process of the physiologic parameters (Seber, 1989). The following parametrization was used

$$f(t) = b - (b - a)\, e^{-k(t - t_0)}, \quad (t = t_0, ..., T)$$

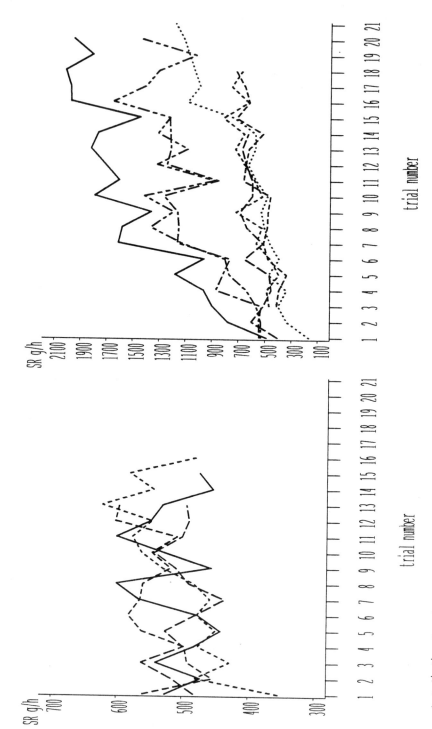

Figure 1: Acclimation of sweat rate

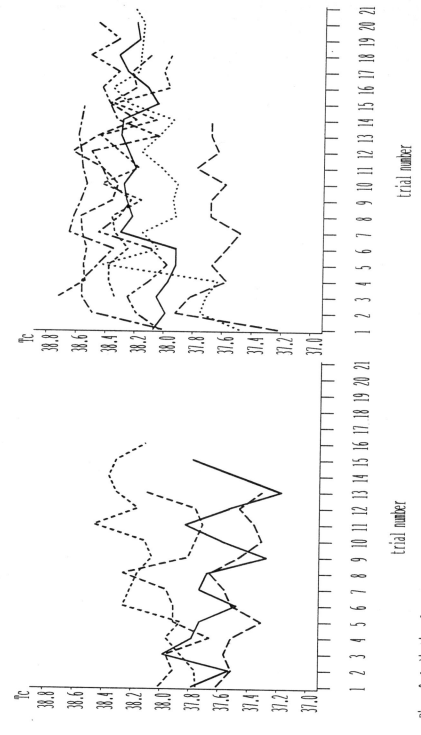

Figure 2: Acclimation of core temperature

with the parameters a (initial size), b (final size), k (scale factor),
t_0 (first experiment), T (number of experiments).

It was assumed that the stochastic process of the acclimation of a physiologic variable y(t)
at the time t can be decomposed into

$$y(t) = f(t) + u(t), \quad (t = t_0, ..., T).$$

where u presents a (weak) stationary, stochastic process with the expected value 0.

To regard the time dependency of the data the stochastic term u(t) is described by an
autoregressive model

$$u(t) = c\,u(t-1) + \varepsilon(t), \quad (t = t_0, ..., T).$$

where $\varepsilon(t)$ are i.i.d. errors of the model:

$$\varepsilon(t) \sim N(0, \sigma^2), \quad (t = t_0, ..., T).$$

If the time dependency of the data are neglected the least squares estimators (LS-estimators)
of the variances of the parameters are biased; these estimators are mostly too small (Glasbey,
1980).

4.3 Determination of the time of acclimation

The time of acclimation T_a is defined as the time t where 90% of the final size are reached.
The deviation of 10% presents the mean error of the model

$$T_\alpha = min(t : f(t) \geq 0.9\,b) = t_0 - \frac{\log(0.1\,\frac{b}{b-a})}{k}$$

4.4 Estimation of the parameters

The following model

$$y_{ij}(t) = b_{ij} - (b_{ij} - a_{ij})e^{-k_{ij}(t_{ij} - t_{0ij})} + u_{ij}(t)$$

with

$$u_{ij}(t) = c_{ij}u_{ij}(t-1) + \varepsilon_{ij}(t), \quad (t = t_{0ij}, ..., T_{ij})$$

and

$$\varepsilon_{ij}(t) \sim N(0, \sigma^2)$$

was fitted to the data $y_{ij}(t)$ of the i-th subject $(i = 1, ..., N_j)$ of the j-th group (j=1,2)
separately for y = SR (sweat rate) and y = T_c (core temperature).

The data of the sweat rate were sufficiently presented by the models. The autoregressive
parameters c_{ij} were estimated from the least squares of the residuals. As this estimated value

439

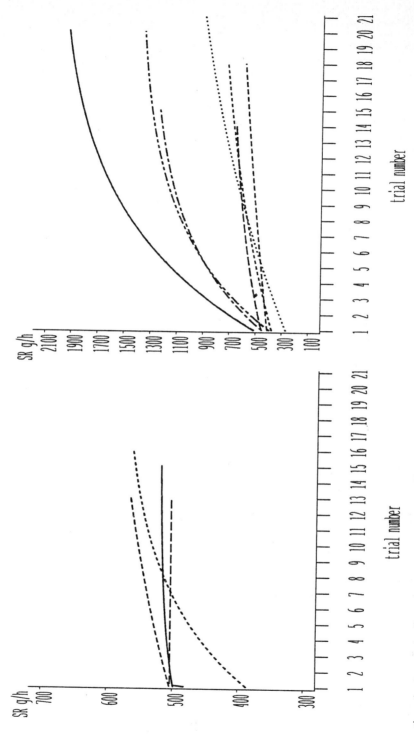

Figure 3: Plots of the fitted curves of sweat rate

Table 2: Least squares estimators and weighted means of the parameters of the sweat rate

group	subj.	$\hat{a}$	$\hat{\sigma}_a$	$\hat{b}$	$\hat{\sigma}_b$	$\hat{\kappa}$	$\hat{\sigma}_\kappa$	$\hat{T}_\alpha$	$\hat{\sigma}_{T_\alpha}$	n	$\hat{\sigma}$	R^2
1	1	498.6	46.2	520.1	50.0	0.21	1.52	-3.2	28.9	15	55.8	0.02
1	2	387.0	34.0	581.2	57.3	0.15	0.12	8.9	6.6	16	43.4	0.64
1	3	504.3	32.0	634.8	705.0	0.05	0.38	14.8	179.7	13	42.4	0.24
1	4	482.0	31.9	503.7	9.2	5.2E7	.	1.0	.	13	31.9	0.04
1	$\bar{x}_w$	459.2	55.9	546.7	30.6	0.14	0.03	8.3	2.6	3		
2	1	514.2	124.1	2054.8	167.6	0.13	0.04	16.0	4.4	19	167.2	0.89
2	2	376.6	108.2	898.7	501.6	0.07	0.11	29.8	50.0	15	78.3	0.66
2	3	411.1	108.0	892.0	3047.2	0.03	0.23	62.8	589.0	16	90.9	0.50
2	4	454.5	53.6	718.5	212.7	0.11	0.18	12.5	21.1	14	68.9	0.51
2	5	450.4	123.0	1326.0	235.9	0.17	0.11	12.5	7.7	13	153.0	0.77
2	6	394.4	145.3	1411.4	149.1	0.16	0.07	13.5	5.5	20	191.2	0.72
2	7	268.8	99.5	1268.1	874.9	0.05	0.07	41.9	58.3	19	152.6	0.75
2	$\bar{x}_w$	419.0	66.8	1441.3	458.3	0.12	0.04	14.7	2.3	7		

was not significantly different from zero it was furtheron neglected. From table 2 which gives the LS-estimators and its weighted averages $\bar{x}_w$ it becomes evident that the estimator for the parameter k is biased for subject 4 of group 1. These data were discarded from averaging.

The plots of the fitted curves are presented in figure 3.

The same model, however, was not suitable for the description of the core temperature. The latter is better presented by a linear regression

$$y_{ij}(t) = a_{ij} + b_{ij}(t - t_0) + u_{ij}(t), \quad (t = t_0, ..., T).$$

4.5   Comparison between the 2 experimental groups

Thereafter a weighted analysis of variance was applied to find out whether the processes of acclimation are different in both the experimental groups. A significant difference was determined for the time of acclimation T_a and for the final size of the sweat rate (see figure 4).

Regarding core temperature a significant difference was determined only for the intercept but not for the slope.

5   Results

The sweat rates during the experiments were sufficiently described by the model. The estimation of the parameters enables to predict where the steady state (acclimation) is reached. Contrary to the hypothesis the descriptive plots revealed ascending values for heart rates, core- and skin temperatures. The expected dependency of the parameters on physical work, air temperature and humidity were proved using the analysis of variance.

Mean curves of sweat rate

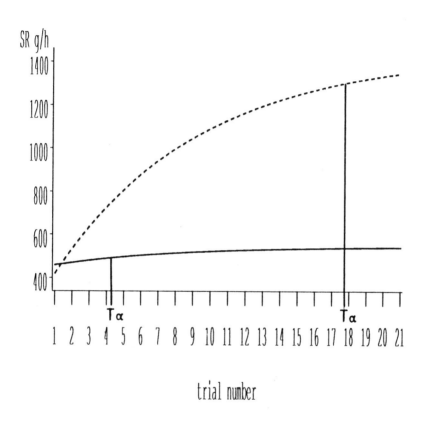

trial number

T_α : time point of acclimation
Group 1: solid line, Group 2: broken line

Figure 4: Plots of the mean curves of sweat rate

442

References

GLASBEY, C.A. (1980), Nonlinear regression with autoregressive time series errors, *Biometrics, 36, pp. 135-140.*

JOHNSON, P. (1983), A simple procedure for testing linear hypotheses about the parameters of a nonlinear model using weighted least squares, *Comm. Statist. - Simula. Computa., 12(2), pp. 135-145.*

SEBER, G.A.F., WILD,C.J. (1989), Nonlinear Regression, *Wiley, New York, pp. 327-328*

Classification of EEG Signals into General Stages of Anesthesia in Real-Time Using Autoregressive Models

R. Bender, B. Schultz

Zentrum für Anästhesiologie, Abteilung IV, Medizinische Hochschule Hannover
Krankenhaus Oststadt, Podbielskistr. 380, W-3000 Hannover 51, FRG

and U. Grouven

Klinik für Abdominal- und Transplantationschirurgie, Medizinische Hochschule Hannover
Dialysezentrum, Stadtfelddamm 65, W-3000 Hannover 61, FRG

Abstract: In anesthesiology an automatic classification of the electroencephalogram (EEG) into general stages of anesthesia is of interest to support narcosis control. The fitting of autoregressive (AR) models to EEG periods reduces the signal information to few parameters, which form a basis for development of suitable discriminant functions. The favourable reclassification results show that autoregressive models are useful tools to describe EEG patterns occuring in anesthesia. The application of the Levinson-Durbin algorithm for the estimation of AR parameters enables a classification of EEG signals into general stages of anesthesia in real-time.

1 Introduction

Computerized analysis of the electroencephalogram (EEG) offered a wide field of medical applications in the last decades. One of them with clinical interest in anesthesiology is an automatic assessment of anesthetic depth. At our institution the Narkograph was developed, an apparatus, which performs an automatic classification of EEG signals into general stages of anesthesia in real-time and is therefore qualified to support an anesthesist in narcosis control. A description of the employed method as well as the necessity of further improvement of the classification procedure was outlined by Bender et al. (1991).

One of the important problems in EEG analysis is the extraction of features to describe EEG signals, because the power of any EEG analysis method depends on the quality of the chosen features (Lopes da Silva, 1987). The classification procedure of the Narkograph works to date with widely used spectral parameters as EEG describing features. As in some papers of late years (Blinowska et al., 1981; Jansen, Hasman and Visser, 1978; Jansen, Bourne and Ward, 1981; Matthis, Scheffner and Benninger, 1981) the superiority of autoregressive (AR) parameters in comparison with spectral parameters was pointed out, the use of AR models promises an improvement of the classification procedure.

In a first study of Bender et al. (1991) it was shown that the extraction of AR parameters from EEG periods of 2 seconds and the application of quadratic discriminant analysis (QDA) has advantages in comparison with the former method using spectral parameters. Although the set of AR parameters is smaller than the set of spectral parameters, the classification results are just as good or better. Especially, the frequently confounded stages "awake" and "slight anesthesia" can better be recognized by AR parameters. In this paper firstly, the method presented by Bender et al. (1991) is extended to a higher number of stages and secondly, the fast Levinson-Durbin algorithm (Pagano, 1972) for estimation of AR parameters is used to make a classification in real-time possible. Further on, the use of the standard deviation of EEG periods as additional feature is studied, which was proposed by Jansen, Hasman and Visser (1978) for recognition of general EEG patterns. As the EEG patterns occuring in anesthesia have absolute differences in signal power in dependence on the stage and AR parameters are only relative values, an amplitude measure like the standard deviation presumably contains supplementary discriminating power.

Therefore, this paper studies the usefulness of AR parameters combined with an amplitude measure for classification of EEG periods into general stages of anesthesia in real-time.

2   Material and Signal Processing

The study has been performed using EEG signals from 40 awake or narcoticized patients between 20 and 50 years of age. Anesthesia was inducted with thiopenthal and maintained with enfluran and a N_2O/O_2-mixture. The EEGs were recorded with the Narkograph using derivation C_3-P_3 of the international 10/20-system (Jasper, 1958). A sampling rate of 128 Hz, a time constant of 0.3 seconds and a low-pass-filter of 70 Hz were used, which form a commonly used EEG recording configuration in anesthesiology (Pichlmayr, Lips and Künkel, 1983).

The EEG signals were devided into epochs of 20 seconds, which were subdivided into periods of 2 seconds. A total of 317 qualified EEG epochs were selected, which contain altogether 2447 artifact-free periods. The EEG epochs were visually classified into 13 stages of anesthesia, which are based upon a division of anesthetic depth proposed by Kugler (1981). Definitions of the stages together with the frequencies in the analysed data set are summarized in Table 1, while in Figure 1 some typical EEG curves are shown.

Tab. 1: *Survey of General Stages of Anesthesia*

No.	Name	Meaning	Number of EEG Epochs	Number of Artifact-free Periods
1	A	awake	25	160
2	B_0		31	273
3	B_1	very slight anesthesia	21	175
4	B_2		17	127
5	C_0		36	259
6	C_1	slight anesthesia	21	124
7	C_2		24	125
8	D_0		26	212
9	D_1	middle anesthesia	24	196
10	D_2		21	160
11	E_0	deep anesthesia	32	293
12	E_1		22	179
13	F	very deep anesthesia	17	164
	Total		317	2447

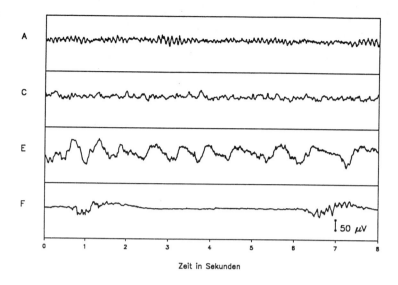

Fig. 1: *Examples of Typical EEG Curves Occuring in Anesthesiology*

3 Theory

3.1 Autoregressive Modelling

An EEG period of 2 seconds recorded and selected in a way described in the previous chapter forms a time series $y=(y_t)_{t=1,...,T}$ with sample size $T=256$. It is assumed that y – after subtraction of the mean – is a realization of a stochastic process Y, which satisfies the equation

$$Y_t = \sum_{j=1}^{p} \varphi_j Y_{t-j} + e_t \quad , \tag{1}$$

where $e=(e_t)$ is a white-noise process with mean zero and variance σ_e^2, i.e., Y forms an autoregressive (AR) process of order p. Thus, the EEG amplitudes are linear combinations of several past values and a random error. To fit models like (1), in practice the approach of Box and Jenkins (1976) is generally accepted, which consists essentially of the three steps identification, estimation and model checking.

3.2 Extraction of Features

In the first study of Bender et al. (1991) the method of conditional least squares (CLS) was used to estimate AR parameters, which is a simple but not very fast procedure. To obtain a procedure which works in real-time a faster method is needed. Therefore the approved fast Levinson-Durbin algorithm (Pagano, 1972) was chosen, which solves the Yule-Walker equations

$$
\begin{pmatrix}
c_0 & c_1 & \cdots & c_{p-1} \\
c_1 & c_0 & \cdots & c_{p-2} \\
\vdots & \vdots & \ddots & \vdots \\
c_{p-1} & \cdots & \cdots & c_0
\end{pmatrix}
\begin{pmatrix}
\varphi_1 \\
\varphi_2 \\
\vdots \\
\varphi_p
\end{pmatrix}
=
\begin{pmatrix}
c_1 \\
c_2 \\
\vdots \\
c_p
\end{pmatrix}
\tag{2}
$$

in a recursive way, where

$$
c_\tau = \frac{1}{T} \sum_{t=1}^{T-\tau} y_t y_{t+\tau}
\tag{3}
$$

is the sample autocovariance function of Y. From c_0 the standard deviation of Y can be estimated by

$$
s = \sqrt{c_0} = \sqrt{\frac{1}{T} \sum_{t=1}^{T} y_t^2} \; .
\tag{4}
$$

If the extracted features describe EEG periods sufficiently, it should be possible to develop suitable discriminant functions based upon $\varphi_1, \ldots, \varphi_p$ and s.

3.3 Development of Discriminant Functions

Assuming that the vector $\vec{x} = (\varphi_1, \ldots, \varphi_p, s)'$ of EEG features, i.e., AR parameters and standard deviation, estimated from an EEG period belonging to cluster k, is multivariate normally distributed with mean $\vec{\mu}_k$ and covariance matrix Σ_k, parametric methods of discriminant analysis can be applied. An univariate analysis using the Shapiro-Wilk test computed with the SAS procedure PROC UNIVARIATE (SAS, 1988a) showed that the assumption of normality does not hold in all cases and therefore it is doubtful that $\vec{x}$ has a multivariate normal distribution. However, because for a classification in real-time a quick and easy procedure is required, a parametric discriminant anlysis method was chosen. Taking this background into account, the quality of the resulting discriminant functions for practical purposes was studied later without using the assumption of normality.

The feature vectors of all considered EEG periods form the learning data set, from which discriminant functions were derived. The use of quadratic discriminant analysis (QDA) leads to the decision rule that an EEG period is assigned to that cluster k, for which the generalized squared distance function

$$
d_k(\vec{x}) = (\vec{x} - \vec{\mu}_k)' \Sigma_k^{-1} (\vec{x} - \vec{\mu}_k) + \ln[\det(\Sigma_k)]
\tag{5}
$$

is at a minimum (Fahrmeir, Häussler and Tutz, 1984). The posterior probability of membership in cluster k can be computed with the formula

$$
P(k|\vec{x}) = \frac{\exp[-0.5\, d_k(\vec{x})]}{\sum_k \exp[-0.5\, d_k(\vec{x})]} \; .
\tag{6}
$$

To apply formulae (5), (6) in practice, it is necessary to estimate the cluster means and covariance matrices from the learning data set and to replace the unknown parameters with their estimates.

3.4 Estimation of Classification Error Rates

To check the adequacy of the evaluated discriminant functions, various methods for estimation of classification error rates exist. As the accuracy of assuming multivariate normality is uncertain, in this analysis only non-parametric methods were applied, which based upon different reclassifications of the learning data set. All estimated error rates have the general form

$$\epsilon = \frac{1}{K} \sum_{k=1}^{K} \frac{a_k}{n_k} \quad , \tag{7}$$

where K is the number of clusters, a_k the number of misclassified EEG periods or epochs belonging to cluster k and n_k is the sample size of cluster k. The misclassification numbers a_k depend on the chosen method and on the definition of "misclassification".

The simplest method to estimate classification error rates is the resubstitution or R method, which has an optimistic bias due to the fact that the sample used to compute discriminant functions is reused to estimate the error. To avoid this disadvantage, a jackknife method can be applied (Lachenbruch and Mickey, 1968). Both methods treat only those observations as correctly classified, if visual and computed classification correspond with each other. For practical purposes in anesthesiology another error rate is of interest. The $K=13$ stages of Table 1 form ordinal categories of a non-measurable continuous variable, namely anesthetic depth. Misclassifications into neighbouring stages have no essential disadvantages for narcosis control and therefore, classification error rates should also be computed by valuing neighbouring misclassifications as correct assessment of anesthetic depth. Consequently, four methods to estimate classification error rates were used, namely the usual resubstitution and jackknife method and the two methods ensued from both by acception of neighbouring misclassifications, abbreviated as R, J, RN and JN method, respectively.

4 Results and Discussion

4.1 Time Series Analysis

The first problem is to choose an order of the AR(p) models adequate for all considered EEG signals, which have quite different patterns in dependence on anesthetic depth. In EEG analysis the model order usually ranges from 3 to 30, but mostly low orders between 5 and 10 are used. If a model is needed, which describes the wide range of possible EEG patterns occuring in anesthesia sufficiently in the sense of Box and Jenkins (1976), a high model order, e.g., $p=30$ is required (Bender et al., 1992). However, the aim of this study is not to describe EEG signals completely, but to select as few features as possible, which are able to distinguish between the stages of anesthesia. For this purpose, based on the study of Bender et al. (1991) and also shown later by the discriminant analysis results, a model order of $p=5$ seems to be adequate.

For the execution of the Levinson-Durbin algorithm to compute AR parameter estimations, a PASCAL program including some subroutines published by Robinson (1967) was used. As the chosen model order is $p=5$, for further computations a $N \times M$ data matrix was obtained with $N=2447$ observations (number of EEG periods) and $M=7$ variables, namely the visually classified stage of anesthesia, the 5 AR parameters and the standard deviation of the EEG periods.

4.2 Discriminant Analysis

For the computation of the discriminant functions the SAS procedure PROC DISCRIM (SAS, 1988b) was used. As the chi-square test for hypothesis of homescedasticity described by Morrison (1976) yields a p-value <0.0001, the homogenity of the cluster covariance matrices had to be rejected. Consequently, no pooled covariance matrix but estimates of the within-group covariance matrices are required, i.e., formula (5) should not be simplified to linear discriminant analysis (LDA), which is commonly used in EEG classification procedures (Jansen, Hasman and Visser, 1978; Jansen et al., 1979; Jansen, Bourne and Ward, 1981). Table 2 contains the reclassification outcome of the learning data set using QDA and the features $\varphi_1, \ldots, \varphi_5, s$.

Tab. 2: *Reclassification of the EEG Periods*

Computed → Visual Class. ↓	A	B_0	B_1	B_2	C_0	C_1	C_2	D_0	D_1	D_2	E_0	E_1	F	Total
A	149	0	0	0	3	2	2	2	0	0	0	0	2	160
B_0	0	224	46	1	1	0	0	0	0	0	0	0	1	273
B_1	0	12	126	27	10	0	0	0	0	0	0	0	0	175
B_2	0	2	19	80	17	8	1	0	0	0	0	0	0	127
C_0	3	1	33	36	122	58	5	1	0	0	0	0	0	259
C_1	2	0	1	2	12	77	20	9	1	0	0	0	0	124
C_2	0	0	1	1	3	32	62	16	9	1	0	0	0	125
D_0	1	0	0	0	2	3	49	132	21	2	1	1	0	212
D_1	0	0	0	0	0	0	14	23	136	14	9	0	0	196
D_2	1	0	0	0	0	1	1	9	28	102	14	4	0	160
E_0	0	0	0	0	0	0	9	8	14	17	222	22	1	293
E_1	0	0	0	0	0	0	0	4	0	4	42	117	12	179
F	0	0	0	0	0	0	0	0	9	14	18	45	78	164
Total	156	239	226	147	170	181	163	204	218	154	306	189	94	2447

The simple R method yields an error rate estimation of $\epsilon=34.4$ %. With regard to the fact that the agreement between visual classification of different EEG experts reaches only about 70 % (Ferber and Künkel, 1979), this first result is of great promise. All four error rate estimations mentioned in Chapter 3 were computed with and without using the standard deviation as discriminating variable to study the effect of s as additional feature. The results are comprised in Table 3.

Tab. 3: *Estimated Error Rates for Classification of EEG Periods (in %)*

Method	Feature Vector $(\varphi_1, \ldots, \varphi_5, s)$	Feature Vector $(\varphi_1, \ldots, \varphi_5)$
R	34.4	42.5
J	37.4	45.1
RN	10.8	13.5
JN	11.6	14.2

From Table 3 the following conclusions can be drawn. The error rates estimated by using s as additional feature are smaller than the others. Together with the results of simple ANOVA and MANOVA models also computed with PROC DISCRIM (SAS, 1988b), which showed clearly that all six features have discriminating power (all p-values < 0.0001), it seemed to be favourable to use the combined feature vector $(\varphi_1, \ldots, \varphi_5, s)$ instead of AR parameters alone.

The differences between the R and J method as well as between the RN and JN method appear to be negligible for practical purposes. As the resubstitution method has disadvantageous properties especially in small samples, this indicates that the sample size of the learning data set is adequate. A great number of misclassifications of the R and J method are classifications into neighbouring clusters, which is shown by the small error rates obtained by the RN and JN method. Both error rate estimations of approximately 11 % show that feature vectors composed of AR parameters and standard deviation are useful tools to describe EEG patterns occuring in anesthesia.

4.3   Classification of EEG Epochs

Although the results of the discriminant analysis are of great promise, attention should be focused on the design of the learning data set. As mentioned in Chapter 2, the 2447 observations (number of artifact-free periods) came from 40 patients and therefore the assumption of independent and identically distributed (i.i.d.) data within the clusters which is necessary for application of QDA does not hold. Instead of this the learning data set has a more complicated unbalanced repeated measurement design which has not been considered in the discriminant analysis, because no standard classification method exists, which takes special dependencies of data into account. Therefore, the results of the discriminant analysis should be interpreted carefully. For example, the reason for the slight difference between R and J methods could be the fact that other periods of the same person with similar patterns as the period which was left out in the jackknife method, were now as before present in the data set. To check the adequacy of the evaluated discriminant functions for anesthesiological practice, they should be applied to EEG signals recorded from patients, which are not a part of the learning data set.

For this purpose, complete EEG epochs of 20 seconds instead of periods are now classified by averaging the feature vectors computed from those periods which are free of artifacts. From anesthesiological point of view a classification of whole epochs is sufficient. On the other hand, a classification procedure based on averaged feature vectors is more robust, because, e.g., periods with not detected slight artifacts have no effect in averaged features of the whole epoch and do not lead to a misclassification. In Figure 2, as an example, the visual and the computed classification of an EEG signal with a duration of 7 minutes is shown, which was recorded during the induction of anesthesia of a 43 year old patient.

From 18 epochs (the epochs 1, 4 and 5 were artifactual) only 4 are correctly classified, i.e., the computed stages of anesthesia agree with the visual assessment only in 22 % of the cases. However, if neigbouring classifications are accepted, the rate of correct classifications increases to 89 %, which shows that most of the computed stages are acceptable for anesthesiological practice. The comparison of visual and computed assessment of anesthetic depth indicates that the whole course of computed stages is useful and reproduces the anesthetic process of the patient. Although one example is of limited scope, it underlines the adequacy

450

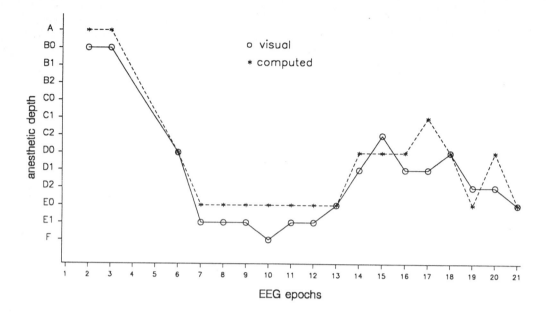

Fig. 2: *Example of a Visual and Computed Assessment of Anesthetic Depth*

of the evaluated discriminant functions for practical purposes in spite of theoretical defects like missing independence of data within clusters. Moreover, the use of the fast Levinson-Durbin algorithm enables a classification in real-time, which is important for an employment in anesthesiological practice. With further developments, e.g., a special recognition of burst-suppression phases (stage F), a classification procedure could be obtained, which provides a useful assistance for narcosis control.

In summary, the results show that autoregressive modelling with subsequent application of quadratic discriminant functions is an adequate way for classification of EEG signals into general stages of anesthesia in real-time.

References

BENDER, R., SCHULTZ, B., SCHULTZ, A. and PICHLMAYR, I. (1991), Identification of EEG Patterns Occuring in Anesthesia by Means of Autoregressive Parameters, *Biomedizinische Technik, 36, 236–240.*

BENDER, R., SCHULTZ, B., SCHULTZ, A. and PICHLMAYR, I. (1992), Testing the Gaussianity of the Human EEG During Anesthesia, *Methods of Information in Medicine, 31, 56–59.*

BLINOWSKA, K, J., CZERWOSZ, L.T., DRABIK, W., FRANASZUK, P.J. and EKIERT, H. (1981), EEG Data Reduction by Means of Autoregressive Representation and Discriminant Analysis Procedures, *Electroencephalography and clinical Neurophysiology, 51, 650–658.*

BOX, G.E.P. and JENKINS, G.M. (1976), *Time Series Analysis – Forecasting and Control (2nd ed.)*, Holden-Day, San Francisco.

FAHRMEIR, L., HÄUSSLER, W. und TUTZ, G. (1984), Diskriminanzanalyse, in: L. Fahrmeir und A. Hamerle (Hrsg.), *Multivariate statistische Verfahren*, De Gruyter, Berlin, 301–370.

FERBER, G. und KÜNKEL, H. (1979), Die Bedeutung der automatischen EEG-Analyse für die klinische EEG-Befundung, *Wissenschaftliche Zeitschrift der Ernst-Moritz-Arndt Universität Greifswald, Medizinische Reihe*, 28, 25–36.

JANSEN, B.H., BOURNE, J.R. and WARD, J.W. (1981), Autoregressive Estimation of Short Segment Spectra for Computerized EEG Analysis, *IEEE Transactions on Biomedical Engineering*, 28, 630–638.

JANSEN, B.H., HASMAN, A., LENTEN, R. and VISSER, S.L. (1979), Usefulness of Autoregressive Models to Classify EEG-Segments, *Biomedizinische Technik*, 24, 216–223.

JANSEN, B.H., HASMAN, A. and VISSER, S.L. (1978), Features to Segmentate an EEG Recording: A Comparative Study, in: D.A.B. Lindberg and P.L. Reichertz (eds.), *Medical Informatics Europe 1978, Proceedings*, Cambridge,Springer, Berlin, 533–544.

JASPER, H.H. (1958); The Ten-Twenty Electrode System of the International Federation, *Electroencephalography and clinical Neurophysiology*, 51, 650–658.

KUGLER, J. (1981), *Elektroenzephalographie in Klinik und Praxis*, Thieme, Stuttgart.

LACHENBRUCH, P.A. and MICKEY, M.R. (1968), Estimation of Error Rates in Discriminant Analysis, *Technometrics*, 10, 1–10.

LOPES DA SILVA, F. (1987), Computer-assisted EEG Diagnosis: Pattern Recognition Techniques, in: E. Niedermayer and F. Lopes da Silva (eds.), *Electroencephalography, Basic Principles, Clinical Applications and Related Fields (2nd ed.)*, Urban & Schwarzenberg, München, 899–919.

MATTHIS, P., SCHEFFNER, D. and BENNINGER, C. (1981), Spectral Analysis of the EEG: Comparison of various Spectral Parameters. *Electroencephalography and clinical Neurophysiology*, 52, 218–221.

MORRISON, D.F. (1976), *Multivariate Statistical Methods (2nd ed.)*, McGraw-Hill, New York.

PAGANO, M. (1972), An Algorithm for Fitting Autoregressive Schemes. *Applied Statistics*, 21, 274–281.

PICHLMAYR, I., LIPS, U. und KÜNKEL, H. (1983), *Das Elektroenzephalogramm in der Anästhesie*, Springer, Berlin.

ROBINSON, E.A. (1967), *Multichannel Time Series Analysis with Digital Computer Programs*, Holden-Day, San Francisco.

SAS (1988a), *SAS Procedures Guide, Release 6.03 Edition*, SAS Institute Inc., Cary, NC.

SAS (1988b), *SAS/STAT User's Guide, Release 6.03 Edition*, SAS Institute Inc., Cary, NC.

Automatic Segmentation and Classification of Multiparametric Image Data in Medicine

Heinz Handels[1]

Institut für Medizinische Statistik und Dokumentation, RWTH Aachen
Pauwelsstr. 30, W-5100 Aachen, FRG

Abstract: In this paper cluster analysis and classification methods in combination with image processing algorithms are presented for automatic segmentation and classification of tissues in medical images. Based on multiparametric image data, which are typically generated in magnetic resonance tomography (MRT), different tissue structures are segmented by using pyramidal histogram analysis methods in combination with a merging algorithm. Though the amount of data is large in the medical images considered, several images can be segmented simultaneously in an efficient manner. After segmentation, the 3-dimensional distribution of tissues in space is visualized. Furthermore, the tissue-specific relaxation parameter values of each segmented tissue are computed to establish a tissue data base. Based on the extracted relaxation parameter values an automatic classification of unknown tissue structures is performed using statistical classifiers. The classified tissues are marked with tissue-specific colours and visualized in tissue class images. For diagnostic support in clinical applications, the algorithms for tissue segmentation, classification and visualization are integrated in the software system SAMSON ('System for Automatic Segmentation and Classification of Tissue in Magnetic Resonance Tomography') [Handels 1992].

1   Introduction

Magnetic resonance tomography is an important and unique imaging technique in medicine, which permits the generation of slice images in every region of a human body. Typically for clinical applications, several MR images of an investigated body slice consisting of 256x256 pixels are generated. Usually, the signals measured in a volume element of a body slice are visualized by grey values (Fig. 1). The measured signals are influenced by several superimposing relaxation processes, which can be characterized by the relaxation times, T_1 and T_2, as well as the spin density, ρ. While the relaxation times T_1 and T_2 describe the relaxation behaviour of the longitudinal and the transversal relaxation processes respectively, the spin density ρ gives information on the spin density of hydrogen protons within one volume element. Using a special multi-echo measurement sequence [Eis et al. 1989], longitudinal and transversal relaxation processes can be observed in a temporal dimension simultaneously in several slices of a human body. Based on the measured signal decays, the relaxation parameters T_1, T_2 and ρ are computed by using a special noise preprocessing algorithm and nonlinear optimization methods [Handels 1992]. After relaxation analysis, a multidimensional MR parameter information is extracted for each volume element from the investigated slices. Interactive analysis and visualization of the relaxation parameter distributions have shown that data driven separation and classification of tissues in MR images can only be achieved by the analysis of combinations of different relaxation parameter informations [Handels et al. 1990].

---

[1] current affiliation: Institut für Medizinische Informatik, Medizinische Universität Lübeck, Ratzeburger Allee 160, D-2400 Lübeck, FRG.

2   Automatic Tissue Segmentation

A pyramidal histogram analysis algorithm, based on multidimensional MR parameter histograms, in combination with a merging algorithm was developed for tissue segmentation. The pyramidal extension of histogram based cluster analysis algorithms leads to a partition of the multiparametric relaxation image data into an unknown number of clusters. The analysis of multidimensional histograms permits the detection of unimodal clusters of any shape, which correspond to different tissue structures in the image. The following merging algorithm improves the segmentation results by merging split tissue structures. For visualization, the extracted tissue structures are represented in cluster matrices, in which every pixel is labeled with a cluster index. Furthermore, the means and standard deviations of the relaxation parameters within each cluster are computed to characterize the relaxation behaviour of the associated tissue structure.

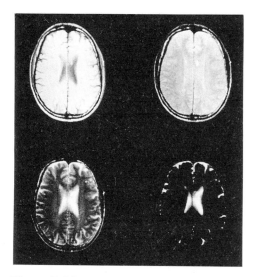
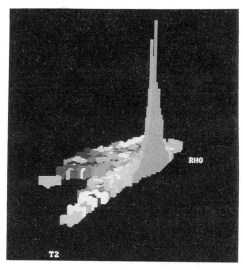

Fig. 1 (left): Four axial MR images of the same slice from a normal human head.

Fig. 2 (right): Pseudo 3-dimensional visualization of a 2-dimensional histogram showing the frequency distribution of the two parameters T_2 and ρ recorded in a supraorbital human head slice. The clusters detected are marked by different colours.

2.1   Histogram Based Cluster Analysis

Histogram based cluster analysis is a non-parametric cluster analysis method, which makes possible an efficient processing of the large amount of multidimensional data occurring in MR image data sets. Histogram based cluster analysis algorithms have been used previously for the analysis of multispectral satellite image data (LANDSAT) [Goldberg and Shlien 1978, Narenda and Goldberg 1977, Wharton 1983/84]. A cluster in a multidimensional histogram is defined as an unimodal histogram hill. Different clusters are separated by histogram valleys (Fig. 2). The analysis of multidimensional histograms is performed without a priori knowledge of the shape or the number of clusters. This is vital for the analysis of medical image data, in which the number of tissues as well as the number of pixels within a tissue change depending upon the investigated patient and the chosen body slice. The basic algorithm for the automatic analysis of multiparametric MR image data is described in the following section.

The algorithm starts with the generation of a d-dimensional histogram ($d \leq 3$). In the application considered here, the 3-dimensional histogram is generated for the parameters T_1, T_2 and ρ evaluated in the investigated images. In the first step, the algorithm detects the peaks (local maxima) in the multidimensional histogram, which are treated as cluster centers. In a second step, the clusters are separated sequentially. Starting at a peak, the cluster is expanded iteratively until the corresponding histogram valleys are determined as cluster boundaries. As illustrated in Figure 3, conflicts between different clusters occur in the histogram valleys, where histogram cells could be assigned to several clusters. In the algorithm developed, these so-called conflict cells are marked automatically during the histogram analysis and a special postprocessing is performed (Figs. 3 and 4). A pixel, whose feature vector occurs within a conflict cell, is called conflict pixel. During postprocessing, the (3x3)- neighbourhood of each conflict pixel is considered separately in the image and the frequencies of the resulting cluster indices are computed. Finally, the conflict pixel is assigned to one of the conflict clusters, whose cluster index occurs with the highest frequency in the (3x3)-neigbourhood of the conflict pixel (Fig. 4).

The computing time complexity of the central algorithm is independent of the number n of pixels or feature vectors, but only linear dependent on the number of histogram cells z with values greater than zero. Therefore, the analysis of the multiparametric image data can be performed rapidly and even the parallel data analysis of several slice images is possible in an interactive mode. This is caused essentially in the repeated occurrence of the same tissues in adjacent slice images in combination with the cluster effect, which leads to small numbers of histogram cells, z, with values greater than zero.

2.1.1 Data Structures for Multidimensional Histograms

While 1-dimensional arrays are suitable for the internal representation of 1-dimensional histograms, the storage complexity of d-dimensional arrays for the representation of multidimensional histograms grows exponentially with the histogram dimension, d. In the application, the representation of the 3-dimensional MR parameter histograms in a 3-dimensional array requires 32 MB storage, but the main part of the histogram cells shows the value zero. To reduce the needed storage a dynamic data structure has been developed for the representation of 3- dimensional histograms. The dynamic data structure consists of three levels, which are described in PASCAL notation as follows:

```
TYPE histogram = ARRAY [0...15,0...15,0...15] OF PointerToSubtreeArray;
                 PointerToSubtreeArray =↑SubtreeArray;
                 SubtreeArray = ARRAY [0...7,0...7,0...7]OF PointerToHistogramBlo
                 PointerToHistogramBlock = ↑HistogramBlock;
                 HistogramBlock = ARRAY [0...1,0...1,0...1] OF INTEGER;
```

At the lowest level ('HistogramBlock'), the 3-dimensional histogram is decomposed into

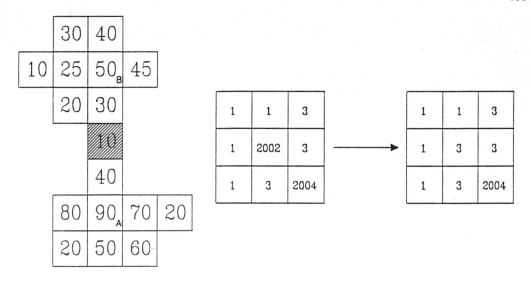

Fig. 3 (left): Two clusters in a 2-dimensional histogram. The conflict cell detected is cross-hatched.

Fig. 4 (right): 3x3-neighbourhood of a conflict pixel (index 2002) in the cluster matrix of an image before (left) and after (right) postprocessing. It is assumed that the conflict pixel corresponds to a conflict cell in the histogram with conflict clusters 2, 3 and 6. Especially cluster 1 is not a conflict cluster. Therefore, the conflict pixel is assigned to the cluster with the index 3, because the cluster 3 is the conflict cluster index occurring with the highest frequency in the 3x3-neighbourhood.

histogram blocks consisting of $2^3 = 8$ neighboured histogram cells. At the second level ('SubtreeArray'), $8^3 = 512$ pointers to histogram blocks are represented, while the third level ('histogram') consists of $16^3 = 4096$ pointers to arrays of subtrees. Using this data structure histogram cells are generated only in those regions of the feature space where feature vectors occur. Furthermore, the neighbourhood connections between histogram cells are preserved in the dynamic data structure, so that neighbour cells of a histogram cell can be addressed easily. This is important for the histogram analysis algorithms, in which the comparison of the values of two adjacent histogram cells is the fundamental operation. The use of the dynamic data structure leads to an essential reduction of the storage needed to represent 3-dimensional MR parameter histograms. In comparison with 3-dimensional arrays the storage required can be reduced to 0,2% - 0,4% on average.

2.1.2   Histogram Pyramid

The histogram pyramid is an extension of histogram analysis algorithms, which leads to a dynamic adjustment of the histogram cell sizes in dependence on the varying data density in different parts of the feature space. In the considered medical image data high variations of the data density in the feature space occur, because on the one hand different tissues are represented by a strongly different number of feature vectors and on the other hand the standard deviations of the relaxation parameters occurring within different tissues vary

456

greatly.

Using the histogram pyramid several multidimensional histograms with histogram cells of different size are generated and analyzed to perform a data driven tissue segmentation (Figs. 5 and 6). During the first analysis step, a multidimensional histogram is generated and all clusters with a peak value larger than a given threshold are analyzed. The analyzed clusters are extracted from the histogram and the corresponding tissue structures are visualized (Fig. 6). In the next step, the histogram cells are enlarged and a new histogram based on the remaining pixels is generated. In practice, a doubling of the the histogram cell size in each dimension of the feature space is performed. Hence, the histogram values of the pyramid level i+1 can be computed in a fast mode using the histogram values of the pyramid level i with i $\in$ {1,2,3}. After several iterations of this process, a so-called histogram pyramid is obtained (Fig. 5).

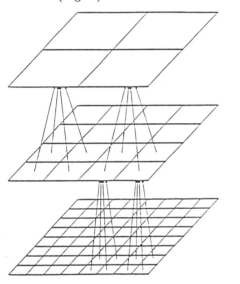 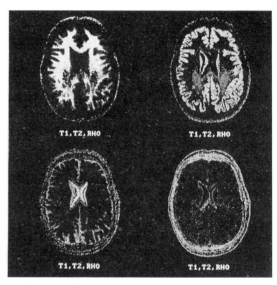

Fig. 5 (left): Schematic illustration of a histogram pyramid with three levels.

Fig. 6 (right): Segmentation results of the histogram pyramid with four levels. The clusters shown upper left, upper right, lower left and lower right are detected at different levels of the histogram pyramid. Especially at the last pyramid level, many outlyers remain (lower right) and are therefore separated automatically from homogeneous tissue structures.

The thresholds used are chosen proportional to the root of the number of the remaining pixels in the considered histogram (proportionality factor = 0,1). The thresholds are computed separately for each level of the histogram pyramid with respect to the changing data situations. Using the histogram pyramid, histogram cells with different sizes are generated dynamically and an automatic adjustment of the cell sizes to the data density in the feature space is achieved. Typically, at the first pyramid levels homogenous tissue structures are extracted, while the most outlyers in the image data, which typically occur in MR image data sets with high frequency, are separated automatically at the last pyramid level (Fig. 6).

2.2 Merging Algorithm

Though most of the tissue structures are segmented by the histogram pyramid algorithm in high quality, some tissues are split into several clusters (Fig. 7). This effect can be induced by variations of the density estimators in the histogram, as well as by experimental artifacts. The merging algorithm has been developed to merge clusters, which represent parts of the same tissue. A combination of feature and image-oriented similarity criteria is used to select split tissue structures in the pre-segmented data set. On the one hand, the euclidian distances between the standardized mean vectors of the clusters are considered. On the other hand, the mean frequency of contacts between the associated tissue segments in the image is computed, which is defined in the following section. During the merging process the basic data structure to describe the clusters corresponding to tissue segments in the image is generated by using the run length encoding technique [Rosenfeld and Kak 1982]. In this data structure a cluster is represented by a set of runs, which can be defined as follows:

A run r is a maximal sequence of pixels of a cluster in an image row. A run $r = (s,i,j,l)$ is determined by the starting point (i,j) in the s'th image and its length $l \in N$. A contact point between the run $r_1 = (s,i,j_1,l_1)$ of the cluster C_1 and the run $r_2 = (s,i,j_2,l_2)$ of the cluster C_2 exists, if

$$j_1 + l_1 = j_2 \quad \text{or} \quad j_2 + l_2 = j_1.$$

Furthermore $n_{C_1,C_2} \in N$ describes the number of contact points between the clusters C_1 and C_2 and r_{C_i} (i=1,2) gives the number of runs in cluster C_i. Then, the mean frequency of contacts f_{C_1,C_2} between the clusters C_1 and C_2 is defined as follows:

$$f_{C_1,C_2} = \frac{n_{C_1,C_2}}{min(r_{C_1},r_{C_2})} \in [0,2]$$

Two clusters C_1 and C_2 with standardized mean vectors m_1 and m_2 fulfill the similarity criteria, if

$$d(m_1, m_2) \leq d_{max} \quad and \quad f_{C_1,C_2} \geq f_{min}.$$

During the merging process all clusters are considered in pairs to check the similarity criteria. The clusters with the smallest distance between the cluster mean vectors (lower equal d_{max}) are merged, which yields a mean frequency of contacts greater equal than f_{min}. The merging process is performed in two phases, in which the thresholds are chosen as follows:

phase 1: $d_{max} = 0,2$ and $f_{min} = 0,1,$ phase 2: $d_{max} = 1$ and $f_{min} = 0,9.$

The combination of feature and image-oriented criteria leads to an automatic selection and fusion of split tissue parts in the analyzed data set. In particular, the fusion of clusters of different tissues is avoided. For visualization and interactive control, the merged clusters are backtransformed into the image space (Fig. 7). Finally, the complete cluster matrices are displayed showing the segmentation result (Fig. 8).

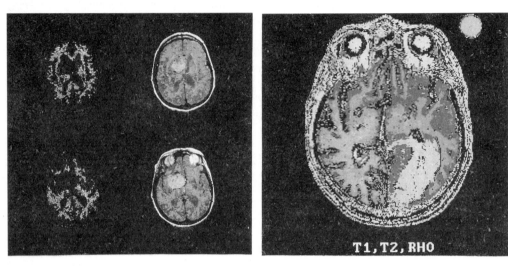

Fig. 7 (left): Merging example: Two clusters, which represent two parts of the white matter are merged in two slices of a human head. The merged clusters are represented by different colours. The clusters are visualized with black background (left) and with an echo image for improved anatomical correlation (right).

Fig. 8 (right): Segmentation result of an axial head slice with a glioblastoma tumor. Each cluster is marked with a specific colour. Especially the tumor is differentiated with high accuracy from normal tissue.

3 Automatic Tissue Classification

After segmentation, the relaxation behaviour of the extracted tissue structures is described by the means of the relaxation parameters. These tissue characteristic values can be stored in a tissue data base to establish a training data set for the classifiers. Based on the tissue data base, an automatic classification of tissues in MR images is performed to support medical diagnostics. In the program system SAMSON, statistical classifiers [Bock 1974, Niemann 1983], such as the maximum-likelihood classifier (ML) and k-nearest neighbour classifiers (k-NN), can be used alternatively for tissue classification. The classification results are visualized in tissue class images, in which every tissue is marked with a tissue specific colour (Fig. 9). Furthermore, weighted lists showing the possible interpretations of the tissue can be displayed. The lists are structured by the used classifier, so that the first element of the list represents the visualized classification result. These lists are of diagnostic interest especially in the case of overlapping tissue specific parameter distributions.

The establishment and the successive extension of the data base is supported by a classification management system integrated in the program system SAMSON. It executes the necessary system routines, for example, to update the tissue specific mean vectors and covariance matrices after the addition to an existing tissue class or the creation of a new tissue class.

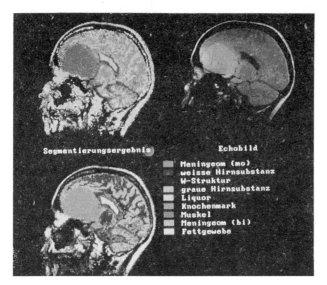

Fig. 9: Tissue class image showing the classified tissue structures in a sagittal head slice with a meningioma. The classification was performed with the k-NN classifier (k=4).

4 Conclusion

The segmentation algorithms described here permit an automatic differentiation of tissues, such as muscle, fat, white and grey matter as well as pathological tissues like meningioma, astrocytoma or glioblastoma in several images simultaneously. The interpretation of the segmented tissue structures is supported by statistical classifiers in combination with a tissue data base. For diagnostic support in clinical applications, the algorithms developed and the tissue data base have been integrated in the software system SAMSON ('System for Automatic Segmentation and Classification of Tissue in Magnetic Resonance Tomography'), which is implemented on a VMS VAXstation 3200 computer. This provides an efficient tool for the analysis and interpretation of multiparametric MR image data, which has already been used to analyze more than 300 image data sets.

Acknowledgement: The author would like to thank Professor Armin Thron and Dr. Jürgen Reul from the Department of Neuroradiology at the Klinikum of the RWTH Aachen for the intensive cooperation and making available the MR images used in this paper. Furthermore, the author thanks Prof. Dr. Walter Oberschelp from the Institut für angewandte Mathematik, insbesondere Informatik at the RWTH Aachen for the support of this work.

References

BOCK, H.H. (1974), *Automatische Klassifikation*, Vandenhoeck & Ruprecht, Göttingen.

DUDA, R.O. and HART, P.E. (1973), *Pattern Classification and Scene Analysis*, Wiley, New York.

EIS, M., HANDELS, H., BOHNDORF, K., DROBNITZKY, M., TOLXDORFF, T. and STARGARDT, A. (1989), A New Method for Combined T_1-Measurement and Multi-Exponential T_2-Analysis in Tissue Characterizing MRI, *Proc 8th Annual Meeting of the Society of Magnetic Resonance in Medicine, Amsterdam, 770.*

FAHRMEIR, L. and HAMERLE, A. (1984), *Multivariate statistische Verfahren*, de Gruyter, Berlin.

GOLDBERG, M. and SHLIEN, S. (1978), A Clustering Scheme for Multispectral Images, *IEEE Trans SMC-8, 86-92.*

HANDELS, H. (1992), *Automatische Analyse mehrdimensionaler Bilddaten zur Diagnoseunterstützung in der MR-Tomographie*, Verlag Shaker, Aachen.

HANDELS, H., HIESTERMANN, A., HERPERS. and TOLXDORFF, T. (1991), Automatische 3D-Segmentierung und Klassifikation von Gewebe in der medizinischen Diagnostik, in: Radig, B. (Hrsg.), *Mustererkennung 1991, 13. DAGM- Symposium, Informatik Fachberichte 290*, Springer, Berlin, 295-303.

HANDELS, H. and TOLXDORFF, T. (1990), A New Segmentation Algorithm for Knowledge Acquisition in Tissue Characterizing NMR-Imaging, *Journal of Digital Imaging, Vol. 3, No 2 (May), 89-94.*

JAIN, A. K. and DUBES, R.C. (1988), *Algorithms for Clustering Data*, Prentice Hall, Englewood Cliffs.

NARENDA, P.N. and GOLDBERG, M. (1977), A Non-Parametric Clustering Scheme for Landsat, *Pattern Recognition 9, 207-215.*

NIEMANN, H. (1983), *Klassifikation von Mustern*, Springer, Berlin.

ROSENFELD, A. and KAK, A.C. (1982), *Digital Image Processing*, Academic Press, New York.

WHARTON, S.W. (1983), A Generalized Histogram Clustering Scheme for Multidimensional Image Data, *Pattern Recognition 16, 193-199.*

WHARTON, S.W. (1984), An Analysis of the Effects of Sample Size on Classification Performance of a Histogram Based Cluster Analysis Procedure, *Pattern Recognition 17, 239-244.*

WIRTH, N. (1979), *Algorithmen und Datenstrukturen*, Teubner, Stuttgart.

Pseudoroots as Descriptors for a Thesaurus Based on Weidtman's Diagnosis Table of Pediatrics

Rudolf-Josef Fischer

Institut für Medizinische Informatik und Biomathematik
der Westf. Wilhelms-Universität Münster
Domagkstraße 9, W–4400 Münster

Abstract: Weidtman's Diagnosis Table enables to encode medical phrases from pediatrics into the International Classification of Diseases (ICD). Due to higher refinement in those sections important for pediatrics it has a widespread use for basic documentation in childrens' hospitals. Thus, testing methods for automated classification of medical phrases into a middlesized thesaurus, Weidtman's Diagnosis Table is a convenient application example.

The descriptors of the method described here are "pseudoroots", i.e. truncated roots of variable lengths. The results show that medical phrases scanned for those descriptors are classifiable with high precision and almost automatically. On the other hand, phrases which ought not to be found by this formal approach are recognized and rejected nearly without exception.

1 Introduction

Classifying medical phrases means mapping them on a given set of codes. The example best known is the International Classification of Diseases (ICD), which has to be used by all medical institutes in Germany. Weidtman's Diagnosis Table (1989) contains a subset of ICD codes from pediatrics, partially extended to five positions, together with describing texts, encoding advices, and cross-references, thus able to be regarded as a thesaurus. In this thesaurus, several thesaurus items, semantically equivalent or not, may belong to the same ICD code.

To support classification, while saving time and reducing the employment of terminology specialists, methods of automated classification are available to map a given medical phrase (input text) on a subset of thesaurus items. If the subset contains only one item, then the related code can be assigned to the input text. In other cases the list of thesaurus items found is presented in a selection menu, usually in decreasing order according to a defined similarity measure. Strictly speaking, the classification methods reported in literature are only semi-automated.

Most of those algorithms use a given set of descriptors which are searched for in an input text. For all descriptors found there are references to some thesaurus items. Gathering these items, their similarity to the input string is calculated. If the similarity exceeds a given minimum level, the thesaurus item belongs to the resulting subset.

One approach to find descriptors in the input text are sophisticated algorithms of word segmentation, using large dictionaries of roots, endings, and other morphemes. The problems related are more complicated than one would imagine at a first glance. While some authors like Dorda (1990) only laconically report, that this work is done, others like Dujols et al. (1991) give a better insight to make the huge amount of efforts obvious.

This led to less ambitious, roughly formal approaches under certain limitations:

- the thesaurus is only middlesized (about 5,000 items)

- no administration of lexicons is necessary beyond the descriptor file

- sufficient storage amount and tolerable response time are achieved already by use of a personal computer

Fischer (1990) presented an algorithm using "4-grams" (substrings of length 4) as descriptors with a recall rate of about 98 % even if trivial cases (input text and thesaurus item identical) were neglected but others with spelling errors included. In a further investigation Fischer (1991) redefined the similarity measure to enable the algorithm to reject not classifiable cases. As a result the recall rate decreased to about 88 %, and 2.4 % of the input texts were misclassified, 1 % even automatically, whereas about 80 % of the not classifiable cases were rejected. Pseudoroots (roots truncated to four characters) resulted in a smaller recall rate (81 %), but 50 % of the input texts could be classified automatically (only 32 % by 4-grams); on the other hand 17.6 % were misclassified (10.4 % automatically). The main reason for this high number of errors was that too often four adjacent characters in the input texts looked like a pseudoroot but were not. Many roots of medical terms are longer than four characters.

The new approach was thus to use variably long pseudoroots searching them in the input text by a simple substring scanning. This is also reported by Pietsch et al. (1992).

2 Problem formulation

The classification algorithm has to recognize input texts as (formally) classifiable or (formally) not classifiable according to the items of a given thesaurus, here Weidtman's Diagnosis Table. The resulting thesaurus item(s) must be semantically equivalent to the input text or they must be its nearest upper term(s) (principle of upper term affinity).

In as many cases as possible a classifiable input text should be mapped on only one thesaurus item (automatically classified), rather than on several items shown in a selection menu.

Storage amount and a acceptable response time should not exceed the capabilities of a personal computer.

3 Problem solution

3.1 The concept of formal classifiability

For every thesaurus item its set of semantic constituents is marked. Primary constituents are independent medicals terms (usually nouns), secondary constituents are semantic modifiers of primary constituents (most of them adjectives). The constituents were truncated and used as descriptors. An input text is formally classifiable, if it contains all descriptors of almost one thesaurus item.

By this concept classification never adds a constituent, i.e. an information, which was not in the original input text. For example: the input text "congenital blindness" must not be encoded as 369.61 = "unilateral congenital blindness" (but it may be classified by the nearest upper term 369 = "blindness or defective sight").

3.2 Creating the help files

Figure 1 shows the thesaurus file and the help files needed. For every thesaurus item its constituents are collected and truncated to descriptors. T_P, the number of primary descriptors, and T_S, the number of secondary descriptors, are stored in the thesaurus file. To match as many variants as possible in input texts, the descriptors are then transformed as usual: only alphanumeric characters and small letters; one blank as only word separator; mapping of "ä", "ö", "ü", "ß" on "ae", "oe", "ue", "ss" respectively; furthermore changing "ph" to "f", and "k" and "z" both to "c". All descriptors are marked as primary or secondary and stored in a reference file, with the number of the thesaurus item, from which they originate. The descriptor file then receives the different descriptors, sorted by their lengths in descending order, and is completed by the content of the synonyms file.

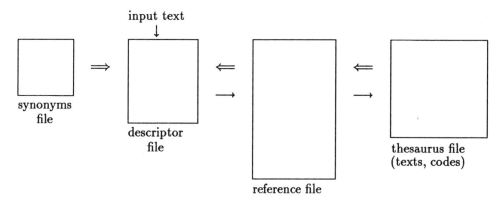

Figure 1: Thesaurus file and help files needed for the classification algorithm. ⟹: Direction of creating/updating, ⟶: direction of the classification process

3.3 Schema of the classification algorithm

First, an input text is subjected to the same transformations as the descriptors and receives one leading and one trailing blank. Second, every item of the descriptor file is searched for as a substring in the input text; if it is found, the related references are collected from the reference file, and the descriptor is deleted in the input text. When the input text or the descriptor file is exhausted, the references collected are ordered to count N_P, the number of primary descriptors, and N_S, the number of secondary descriptors, found for any thesaurus item.

An input text is thus (formally) classifiable, if

$$D_P := N_P - T_P = 0 \text{ and } D_S := N_S - T_S = 0$$

If any of the differences D_P and D_S is negativ, there is at least one constituent in the thesaurus item not occurring in the input text.

The set of referred thesaurus items is now sorted in descending order, by D_P, N_P, D_S, and N_S successively, thus obtaining upper term affinity.

Let I be the input text, T_i the i-th thesaurus item in the ordered set. Then I is said automatically classifiable by T_1, if I is classifiable by T_1 (i.e. $D_P = D_S = 0$ for T_1) and if at least one of the four items D_P, N_P, D_S, N_S of T_2 is less than the corresponding one of T_1. Figure 2 shows an example. The input text "malignant neoplasm of the kidneys" is automatically classifiable by "Kidney, malignant neoplasm" and thus encoded to 189.09.

D_P	N_P	D_S	N_S	Code	Thesaurus item
0	2	0	1	189.09	Kidney, malignant neoplasm
0	2	-1	0	223.09	Kidney, benign neoplasm
0	1	0	1	199.19	Neoplasm, malign (loc. not indic.)
0	1	0	0	593.99	Disease of the kidneys

Figure 2: First four referred thesaurus items for the input text "malignant neoplasm of the kidneys". D_P, D_S: difference between number of primary/secondary descriptors in the input text and the thesaurus item; N_P, N_S: number of primary/secondary descriptors found in the input text

If I is only classifiable (but not automatically), i.e. if there is at least one thesaurus item with $D_P = D_S = 0$, all thesaurus items with $D_P = 0$ are shown in a selection menu. Otherwise I is not classifiable.

4 Creating the descriptors from Weidtman's Diagnosis Table

4.1 Some special problems

When creating descriptors for the texts of Weidtman's Diagnosis Table, there arise many difficulties. Some of them are mentioned here.

- A medical phrase may appear in many variants. Especially in German texts the constituents may be in any combination; for example: "verzögerte Sprachentwicklung" (retarded development of speech) versus "sprachliche Entwicklungsverzögerung".

Therefore descriptors are truncated ("verzög", "sprach", "entwick") and searched as substrings.

- The order of constituents is significant in some cases; for example "Beckenniere" (pelvic kidney) in opposition to "Nierenbecken" (renal pelvis).

Here it is helpful to create a longer descriptor "beccenniere", which is searched in the input text prior to the shorter descriptors "beccen" and "niere" (principle of longest match).

- Unavoidably short constituents (abbreviations, names) which formally are common substrings of input texts.

To reduce this cases, descriptors may be extended by a leading and/or trailing blank (" lichen ") or consist of more than one word (" in situ ").

- Terms with negative prefixes as "non-", "un-", "a-"

They can be distinguished by the classification algorithm, if one uses pairs of negative/positive descriptors ("abacterial"/"bacterial"). Then the negative descriptors are longer than the positive ones and thus found first.

- Semantically too general phrases

Weidtman's Diagnosis Table contains very general phrases in those sections, where diseases are seldom to be found in pediatrics. These are hardly approachable by the concept of constituents. Too general terms as "with complications" are neglected; in the worst case, this leads to a selection menu and does not prevent the classification (see figure 3 for an example).

D_P	N_P	D_S	N_S	Code	Thesaurus item
0	2	0	1	941.11	Burn,neck,1st degree without complications
0	2	0	1	941.12	Burn,neck,1st degree with complications
0	1	-1	0	946.	Burn, several specified locations

Figure 3: Selection menu from input text "first-degree burn of neck". (For D_P, N_P, D_S, N_S see figure 2.) The encoding person has to decide, which code is to be selected, eventually by further informations, or to postpone the classification

4.2   Adaptation of the thesaurus texts

Due to the problems mentioned, the texts of Weidtman's Diagnosis Table (1989) had to be adapted. One version of some synonymous phrases could be dropped, because they contained the same constituents. In many other cases synonymous texts with different constituents had to be added. A dialogue program helps to create descriptors for a new thesaurus text, allowing the user to browse in the descriptor file, while showing the referred thesaurus items at the same time.

The actual version of the adapted thesaurus contains 4,245 items with 10,707 references. Figure 4 shows the distribution of the 2,770 different descriptors.

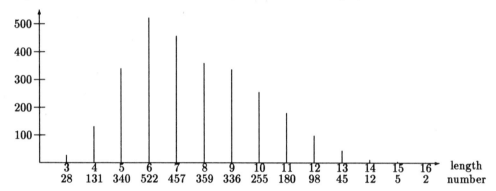

Figure 4: Distribution of the 2,770 different descriptors

5   Results

For a trial a sample of 1,018 input texts from the Childrens' Hospital of the University of Münster was processed by the classification algorithm described above. 507 texts were known to be (formal) classifiable by Weidtman's Diagnosis Table; the remaining 511 texts were not classifiable and should be rejected. Table 1 shows the results.

input texts	thesaurus texts			sum
	just one item found	several items found	no item found	
formal classifiable	472 93.1 %	35 6.9 %	0	507
not formal classifiable	9 1.8 %	0	502 98.2 %	511

Table 1: Results achieved by a sample of 1,018 input texts

All 507 classifiable input texts were classified correctly, 472 among them even automatically. Misclassifications occurred only in 9 out of the 511 cases, which ought to be rejected, but were not (see table 2).

input text:	misclassified by following thesaurus text:
"atypical vaccination reaction"	V05.0 Vaccinations, other
"Pseudo-Cushing syndrome"	255.00 Cushing syndrom
"Zystinspeicherkrankheit" (Germ.) (aminoaciduria)	216.81 Cyst
"cough syrup poisoning"	786.2 cough
"Mohr Claussen syndrome"	388.9 Ohrerkrankung (ear disease)
"pseudo-achondroplasia"	756.40 Achondroplasia
"cysticercosis"	216.81 Cyst
"plant protecting poisoning"	988.99 poisoning by harmfull plants
"Milkman syndrome"	289.59 Milzerkrankung (spleen disease)

Table 2: The 9 cases which were misclassified, rather than to be rejected

6 Discussion

The classification algorithm described here is limited to a middlesized thesaurus and thus not in competition with more global methods reported in literature. For example, indexing based on the Systematized Nomenclature of Medicine (SNOMED) Wingert (1986) hardly uses a textual approach, but tries to map the results of his sophisticated word segmentation on the SNOMED code directly. A critical point of most of the methods is to dismantle compound terms into its component parts. Though just compound terms are typical for medical texts, Debold et al. (1986) describing the system DIACOS only mentions roots and endings, while Bartowksi and Graubner (1991) obviously treat a compound term as one "root". Neither by a fixed number of leading characters of every word nor by pseudoroots of fixed length an input text can be evaluated successfully, as Fischer (1991) showed. But searching variably long pseudoroots in the input text by substring scanning led here to satisfying results.

Table 1 proved that the classification algorithm tends to a significant improvement compared with the approach by 4-grams reported by Fischer (1991). In particular pseudoroots as descriptors cause a surprising precision in distinguishing classifiable input texts from not classifiable ones. That is very important in practice, for even complete code systems as the ICD become incomplete by every formal approach using textual thesaurus items. Thus, the connectionistic method of Pietsch, Ruppel, and Schneider (1992) yet lacks a criterion to reject an input text; a lower threshold of the cumulated similarity measure would enable this, but its definition may be difficult.

Longer input texts, possibly formulated with redundancy, cause a general risc to be classified by a short, but semantically not matching thesaurus item (see examples in table 2). Therefore, an exclusively automated (correct) classification is not conceivable. For some cases there will always result a selection menu, and it is advisable to include those thesaurus items, for which yealds $D_P = -1$ (one primary descriptor not found in the input text). The encoding person may then in reply reformulate the input text such that it becomes classifiable. If the same, not classifiable texts occur several times, they may be inserted into the

468

thesaurus. If necessary, it is easy to update the help files (see figure 1) subsequently.

As shown, compared with n-gram methods pseudoroots as descriptors enable a higher percentage of automatically classifiable cases. But there is one disadvantage: most of the spelling errors in the input texts cause that the related pseudoroot is afflicted and thus unrecognizable, while n-gram methods are very insensitive in regard to spelling errors. Nevertheless the transformations applied here (see section 3.2) are designed only to consider real variants of medical terms. Any further phonetic projections would decrease the descriptors' selectivity. To resume, under the given limitations (see section 1) pseudoroots proved to be very efficient descriptors to fulfil an almost automatical classification with high precision.

References

BARTKOWSKI, R.; GRAUBNER, B. (1991), Automatische Klartextverschlüsselung histologischer Tumordiagnosen mit dem Personalcomputer. In: Überla,K., Rienhoff,O., Victor,N. (eds.), *Quantitative Methoden in der Epidemiologie, Medizinische Informatik und Statistik 72*, Springer-Verlag, 258-62

DEBOLD, P.; DIEKMANN, F.; MÜLLER, U. (1986), DIACOS - Programmsystem zur Diagnosecodierung im Dialogverfahren. In: Köhler,C. (ed.), *Medizinische Dokumentation und Information: Handbuch für Klinik und Praxis*, ecomed-Verlag, chapter III-11

DORDA, W.G. (1990), Data-Screening and Retrieval of Medical Data by the System WAREL, *Methods of Information in Medicine 29*, Springer-Verlag, 3-11

DUJOLS, P.; AUBAS, P.; BAYLON, C.; GRÉMY, F. (1991), Morphosemantic Analysis and Translation of Medical Compound Terms, *Methods of Information in Medicine 30*, Schattauer-Verlag, 30-35

FISCHER, R.-J. (1990), Semi-automated classification of medical phrases using a personal computer. In: Bock,H.-H., Ihm,P. (eds.), *Classification, Data Analysis, and Knowlegde Organisation*, Springer-Verlag, 270-76

FISCHER, R.-J. (1991), Vergleich verschiedener Methoden der Abbildung medizinischer Aussagen auf Texte einer standardisierten Terminologie. In: Überla,K., Rienhoff,O., Victor,N. (eds.), *Quantitative Methoden in der Epidemiologie, Medizinische Informatik und Statistik 72*, Springer-Verlag, 253-57

PIETSCH, W.; RUPPEL, A.; SCHNEIDER, B. (1992): Effizientes Lernen von ICD-Klassifikationen anhand von Diagnosetexten, *Proceedings der TAT '91 in Aachen*, Springer-Verlag (in print)

WEIDTMAN, V. (1989), Diagnoseschlüssel für die Pädiatrie, Springer-Verlag

WINGERT, F. (1986), An Indexing System for SNOMED, *Methods of Information in Medicine 25*, Schattauer-Verlag, 22-30

An Approach to a Space Related Thesaurus

E. Weihs

Bavarian Ministry for State Development and Environmental Affairs
Munich, FRG

Abstract: For classifying the spatial relations of coordinated numerical data there are topological and object oriented models with which nontrivial questions can be researched while accounting for topic related contents, such as "what borders on", "contains", "lies within" etc. For classifying text related data in which the spatial relation is only implicit, e.g. determined by names, no topological classification models are known for researching the above questions.

Initially, therefore, we shall derive an object oriented topological model for classifying text related data with a spatial reference from a coordinate oriented approach. The result is a thesaurus with necessary and adequate topological relations. In this process we notice that an object oriented approach has considerable advantages as it avoids redundant data retention. On the basis of a specific implementation we shall deal with search possibilities and the underlying thesaurus.

1 Defining the Problem

Space related data are more important in environmental protection and prevention than in most other fields. Accordingly, the objective and central issue of every environment information system is to provide corresponding data collections for legislation on the one hand, and for daily administration work on the other. Environment information systems usually present themselves as an organizational pool of data collections from specialized information systems (= specialized information centers). It constitutes an organizational pool because the required data should be available but not transferred. This type of process would not be viable from a technical, organizational and legal viewpoint.

In the light of the above situation, various recommendations including the EXPERT Ground Information Working Group (1989) have led to the choice of an organizational form which distinguishes between the independent specialized information systems and what is known as the core system. The specialized data are kept separate (usually spatially and organizationally) in the specialized information system, while the information on data, methods and access paths is taken from the core system (EXPERT (1989), fig. 11)

The core system is the pivotal point of the system; the necessary "common denominator" is the common vocabulary and its relations (= thesaurus) with which comparable and linkable search results on specialized information and the various references are obtained. The more heterogeneous the distribution of the specialized information systems, the greater the significance of the core system.

One crucial part of the core system has to be the reference system which implements the access to the desired data. Because, as we ascertained above, most data on protection of the environment is available in space related form, we need not only topic related searches but also space related searches for place names, certain geographical units, spatial relationships or coordinates. Beside (fuzzy) geographical location, spatial relationships also mean associations such as "neighboring", "bordering on", "located within" etc. The underlying data can be available in the form of text or tables, and the spatial relationship itself can be defined explicitly by coordinates and references or implicitly by unformatted descriptions.

It is evident that a uniform data model of space related data which describes the topology of the spatial relationship independently of the storage methods is of considerable advantage when it comes to constructing search algorithms. Beside this, it also provides technical advantages in the design of databases, but that is not the subject of this paper. In literature we find no space related approach which could serve as the basis of a text related reference system. Only the problems of standardizing the terminology of cartographic expressions are treated, for example Haller (1976, 1987), Schnabel (1987), the International Federation of Library Associations and Institutions (1977), STIBOKA (1982) and others.

In the following we will assume a basis of formatted data to develop a space related data model, and then derive from it a model for unformatted data. We will see that the spatial relation, accounting for a methodically conditioned fuzziness, can be mapped and put to use in research. Finally we will outline the structure of a thesaurus like the one used for the text data of the Bavarian Ground Information System.

2 Representing Space Related Data

2.1 The space related model for formatted data

Normally, a discussion of space related data assumes that the data are formatted. Let us take formatted data to mean data in which the place of their storage is unique in the data record. Furthermore, we will deal only with *focussed surface* data. (One can easily see that the model can be ported to a three-dimensional case; grid related data can be treated as a special case.) Focussed surface means that the border of the surface - or, more generally, the border of a geometrical object - can always be specified with sufficient accuracy. It is fully conceivable that the border of the object could be described by a set of functions. But following practice, we shall assume that the representation of the object border is usually provided by specifying coordinates (polygons). We shall treat points, lines and networks or other objects as special case surfaces. We shall also choose to neglect some other technical requirements of designing a data model, such as those posed by the technical aspects of saving the data, unless they contradict the basic premises outlined here. The same applies to the attributes assigned to the objects, with their topic related characteristics.

Unlike graphical data, whose contents can only be interpreted by means of visualization (so that the requirements of the data model can be limited to specifying coordinates, allowing priority to be given to things like optimizing technical implementation issues), the topology of the spatial relationship plays the decisive role in space related data. Each surface border line always divides two surfaces; each point x_i where $i = 1, 2$, which is not a surface border, is always associated with one, and only one, surface. The surfaces themselves can contain further surfaces or other geometrical objects, or be contained within them. Beside the coordinates, further content data (attributes) are required to describe the geometrical objects and their locations in space. With certain restrictions, a case-by-case derivation of the spatial relationship from graphical data is (perhaps after vectorizing) basically possible using methods of linear algebra or analysis (as long as surface equations $\mathbf{F}(x)$ are known), but not technically feasible with large quantities of data.

The topology of the spatial relationship is discussed exhaustively in literature by such people as Bill and Fritsch (1991) and Molenaar (1989, 1991). Here we shall limit our attention to the parts that are most important to our considerations, and shall also omit further details of graph theory for the sake of our further considerations:

There are three discrete functions that are necessary and adequate

$$\{O\}_{z,r} = \mathbf{V}(z,r) \tag{1}$$

where $\{O\}$ = result set of the elements $o_n, n = 1, 2, \ldots, N_z$, as unique names of the geometrical objects (cf. fig. 1). $z = e, k, n$ identifies the function type, where e = "is contained", k = "contains" and g = "borders on" and finally, $r = 1, 2, \ldots, s, \ldots, S, \ldots R; R$ = number of all objects in the area which describe the spatial relationships between the geometrical objects O (e.g. surfaces, lines and points, networks). The functions $\mathbf{V}$ are always unique, as we specify that each surface point (result set of the equation of the surface $\mathbf{F}(x)$) is assigned to one surface only:

$$\{\mathbf{F}_r(x)\} \neq \{\mathbf{F}_s(x)\} \tag{2}$$

So "overlapping" surfaces (general objects) are not allowed. Thus the discrete function

$$O_r = \mathbf{V}e(r) \tag{3}$$

can be specified, which determines the objects O_r which "contains" a surface $\mathbf{F}$ (islands, "children"). We also define the function

$$O_r = \mathbf{V}k(r) \tag{4}$$

whose result set references the objects O_r in which the object r is itself contained ("parents") and which can be derived from (3) ($\mathbf{V}^{-1}$).

Finally, the function

$$O_r = \mathbf{V}g(r) \tag{5}$$

describes the direct neighbor relationships between the geometrical object r and its adjacent neighbor $s = 1, 2, \ldots, S, s \neq r$ with the following properties:

1. adjacent means that parts of the object borders are superimposed on those of other objects: we split the set of values of the object's border

$$\mathbf{F}_r'(x) = 0 \tag{6}$$

into m intervals open at both sides, in such a way that for each interval formed,

$$\mathbf{F}_r'(x) = \mathbf{F}_s'(x) \quad \text{for } x \text{ from } (g_1, g_2) \tag{7}$$

This defines the common border sections between the objects and also meets the condition (2) which requires that a surface segment dx/dy is assigned to only one surface.

2. The result set of the discrete function $\mathbf{V}g(r)$ contains the names of the objects that are adjacent within the m intervals $[G]$. It is evident that (5) is not necessarily reversibly unique because each interval limit G, provided we are dealing with surfaces, can reference the names of two surfaces and/or $s = 1, 2, \ldots t, \ldots, S$ objects.

Figure 1 shows an example of the relationships. The map page Z shows a community area X containing a lake. Running along the southern boundary of the community is a road S. The names of the objects are, following (1), described by way of example as COMMUNITY X, FOREST Y etc.; the names of the border sections (lines) are numbered 1 to 5. The arrows represent the relationships of the functions $\mathbf{V}$. The functions $\mathbf{V}e$ and $\mathbf{V}k$ are indicated by double arrows, the ambiguous function $\mathbf{V}g$ by single arrows. Thus, for example, we have the following functions:

472

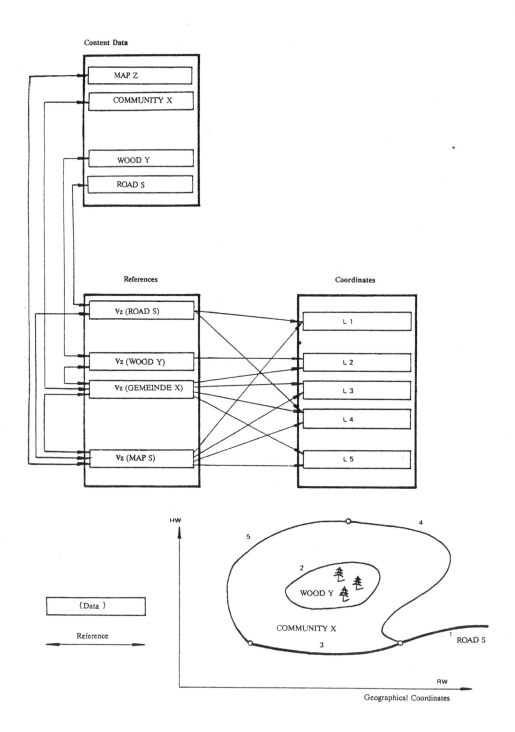

Figure 1: Topology of the formatted space related data

$$\mathbf{V}e(\text{COMMUNITY } X) \;\; <=> \;\; \{\text{FOREST } Y\},$$
$$\mathbf{V}k(\text{COMMUNITY } X) \;\; <=> \;\; \{\text{MAP } Z\}$$
$$\mathbf{V}g(\text{COMMUNITY } X) \;\; <=> \;\; \{2,3,4,5\}$$
$$\mathbf{V}k(\text{ROAD } S) \;\; <=> \;\; \{\text{MAP } Z\}$$
$$\mathbf{V}g(\text{ROAD } S) \;\; <=> \;\; \{1,3\}$$
$$\mathbf{V}e(\text{MAP } Z) \;\; <=> \;\; \{\text{ROAD } S, \text{FOREST } Y\}$$

and we obtain, for example, by inserting

$$\mathbf{V}e(\text{MAP } Z) = \{\mathbf{V}e(\text{COMMUNITY } X), \text{ROAD } S\} = \{\text{FOREST } Y, \text{ROAD } S\}$$

the "grandchild" FOREST Y of the object MAP Z etc. The coordinate related representation is produced by inserting the results in the functions $\mathbf{V}g$.

2.2 The space related model for unformatted data

By unformatted data we mean data which cannot be uniquely located in the data record (i.e. by place or field name). The data can only be interpreted from the overall context (of their grammar). Also, we shall only be dealing with text related data. Text **related** in the sense of a spatial allocation means that the geometrical object, unless a functional context is described, always has fuzzy borders by definition, but **can** always be specified with any degree of accuracy by providing a more exact text description in accordance with the requirements given under 2.1. A separation of the data into those which describe the objects in more detail by means of attributes (e.g. nature area description of a certain type with certain species of animals, usages etc.) and those which describe the spatial relationship is not possible because of the nature of unformatted data.

The text data describing the spatial relationship, beside attributes such as "near", "between" etc., are usually explicit statements of location which are embedded in the context of the text, or ones which refer indirectly to a certain area. Such areas can be politically defined, such as counties, districts etc., or topic related such as nature reserves, forestry areas, geological units etc. Just as important in practice are the map page names. Map pages of different scales from 1:5000 to 1:1 million usually have known page formats and standardized page names. In practice these are often used more frequently for delimitation than are place names. Thematic maps are also usually drawn up by such page sections.

In the concept of a data model, which is to be the basis of the method of uniform space related research, we must therefore take account of the requirements of 2.1, and also account for the requirements of unformatted storage. One of the methods of text related research is to use a thesaurus in conjunction with inverted text. In accordance with the subject matter of this article, we shall in the following examine the conditions under which the model approach outlined below can be applied to a thesaurus:

The space related place name (in the real, tangible world) can essentially be specified with sufficient accuracy in both formatted and unformatted text data. While the accuracy can be described constantly in numerical data, text data require inconstant modification of the text to produce an accurate description of the location.

Thus, because no surface equation as per (2) is derived, but rather is defined by text for the objects represented by the descriptions from the text data, the condition of focussed borders as per (5) cannot be met without further effort:

The border must be classified implicitly via the assignment of the place names in accordance with the represented surface to a specified smallest spatial unit (in the example below the "common denominator" is the map grid 1:5000). This spatial unit determines the maximum resolution of the spatial relation during the search, and must meet condition (2).

Consequently, surfaces below the **terminologically** determined resolution limit must be viewed as **spatially identical**; the different terms describing them (as regards the spatial relation) must therefore be regarded as synonyms. Here, (5) refers to the synonym terms.

Space related superordinate and subordinate classes are defined according to (3) and (4). Because algebraic determination of the superordinate and subordinate classes (grandchildren, children - see above) is not possible, they must be explicitly specified.

The spatial relation is represented by roots. Assigned beneath a root are the place terms in accordance with functions (3) and (4), which also lie spatially beneath the root described by the corresponding term. As (4) can be derived from (3) due to the uniqueness of (4), the relationships represented by the roots can also be uniquely formulated (and (4) can be automatically generated by database software).

The space related accuracy of the search is thus determined by specifying the maximum resolution set in the thesaurus; the success of the search is determined by the definition of the spatial relationships specified by the roots.

In the subsequent search, the texts include place names that are in no context with the spatial accuracy requirements. The different accuracy requirements are now reflected in a **differing usage and knowledge of space related terms** in the search. The access to the text can therefore only be achieved in the nontrivial case by linking the terms used in the search and by making use of the relations formed by the roots.

Figure 2 shows a spatial relation, corresponding to figure 1, for text related applications. The aforementioned common denominator is schematically shown by the broken map grid.

3 The Thesaurus of the Bavarian Ground Information System

The significance of space related searches in environmental data becomes directly evident, as by nature we are always dealing with conditions that interact with their environment. The development of a thesaurus as per 2.2, which is the prerequisite for a nontrivial search in text data, must therefore be given special attention. As mentioned above, the accuracy requirement is determined by the use of various terms. For example, at the "lowest" level in Bavaria alone, some 100,000 place names provided on official maps are relevant. The "highest" level is formed by the political districts or regions. Between the two are the delimited communities. On the "middle" level, standardized maps on a scale between 1:5000 and 1:100000 are used. Scales between 1:5000 and 1:25000 are frequent. These maps reflect the space related thinking of the users; the map page names are just as familiar to the user as the community names or other place names. The texts themselves contain space related designations of arbitrary order. Just as important are relevant topical divisions such as nature reserves boundaries, forest location mapping etc.

In accordance with the practical requirements, the grid of the 1:5000 scale non-overlapping field maps according to (2) were chosen for space related resolution by coordinates. The lowest resolution here is about 2 km. The place names (often settlements with only a few houses) are declared as synonyms to the field map name and to the name of the community. This produces two spatial reference systems at the lowest level. Via the definition of roots, the relationships with superordinate spatial units such as region, legislative district or map

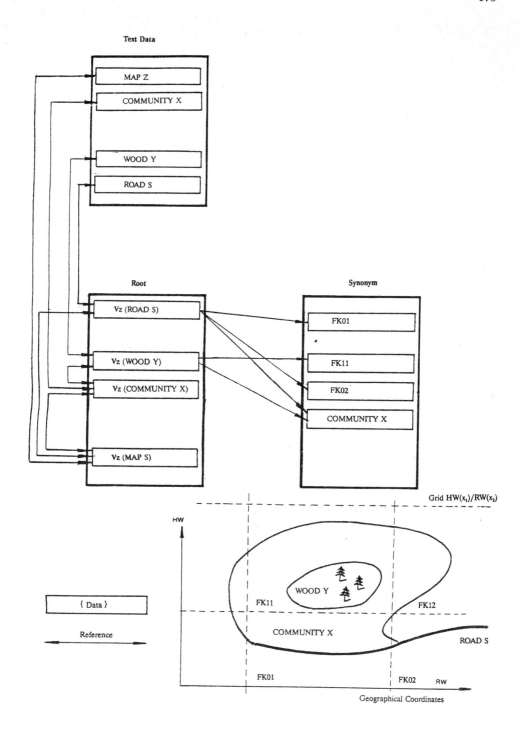

Figure 2: Topology of the unformatted space related data

page name are created on a scale of 1:25000 and 1:50000. In the thesaurus, roots and synonyms are therefore defined by the following system:

Word: **WEINBERG**
Category: PLACE NAME, TERRESTRIAL, DESCRIPTION, MAP CATALOG, ...
Root: L7734, PETERSHAUSEN, RG14, *AK24* ...
Synonyms: *FK24NW1304*, GS174136

and

Word: *FK24NW1304*
Category: FIELD MAP
Root: *AK24*
Synonyms: ZIEGELBERG, WASENHOF, ASBACH-PETERSHAUSEN

The division into categories allows a further narrowing of the search: The string WEIN-BERG describes not only a place (place name category), but also an area of high ground (terrestrial category). "Weinberg" also occurs in documents of the map catalog etc. Various "parents" were determined as roots in accordance with (4); the narrowing during the search is done by explicitly specifying the category, e.g. region, the top map or the root. In accordance with (5), the best resolution was set as the 1:5000 field map. In the above example it (field map *FK24NW1304*) also contains Ziegelberg, Wasenhof and Asbach-Petershausen.

In accordance with (3), under the term of the described root (e.g. L7734, the abbreviation of a topographic map page), a reference is made to all terms ("children") produced by the relation from (4).

The final spatial relationships required for the search are created by using (or negating) suitable synonymous terms and roots (the linked search in the topical text data follows the usual rules). As the link between the place names and the field maps representing a spatial grid is made in the thesaurus, consistent relationships can be created between the place names and ambiguities can be excluded. As the spatial coordinates are also known with the field maps, we have a link to the formatted data as per 2.1.

In this manner it is possible to gain access to the data both via spatial order terms and by marking an arbitrarily defined area.

One of the main advantages of this object oriented model approach is that text can be searched for which has a relationship with the place names you are searching for but which is not necessarily included explicitly in the text (and thus cannot be inverted). Merely adding new keywords (= constants) to the document would be redundant (with the associated problems). It must also be remembered that the number of assigned keywords would hit a natural limit: a surface area of approximately 30 km × 30 km can refer to several hundred place names.

References

BILL, R. and FRITSCH, D. (1991), *Grundlagen der Geoinformationssysteme*, Bd. 1, Karlsruhe, 220–238.

HALLER, K. (1976), Titelaufnahme nach RAK, *Einführung in die "Regeln für die alphabetische Katalogisierung"*, Generaldirektion der Bayerischen Staatlichen Bibliotheken, München.

HALLER, K. (1987), Sonderregeln für kartographische Materialien: *"Regeln für die alphabetische Katalogisierung"*, Bd. 4 RAK-Karten, Kommission d. Dt. Bibliotheksinstitute f. Alphabet. Katalogisierung unter Vorsitz von Klaus Haller, Wiesbaden.

INTERNATIONAL FEDERATION of LIBRARY ASSOCIATIONS and INSTITUTIONS (1977), ISBD (CM): *International Standard Bibliographic Description for Cartographic Materials*, IFLA, London.

MOLENAAR, M. (1989), Single Valued Vector Maps., A Concept in Geographic Information Systems, *Geo-Informationssysteme (GIS), 2, Karlsruhe 18-26.*

MOLENAAR, M. (1991), Terrain Objects, Data Structures and Query Spaces, *Geo-Informatik, Berlin 53–70.*

SCHNABEL, W. (1987), GEOKART - Ein Informationssystem für thematische Karten, *Raumplanung in Österreich 14,* Bundeskanzleramt (Hrsg.), 233–236, Wien.

STIBOKA (1982), *Kartencatalogue van de Stiching voor Bodemkartering*, Wageningen.

SONDERARBEITSGRUPPE BODENINFORMATIONSSYSTEM (1989), *EXPERT länderübergreifendes Bodeninformationssystem*, Hannover 1989, Hrsg. Niedersächsisches Umweltministerium.

Classification of Archæological Sands by Particle Size Analysis

E.C. Flenley
Department of Probability & Statistics
The University, P.O. Box 597, Sheffield, U.K.

W. Olbricht
Fakultät für Mathematik, Ruhr-Universität Bochum
Postfach 102148, D-4360 Bochum, Germany

Abstract: Particle size data have long been a useful tool for classification studies. Early work used *ad hoc* techniques such as simple scatterplots of summary statistics based on sample moments, but the modern approach is to use multivariate classification techniques on either the raw data or estimated parameters of fitted distributional models. This latter method will be described with reference to an archæological example in which sands from a mesolithic shell midden were to be classified as "beach" or "dune". The overall aim was a palæogeographic reconstruction of the mesolithic coastline of the island of Oronsay in the Inner Hebrides. The paper initially addresses the problem of which parametric model should be used. We then turn to the question of how to use the parameter estimates in classification. In both problems contextual knowledge plays an important rôle.

1 Introduction

The use of particle size distributions has a long history in Geoarchæology. This is mainly due to the fact that they provide an objective, relatively easily measurable, quantity which is not prone to changes even on the archæological time scale. The latter severely impairs the use of chemical measurements on sediments since they are often contaminated by later overprinting. But how can particle size data be used and to what purpose? In this paper we report on a study in which sands of unknown origin from a well-known archæological site were to be classified as "dune" or "beach" sand on the basis of their particle size distribution. The ultimate aim was to reconstruct the position of the site with respect to the mesolithic coastline. We briefly describe the background of the problem and the main approaches to its solution. The fitting of log-skew-Laplace distributions turns out to be an appropriate way to condense the information contained in a mass-size distribution into three summary parameters. If fitted distributions are available for a large number of sand samples, their parameter vectors can be used for clustering and classification. However, whilst a routine statistical analysis may be suitable initially, it is clear that such data often have particular features which should be taken into account. We give some examples with possible solutions.

2 Background

2.1 The Problem and the Oronsay Data Set

The mesolithic shell middens on the island of Oronsay (Inner Hebrides) are one of the most important archæological sites in Britain (cf. Mellars, 1987). It would be of considerable interest to determine their position with respect to the mesolithic coastline. An essential step in this direction would be the classification of sand samples from the sites as "beach" or "dune". If, for example, the sand below the midden were beach sand and the sand from the upper layers were dune sand, this would indicate a seaward shift of the beach-dune interface. Thus 226 sand samples were taken from the two locations Cnoc Coic ("CC") and Caisteal nan Gillian ("CNG"). They include 77 archæological samples of unknown (but to be determined) sand type taken vertically through the middens themselves and samples from known beach (110) or dune (39) environments taken horizontally along transects from the sea towards the middens which are located today about 150m inland. A full description of the archæological background is given in Fieller *et al.* (1984, 1987). From each sample 60g or 70g of sand were sieved through a stack of 11 sieves of mesh sizes (in mm) 0.063, 0.09, 0.125, 0.18, 0.25, 0.355, 0.5, 0.71, 1.0, 1.4 and 2. This yielded (after exclusion of the extreme classes) a histogram with 10 classes for the mass-size distribution of every sample. Our raw data set consists of 226 such mass-size histograms. The questions are whether the histograms for the modern sand samples show a clear difference between beach and dune environment (and hence can be separated) and how, if at all, the ancient sand samples from the middens can be classified.

2.2 Brief Historical Survey of Particle Size Analysis

Initially, particle size data of the above type were summarized by their sample moments, which were then in turn used for grouping and classification. In particular, mention should be made of Friedman's work (1961, 1979) where scatterplots of specific low order moments (sorting coefficient, skewness coefficient) are suggested for the discrimination of beach and dune sands. Many of these quantities are commonly read from graphs and are based on the implicit assumption of log-normality of the particle size distribution. This has led to some doubts about the validity and the accuracy of the procedure (Ehrlich, 1983). A more computational approach is to regard the mass-size histogram as a multivariate observation of dimension equal to the number of classes and to employ standard multivariate techniques (e.g. Syvitski, 1984). These ideas are not applicable if different sets of size classes are used for different samples, neither do they make use of the implicit ordering of the size classes. Finally, the mass-size histogram can be described by a distributional model, i.e. one can fit a probability distribution to the data and use the estimated parameters as input for further analyses. This method was developed into a coherent statistical approach mainly by Barndorff-Nielsen (1977) and Bagnold and Barndorff-Nielsen (1980) building on earlier observations by Bagnold (1941) and will be described in the next section.

3 Fitting of Particle Size Distributions

A variety of different distributional models is used in particle size analysis. Some of these are mainly based on empirical evidence (Weibull distribution following Rosin and Rammler,

1933), some have theoretical support from breakage models (log-normal distribution), some are chiefly pragmatic choices (normal distribution), some are based on specific features of particle size distributions such as their exponential tails which were first noticed by Bagnold (1937) (log-hyperbolic distribution, log-skew-Laplace distribution). For sand samples it has always been common to use a logarithmic scale on the abscissa. Bagnold (1941) suggested using a logarithmic scale on the ordinate as well and obtained mass-size histograms which looked approximately like hyperbolæ. Barndorff-Nielsen (1977) took up these ideas and introduced the log-hyperbolic distribution as a standard model for particle size analysis. However, attempts to apply this model to the Oronsay data set were not satisfactory (Fieller *et al.* 1984, 1992), since the log-hyperbolic distribution could not be fitted to a large number of our samples. The chief reason for this appears to be its computational complexity which is indicated by the formula for the density of the hyperbolic distribution:

$$f(x; \phi, \gamma, \delta, \mu) = \frac{\sqrt{\phi\gamma}}{\delta(\phi + \gamma)\mathcal{K}_1(\delta\sqrt{\phi\gamma})} \times \exp\left\{-[(\phi + \gamma)\sqrt{(\delta^2 + (x - \mu)^2)} + (\phi - \gamma)(x - \mu)]/2\right\}$$

where $\mathcal{K}_1(.)$ is the modified Bessel function of the third kind and ϕ, γ, $\delta > 0$. Two other contenders, the skew-Laplace and the normal distribution are limiting cases of the hyperbolic but have only three and two parameters, respectively, with densities:

$$f(x; \alpha, \beta, \mu) = \begin{cases} (\alpha + \beta)^{-1}\exp\{(x - \mu)/\alpha\} & x \leq \mu \\ (\alpha + \beta)^{-1}\exp\{(\mu - x)/\beta\} & x > \mu \end{cases}$$

where α, $\beta > 0$, and

$$f(x; \mu, \sigma) = \frac{1}{\sqrt{2\pi}\sigma} \exp\left\{-\frac{(x - \mu)^2}{2\sigma^2}\right\}$$

where $\sigma > 0$. A detailed comparison of these last three models as applied to the Oronsay data set can be found in Fieller *et al.* (1992). It turns out that, in particular with respect to discriminatory power, the log-hyperbolic model did not give better results than the simpler log-skew-Laplace and log-normal models. A similar conclusion was reported by Wyrwoll and Smyth (1985). We therefore concentrate on the 226 three-dimensional data vectors of estimated parameters for the log-skew-Laplace distributions and try to base a classification of the midden sands on these. We note that the three parameters of the log-skew-Laplace model can be interpreted geometrically as the reciprocals of the slopes of the two "legs" of the distribution and the abscissa of their point of intersection (when the density is plotted on a logarithmic vertical scale) and geologically by their correspondence to the three quantities "small-grade coefficient", "coarse-grade coefficient", and "peak diameter" defined by Bagnold.

4 Classification Considerations

4.1 First Steps

The problem under consideration appears at first sight like an archetypal discrimination-classification problem:

- first a function has to be established to distinguish between beach and dune samples on the basis of the sands from known environments

• then the samples from unknown environments can be attributed to one of these types.

There are, however, a few peculiarities. First of all, the sand samples are by no means a random sample. Quite the contrary, they were taken according to a very elaborate plan horizontally along transects towards the middens from the modern environment and vertically through the middens for the ancient sands. This violates some of the assumptions for standard procedures, but yields additional information about a potential "seriation" or "sequencing" in the data, i.e. stratigraphic coherence. Moreover, the "objects of unknown type" do not necessarily belong to one of the "groups", since the ancient sands differ in time and location from the modern ones. On *a priori* grounds it seems plausible that this does not affect classification with respect to the types "beach" and "dune", but it is not certain. In view of these properties, a deliberately cautious perspective was adopted for the initial analyses of the Oronsay data set in Fieller *et al.* (1984, 1987). It consisted mainly of the following steps:

1. Use of two-dimensional scatterplots. Without interactive graphics and in view of the obvious clustering in the data this appeared a promising tool.

2. Use of the original parameters (α, β, μ) rather than derived quantities. This is justified because of Bagnold's ideas about the physics of sand transport and facilitates "thinking in the model". It also turned out that the traditional suggestions (Friedman, 1961) provided no improvement, but rather blurred the picture.

3. Use of Fisher's linear discriminant function as a simple unequivocal tool for classification rather than the earlier eye-fitted curves which have a distinctly *ad hoc* flavour.

The analyses demonstrated that classification is indeed possible. μ was recognized as a good discriminator between location (CC vs CNG) and α in combination with μ as a good discriminator for the type (beach vs dune). Figure 1 shows a plot of α vs μ for the 226 samples. The main groups are strongly clustered and, furthermore, stratigraphic coherence was seen to hold throughout and in great detail as evidenced by seriation along transects. In summary, the simple tools employed seemed to lead to a reliable analysis with interesting archæological conclusions.

4.2   More Elaborate Techniques

How useful are two-dimensional scatterplots of the original parameters? It was mentioned above that our attempts to use Friedman-type quantities instead did not help. However, it is not clear that they cannot be improved upon by other projections. This has to be balanced against increasing difficulties of "thinking in the new parameters" if they do not admit a sensible subject-matter interpretation. An obvious approach in this context is Principal Component Analysis, which raises two immediate questions:

• Should we base PCA on the covariance matrix or on the correlation matrix (i.e. standardize)?

• Should we use transformed parameters, in particular should we use the logarithms of α and β (cf. Wyrwoll and Smyth, 1985)?

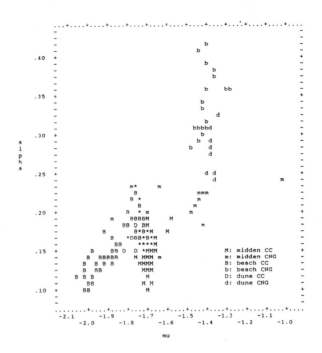

Figure 1: Plot of alpha versus mu

We have carried out all four variants and found no substantial differences between them so we only refer to a PCA on the unstandardized, untransformed data. It appears that the measurement scales for our three parameters are close enough not to require any other treatment. The first two principal components for the 149 data points of known sand type are

$$pc1 = 0.3215\alpha + 0.0230\beta + 0.9466\mu$$

$$pc2 = 0.1679\alpha + 0.9825\beta - 0.0808\mu.$$

They explain cumulatively 80.9% and 98.7% of the total variance. Their correlations with the original variables are given in the following table:

	α	β	μ
pc1	0.9129	0.0496	0.9983
pc2	0.2238	0.9955	-0.0400
ex. variation	0.8835	0.9935	0.9942

The last row contains the proportion of variability for each single variable which is explained by pc1 and pc2 jointly. Figure 2 shows a plot of the data (with the midden samples included) in terms of the first two principal components. In general, PCA seems to work quite well for the Oronsay data set as opposed to, for example, the crab data treated in Huber (1990). (This is particularly visible if one compares Figure 2 with Figure 3 which was obtained by

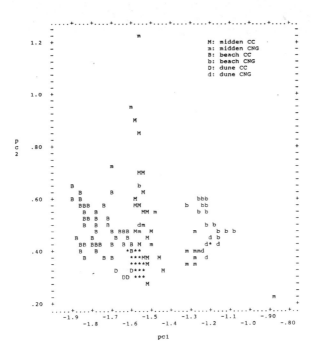

Figure 2: PCA for the modern sands

Canonical Correlation Analysis and displays the best way of separating the four clusters for sand type and location.)

On the other hand the achieved separation of the clusters is not decisively better than in Figure 1 and the interpretation of pc1 and pc2 is not as obvious as the original meaning of α and μ. There are two main things that PCA teaches us in this situation. Firstly, it brings out outlying values of the midden samples more clearly (note that the midden samples were not used in the calculation of pc1 and pc2). Secondly, and perhaps more importantly, it emphasizes that the main separation in the modern environment is between location (pc1) and only the second is between sand type which we are interested in. This may have some repercussions for a more refined analysis. Consider, for example, the situation depicted in the following diagram:

beach1	dune1
dune2	beach2

If we only want to discriminate between sand type, Fisher's linear discriminant function is bound to do badly. If, however, we knew from contextual knowledge that values in the upper semi-plane are from location 1 and those in the lower from location 2, we could easily obtain a perfect separation by classifying into four groups (the four quadrants) and "ignoring" the location information afterwards. We now compare the two-group separation with the four-group separation for the modern environment. In both cases Fisher's approach was used (with one canonical variable in the first case and three in the second) and all variables were

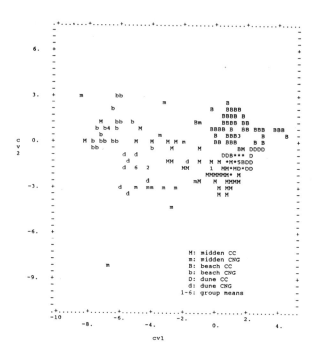

Figure 3: CCA for all samples

included. The following tables give the raw (not jackknifed) classification matrices:

	beach	dune
beach	84	26
dune	3	36

	beach CC	beach CNG	dune CC	dune CNG
beach CC	64	0	26	0
beach CNG	0	19	0	1
dune CC	0	0	29	0
dune CNG	0	0	0	10

Thus the misclassifications go down from 26 + 3 to 27 + 0 for the "pooled" second table. In particular the samples from the "base of dune" are now correctly classified. The improvement is certainly not very substantial, but it serves to illustrate a way in which additional stratigraphic information might be incorporated. We note that amongst the samples that were wrongly classified as dune, we find in both cases entire "subclusters" of "upper beach" samples. This illustrates again the vital role of stratigraphic coherence in this data set. One might consider allowing for this by similar considerations to those above but we shall not pursue this here. However, there is some need to take this fact into account again when classifying the unknown samples.

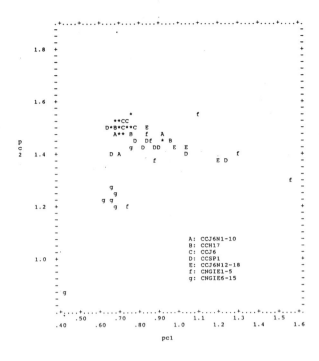

Figure 4: PCA for the middern sands with subclusters

4.3 A Remark on Classifying the Midden Samples

Normally, there should not be any problem applying the classification rules to the samples of unknown type. However, again we should watch out for potential complications or improvements due to the additional contextual information. Consider, for example, the following sketch where only one subcluster is shown:

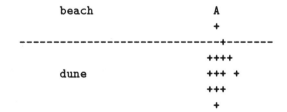

Most analysts would probably group point A with its whole subcluster as "dune" because the subcluster seems simply to dip into the beach area and the "cluster-cohesion" would overrule the demarcation line. In an informal way this is done anyway when the classified individual points are later interpreted in the subject-matter context. More formally, one could attempt to classify subclusters (rather than single points) right away. Again, we might expect that some sort of stratigraphy holds. This is also the reason why the sands from the middens were split *a priori* into several groups according to the sampling scheme ("below midden", "above midden" etc.). The situation is however even more complicated. It is well known that archæological sites can contain fair amounts of different sand which was introduced by human activities (anthropogenic contamination). Essentially, we find ourselves with respect to the midden sands in a position where we wish to look for all sorts of subclusters and interesting undetected (and unspecified) structure. In spite of the seemingly well-defined original problem, this is the starting point for an unstructured data analysis problem. As a first step we have again tackled this problem using Principal Component Analysis; this time for the midden sands. Figure 4 shows the result (the numbers there refer to the sampling subgroups, not necessarily to any relevant subclusters). There appears indeed to be some obvious grouping, which tallies fairly closely with the subsample coding in fact. This subclustering could be used in a sophisticated classification procedure, as described above. It is also interesting to see how Projection Pursuit methods fare here, since they sometimes seem able to detect clustering without external information (cf. Huber, 1990). We follow the suggestions in Huber (1987, 1990), the theoretical background is given in Huber (1985). We used the χ^2 index on the (sphered) data, to determine the two "least normal" projection directions. For the first direction a grid search on the upper half of the unit sphere was carried out in spherical coordinates with steps of one degree. The second direction was then found by grid search in the plane orthogonal to the first. Figure 5 shows the result.

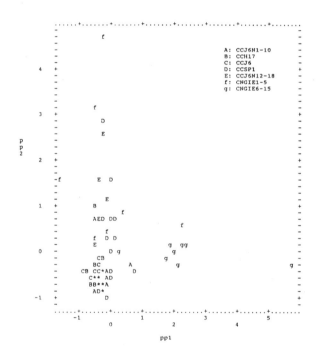

Fig. 5: Projection Pursuit for midden sands

A fairly similar clustering to that of PCA is observed. Projection Pursuit was also used on the (sphered) modern data to see what structure was revealed there (see Figure 6). Some interesting clustering has been detected. In particular, the first direction (pp1) discriminates well between the sites. For modern sands the second direction (pp2) appears to separate

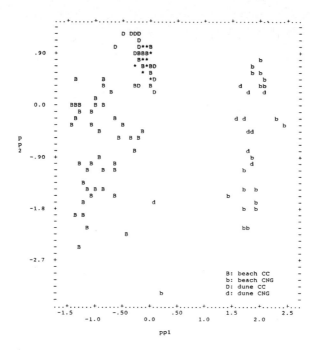

Fig. 6: Projection Pursuit for modern sands

beach and dune sands less clearly than the second principal component. However, it brings out the distinction between various types of beach sand ("mid beach", "upper beach") more distinctly. This is clearly seen in the three main clusters on the left of Figure 6 if a more detailed grouping is superimposed.

5   Conclusions

We have seen above that even in comparatively straightforward situations additional problems and additional information can come to light once we take a closer look. Today's computer facilities have greatly increased our ability to deal with such questions in a more specific tailor-made way rather than simply applying a standard analysis. In the foregoing, we have moved from basic subject-matter orientated considerations to increasingly more elaborate techniques. Our conclusions can be summarized as follows:

1. A sound basic analysis is always recommended and often reaches the most important conclusions in a direct manner. A suitable model, such as Bagnold's ideas, can be of great value.

2. For a well-structured data set, such as the Oronsay data, Principal Component Analysis works satisfactorily. One should, however, always watch out for additional features that might complicate or help to improve the analysis.

3. Even in well-structured problems, less structured subproblems may be encountered. Projection pursuit techniques can be of great use here.

4. Specifically, with respect to the classification of sands, the fitting of log-skew-Laplace distributions seems to be a feasible approach. The classification can then be done on the basis of the parameters and take into account any contextual information that might be available.

Acknowledgements

We are grateful to Yvonne Timmins, David Gilbertson and Nick Fieller for provision of the Oronsay data and for help in its analysis and and interpretation.

References

Bagnold, R.A. (1937), The size grading of sand by wind, *Proc. R. Soc. A, 63, 250–264.*

Bagnold, R.A. (1941), *The Physics of Blown Sand and Desert Dunes*, Methuen & Co. London.

Bagnold, R.A. and Barndorff-Nielsen, O. (1980), The pattern of natural size distributions, *Sedimentology 27, 199–207.*

Barndorff-Nielsen, O. (1977), Exponentially decreasing distributions for the logarithm of particle size, *Proc. R. Soc. A, 353, 401–419.*

Ehrlich, R. (1983), Size analysis wears no clothes or Have moments come and gone?, *J. sedim. Petrol., 53, 1.*

Fieller, N.R.J., Gilbertson, D.D. and Olbricht, W. (1984), A new method for environmental analysis of particle size distribution data from shoreline sediments, *Nature, 311, 648–651.*

Fieller, N.R.J., Gilbertson, D.D. and Timmins, D.A.Y. (1987), Sedimentological studies of the shell midden sites, in: P.A. Mellars (ed.), *Excavations on Oronsay: prehistoric human ecology on a small island*, Edinburgh, E.U.P., 78–90.

Fieller, N.R.J., Flenley, E.C. and Olbricht, W. (1992), Statistics of particle size data, *Appl. Statist., 41, 127–146.*

Friedman, G.M. (1961), Distinction between dune, beach and river sands from their textural characteristics, *J. sedim. Petrol., 31, 514–529.*

Friedman, G.M. (1979), Address of the retiring president of the International Association of Sedimentologists: Differences in size distributions of populations of particles among sands of various origins; and addendum, *Sedimentology, 26, 3–32, 859–862.*

Huber, P.J. (1985), Projection Pursuit (with discussion), *Ann. Statist., 13, 435–525.*

Huber, P.J. (1987), An illustrated guide to projection pursuit, *unpublished manuscript.*

Huber, P.J. (1990), Data Analysis and Projection Pursuit, *Technical Report, M.I.T., Cambridge MA.*

Mellars, P.A. (ed.) (1987), *Excavations on Oronsay: prehistoric human ecology on a small island*, Edinburgh, E.U.P.

Rosin, P. and Rammler, E. (1933), The laws governing the fineness of powdered coal. *J. Inst. Fuel, 7, 29–36.*

Syvitski, J.P.M. (1984), Q-mode factor analysis of grain size distributions. *Geological Survey of Canada: Open file report, 965.*

Wyrwoll, K.-H. and Smyth, G.K. (1985), On using the log-hyperbolic distribution to describe the textural characteristics of eolian sediments. *J. sedim. Petrol., 55, 471–478.*

The Analysis of Stratigraphic Data with Particular Reference to Zonation Problems

Joanne Padmore

Management School, Conduit Road, University of Sheffield, UK

Abstract: Stratigraphic data arise commonly in archæological contexts and are frequently multivariate in nature. Often, one aim of their analysis is to divide the original sequence into subsequences or 'zones' (in a palynological context) where each zone comprises samples of similar composition. This paper will review some of the multivariate techniques which may be employed in such analyses and propose some enhancements.

A commonly used technique is that of constrained cluster analysis (Birks & Gordon, 1985) which is an adaptation of standard cluster analysis. A stratigraphic constraint is imposed which ensures that the sequential ordering of the samples is maintained during the clustering process. Thus only clusters consisting of stratigraphically adjacent samples are formed.

The potential of change-point techniques in the analysis of such data is also considered. The problem of zonation is formulated in a hypothesis testing framework and several models of use in the analysis of stratigraphic data are considered.

The various techniques described are applied to a data set originating from the Outer Hebrides and their results are compared. The analysis which best summarises the structure of the data is found to be a multinomial change-point analysis.

1 Introduction

This paper is concerned with the analysis of stratigraphic data sets of a multivariate nature. Within such data sets the samples follow a natural ordering typically determined by the depth at which the samples are taken from the core of sediment. This unique ordering of samples with respect to depth and hence, time allows the data to be used in determining changes in past populations through time.

The data may be presented in the form of an $n \times p$ data matrix, n representing the number of levels (from each of which a single sample was taken) and p, the number of individual pollen taxa encountered in at least one of these levels. These data are often so complex in the number of taxa represented, stratigraphic levels analysed and individual fossils counted at each level that it is generally necessary to subdivide each stratigraphic sequence into smaller units for ease in:

- the description;

- the discussion, comparison and interpretation;

- the correlation in both time and space of the sequences.

The most useful unit of subdivision of the vertical or time dimension of stratigraphic pollen data is the pollen zone. Cushing (1964) defines a pollen zone as

> '...a body of sediment distinguished from adjacent sediment bodies by differences in kind and amount of its contained fossil pollen grains and spores which were derived from plants existing at the time of deposition of the sediment.'

A variety of numerical techniques have been applied to this problem of delimiting pollen zones from a stratigraphic sequence of data. These range from plotting taxon frequencies and determining zones by eye (Flenley *et al*, 1990; Mills, 1984), to computer intensive classification techniques (Adam, 1974; Dale & Walker, 1970; Birks & Gordon, 1985). Standard zonation techniques involve the use of exploratory multivariate techniques such as clustering and scaling and some of these are considered briefly in this paper.

An alternative approach incorporating change-point methodology is also considered and it is shown how such an approach may provide a more objective zonation of the data than that provided by the exploratory multivariate techniques.

2 Exploratory multivariate techniques

As is the case with any multivariate data set, the patterns of variability within pollen data are often difficult to assess by a visual inspection of the data alone. The data are multivariate and exploratory multivariate techniques which exploit, accommodate and display the structure within the data are particularly useful. The aim of such techniques is to summarise a complex multivariate data set in such a way that any structure within it is revealed whilst the information within it is retained.

The major aim of zonation is to group samples, hence clustering procedures are particularly useful in the analysis of stratigraphic data. Since the data possess a stratigraphic ordering it would appear appropriate to analyse them employing some stratigraphic constraint in order to ensure that the sequential ordering of the objects (samples) is maintained during the clustering process. Thus a stratigraphic constraint may be imposed, ensuring that only clusters consisting of stratigraphically adjacent samples are formed and leading to a constrained cluster analysis. The results of this form of analysis provide an insight into the structure of the data whilst maintaining its natural ordering and the formation of groups acts as a zonation of the data. These clusters can then be interpreted as indicators of pollen zones. Such analyses are provided by the computer programs ZONATION (Birks & Gordon, 1985) and SHEFPOLL (Padmore, 1987).

Other exploratory multivariate techniques which have been applied to pollen data include Principal Component Analysis, Correspondence Analysis and the Biplot.

There is always a problem in assessing the validity of the structure displayed by these techniques because different methods are likely to give slightly different results. Everitt (1980) considers this problem and proposes several approaches to it. These include the use of several techniques to analyse the same data set, the omission of samples or variables and the use of sub-sequences of the full data set. In each case the results obtained should be examined for consistency. There remains, however, no fully satisfactory solution to the problem and exploratory techniques, whilst giving an insight into data structure, remain, to a certain extent, subjective.

3 The change-point approach

The problem of testing for a change-point in a sequence of independent random variables: $X_1, X_2, \ldots, X_n$; is well recognised. It is usually assumed that the sequence of random variables arise from the same family of distributions, specified by some parameter, θ, and that at some unknown point in the sequence, r say, there is a change in this parameter value. The problem of a single change-point in a sequence of n random variables, X_i, can be expressed more formally in terms of a test of hypothesis: H_0 vs H_1

$$X_i \sim f(x_i, \theta_i)$$

$$H_0 \quad : \quad \theta_1 = \ldots = \theta_n$$

$$H_1 \quad : \quad \theta_1 = \ldots = \theta_r \neq \theta_{r+1} = \ldots = \theta_n \quad (1 \leq r < n)$$

In a test for a single change-point the general procedure is to select some test statistic, T_r say, which tests for a change in the parameter vector, θ, at some particular point, r, in the sequence and to maximise this over the range of possible change-point positions to give an overall test statistic, T, and an estimate of change-point position, $\hat{r}$, where

$$T = \max_{1 \leq r < n} T_r = T_{\hat{r}}$$

The univariate case has been studied by many authors (Hinkley, 1970; Hinkley & Hinkley, 1970; Hawkins, 1977; Worsley, 1979; James et al, 1987). The multivariate case has been considered less widely. Srivastava & Worsley (1986) investigate the properties of a likelihood ratio test statistic for testing for a change in multivariate normal mean and extend this to testing for a change in row probabilities of a contingency table with ordered rows.

Stratigraphic pollen data take the form of an ordered sequence of multivariate observations. If it can be assumed that the data form a sequence of independent random variables from a specified family of distributions and that changes in zones may be regarded as changes in the parameter value of these distributions, then the zonation problem may be regarded as a multiple change-point problem where change-point positions represent zone boundaries and a sequence of samples between consecutive change-points may be thought of as a pollen zone.

Two multivariate distributions of potential use in zonation are the multivariate normal and multinomial distributions. Srivastava and Worsley (1986) propose test procedures for a single change-point based on changes in multivariate normal mean and multinomial proportions. In each case the appropriate test statistic is calculated for a change-point at each sample in the sequence. The change-point is identified as the point at which the highest value of the test statistic occurs and its significance is assessed using a Bonferroni bound.

A wide variety of palynological data sets may be analysed by testing for a change in the mean of a sequence of multivariate normal random variables. Absolute abundances consisting of concentrations or accumulation rates may be analysed directly whilst data comprising relative abundances may be analysed via the additive logratio transformation described by Aitchison (1986). A further possibility is to apply this test procedure to the transformed data rotated onto their principal components. This is of potential use when, as is often the case in palynological studies, the dimensionality of the observations (i.e. number of taxa encountered) is large and places a lower bound on the length of sequence which can be searched. In such a case it would be useful to reduce the dimensionality of the data set whilst retaining as much of the information within it as possible. If it is assumed that the transformed data are multivariate normal then any linear combination of them should also be multivariate normal, in particular that combination defined by the principal components. Thus they may be modelled by the multivariate normal model with lower dimensionality. Counts to a pollen sum may be modelled by the multinomial distribution (Mosimann, 1965).

4   Multiple change-points

When analysing pollen data it is usually the case that several zone boundaries exist. A core of sediment will typically cover a 10,000 - 100,000 year span incorporating several distinct ecological phases and the resulting sequence of samples will contain several zones with associated boundaries. Each zone boundary may be thought of as a change-point and hence, in many situations, it will be unknown whether there is no change-point, one change-point or more than one change-point in the sequence under study. Thus the problem of zonation is concerned with estimating the number of change-points as well as their position.

One approach to this might be to examine all sets of possible change-point positions for that which optimises some appropriate test statistic. This is computationally infeasible, particularly since it is not known *a priori* how many change-points there are. The objective of any multiple change-point technique is to identify the correct number of change-points and their correct locations.

In practice problems may arise. Where several change-points exist the test procedure may fail to locate any as the change-points may have a masking effect on one other. The masking effect being the result of a false null hypothesis being tested against a false alternative (i.e. testing for a single change-point when several are present). This may lead to the non rejection of the false null hypothesis. There may also be a problem with the identification of spurious change-points whereby a change-point may be located at an incorrect position.

In this paper two approaches are considered, both of which are extensions of the tests for a single change-point:

1. A binary segmentation procedure (Vostrikova, 1981)

2. A window search technique

4.1   Binary segmentation

Vostrikova (1981) proposes that an unknown number of change-points can be estimated by a binary segmentation procedure as follows:

The full sequence is first tested for a single change-point; if a change is found and assessed as significant at a prespecified level, α say, then the 2 sequences before and after the change-point are tested separately for a change. The process is repeated until no further sub-sequences are found to have significant changes at level α. Diagrammatically the process may be described thus:

SEQUENCES SEARCHED	CHANGE – POINTS
$x_1 \longrightarrow x_n$	r_{11}
$x_1 \to x_{r_{11}}$ $\quad$ $x_{r_{11}+1} \to x_n$	r_{21} r_{22}
$x_1 \to x_{r_{21}}$ $\;$ $x_{r_{21}+1} \to x_{r_{11}}$ $\;$ $x_{r_{11}+1} \to x_{r_{22}}$ $\;$ $x_{r_{22}+1} \to x_n$	r_{31} r_{32} r_{33} r_{34}
$\vdots$	$\vdots$

The major problem arising in searches of this nature is that of masking. Depending on the data involved this may lead to the non-rejection of the false null hypothesis over the false alternative or, alternatively, the change-point $\hat{r}_{11}$ may be estimated at an incorrect position due to the masking effect of other change-points, so affecting the identification of further change-points.

4.2 Window search

In this approach windows are used to selectively examine the sequence. A window width, w, is specified at the beginning of analysis. This is dependent upon the size of the sequence, dimensionality of the data set and any other information available such as typical zone length. A pragmatic choice of window width which has been found reasonable in practice is 10 samples.

Having specified the window width every ordered sub-sequence of this length is tested for a single change-point. A cumulative total is kept of the number of times a particular sample is judged to be a significant change-point at a prespecified significance level of α, typically 5% or 1%. If a sample is judged a significant change-point in several of the window searches then this indicates that the sample is a valid change-point for the full sequence.

In the analysis which follows both methods are used as it is felt that they are complementary and should be used in conjunction with one another in multiple change-point analysis.

5 An application of change-point analysis

In order to compare the results of multiple change-point analysis with those of standard constrained clustering techniques, an analysis was made of a palynological data set originating from Lochan na Cartach, Barra, Outer Hebrides (henceforth referred to as the Barra data set). A 5.24 metre core was taken from the infilled edge of the loch which is thought to cover the period 11,000BP to the present. Samples were taken at 1cm intervals from 524-500cm and at 5cm intervals from 500cm upwards. The full data set comprises 125 samples and 44 taxa.

Constrained versions of single link analysis (city block metric) and Ward's analysis and a hierarchical divisive routine, based on total within group sum of squares, were applied to the data. Full descriptions of these techniques are available in Birks & Gordon (1985). The results of these analyses are summarised in Table 1 and were determined on the basis of diagrammatic and numerical output.

The data were also analysed using a multinomial change-point model. The binary segmentation technique was implemented to a factor of four, so that a maximum of 15 $(1+2+4+8)$ change-points would be identified. For the window searches, window widths of 10 and 20 were selected. The significance level was established as 5%. In order to qualify as a potential change-point a sample had to be recognised at least 5 (window width 10) or 9 (window width 20) times, although the full set of results was examined in order to gain a richer insight into the structure of the data.

Table 2 presents the results of the binary segmentation procedure. A primary change-point is recognised at sample 101, a result consistent with that of Ward's analysis. Secondary change-points are identified at samples 43 and 107. Sample 107 has been identified consistently as a zone boundary by the clustering techniques. Some change around sample 90 is identified by all three clustering techniques and this is identified as a tertiary change-point.

Single Link	Ward's	Divisive SSquares
1–14	1–33	1–14
15–20	34–51	15–33
21	52–87	34–47
22–33	88–92	48–89
34–50	93–101	90–93
51	102–108	94–107
52–90	109–125	108–125
91–92		
93–107		
108–125		

Table 1: Summary of Unweighted Zonation of Barra data set

1	2	3	4
			14
		22	
			33
	43		
			51
		90	
			92
101			
			-
		105	
			-
	107		
			110
		117	
			-

Table 2: Results of multinomial binary segmentation of Barra data set

At the fourth level of division the change-points identified at samples 14, 33 and 51 show consistency with the results of the constrained clustering solutions.

Table 3 presents the results of the window searches. Any sample which was identified as a significant change-point by at least one of the window searches is listed together with the number of times it was identified as an individual change-point. It is apparent that the window width of 10 provides a much more sensitive analysis compared with that of window width 20. However, both window search analyses do show consistency with one another. Furthermore, they show consistency with the results of the binary segmentation and clustering procedures.

The primary and secondary change-points identified by the binary segmentation analysis are also identified by the window searches (samples 43, 101 and 107). It is interesting to note the identification of change-points at samples 14 and 33. This result shows consistency with the results of the clustering procedures. At the third level of division the binary segmentation

Sample	Window 10	Window 20
14	8	9
33	7	9
42	6	2
43	4	14
51	6	8
60	7	0
81	6	7
90	5	9
101	7	7
107	8	14
117	5	0

Table 3: Results of multinomial window search of Barra data set

Zone	
Numerical	Archaeological
1–14	LNC
15–33	4
34–43	LNC
44–51	3
52–90	LNC
91–101	2
102–107	LNC
108–117	1
118–125	

Table 4: Archaeological significance of numerical zonation

technique examines the sequence consisting of samples 1–43 and identifies a change-point at sample 22. It may be that the nature of the binary segmentation procedure, seeking a single change-point in a sequence when in fact there are two, has resulted in the two true change-points being masked and a spurious change-point being identified in their place. At the fourth level the change-points at samples 14 and 33 are recognised. The change-point identified by the binary technique at sample 90 is confirmed and some instability at sample 117 is identified. The standard clustering analyses appear to identify the full set of zone boundaries with the exception of possible zone boundaries at samples 43 and 117. Zone boundaries are identified consistently at samples 14, 33, 50/51, 89/90, 101, 107/108 by both clustering and change-point analyses.

The multinomial change-point solution using the window search technique summarises the location of zone boundaries very well and may be regarded as the 'best' technique in that all zone boundaries (change-points) identified consistently by the other techniques are present in this solution.

The final zonation of the Barra data set, based upon the multinomial window search analysis, was examined in order to determine the archaeological significance of the zones indicated. Four major zones were identified in this way and these are shown together with their corresponding numerical zones in Table 4.

Zone LNC1 (samples 102–125) is thought to represent vegetation response to fluctuating climatic conditions during the Devensian Late glacial and up to the start of the Holocene. Within this major zone, three sub-zones were identified by the numerical techniques: LNC1a comprising samples 118–125, LNC1b comprising samples 108–117 and LNC1c comprising samples 102–107. LNC1a represents the establishment of vegetation cover as the climate warmed during the Devensian Late-glacial interstadial. LNC1b is thought to represent a change back to colder climatic conditions during the Loch Lomond stadial (C.11,000 B.P.). LNC1c is thought to represent the beginning of the Holocene.

The major zone LNC2, comprising samples 52–101, is thought to represent early woodland history and can be divided into two sub-zones on the basis of changes in the pollen representation of the woodland taxa: LNC2a comprising samples 91–101 and LNC2b comprising samples 52–90.

The third zone, LNC3, is thought to represent mid to late Holocene change. It may be subdivided into two sub-zones: LNC3a comprising samples 44–51 and LNC3b comprising samples 34–43.

The final zone, LNC4, is thought to represent blanket peat communities.

6 Discussion

This paper briefly reviews the exploratory multivariate techniques commonly used in the analysis of stratigraphic data and goes on to formulate zonation as a change-point problem. Although multivariate change-point techniques do not appear to have been applied to palynological data, a univariate approach was adopted by Walker & Wilson (1978) termed 'sequence splitting' and used by Green et al (1988). They test hypotheses using presence/absence data and pollen accumulation rates, both for individual taxa. The presence/absence data for individual taxa are treated as Bernoulli random variables and a change in the parameter π_i, the probability of 'success' (in this case presence of that taxon in sample i) is sought using a binary segmentation procedure. The pollen accumulation rates are modelled by the univariate normal distribution and a point at which there is a change in μ and σ^2 is sought using a likelihood ratio test statistic. Again a binary segmentation procedure is employed. Both of these analyses treat different pollen taxa as separate entities and no assumption of synchroneity in changes in the abundance of individual taxa is made.

In this paper, multivariate models for the data are proposed. These allow the multivariate data to be treated as a complete entity, hence changes in the overall composition of the pollen content of the core are sought rather than changes in individual species. Since the majority of pollen data collected takes the form of relative data, rather than 'absolute' pollen accumulation rates, this form of change-point model is probably of more practical use. In the analysis of the Barra data the multinomial change-point model with window searches produced a comprehensive summary of the results of the constrained clustering techniques, yielding an interpretable zonation of the data.

The change-point techniques developed here have a variety of practical applications in the fields of geology and archæology where data are often stratigraphic and multivariate. If an appropriate model can be found for a particular data set then the use of change-point techniques may provide the analyst not only with added information about the structure of the data but also a quantitative assessment of the degree of confidence to be placed in a zone boundary.

References

Adam, D.P. (1974). Palynological applications of principal component and cluster analyses. *Journal of Research of the United States Geological Survey* **2**, 727–741.

Aitchison, J. (1986). *The Statistical Analysis of Compositional Data.* Chapman and Hall, London.

Birks, H.J.B. & Gordon, A.D. (1985). *Numerical Methods in Quaternary Pollen Analysis.* Academic Press, London.

Cushing, E.J. (1964). Application of the Code of Stratigraphic Nomenclature to pollen stratigraphy. (Unpublished manuscript).

Dale, M.B. & Walker, D. (1970). Information analysis of pollen diagrams. *Pollen et Spores* **12**, 21–37.

Everitt, B.S. (1980). *Cluster Analysis* (Second edition). Heinemann, London.

Flenley, J.R., King, A.S.M., Teller, J.T., Prentice, M.E., Jackson, J. & Chew, C. (1990). The Late Quaternary vegetational and climatic history of Easter Island. (Unpublished).

Green, D., Singh, G., Polach, H., Moss, D., Banks, J. & Geissler E.A. (1988). A fine resolution palaeoecology and palaeoclimatology from South-eastern Australia. *Journal of Ecology* **76**, 790–806.

Hawkins, D.M. (1977). Testing a sequence of observations for a shift in location. *Journal of the American Statistical Association* **72**, 180–186.

Hinkley, D.V. (1970). Inference about the change-point in a sequence of random variables. *Biometrika* **57**, 1–17.

Hinkley, D.V. & Hinkley, E.A. (1970). Inference about the change-point in a sequence of binomial variables. *Biometrika* **57**, 477–488.

James, B.A., James, K.L. & Siegmund, D. (1987). Tests for a change-point. *Biometrika* **74**, 71–83.

Mills, C. (1984). The environmental history of the Gordano Valley, Avon — a palynological investigation. Unpublished M.A. Thesis, Department of Prehistory and Archæology, University of Sheffield.

Mosimann, J.E. (1965). Statistical methods for the pollen analyst: multinomial and negative multinomial techniques. In *Handbook of Paleontological techniques* (ed. Kummel, B. & Raup, D.). Freeman, London.

Padmore, J. (1987). Program Sheffpoll — a program for the zonation of stratigraphical data sets. Dept. of Probability & Statistics, University of Sheffield, Research Report No. 304/87.

Srivastava, M.S. & Worsley, K.J. (1986). Likelihood ratio tests for a change in the multivariate normal mean. *Journal of the American Statistical Association* **81**, 199–204.

Vostrikova, L.Ju. (1981). Detecting 'disorder' in multidimensional random processes. *Soviet Mathematics Doklady* **24**, 55–59.

Walker, D. & Wilson, S.R. (1978). A statistical alternative to the zoning of pollen diagrams. *Journal of Biogeography* **5**, 1–21.

Worsley, K.J. (1979). On the likelihood ratio test for a shift in location of normal populations. *Journal of the American Statistical Association* **74**, 365–367.

Classification Criterion and the Universals in the Organization of a Musical Text

M. G. Boroda

Sprachwissenschaftliches Institut, Ruhr-Universität Bochum,
D-4630 Bochum 1, FRG

Abstract: The paper deals with the role of classification criterion in the organization of a musical composition resp. in the development processes in the European composed music of the last three centuries. It is shown that in the period under study qualitative changes of the classification criterion on the motif level took place, and that these changes were one of the moving forces in the development of a European musical language. The study is based on the unambiguously defined melodic units – the F-motif and the mr-segment – suggested in (Boroda 1973, 1988).

1   The Problem

The trends and moving forces in the European composed music of the last three centuries attract for a long time attention in musicology. The general integrity of the European musical language, which can be observed notwithstanding all its deep changes in this period, allows one to suggest that its development in the 18th to the 19th and partly to the 20th century was subject to very general principles. However, the questions:

— *what invariants* in a musical composition underlay this development?

— to what extent is it a *self-development* and what internal processes trigger the "*stylistic revolutions*" in the period in question?, and

— what is the role of the *classification criterion* of units in these processes?

– these questions remain open.

The latter question, generally neglected in musicology, is very important: the difference between "repetition" and "alteration" – that is to say, between units *already occurred* in a composition, and those *new* ones in it – is fundamental for a musical form.

The present paper reports a study of the organization of repetition of the micromotifs and motifs in musical compositions of the 18th to the 20th century, based on various classification criteria. It is shown that

• in this organization both *invariants and shifts* can be observed,

• both of them are connected with the definite *classification criteria* of units in question, and

• changes in the classification criterion over the course of time have led to certain *general changes in the European musical language* in the period under study.

As a micromotif, resp. motif type unit the F-motif, resp. the mr-segment – the unambiguously defined melodic units suggested in (Boroda 1973, 1988) – are used.

2 *F*-motif / *mr*-segment: Definitions

The **F**-*motif* was defined as a melodic fragment within the framework of one of the following elementary metrorhythmic groups:

a) a group M of two or three successive tones of equal duration, the first of which is metrically stronger than the rest – e. g. ♫; ¾|♩ ♩ ♩|;

b) a group I of $k \geq 2$ tones with increasing durations, the final tone of which is not shorter in duration than the following one – e. g. ♪ ♩ ♩ ♩. ♩ ;

c) a single tone • which does not built an M- or an I-group with the preceding or the following tone e. g. – ♩ ♩ ♩ ♪♩ ;

d) a chaining of the M-group with the following I-group beginning with the final tone of this M-group – e. g. ♫ ♩; ♫♫♩[1]

A melody is segmented into F-motifs (see examples in Fig.1) in its tone-by-tone examination according to the following two rules:

R1. A given tone $x^{\bullet}$ closes an F-motif X under formation iff. $x^{\bullet}$ is **(a)** longer in duration than the following tone or **(b)** equal to it in duration and either metrically weaker than it or longer in duration than the previous tone.

R2. The F-motif X_{+1}, following X in a melody, begins with its final tone $x^{\bullet}$ iff. $x^{\bullet}$ is metrically stronger than the following tone $x^{\bullet}_{+1}$ and equal to it in duration (that is to say, if these two tones build an M-group). Otherwise X_{+1} begins with $x^{\bullet}_{+1}$.

The **mr**-segment is defined in (Boroda 1988) as a chain of successive F-motifs connected together by the "tendencies" and "pushing off" similar to those mentioned in footnote 1 above. The definition of the *mr*-segment, which is much more compilcated than that of the F-motif, is not given here due to the space limitation (cf. examples of the *mr*-segmentation of a melody in Fig. 1).

3 *F*-motif Level: Invariants

The study of the F-motif repetition in musical compositions of the 18th to the 20th centuries, based on the *sequence classification criterion* of the F-motifs (cf. section 3 below), has revealed the following general regularities:

---

[1]The F-motif was defined on the basis of the following principles:

(a) A tone *rhythmically tends* to the following tone longer in duration and is *rhythmically pushed off* from the following tone shorter in duration.

(b) A tone *metrically tends* to the following metrically weaker tone and is *metrically pushed off* from the following metrically stronger tone.

(c) The *rhythmic tendency between tones* mentioned in p.1 is *stronger than their metric push off*; the *metric tendency between tones* (p.2) is *weaker than their rhythmical push off*.

502

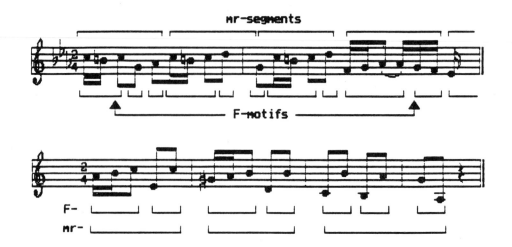

Fig.1

The note examples are arranged with the help of the Program "ESAC" elaborated by Professor H.Schaffrath and Mr. U.Franzke at the University of Essen.

a) In each composition the greatest part of its F-motif vocabulary was occupied by the F-motifs *rarely occurred in it*. This part was the greater the smaller the frequency of occurrence was.

b) The F-motifs occurred only once in a composition – i. e. those unrepeated in it – occupied about a half of its F-motif vocabulary. That is to say, in musical compositions under study a *balance between the repetition and the renewal* on the F-motif level was observable.

c) If the set of all frequences of occurrence of the F-motifs in a composition was arranged in diminishing order then the relation between the frequency f_i, $i = 1, 2, \ldots, V$ and its *rank* i in this ordered set proved to be subject to the *Zipf-Mandelbrot law*:

$$f_i = \frac{L \cdot K}{(B + i)^c}, \quad i = 1, 2, \ldots, V \tag{1}$$

(where L is the total number of the F-motifs in a compositioin, and V is its F-motif vocabulary size) with $c = 1$, $K = 1/ln F_1$, and $B = KL/F_1$ (cf. Fig. 2)[2]. The corollaries of the ZM law prognosed the F-motif vocabulary size V of the composition, as well as the size V_F of the group of F-motifs each of which occurred f times in it[3]

---

[2] Note that the Zipf-Mandelbrot Law with the same values of parameters described the rank-frequency relations in literary texts on the level of words – cf. Altmann 1988, Orlov 1976, 1982, a.o.).

[3] The corresponding relations were $V = L(F_{max} - 1)/(F_{max} ln F_{max})$, and $V_F = V/(F(F+1))$ respectively.

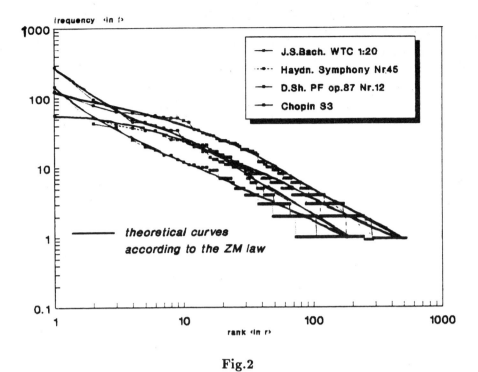

Fig.2

d) The regularities mentioned in a)-c) were characteristic of *complete compositions*; they violated on fragments. Even if a fragment and a complete composition belonged to the same author and the fragment was larger, it did not follow the ZM law, whereas the complete composition did.

That is to say, the organization of repetiton of the F-motifs in a musical composition proved to be stable in time, and subject to metastylistic principles.

4   Classification Criterion: Uniquity

The results reported above were based on the following classification criterion:

> two F-motifs A, B were considered to be equal iff. they coincided with each other in the interval structure and rhythm – in other words, if one could be obtained from the other by a parallel transfer of pitch (Fig. 3a).

Such transformation, known in musicology as *(strict) sequence*, is a very popular means of repeating small units in music of a broad group of styles.

It was reasonable to question, to what extent a "sequence" classification criterion is decisive for the results above.

To answer this question, in the same musical compositions as above the organization of the F-motif repeptition was studied under other musically natural classification criteria.

504

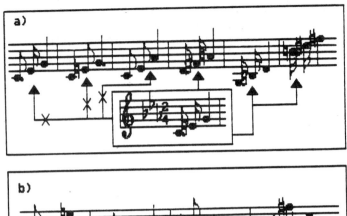

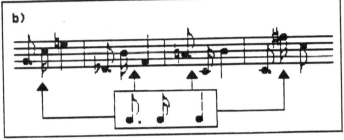

Fig.3

In particular, a criterion has been used considering two F-motifs equal iff. they coincided with each other as to their *rhythmic structure only* (let us call such "rhythmic F-motifs" the F_R-*motifs*) – Fig. 3b.

The study has shown that with this classification of the F-motifs the regularities above can not be observed.

The F_R-motifs rarely occurred in a composition did *not* form the greatest part of its vocabulary; this place was occupied rather by the medium frequent and the frequent F_R-motifs. Consequently, the frequencies in the set of all F_R-motif frequencies in a composition, arranged in a diminishing order, did not decrease rapidly. In certain cases there were even two most frequent F_R-motifs in a composition. The organization of repetitions of the F_R-motifs in a composition did not follow the Zipf-Mandelbrot law above (Fig. 4). Moreover, it was not possible to state certain definite general regularity to which the frequency structure of the F_R-motifs was subject.

The study of the F-motif repetition by other musically natural criteria of classification – e. g. by considering two F-motifs equal to each other if they coincided with each other as to the *interval structure* only (i. e. by neglecting their rhythmic differences) has shown similar picture as to the general principles of repetition and the ZM law.

In other words, the study makes it possible to state that the metastylistic regularities of the organization of repetition of the F-motifs in a musical composition described above (in particular, those connected with the Zipf-Mandelbrot law) are based on a definite classification criterion – specifically, on the "*sequence classification*" of the F-motifs. This criterion proved also to be very stable in time.

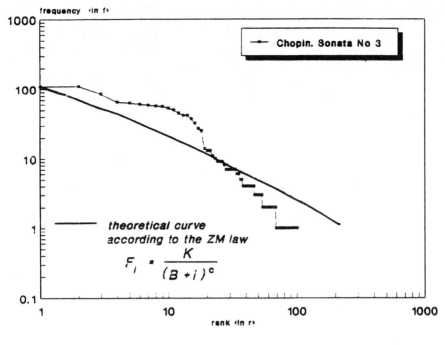

frequency ‹in f›

theoretical curve
according to the ZM law

$$F_i = \frac{K}{(B+i)^c}$$

rank ‹in r›

Fig.4

5 The Classification Criterion and the Level of a Composition: F-motifs

The results above led to the following question:

> The repetition of the "full" F-motifs in a musical composition proved to be based on the *sequence classification*, uniting rhythm and pitch. What are the relations between rhythm and pitch in the F-motif *vocabulary* of a composition?

Such a question seemed reasonable since not rarely different F-motifs coincided with each other as to the rhythmic structure. In other words, in the F-motif vocabularies of the musical compositions the *pitch variation* of the F_R-motifs was observable.

It was to question, what principles this variation underlays.

To answer this question, a study has been carried out on the F-motif vocabularies of the musical compositions where the organization of repetitions of the "full" F-motifs, resp. of the F_R-motifs was investigated (cf. above).

The study has revealed that the pitch variation of the F_R-motifs in the vocabulary of a musical composition *is subject to the same principles as the repetition of the "full" F-motifs in the whole composition.*

Specifically, the group of the F_R-motifs, each of which had only few or no pitch variants in a vocabulary under study, was considerably large – the larger the smaller the number of the variants was. The relations between the basic characteristics of the "variation activity" of the F_R-motifs in a vocabulary were subject to the Zipf-Mandelbrot law with the same parameters values as for the "full" F-motifs in a musical composition (cf. above).

The same study of the F_R-motif variation in a vocabulary with *other classification criteria* (e. g. by considering the *rhythmic variation* of the *intervalic structures* of the F-motifs) led to the suggestion that the classification of the F-motifs according to their rhythmic structure is just as important to the pitch variation in the F-motif vocabulary as the "sequence" classification to the repetition of the F-motifs in the whole composition. In both cases the classification criterion proved to be stable in time.

6   Classification Criterion: Motifs. Changes and Shifts

The results above lead to the following questions:

— what principles regulate the repetition in a musical composition on the level of *motifs*, considered by the musicological tradition as basic units of a musical language?

— to what extent are these principles stable in time?

The study of the repetition of the motif type units **mr-segments** in a musical composition (cf. section 2), based on the same "sequence" classification as in the study of the F-motif repetition above, revealed quite different picture from that characteristic of the F-motifs.

First it occurred that the part of the *mr*-segments, rarely occurring in a composition, in its vocabulary increases from the 18th to the 20th century.

Second, in contrast to the F-motif case, where many F-motifs differing from each other in pitch could have one and the same rhythmic structure, the different *mr*-segments often differed also rhythmically. Moreover, in the period under study a tendency could be observed to *growing rhythmic individuality of the mr-segments*.

That is to say, if the "sequence" classification of the *mr*-segments remained actual in musical compositions in the period in question then alteration should have played in a composition more and more significant role; the balance between it and the repetition mentioned above for the F-motifs should have been no more the case.

It seemed however more likely that another process took place in the period in question – namely, such a **change of the classification criterion** on the motif level that a number of the variation transformations of motifs were considered as (varied) *repetitions*.

It has also been suggested that in the period in question the varied repetition became more and more "typicised" – i. e. that certain types of rhythmic variation, occurring frequently in each composition, became over the course of time a kind of "norms".

Recent study of the rhythmic variation of the *mr*-segments in musical compositions of large forms (Boroda 1991) has revealed such "norms" characteristic of music of a broad group of styles (cf. some of them in Fig. 5).

On the other hand, a study has shown that over the course of time the rhythmic variation of motifs which changed *only the duration of tones* (e. g. ♪♫ ♩ $\Longrightarrow$ ♪ ♫ ♩.), occurs less and even less frequently. On the contrary, variation which changed the *structure of the mr-segments* (e. g. ♫ ♩ ♩. or ♫♫ ♩ $\Longrightarrow$ ♫ ♫♫ ♩♩, etc.) – a variation in certain respect similar to the *addition or subtraction of suffixes* to a word, resp. to its *internal flexion* – occurs more and more frequently.

It can be suggested that such changes in classification criterion would lead to the isolation of "kernel" *mr*-segments – i. e. of those **simplest and shortest ones** –, which

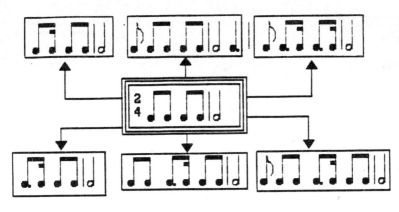

Fig.5

would be then used as a "basis" in a musical composition, on which the rhythmic/pitch variation can be build. Such a process would lead over the course of time to the general **shortening of the mr-segment length**, and cause, in order to avoid the information lost due to the shortening of the mr-segment length, an **increase of complexity of the mr-segments** in rhythmic, interval, and tonal aspects. This "length shortening ⟹ complexity increasing" process would influence also the F-motifs, the length of which would also decrease. As a general effect, the length of a musical composition within the framework of a given genre would (slowly) grow.

The study of the length distribution of the mr-segments and the F-motifs in musical compositions of the 18th to the 20th century proved these suggestions. The study revealed the shortening of the mr-segment, resp. the F-motif length in this period (Fig. 6). These changes for both the mr-segment and the F-motif were accompanied with a tendency to growing their interval and tonal complexity: the study revealed the *growing occurrence in a musical composition of the mr-segments (F-motifs)* containing chromatic scale degrees and large melodic intervals.

The entropy H_1 per F-motif, as well as the relation V_1/L of the number of the unrepeated F-motifs to the F-motif length of a composition proved to be slowly growing over the course of time. The values of corresponding correlation coefficients proved to be: $r(H_1, T) = 0.892$; $r(V_1/L, T) = 0.727$ $(P < 0.001)$.

The study of the compositions of the 18th to the 20th centuries in large forms has also proved the suggestion above about the changes in length of a musical composition. It has shown high positive correlation between the length of a composition and time: $r(lnL, lnT) = 0.958$ (cf. here the musicological observations about the tendency for symphonic movements to grow in scope and length during the nineteenth century – cf. e. g. *Leonard Meyer*'s notion in his book "Explaining music"); cf. also the increase of length of a symphony from, say Mozart to Maler).

Thus the changes in the classification criterion of motifs in the three last centuries (which in their turn were caused by the growing individualization of rhythm in music of the 18th to the 20th century, with its emancipation from a role of a basis for melodic variation to the self-dependent factor) have led to the series of changes in the basic elements of language of the European music, and have influenced such its fundamental characteristics as the tonal

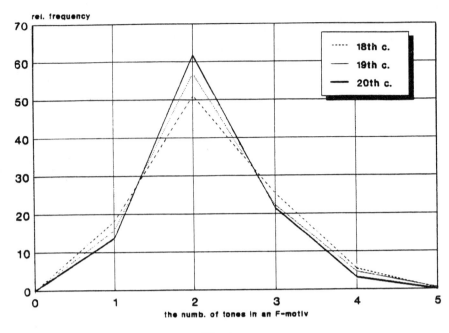

Fig.6

structure of motifs, and finally, the length of a musical composition.

The system "musical language – musical composition" has acted in this process as a **self-regulated system** where the length and the complexity of the motifs/micromotifs were regulated by the growing, resp. decreasing role of variation – especially, of the rhythmic one.

A hypothesis can be suggested that one of the very important internal prerequisites for "stylistic revolutions" in the European music of the last three centuries was the **"scooping out" of the "kernel" motifs** considered above, so that they became in the end too "formulaic" (that is to say, they were interpreted as rather melodic/rhythmic schemes). The new motifs were in this case build out of the "kernel-motifs" by adding a note or two to the corresponding "kernel" or by its "internal flexion" (for example by puncturization of its certain tone).

Such revolutions, resulting in particular in birth of new intonations, lead to certain increase of the motif length and to certain general decrease of the length of a composition (cf. e. g. the wide-spreading of the instrumental miniatures in the 19th century). This leads over the course of time to a new "wave" in the direction "growing complexity $\Longrightarrow$ decreasing length of the motifs, resp. micromotifs $\Longrightarrow$ increasing length of a composition", and so on. Such processes, if they indeed take place, should be one of the main internal **"belt drives"** of the development of the European musical language in (at least) the period of the 18th to the 20th century.

Proving such a hypothesis is a task of a special study – perhaps, of a series of them. The results reported here demonstrate that the investigation of changes in the classification criterion can form a basis for such studies.

References

BORODA, M.G. (1982). Die melodische Elementareinheit. *Quantitative Linguistics, vol.15.* Bochum: Brockmeyer, 205–221.

BORODA, M.G. (1990). The organization of repetitions in a musical composition: Towards a quantitative-systemic approach. *Musikometrika 2*, Bochum: Brockmeyer, 53–106.

KÖHLER, R. (1990). Linguistische Analyseebenen. Hierarchisierung und Erklärung im Modell der sprachlichen Selbstregulation. *Glottometrika 11*, Bochum: Brockmeyer, 11–18.

ORLOV, Ju.K. (1976). Obobščennyj zakon Zipfa-Mandel'rota i častotnye struktury informacionnykh edinits razliɲakh urovnej. In: *Vyčislitel'aya lingvistika*, Nauka, Moskva, 179–202.

ORLOV, Ju.K. (1982). Dynamik der Häufigkeitsstrukturen. In: Orlov, Ju.K., Boroda, M.G., Nadarejšvili, I.Š. *Sprache, Text, Kunst: Quantitative Analysen.*, Brockmeyer, Bochum, 82–117.

A Two-Mode Clustering Study
of Situations and Their Features

Thomas Eckes

Fachbereich Gesellschaftswissenschaften, Bergische Universität Wuppertal

Gauss-Str. 20, W-5600 Wuppertal 1, FRG

Abstract: In order to examine categorical beliefs about situations typically occuring within the university milieu, 48 students were asked to keep a written record of the situations they encountered over a one-week period. The 30 most frequently mentioned situation labels were presented to an independent group of 34 students for rating on 18 bipolar scales. Mean ratings were arranged in a situation × feature matrix and submitted to a two-mode hierarchical clustering analysis. This analysis aimed at a common representation of situations and features. Nine two-mode clusters were identified showing how subjects gave meaning to classes of social situations. More generally, the two-mode clustering solution was found to illuminate the prototype organization of culturally shared knowledge about everyday situations.

1 Introduction

Traditionally, the bulk of personality psychology has been a search for the characteristics of persons more or less regardless of the situations or contexts in which they function. In the 1960s, recognition of the theoretical and practical limitations of a situation-free conceptualization of persons led to a focus on psychological situations as potential sources for a more accurate account of individual behavior (see, for a recent review, Kenrick & Funder, 1988). Currently, situations are receiving much attention from researchers in personality and social psychology, with a vigorous search for methods suited to the analysis of how situations impact on the perceiver and the actor (see, e.g., Argyle, Furnham & Graham, 1981; Krahé, 1992; Snyder & Ickes, 1985). Foremost among these methods are *clustering* and *multidimensional scaling (MDS) techniques* by which the conceptual structures (i.e., categories or dimensions) underlying the perception of situations may be revealed. Using primarily MDS procedures, a number of studies have found that a relatively limited set of dimensions, such as self-confidence, involvement, evaluation, formality, and task vs. socio-emotional orientation, implicitly serve as the basis for cognizing everyday situations across different cultural groups and milieus (see, e.g., Battistich & Thompson, 1980; Forgas, 1976, 1983; King & Sorrentino, 1983; Magnusson, 1974; Pervin, 1976).

However, the pure dimensional approach is subject to several criticisms. First, it has been argued that the high convergence of dimensional structures across different subject samples and cultures may indicate a relatively *low sensitivity* of MDS procedures toward culture-specific ways of construing situations (Eckes, 1990). Second, the appropriateness of the *dimensional* and *metric assumptions* inherent in this approach have been questioned. Thus, in a review of several methodological routes to the study of situations, Argyle, Furnham and Graham (1981) have come to the conclusion that "... situations fall into types rather than along dimensions" (p. 11). Similarly, in a discussion of Tversky's (1977) contrast model of similarity as applied to the analysis of situations, Bem (1981) states that "... the similarity of two situations would seem to be a function of the number of salient features they share with one another, the number of features unique to one or the other of them, and the distinctiveness of both their shared and unshared features within the context of all the situations in the comparison set" (p. 256). Third, recent studies in the closely related field of person perception have shown that *classificatory methods* can yield structural information usually not contained within dimensional representations (Anderson & Sedikides, 1991; Powell & Juhnke, 1983).

Viewed from a cognitive-anthropological perspective (see D'Andrade, 1989; Quinn & Holland, 1987), what is needed is a fine-grained categorical analysis of cultural models of situations. A *cultural situation model* may be defined as the body of shared implicit knowledge about everyday situations and about ways to talk about these situations. It is further assumed that this culturally shared knowledge is organized into schemata or prototypes. A *situation prototype* is defined as an abstract set of features commonly associated with members of a situation category (Cantor, 1981).

To date, studies on the conceptual structures of situation cognition rely exclusively on procedures constructing a representation of only one *mode* or set of entities (Tucker, 1964). For example, Battistich and Thompson (1980) had subjects rate (a) the similarity between 30 social situations and (b) each situation along 38 bipolar scales. The set of situations constituted the one mode and the set of rating scales the other. While the similarity data were analyzed using nonmetric MDS, the mean property ratings were *separately* submitted to linear multiple-regression analysis to aid interpretation of the obtained scaling solution. The main drawback of such a sequence of one-mode analyses is that it cannot adequately take into account the *relationships between elements from both modes*. As a consequence, it is often difficult to find an adequate interpretation of a given dimension or cluster. In contrast, two-mode procedures allow the simultaneous representation of the similarity structure within each mode, as well as between both modes. In the present paper, a *two-mode clustering approach* to the analysis of culturally shared beliefs about social situations is suggested as a viable alternative and/or complement to the prevailing use of one-mode dimensional techniques.

2 Two-Mode Model and Algorithm

Several two-mode clustering models and algorithms for constructing hierarchical and/or nonhierarchical classifications are currently in use (see, for a review, Eckes, 1991). The model employed here aims at a two-mode hierarchical clustering representation of the input data. Recently, Eckes and Orlik (1991, in press) suggested an algorithm for the implementation of such a model, the *centroid effect method*. This method constructs a two-mode hierarchical clustering system with clusters having minimal internal heterogeneity. Specifically, a two-mode cluster is said to have low internal heterogeneity to the extent that its elements show strong inter-mode relations with as small a variance of corresponding numerical values as possible. Presuming that the input data are scored or normalized such that larger matrix entries indicate a stronger relationship between the corresponding row and column elements (i.e., the data are interpreted as "similarities"), the internal heterogeneity measure of a two-mode cluster $\mathcal{C}_r$ is expressed as

$$MSD_r = \frac{1}{n_r m_r} \sum_{\substack{A_{i'} \in A' \\ B_{j'} \in B'}} (x_{i'j'} - \mu)^2, \tag{1}$$

where n_r is the number of entities $A_{i'}$ in cluster C_r belonging to mode $A = \{A_i\}$, m_r is the number of entities $B_{j'}$ in cluster C_r belonging to mode $B = \{B_j\}$, $x_{i'j'}$ is the numerical value assigned to the elements $(A_{i'}, B_{j'})$ of a two-mode array $A \times B$, and μ is the maximum entry in the input matrix $\mathbf{X}$, that is $\mu = max_{i,j}(x_{ij})$. Thus, MSD_r is the mean squared deviation of entries $x_{i'j'}$ in the submatrix $\mathbf{X}_r$ corresponding to cluster C_r from the maximum entry μ in $\mathbf{X}$.

The fusion rule may now be specified as follows. At each step of the agglomerative algorithm, a particular subset of mode A is merged with a particular subset of mode B such that the increase in the internal heterogeneity measure of the resulting two-mode cluster is as small as possible. To accomplish this objective, several heuristic criteria are employed that are closely related to the MSD index. Which criterion will be used at any particular step in the agglomerative process depends on the subsets considered. Three general cases can be distinguished. In each case, those two subsets yielding the smallest criterion value will be merged.

Case I. A single-element subset $\{A_{i'}\}$ is to be merged with another single-element subset $\{B_{j'}\}$:

$$MSD_{i'j'} = (x_{i'j'} - \mu)^2. \tag{2}$$

Case IIa. A single-element subset $\{B_{j'}\}$ is to be merged with an existing two-mode cluster C_r:

$$MSD_\alpha = \frac{1}{n_r} \sum_{A_{i'} \in A'} (x_{i'j'} - \mu)^2. \tag{3}$$

Case IIb. A single-element subset $\{A_{i'}\}$ is to be merged with an existing two-mode cluster C_r:

$$MSD_\beta = \frac{1}{m_r} \sum_{B_{j'} \in B'} (x_{i'j'} - \mu)^2. \tag{4}$$

Case III. A two-mode cluster $C_p = A' \cup B'$ is to be merged with a two-mode cluster $C_q = A'' \cup B''$ into a new cluster C_t:

$$MSD_{\alpha\beta} = \frac{1}{n_p m_q + n_q m_p} \left[\sum_{\substack{A_{i'} \in A' \\ B_{j''} \in B''}} (x_{i'j''} - \mu)^2 + \sum_{\substack{A_{i''} \in A'' \\ B_{j'} \in B'}} (x_{i''j'} - \mu)^2 \right]. \tag{5}$$

As a conventional criterion for determining the number of two-mode clusters present in a given data matrix, a marked increase in fusion values can be considered indicative of the formation of a relatively heterogeneous cluster.

3 Data Collection and Analysis

Subjects participating in the situation description study were 48 (female and male) German university students. They were asked to keep a written record of the situations they encountered over a one-week period. For each situation described, subjects supplied a commonly-used label and a list of its most characteristic features.

On the basis of frequency of mention and range, 30 situation labels were selected (see Table 1). The features listed were used as the basis for constructing rating scales. Eighteen features were chosen from the descriptive protocols according to (a) overall frequency of usage across subjects and situations and (b) diversity of usage across situations. The bipolar scales were derived by pairing the 18 top-rated features with their appropriate antonyms (see Table 2).

Table 1: Situation Labels

at the supermarket	at the movies
going in for sports	in an elevator
going by train	preparing a lecture presentation
at a dentist	at a store
getting to know someone	having breakfast
at the swimming pool	watching TV
at the university cafeteria	studying
at daybreak	at a bar
going by bus	working part-time
going window shopping	at a private party
meeting a friend	at an accident
at a city festival	family get together
at a lecture	driving a car
at a cross-country run	a day at the university
hitchhiking	at a café

Table 2: Features and Their Antonyms

exciting–boring
constrained–free
complicated–uncomplicated
frightening–not frightening
formal–informal
relaxed–tense
frustrating–not frustrating
alone–with others
emotionally involved–emotionally uninvolved
pleasant–unpleasant
interesting–uninteresting
intimate–nonintimate
knowing how to behave–not knowing how to behave
familiar–unfamiliar
clear–unclear
noisy–quiet
serious–not serious
competitive–cooperative

The seven-point, bipolar scales were presented to an independent group of 34 (female and male) subjects from the same university in a fixed, random order. The order of situations varied randomly between subjects. Ratings were averaged across subjects and arranged in a 30 × 18 data matrix.

Table 3: Nine Cluster Solution Representing Situations and Their Features

Cluster	MSD	Elements
A	0.58	DENTIST *constrained, frightening, alone*
B	0.68	GETTING TO KNOW SOMEONE, MEETING A FRIEND, PRIVATE PARTY *exciting, interesting, intimate*
C	0.84	MOVIES, BREAKFAST *free, uncomplicated, not frightening, relaxed, not frustrating, pleasant, knowing how to behave, familiar, clear, quiet, not serious, cooperative*
D	1.53	LECTURE, PREPARING A LECTURE PRESENTATION, STUDYING, WORKING PART-TIME *competitive*
E	1.74	GOING IN FOR SPORTS, SWIMMING POOL, AT DAYBREAK, GOING WINDOW SHOPPING, CROSS-COUNTRY RUN, HITCHHIKING, WATCHING TV, DRIVING A CAR *informal*
F	1.98	CITY FESTIVAL, BAR, FAMILY GET TOGETHER, A DAY AT THE UNIVERSITY, CAFÉ *with others*
G	1.98	SUPERMARKET, GOING BY TRAIN, GOING BY BUS, IN AN ELEVATOR *nonintimate*
H	2.15	ACCIDENT *complicated, tense, frustrating, emotionally involved, unpleasant, not knowing how to behave, unfamiliar, unclear, serious*
I	2.70	UNIVERSITY CAFETERIA *emotionally uninvolved, noisy*

Note. MSD = mean squared deviation. One situation element (AT A STORE) and three feature elements (*boring, formal, uninteresting*) remained unclassified at the level of nine clusters.

In a preparatory step of the analysis, the mean ratings were centered by subtracting the grand mean from each entry. Then, the columns (representing the scales) were reflected (i.e., entries were first duplicated in a columnwise fashion, and then the duplicated entries were rescored by multiplying with -1). This reflection of columns elements assured that both poles of a given scale represented a single feature each, and thus could be clustered separately from each other. The augmented 30 × 36 data matrix was submitted to the centroid effect algorithm outlined above. A nine cluster solution was deemed appropriate according to the increase in fusion values from one hierarchical level to the next. Table 3 gives a summary presentation of this solution. The nine clusters are ordered according to degree of internal heterogeneity as measured by the MSD index.

Two-mode clusters A, H, and I contain one situation element each whose culturally shared meaning is readily described by referring to the respective list of feature elements. Thus, AT A DENTIST is consensually characterized as *constrained, frightening*, and being *alone* (see Cluster A); this cluster has very low internal heterogeneity ($MSD = 0.58$), showing that the intermode relations are fairly strong. A similarly *unpleasant* situation is AT AN ACCIDENT to which the subjects additionally ascribe features such as *complicated, tense*, and *not knowing how to behave* (see Cluster H). Although seen as *noisy*, AT THE UNIVERSITY CAFETERIA is not associated with *emotional involvement* (see Cluster I); however, it should be noted that this cluster has relatively high internal heterogeneity ($MSD = 2.70$), indicating that the intermode relations are not very strong. Clusters D through G contain a single feature element each that binds together many diverse situations. Thus, Cluster D comprises *competitive* situations, Cluster E contains a large number of *informal* situations, Cluster F consists of situations that are characterized by familiar *social* settings, and Cluster G refers to *nonintimate* situations. In contrast, the situation elements of Cluster B involve social encounters or interactions that are viewed as *exciting, interesting*, and *intimate*. Finally, the common features of AT THE MOVIES and HAVING BREAKFAST show that these situations are consistently perceived as *pleasant* (see Cluster C). Due to space limitations, an overlapping clustering solution taking into account the cross-classifiability of situations and features is not presented here.

4 Discussion

The results of the present study provide strong support for the claim that a two-mode classificatory approach is particularly well suited to the analysis of the conceptual structures underlying the cognition of everyday situations in a specified cultural group. Specifically, the clustering solution considered here showed that university students share a richly articulated system of knowledge about situations typically occurring in their milieu. Each of the nine clusters was found to be readily interpretable on the basis of its situation and feature elements. The major clusters referred to intimate social encounters, familiar social settings, as well as to relaxed, competitive, informal, and nonintimate situations.

It should be noted that the clustering of elements at a particular level of a two-mode hierarchy need not be exhaustive, that is there may be elements of one or the other mode that do not belong to a two-mode cluster. In the present study, one situation (AT A STORE) and three features (*boring, formal, uninteresting*) remained unclassified at the level of nine clusters. These elements did not have sufficiently strong relations to any of the other elements, that is, none of the situations under consideration was described as *boring, formal*, or *uninteresting*, and none of the features included here was seen as descriptive of AT A STORE. Obviously, the existence of such isolated elements may be of some interest in its own right.

The prototype view of situation cognition holds that a particular category of everyday situations is mentally represented as an abstract set of features commonly associated with members of the category. According to this view, an individual who is confronted with a particular situation compares the relevant features of that situation with those of existing prototypes of situations. The higher the match between the situation and the activated situation prototype, the higher the individual's confidence that behaviors associated with that prototype will be appropriate in the specific situation encountered. For example, a situation that is cognized as being highly similar to the MOVIES prototype will suggest displaying various behaviors appropriate in relaxed, clear, and pleasant situations. A two-mode clustering representation of situations and their features can depict the prototype organization of culturally shared situation knowledge and thus advance our understanding of the relation between situation cognition and behavior in social situations.

References

ANDERSON, C.A., and SEDIKIDES, C. (1991), Thinking about People: Contributions of a Typological Alternative to Associationistic and Dimensional Models of Person Perception, *Journal of Personality and Social Psychology, 60, 203–217.*

ARGYLE, M., FURNHAM, A., and GRAHAM, J.A. (1981), *Social Situations*, Cambridge University Press, Cambridge.

BATTISTICH, V.A., and THOMPSON, E.G. (1980), Students' Perceptions of the College Milieu: A Multidimensional Scaling Analysis, *Personality and Social Psychology Bulletin, 6, 74–82.*

BEM, D.J. (1981), Assessing Situations by Assessing Persons, in: D. Magnusson (ed.), *Toward a Psychology of Situations: An Interactional Perspective*, Erlbaum, Hillsdale, 245–257.

CANTOR, N. (1981), Perceptions of Situations: Situation Prototypes and Person-Situation Prototypes, in: D. Magnusson (ed.), *Toward a Psychology of Situations: An Interactional Perspective*, Erlbaum, Hillsdale, 229–244.

D'ANDRADE, R.G. (1989), Cultural Cognition, in: M.I. Posner (ed.), *Foundations of Cognitive Science*, MIT Press, Cambridge, 795–830.

ECKES, T. (1990), Situationskognition: Untersuchungen zur Struktur von Situationsbegriffen, *Zeitschrift für Sozialpsychologie, 21, 171–188.*

ECKES, T. (1991), Bimodale Clusteranalyse: Methoden zur Klassifikation von Elementen zweier Mengen, *Zeitschrift für Experimentelle und Angewandte Psychologie, 38, 201–225.*

ECKES, T., and ORLIK, P. (1991), An Agglomerative Method for Two-Mode Hierarchical Clustering, in: H.H. Bock and P. Ihm (eds.), *Classification, Data Analysis, and Knowledge Organization*, Springer-Verlag, Berlin, 3–8.

ECKES, T., and ORLIK, P. (in press), An Error Variance Approach to Two-Mode Hierarchical Clustering, *Journal of Classification.*

FORGAS, J.P. (1976), The Perception of Social Episodes: Categorical and Dimensional Representations in Two Different Social Milieus, *Journal of Personality and Social Psychology, 34, 199–209.*

FORGAS, J.P. (1983), Episode Cognition and Personality: A Multidimensional Analysis, *Journal of Personality, 51, 34–48.*

KENRICK, D.T., and FUNDER, D.C. (1988), Profiting from Controversy: Lessons from the Person-Situation Debate, *American Psychologist, 43, 23–34.*

KING, G.A., and SORRENTINO, R.M. (1983), Psychological Dimensions of Goal-Oriented Interpersonal Situations, *Journal of Personality and Social Psychology, 44, 140–162.*

KRAHÉ, B. (1992), *Social Psychology and Personality: Towards a Synthesis,* Sage, London.

MAGNUSSON, D. (1974), The Individual in the Situation: Some Studies on Individuals' Perceptions of Situations, *Studia Psychologica, 2, 124–131.*

PERVIN, L.A. (1976), A Free-Response Description Approach to the Analysis of Person-Situation Interaction, *Journal of Personality and Social Psychology, 34, 465–474.*

POWELL, R.S., and JUHNKE, R.G. (1983), Statistical Models of Implicit Personality Theory: A Comparison, *Journal of Personality and Social Psychology, 44, 911–922.*

QUINN, N., and HOLLAND, D. (1987), Culture and Cognition, in: D. Holland and N. Quinn (eds.), *Cultural Models in Language and Thought,* Cambridge University Press, Cambridge, 3–40.

SNYDER, M., and ICKES, W. (1985), Personality and Social Behavior, in: G. Lindzey and E. Aronson (eds.), *Handbook of Social Psychology,* Random House, New York, Vol. 2, 883–947.

TUCKER, L.R. (1964), The Extension of Factor Analysis to Three-Dimensional Matrices, in: N. Frederiksen and H. Gulliksen (eds.), *Contributions to Mathematical Psychology,* Holt, Rinehart, and Winston, New York, 109–127.

TVERSKY, A. (1977), Features of Similarity, *Psychological Review, 84, 327–354.*

Printing: Druckhaus Beltz, Hemsbach
Binding: Buchbinderei Schäffer, Grünstadt

5